**Books are to be returned on or before
the last date below**

−2. OCT. 1978		
27. NOV. 1978		
0 8 JAN 1992		
1 6 MAR 1992		

LIBREX —

Vacancies '76

Vacancies '76

Edited by

R. E. Smallman and J. E. Harris

Proceedings of a conference on 'Point defect behaviour and diffusional processes' organized by The Metals Society and held at The Royal Fort, University of Bristol, on 13–16 September, 1976

The Metals Society

Book 186 published by

The Metals Society
1 Carlton House Terrace
London, SW1Y 5DB

ISBN 0 904357 10 4

Text set in 10/11 pt Linotron Times and titles in 28 pt Linotron Univers
Printed by J. W. Arrowsmith Ltd, Bristol

Contents

Session IV   Oxidation

Foreword

Since the introduction into the scientific literature of the Frenkel defect 50 years ago and the Frank sessile dislocation loop almost 30 years ago, these concepts have been applied in many branches of materials science from reactor physics to semiconductor technology. The last major conference in the UK on 'Vacancies and other point defects' was organized almost 20 years ago by The Institute of Metals as it then was. That conference, held at Harwell in 1957, marked a definite landmark in the subject of point defects and the published conference volume is still widely consulted as a useful reference work. Even allowing for the fact that scientific progress is often measured in decades, rather than months or individual years, the tremendous developments in the subject since 1957 indicated that another major conference on vacancies and other point defects seemed timely: hence 'Vacancies '76'.

Following the successful organization of the '57 conference it is particularly appropriate that the '76 conference should have been organized by The Metals Society, the progeny of The Institute of Metals, and that it should have been held in the Physics Department at Bristol University where much of the pioneering work in the field was carried out.

The 1976 conference was divided into four sessions: I Fundamental Properties of Point Defects; II Point Defect Aggregates and their Effects on Properties; III Microstructure and Creep; and IV Oxidation. The overall programme and selection of invited speakers was the responsibility of the programme committee but the organization of each session was the sole responsibility of the individual technical secretary and these are acknowledged at the beginning of each section of the book.

All the papers were by invitation and it was intended that each should form a major (though short) critical review of a particular field by acknowledged experts. With a few distinguished exceptions the contributions were all from the UK. This was not chauvinism but simply reflected the fact that in these times of economic stringency and expensive travel it is often not possible for an overseas researcher to make a firm commitment to appear in person at some future time to present a paper. This is a disadvantage but in this case not disastrously so for this particular branch of metal physics continues to be a well-supported and healthy activity in Britain, following a long-established tradition. It is to be hoped that this important field continues to receive the support necessary for its advancement.

A conference consisting of only invited papers suffers from the potential disadvantage that the recent findings of the young active research worker may be excluded.

To guard against this possibility a Poster Session was organized. This consisted of the individual research worker setting up a display board illustrating his work and being available for informal discussion alongside the display during a specially allocated session. This was the first time such a session had been organized at a Metals Society Conference and it proved to be a highly successful innovation. The inclusion of Poster Sessions at future conferences of The Metals Society is strongly recommended.

Conferences are much warmer affairs than appears through the cold prose of the published Proceedings. The social activities organized in association with Vacancies '76 were in fact a huge success and provided a valuable forum for informal contact and discussion. Following the conference dinner at the University Senate House, Professor Norman Thompson and Professor Jock Eshelby gave a fascinating account of research in metal physics at Bristol in the immediate pre- and post-war years. Unfortunately, a written account of their reminiscences is not available. In addition to this the City of Bristol generously organized a Civic Reception for the conference delegates which was held at St. Nicholas Church Museum. In response to a short speech of welcome from the Lord Mayor, Dr J. E. Harris, on behalf of The Metals Society, spoke at some length describing the background to the conference, why it was held in Bristol, and discussed some of the consequences of the communications gap between scientists and the general public. A number of delegates have asked that this speech be written up and provided in a more permanent form and, accordingly, it is reproduced in the following pages.

To summarize the scientific highlights of the conference is too large a task to attempt here; the published papers must speak for themselves. It is, however, pleasant to record that in spite of the evident financial stringency, research in the various areas is of a very high quality and new knowledge is increasing rapidly. Indeed, development in the understanding and application of point defects in many technological areas, e.g. oxidation, creep, sintering, is so fast that it is tempting to suggest that a conference report on progress is shortened from an interval of 20 years to perhaps 5 years and we are already looking forward to 'Vacancies '81.'

R. E. Smallman
Chairman, Organizing Committee

Organizing Committee

R. E. Smallman (Chairman)
K. H. G. Ashbee
W. Beeré
B. Burton
B. L. Eyre
P. T. Heald
J. E. Harris
J. Stringer
Mrs J. Hermann
Mrs R. Lambert

Scientific images and the image of scientists

J. E. Harris*

My Lord Mayor, Lady Mayoress, and fellow delegates, it is my privilege this evening to thank, on behalf of The Metals Society, the Rt. Hon. Lord Mayor and the City of Bristol for their hospitality in launching the social side of our activities with this excellent Reception.

Those of us who are fortunate enough to live close to Bristol know very well the beauty of the city and its surrounding countryside, its fascinating history, and the hospitality of its people. However, all this may come as a pleasant surprise to those delegates who come from further afield. Some have come from very distant fields indeed, from Japan, America, Russia, and many countries on the continent of Europe. I am sure they will all take back with them many happy memories of their short stay in Bristol.

The Metals Society is the youngest of the major learned societies so perhaps it might be in order, Lord Mayor, if I described to you its brief history and outlined its terms of reference. I am sure you will also be interested to learn why the Society decided to organize its first major conference with a metal physics theme at the University of Bristol.

The Metals Society dates only from 1974 but it was formed by the amalgamation of the two well established metallurgical societies, The Iron and Steel Institute and The Institute of Metals. The former had celebrated its centenary before the union so our roots go back well into the last century and number many distinguished nineteenth century metallurgists amongst its members.

The Society has unusually wide terms of reference for it covers all aspects of extraction of metal from its ores, melting and casting, fabrication into useful articles, and the marketing of these products. It embraces chemical metallurgy, engineering metallurgy, management studies, and sales and at one end of the spectrum the study of the physics of metals at the atomic level. It is the latter aspect which is the subject of this Conference.

I am afraid that I must plead guilty to having proposed the rather jazzy title of 'Vacancies '76'. It has been suggested to me that it rather indicated a convention of employment officers rather than metal physicists so perhaps at this point I should explain that a vacancy is simply the space which should have been occupied by an atom in a metal lattice. It must then appear curious to the layman how it is that scientists can organize a four-day meeting on absolutely nothing! A vacancy is in fact a defect which is much beloved by physicists. It is pictured as a nice round object whose motion can be predicted quite accurately, like the path of a ball of wool projected across a room by a playful kitten. In this respect it is quite different from its linear counterpart, the dislocation, whose movement is more complex as it meets and interacts with many more obstacles. It can be likened to an unravelled ball of wool which tangles with the legs of all the tables and chairs as it tries to move across the room.

So much for The Metals Society and the theme of our Conference, but why hold it in Bristol? The reason is simply that the Physics Department of the University of Bristol is recognized internationally as a centre of excellence for the study of the science of metals. Indeed it was here in the Royal Fort in the late 1940s and early 1950s that enormously important advances were made providing the foundation stones upon which has been built the modern theory of defects in metals which is the subject of our Conference. Many of the Bristol pioneers are still active in the field though not all have remained in the city. One thinks most obviously of the work of Mott, Frank, Nabarro, Thompson, Eshelby, and Nye and the able assistance of many of their students. These spearheaded the revolution in our thinking but like all revolutions it had its traumas. Old ideas and prejudices had to be discarded and new theories substituted. New pictures had to be painted over the old canvasses. But it was a quiet revolution to all but the specialists. The man in the Bristol omnibus knew nothing of this renaissance which was taking place right in the centre of the city. Sadly he knows nothing about it to this very day. It is as though Mott and his merry men were repainting the Sistine chapel on the walls of the Royal Fort, yet no one bothered to pop in off the street to have a look.

Bristol is of course rightly famous for many feats of engineering and developments in industrial metallurgy. Innovations in zinc smelting at Avonmouth spring to mind and the construction of Concorde and the Severn Bridge, but most especially the monumental works of Brunel capture our imagination. His architectural and engineering works at Temple Meads Station and at Bristol Docks occupied much of his time and one thinks also of the launching of the SS Great Britain (happily now returned to Bristol) and the design of the Clifton Suspension Bridge where Brunel's design was preferred to one by Thomas Telford himself. Brunel's versatility truly made him the Michaelangelo of the engineering world.

If I may be allowed a small digression on the subject of the Clifton Bridge: Brunel, like many modern engineers, knew little physical metallurgy (there was in fact little to know in those days, which is not the case today) so inevitably his structures were overdesigned and expen-

* In response to an address of welcome by the Lord Mayor of Bristol.

sive. So expensive in fact did the Clifton bridge turn out to be that the people of Clifton could not afford to complete it until after Brunel's death and then only by employing second-hand suspension chains from a bridge which had been demolished in London. I am happy to report that recently fatigue tests have been conducted on specimens cut from the bridge and the metal is in as good condition today as the day it was constructed. I wonder if one will be able to say the same of segments of the Severn Bridge tested 100 years hence?

The SS Great Britain, Clifton Suspension Bridge, and Concorde are undoubtedly great feats of innovatory engineering. Many screw-propelled iron ships followed SS Great Britain; many suspension bridges followed the example and principles laid down at Clifton culminating ten years ago in the construction of the Severn Bridge. For good or ill undoubtedly many supersonic airliners will follow Concorde.

Although innovatory, these feats of engineering are in another sense somewhat specific. While not wishing to denigrate them in any way it must be said that they do not provide new insight or universal facts which change fundamentally our view of each other and our understanding of the physical world. A ship, a bridge or an aeroplane allows one to move more conveniently from one point on the earth's surface to another, but makes no comment on whether the journey is really necessary or is part of some mad modern random Brownian movement leading nowhere and satisfying only those who mistake frantic activity for steady advancement. They are landmarks to our material progress rather than to our spiritual or artistic progress. They are products of distinctly different thought processes than those which have led to, shall we say, a novel by James Joyce, a building by Le Corbusier or a painting such as Les Demoiselles d'Avignon by Picasso. Moreover, evolution in engineering is moving away from control by the individual genius to management by working parties and committees with consequent loss of individuality and creativity. There are no contemporary Brunels.

On the other hand creativity, and universality, in literature, art, music, and science remain essentially the same phenomenon as has been recently argued so persuasively by Koestler. It is entirely possible that when the history of our times is written some of the Bristol physicists will find themselves in the select company of the great creative artists of the first half of the twentieth century. They may well occupy a place in history above those of the designers of the Severn Bridge and Concorde and stand alongside the great Brunel himself. Their contemporary lack of public recognition may simply be due to the fact that a vacancy is so much smaller than a bridge.

At this point I must emphasize that my purpose tonight is not to drive a wedge between science and engineering, but on the contrary to stress that the two activities are completely interdependent. It is possible through bad organization (as we in this country should be all too aware) to have bad engineering being carried out alongside good science, but it is simply not possible in modern times to have good engineering based on bad science. So in many senses the good science must come first.

The next question to ask is: does an almost complete lack of awareness by the general public of exciting developments in our University Science Departments matter? I think it does, although it is as well to recognize immediately that whatever efforts we make to publicize these achievements the man in the street will continue to be more interested in the exploits of Whitehead, Ritchie, and Cheesely rather than those of the contemporary equivalent of Eshelby, Frank, and Nabarro (perhaps I should explain that the former trio are the front line of Bristol City FC which has recently been promoted to the First Division, whereas the latter are the front line in a different First Division). However, what is of real concern is that we are failing to sell the excitement of science to our school kids. The image of the scientist amongst our children is deplorable; we are seen as uncreative, unimaginative despoilers and polluters. A consequence of this is that our social science, law, medicine, and arts faculties are overwhelmed with applicants from potential students, whereas our science and engineering departments are less than half full. There are in fact thousands of places unfilled in science and engineering in our universities and polytechnics. Yet in a sense the oversubscribed disciplines of social science, law, arts, and medicine are parasitic upon science and engineering. The wherewithall, the money, to build and run hospitals, libraries, schools, and galleries and to support the social services can only come from the profits of a scientifically based manufacturing and engineering industry. And yet we are failing to provide the scientifically trained personnel to man these industries and to carry out the associated basic research. Perhaps this failure is an important contributory cause to our present lack of competitiveness in world markets in advanced technology. If so the prognosis is bleak indeed for there are no signs of the popularity of science picking up in the near future.

What can we as a learned society and as individual scientists do to improve the situation? Firstly, we must spend more time in schools selling science as a 'fun' activity; to try and convey the excitement of scientific discovery and technical innovation; to persuade our brightest children that science is socially useful, that our very survival depends upon it; to argue to those children who are equally interested in the arts and science that it is probably better to concentrate on the latter while they are still young. It is easier for a science graduate to learn to speak say French than for a graduate in modern languages to learn science in later life. More imagination is needed on our part. A few years ago I was involved in the organizing of an international conference on nuclear metallurgy at the Berkeley Laboratories. As part of the Conference Proceedings we hired the Colston Hall and invited Sir Alan Cottrell to give a public lecture on world energy resources. I was extremely nervous knowing the size of the Hall and wondered if we could possibly attract a reasonably sized audience. In the event I need not have worried, the Hall was packed with school children and interested members of the general public and we had a most enjoyable and informative evening. Perhaps the organizers of this conference should have been equally bold and organized a public lecture. The audience is there, let us go out and speak to them. Let us grab them by the arm and show them the pictures on the walls of the Royal Fort.

My Lord Mayor thank you once again for your hospitality and attention. We look forward with keen anticipation to spending a few more days in this pleasant and amiable city of Bristol.

Session I
Fundamental properties of point defects

Chairman: B. A. Bilby (University of Sheffield)
Technical Secretary: P. T. Heald (CEGB)

Interaction and diffusion of point defects

J. D. Eshelby

Simple derivations are given for some results in the continuum theory of point defects; in particular the interaction between defects, both directly and through image (surface) effects. It is argued that these results have a wider validity than appears at first sight. The jump frequency for particles diffusing from one minimum to another of a rigid one-dimensional potential is derived by a simple argument which shows clearly the origin of the frequency factor, which turns out not to be, in any reasonable sense, a measure of the number of attacks on the barrier per unit time, as commonly stated. The method can quite easily be extended to give the jump frequency of a point defect or vacancy in three dimensions both when the effect of the host lattice is simulated by a rigid set of potential wells and also when, more realistically, it is not (Vineyard formula).

The author is at the University of Sheffield

Before treating, by request, the continuum theory of defects, the author would like to recall in defence of its obvious shortcomings an earlier remark[1] to the effect that 'the theory perhaps suffers from the disadvantage that its limitations are more immediately obvious than are those of other approximate methods which have to be used in dealing with the solid state', and add that a simple treatment at cottage-industry level may sometimes provide a certain degree of insight into some phenomenon or other when an exact calculation, based on a theoretical model which apes reality precisely, will give accurate answers but may be too unsurveyable, or too numerical, to give much insight.

In the next section the elastic interaction of defects is reviewed and in the following section some views are given on the right way to look at jump frequencies.

DEFECT INTERACTIONS

The interaction energy of a defect with a stress field can be worked out in macroscopic terms by considering the result of introducing a uniform distribution of many such defects into a stressed specimen. We suppose that the solution of defects is dilute enough for the interaction between them to be neglected.

Suppose that in a perfectly rigid block of material a badly made cavity has been cut which was meant to be a unit cube but is actually slightly too small and is also slightly parallelepipedal. If we force a unit cube of the elastic material we are interested in into the cavity the resulting contraction and shear will give it an elastic energy

$$W = W_d + W_s \qquad (1)$$

Here W_d, the dilatational part of the energy density, is $\frac{1}{2}Ke^2$ where K is the bulk modulus and e the dilatation (actually negative) induced by the too-small cavity. Similarly the shear part of the energy density W_s induced by the small departure from cubic shape is $\frac{1}{2}\mu s^2$ where μ is the shear modulus and s a suitable measure of distortion which we need not write out. Next insert n defects, uniformly scattered, into the deformed cube in the cavity. The work done is nE_s, where E_s is the self-energy of one defect, plus n times the interaction energy E_{int} of a single defect with the imposed stress field.

Now start again with the undeformed cube. Insert the defects in it; the work done is nE_s. Force the cube into the cavity. The work required to do this will differ from (1) for two reasons. First, the introduction of the defects into the unconstrained unit cube changes its volume by, say, e' without producing any corresponding stress (compare the case of thermal expansion), so that now we have to write $W_d = \frac{1}{2}K(e-e')^2$. There would be a corresponding change in W_s if, contrary to our assumption, the defects changed the shape as well as the size of the cube (carbon or nitrogen in ferrite). Secondly, the macroscopic K and μ of the material are changed if the defects are in some sense elastically harder or softer than the matrix. Hence in all W increases by

$$C\frac{\mathrm{d}}{\mathrm{d}C}W = -CKe\frac{\mathrm{d}e}{\mathrm{d}C} + \frac{1}{2}\frac{\mathrm{d}K}{\mathrm{d}C}e^2 + \frac{1}{2}\frac{\mathrm{d}\mu}{\mathrm{d}C}s^2$$

$$= CP\frac{\mathrm{d}e}{\mathrm{d}C} + \frac{1}{K}\frac{\mathrm{d}K}{\mathrm{d}C}W_\mathrm{d} + \frac{1}{\mu}\frac{\mathrm{d}\mu}{\mathrm{d}C}W_\mathrm{s}$$

where P is the imposed hydrostatic pressure and C is the concentration of defects. Both this quantity and $nE_\mathrm{int.} = CE_\mathrm{int.}/\Omega$ represent the work, over and above nE_s, required to establish the defect-filled cube in the cavity, and they must be equal; otherwise we should be in the energy-making business by loading up the easier way and unloading in the reverse of the harder way. Consequently we must have

$$E_\mathrm{int.} = E_\mathrm{mis.} + E_\mathrm{inh.} \tag{2}$$

where the first term is

$$E_\mathrm{mis.} = P\frac{\mathrm{d}e}{\mathrm{d}C}\Omega = P\cdot\frac{3}{a}\frac{\mathrm{d}a}{\mathrm{d}C}\Omega = P\,\Delta V \tag{3}$$

and the second is (5) below. The component (3), which is associated with the misfit in the sphere-in-hole model, can be written in terms of the imposed hydrostatic pressure P and either the rate of change of lattice parameter with defect concentration or the volume change ΔV associated with the introduction of a single defect.

This is perhaps the point to comment on the fact that in his original calculation Cottrell[2] found the misfit interaction by calculating the work done in 'blowing up' the defect against the pressure P and arrived at

$$E_\mathrm{mis.} = P\,\Delta V^\infty \tag{4}$$

instead of (3), where ΔV^∞ is a certain volume less than ΔV and related to it by (7) and (10) below. This is plausible because, as we shall see shortly, ΔV^∞ is the volume change of any surface *closely* surrounding the defect. The standard calculation of Bilby[3] used a sphere elastically homogeneous with the matrix whereas Cottrell used a rigid sphere, and so the difference between (3) and (4) is commonly said to be due to the elastic energy in the defect. This is untrue. Attention to two quite unsophisticated points, both of which increase the work of insertion, will raise (4) by the necessary factor γ. The first is that the volume of a hole is reduced by a factor $(1 - P/K)$ when the body containing it is subjected to an external pressure P, supposing that there is also a pressure P inside the hole. (If not, the volume is even smaller.) The second is that the internal pressure has to be increased somewhat above P to expand the hole so that the (rigid) sphere may be inserted. (The 'bulk modulus for blowing up a hole' is, in fact, $4\mu/3$.)

The second term in (2),

$$E_\mathrm{inh.} = \frac{1}{K}\frac{\mathrm{d}K}{\mathrm{d}C}\Omega W_\mathrm{d} + \frac{1}{\mu}\frac{\mathrm{d}\mu}{\mathrm{d}C}\Omega W_\mathrm{s} \tag{5}$$

takes account of the defect in its role as elastic inhomogeneity.

To find the interaction between two defects we must calculate the elastic fields which have to be inserted in (3) or (5). If we require the displacement around a defect to be purely radial, spherically symmetric, and to decrease with increasing distance r from the defect, the equations of isotropic linear elasticity force it to be proportional to $1/r^2$,

$$U_\mathrm{R} = \frac{c}{r^2} \quad \text{or} \quad u_i = c\frac{x_i}{r^3} = -c\frac{\partial}{\partial x_i}\left(\frac{1}{r}\right) \tag{6}$$

where the constant c measures the strength of the defect. Equation (6) has the same form as the electric field of a point charge and so its divergence is zero, that is to say the dilatation $e = \mathrm{div}\,\mathbf{u}$ around the defect is zero. However this does not mean that there is no volume change associated with the defect. In fact a sphere of radius r suffers an increase in volume

$$\Delta V^\infty = 4\pi r^2 . c/r^2 = 4\pi c \tag{7}$$

when the defect is introduced at its centre. If the spherical surface is deformed into a new surface of arbitrary shape embracing the defect the volume change is still (7) because throughout the volume between the original sphere and the deformed surface the dilatation is zero.

Although (7) is the volume change associated with any surface surrounding the defect it is not the whole of the volume change when the defect is introduced into a finite body with a stress-free surface. At the surface of a sphere round the defect there is a radial strain $\mathrm{d}U_\mathrm{R}/\mathrm{d}r$ accompanied by two equal transverse strains of half this magnitude and opposite sign so as to give zero dilatation. Thus the radial and transverse strains at distance r are

$$-2c/r^3, \qquad c/r^3, \qquad c/r^3 \tag{8}$$

and the corresponding stresses are, by Hooke's law,

$$-4\mu c/r^3, \qquad 2\mu c/r^3, \qquad 2\mu c/r^3 \tag{9}$$

Thus if a sphere of radius r is cut out a normal radial inward force, i.e. a pressure, of amount $4\mu c/r^3$ must be supplied if the original elastic field (6) is to be maintained. If this unwanted pressure is relaxed to give a stress-free surface there will be an additional 'image' volume increase

$$\Delta V^\mathrm{I} = \frac{4\pi}{3}r^3 . \frac{4\mu c}{r^3} . \frac{1}{K} = 4\pi c . \frac{4\mu}{3K} \tag{10}$$

or in all a volume change

$$\Delta V = \Delta V^\infty + \Delta V^\mathrm{I} = 4\pi c\gamma$$

with

$$\gamma = 1 + \frac{4\mu}{3K} = 3\frac{1-\nu}{1+\nu}$$

where ν is Poisson's ratio. Like (7) the extra volume change (10) is independent of the shape and size of the solid or the position of the defect in it, though this is rather more difficult to show. One way is to use the sphere-in-hole model plus the general result,[4] obvious to some, that the average, or volume integral, of any stress-component over a self-stressed body with a stress-free surface is zero. Applied to the hydrostatic pressure this means that the self-stress produces no change in the volume of the material if Hooke's law holds and the material is elastically homogeneous. Let the body be the sphere plus the solid with the hole to receive it (the theorem does not require the body to be all in one piece). Before assembly the volume of material is the volume inside the outer surface, minus the volume of the hole, plus the volume of the sphere, that is, the volume inside the outer surface plus the misfit volume $V_\mathrm{mis.}$ by which the sphere exceeds the hole. After the sphere has been forced into the hole the volume of material is just the volume inside the outer surface. As the volume of material has not changed, the increase in the volume inside the outer surface must be equal to V_{mis}. In addition to showing that ΔV and its parts ΔV^{∞}, ΔV^I are independent of the shape and size of the body and the position of the defect within it, this argument shows that $\Delta V = 4\pi c\gamma$ is actually equal to the misfit volume V_{mis}. For a defect at the centre of a sphere the extra 'image' elastic field associated with ΔV^I is a uniform hydrostatic pressure

$$\bar{P} = -K\frac{\Delta V^I}{V} = -16\pi\mu c/3V \qquad (11)$$

which is actually a tension for positive c or ΔV, inversely proportional to the volume of the specimen.

In considering the interaction between defects we can usually ignore the modification of the defect elastic fields by image terms (but see below). Then, since each defect produces no hydrostatic pressure the misfit interaction (3) between them is zero, a result apparently first noticed by Bitter.[5] To get the inhomogeneity interaction (5) we need to know the effective elastic constants of a dilute suspension of spheres of elastic constants K', μ' in a matrix with constants K, μ. They have been worked out by many people; the first to arrive at both K and μ correctly seems to have been Dewey.[6] If the Poisson's ratio of the matrix is $1/5$ we have the easily memorized results

$$\alpha_K \equiv \frac{1}{K}\frac{dK}{dC} = \frac{K'-K}{\frac{1}{2}(K'+K)}, \qquad \alpha_\mu \equiv \frac{1}{\mu}\frac{d\mu}{dC} = \frac{\mu'-\mu}{\frac{1}{2}(\mu'+\mu)},$$

difference over average; though $\nu = 0.20$ is rather small to be typical these expressions are not far out for, say, $\nu = \frac{1}{3}$ or $\nu = \frac{1}{4}$. For the elastic field (6) W_d is zero and W_s is half the sum of the products of the three quantities (8) by the three quantities (9), each to each, which gives

$$W_s = 6\mu c^2/r^6 \qquad (12)$$

and hence

$$E_{inh.} = 6\alpha_\mu \mu c^2 \Omega/r^6$$

for the interaction of one purely misfitting defect with one purely inhomogeneous one. If they are both inhomogeneous and misfitting the result is doubled. Generally, for two different defects 1 and 2 we have

$$E_{inh.} = 6(\alpha_{\mu 1}c_2^2 + \alpha_{\mu 2}c_1^2)\Omega/r^6 \qquad (13)$$

According to the estimate above, α_μ is -2 for a hole and $+2$ for an unshearable ('rigid') inclusion. It is not unreasonable to expect a vacancy to behave like a hole, not necessarily of volume Ω, or at any rate as an elastically soft spot. One might expect interstitials to act as hard spots and put up the bulk modulus (which does not interest us here), but it is not intuitively obvious that they would also put up the shear modulus. Indeed one might perhaps feel that the excessive distortion around the interstitial would allow the neighbouring atoms to as it were slip round it under shear. In that case a more appropriate boundary condition might be a freely slipping interface rather than bonding between sphere and hole. The bulk shear modulus has been worked out for this case, but only for a Poisson's ratio $\frac{1}{2}$.[7] It gives $\alpha_\mu = 1$

in place of the value $\alpha_\mu = 2.5$ for bonding, so that, on this model also, an interstitial is still a 'hard' defect. What we have said suggests that vacancies should attract one another, and interstitials repel. For the interaction of an interstitial 1 with a vacancy 2 we should expect to have $\alpha_{\mu 1}$ moderate and positive, c_1^2 large, $\alpha_{\mu 2}$ moderate to large in magnitude and negative, c_2^2 small, leading to attraction between the pair since the second term in (13) will dominate.

The result that the misfit interaction (3) between two defects is zero because each looks for a hydrostatic pressure which its colleague does not provide is peculiar to the isotropic misfitting sphere model and can easily be upset. One way is to make the material anisotropic. Another is to make the defect less symmetrical, a misfitting ellipsoid in a spherical hole say, representing a split interstitial or carbon and nitrogen in alpha-iron. Either modification gives an interaction energy proportional to r^{-3} times the appropriate angular factor. However, we shall only consider here defects with cubic symmetry in a cubic crystal.

Before going over to anisotropy we shall see if anything fresh can be obtained from an isotropic model. As an alternative to the misfitting inclusion model we can imagine that the elastic field of the defect is produced by a small cluster of point forces, with zero resultant and couple, representing, approximately, the force which neighbouring atoms would feel if they refused to move when the defect was introduced. The i-component of displacement produced by unit force at the origin, directed parallel to the x_k-axis is

$$U_{ik} = \frac{1}{4\pi\mu}\frac{\delta_{ik}}{r} - \frac{1}{16\pi\mu(1-\nu)}\frac{\partial^2 r}{\partial x_i \partial x_k} \qquad (14)$$

Hence if, say, a cluster of six forces, all of magnitude f, act radially outwards at the face centres of a cube of side a_0, forming three crossed double forces, the displacement for large r/a_0 is

$$u_i = -fa_0\frac{\partial U_{ik}}{\partial x_k} \qquad (15)$$

which is identical with (6) if the moments fa_0 of the double forces are related to the defect strength c by

$$fa_0 = 4\pi K\gamma c \qquad (16)$$

Equation (15) is strictly valid for all r only if we let the force dipoles become genuine point dipoles, that is, if we make the transition $a_0 \to 0$, $f \to \infty$ in such a way that (16) is satisfied even in the limit. However, if we take the model seriously we ought to let a_0 be finite and of atomic dimensions. As all we shall need is the dilatation we might as well find it directly. The dilatation corresponding to (14) is

$$e = \frac{1}{4\pi K\gamma}\frac{\partial}{\partial x_k}\left(\frac{1}{r}\right)$$

Thus the forces $\pm f$ at $(x_1 \mp \frac{1}{2}a_0, x_2, x_3)$ contribute an amount

$$\frac{f}{4\pi K\gamma}\frac{\partial}{\partial x_1}\{|r - \tfrac{1}{2}a_0\mathbf{i}|^{-1} - |r + \tfrac{1}{2}a_0\mathbf{i}|^{-1}\}$$

to the dilatation, where $\mathbf{i}$ is the vector $(1, 0, 0)$. If we expand the quantities in { } in a Taylor series and add the contributions from the other two dipoles we get

$$e = -c\left[\nabla^2 + \frac{1}{24}a_0^2\left(\frac{\partial^4}{\partial x_1^4} + \frac{\partial^4}{\partial x_2^4} + \frac{\partial^4}{\partial x_3^4}\right) + \cdots\right]\frac{1}{r}$$

for the dilatation due to the defect, or, dropping the term $-c\nabla^2(1/r) = 4\pi c\delta(r)$ which just represents a blob of expansion at the origin, and the unwritten higher terms,

$$e = -\frac{35}{8}ca_0^2\Gamma r^{-5} \tag{17}$$

where the angular factor

$$\Gamma(l, m, n) = l^4 + m^4 + n^4 - \tfrac{3}{5} \tag{18}$$

depends on the direction cosines

$$l = x_1/r, \qquad m = x_2/r, \qquad n = x_3/r$$

It is easy to see that any centrally symmetric cluster of forces will give a dilatation proportional to r^{-5}. Since e is harmonic the angular factor Γ must then be a surface harmonic of order 4 and there is only one such which has cubic symmetry, namely (18). The original sphere-in-hole model can be modified to give (17). Replace the misfitting sphere by a water-worn cube or octahedron which has become nearly, but not quite, a sphere. Apply surface forces to make it a sphere the size of the hole, and cement it into the hole. At this stage there is no stress outside the defect, but there is an unwanted cubically symmetric layer of body force at the interface. To remove it we may apply an equal and opposite layer which, as we have seen, induces a dilatation of the form (17) with some value for the constant ca_0^2. Thus (17) gives the first non-vanishing term in the expansion of the dilatation in negative (odd) powers of r for any model of the defect which allows it to manifest its finite size and cubic symmetry. Such a model involves a pair of constants, say the volume change $\Delta V = 4\pi\gamma c$ and ca_0^2 which controls the magnitude of (17), and we can characterize the defect by specifying the pair of numbers $(\Delta V, a_0^2)$. We choose a_0^2 rather than a_0 because in some cases a_0^2 may be negative. An example is a set of forces directed outward from the corners rather than the face centres of the small cube. To see this apply the corner and face forces together. We are approaching the spherically symmetric situation of the misfitting sphere with zero dilatation. Hence the dilatations produced by the corner and face forces must partly cancel. With the $(\Delta V, a_0^2)$ notation our original spherically symmetric defect evidently has to be denoted by $(\Delta V, 0)$.

According to (3) the interaction energy of an ordinary $(\Delta V, 0)$ defect with the dilatation (17) is $-K\Delta V$ times (17). If both defects are of type $(\Delta V, a_0^2)$ each feels the dilatation of the other and the effect is doubled, giving the interaction energy

$$E_{\text{mis.}} = \frac{35}{16\pi}\frac{K}{\gamma}(\Delta V)^2 a_0^2 \frac{\Gamma}{r^5} \tag{19}$$

An interaction energy with this radial and angular dependence and about the same magnitude, was originally presented by Hardy and Bullough[8,9] on the basis of a lattice calculation. As a result the Γ/r^5 interaction is commonly said to be an essentially lattice effect, described variously as involving non-local forces, phonon dispersion curves, or the lattice Green's function.

However, to derive (19) the only concession we had to make to the lattice was to admit that the defect was of finite size and had cubic symmetry. To make our calculation comparable with Hardy and Bullough's we should in addition have to replace (14) by the expression for the displacement of one atom due to a force applied to another. According to Siems[10] the effect of doing this would be (for the case considered in ref. 8) to reduce (19) by 25%, so that the Γ/r^5 interaction in a lattice could fairly be described as an essentially continuum effect somewhat reduced by the lattice structure of the crystal.

To finish our discussion of the direct interaction between defects we drop the refinements which led to (19) and ask what effect the introduction of cubic anisotropy has on the original model. For a cubic crystal one of the elastic equilibrium equations is

$$c_{44}\nabla^2 u_1 + (c_{12} + c_{44})\frac{\partial e}{\partial x_1} + d\frac{\partial^2 u_1}{\partial x_1^2} = 0 \tag{20}$$

where

$$d = c_{11} - c_{12} - 2c_{44}$$

is zero for isotropy. If we differentiate (20) with respect to x_1 and add to it the two similar equations in u_2, u_3 we obtain

$$(c_{12} + 2c_{44})\nabla^2 e + d\left(\frac{\partial^3 u_1}{\partial x_1^3} + \frac{\partial^3 u_2}{\partial x_2^3} + \frac{\partial^3 u_3}{\partial x_3^3}\right) = 0$$

so that e is no longer necessarily harmonic. If d is small (nearly isotropic material) we may, with an error of order only d^2, use the isotropic value of u_i, $-c\partial(r^{-1})/\partial x_i$ in the present case, in the terms involving d. If in addition we write r^{-1} as $\nabla^2(\tfrac{1}{2}r)$ we get

$$\nabla^2\left\{(c_{12} + 2c_{44})e - \tfrac{1}{2}cd\left(\frac{\partial^4}{\partial x_1^4} + \frac{\partial^4}{\partial x_2^4} + \frac{\partial^4}{\partial x_3^4}\right)r\right\} = 0 \tag{21}$$

which says that { } is harmonic, i.e.

$$e = \frac{\tfrac{1}{2}cd}{c_{12} + 2c_{44}}\left(\frac{\partial^4}{\partial x_1^4} + \frac{\partial^4}{\partial x_2^4} + \frac{\partial^4}{\partial x_3^4}\right)r + h(x_1, x_2, x_3)$$

where the harmonic function h must have no singularities, not even at $r = 0$ (one there would alter the magnitude or nature of the elastic singularity) and so we can leave it out, as merely being some applied field unrelated to the defect. Hence, working out the derivatives,

$$(c_{12} + 2c_{44})e = -(15/2)cd\Gamma(l, m, n)r^{-3} \tag{22}$$

with the Γ of equation (18). The angular factors in (17) and (22) are the same not for any deep physical reason, but merely because of the dearth of suitable functions. From (21) one can see that $\nabla^2\nabla^2 e = 0$, i.e. e is a biharmonic function, so that by a standard theorem it has the form $h_1 + r^2 h_2$ with harmonic h_1, h_2. With an r^{-3} dependence we must have $h_1 = A(l, m, n)r^{-3}$, $h_2 = B(l, m, n)r^{-5}$, and we have already seen that, with cubic symmetry, $A = 0$, $B = $ const. Γ is the only possibility. According to (3) the interaction between two defects via their dilatation fields is

$$E_{\text{mis.}} = 30\pi dc^2\frac{\Gamma}{r^3} = \frac{15}{8\pi\gamma^2}(\Delta V)^2 d\frac{\Gamma}{r^3} \tag{23}$$

If d is not small the interaction is still proportional to r^{-3} and to $(\Delta V)^2$ but $d\Gamma$ is replaced by a power series in d whose successive coefficients are increasingly complicated functions of l, m, n.

The finite-size interaction (19) beats the anisotropy interaction (23) if, roughly,

$$r^2/a_0^2 < d/2K$$

For highly isotropic aluminium with $d/K \sim 0 \cdot 06$ this gives $r < 6a_0$. For most other cubic crystals where d/K is of the order of unity we must have $r < a_0$, so that if a_0 is in fact something like a lattice constant the anisotropy interaction will always win. (Of course we are cheating a little here—for $d/K \sim 1$ equation (23) is not a very good approximation.)

The indirect interaction between defects via their image fields, which we have so far ignored, may become important when there are a number of them, when say the 'defects' are actually the minor constituent in a binary alloy. In that case the image term makes a (negative) contribution to the strain energy per defect which is comparable with the elastic self-energy of a defect. To see this we must first calculate the self-energy. According to (12) the energy in the matrix round a defect is

$$\int_r^\infty \frac{6\mu c^2}{r^6} \cdot 4\pi r^2 \, dr = \frac{8\pi\mu c^2}{r^2} \tag{24}$$

On the other hand, since the matrix expands to partially accommodate it the sphere suffers a volume change $V_{\text{mis.}} -4\pi c = 4\pi c(\gamma - 1)$, so that the elastic energy of the sphere is

$$\tfrac{1}{2}K[4\pi c(\gamma - 1)]^2/\Omega \tag{25}$$

The sum of (24) and (25) gives E_s, the self-energy of a defect:

$$E_s = (8\pi\mu c/3)\,\Delta V$$

By the time a concentration C has been introduced an image pressure (actually a tension for positive c)

$$P(C) = -(16\pi\mu c/3)C \tag{26}$$

equal to (11) times the number of defects in the volume V, will build up linearly with C, so that by (3) the average energy expended in inserting all the defects introduced so far is $E_s + \tfrac{1}{2}P(C)\,\Delta V$ per defect, which is just $E_s(1-C)$ because of the similarity of the coefficients in E_s and $P(C)$. Consequently the strain energy per atom of alloy takes the simple form

$$E = E_s C(1-C) \tag{27}$$

characteristic of the heat of solution of a regular solution. It is not surprising that (27) looks sensible not only near $C = 0$ but also near $C = 1$, for there it is just what we should have got from a model in which atoms of the former major constituent were inserted, with minus the former volume misfit, into a matrix of the former minor constituent, but that it also looks sensible near $C = \tfrac{1}{2}$ is perhaps a matter of luck. However, (27) or some complication of it, has been used to discuss strain energy in alloy theory. One can, for example, 'derive' Hume-Rothery's 15% rule from it.[1] Here we shall just use it to talk about the strain energy associated with non-uniform solute distribution as, for example, in spinodal decomposition. Begin with an AB alloy with a concentration C

of A. At first it is homogeneous but then it segregates into, say, cubical regions which are alternatively A-rich and A-lean. If we cut the specimen up along the cube boundaries the A-rich cubes will expand and the others will contract, supposing that da/dC is positive. With sufficient care we could get useful work from these expansions and contractions, and so there is strain energy associated with the non-uniform solute distribution. On the other hand, one can see that it requires no work (beyond the osmotic work done in changing the configurational entropy) to rearrange the solute. In fact as we have seen the defects only interact via their image fields. Clearly the fine-scale rearrangement we have postulated will leave the image hydrostatic tension uniform and unaltered in magnitude in the bulk of the material, and so there is no change in the image interaction. The same is true for any other type of segregation (blobs, stripes etc.) which does not produce macroscopic deformation of the specimen. We can use these facts to avoid the difficulties of a direct calculation of the strain energy. As the energy of the specimen with fluctuating composition and coherent lattice is unaffected by redistributing the solute uniformly, we must have, by (27)

$$E_{\text{coh.}} = E_{\text{hom.}} = E_s\bar{C}(1-\bar{C}) \tag{28}$$

where $\bar{C}$ is the average composition. On the other hand, if we divide the specimen into cubes small enough for the composition to be substantially constant and the stress to be relaxed in each, the energy is

$$E_{\text{relax.}} = E_s\overline{C(1-C)} \tag{29}$$

The elastic strain energy released is (28) minus (29):

$$E_{\text{strain}} = -E_{\text{incoh.}} = E_s(\overline{C^2} - \bar{C}^2) = E_s\overline{(C-\bar{C})^2} \tag{30}$$

On the other hand, the energy required to make an incoherent block (or a collection of loosely assembled infinitesimal cubes) from a uniform block with the same mean composition, or from pure A and B, contains a term equal to (29) minus (28), in its role as $E_{\text{hom.}}$ rather than $E_{\text{coh.}}$, that is, as indicated, (30) with reversed sign. (This decrease is of course exactly cancelled by the strain energy if the blocks are forced to fit together coherently.) Rather paradoxically the energy required to make an inhomogeneous incoherent alloy from pure A and B depends not only on the mean composition $\bar{C}$, but also on its distribution, whereas the energy required to make an inhomogeneous coherent alloy depends only $\bar{C}$.

Since (3) and (5) were derived for the interaction of a defect with a uniform externally applied elastic field it is not entirely obvious that they will work when the 'applied' field is due to another defect, unless we take the anthropomorphic view that the defect cannot distinguish between one field and another. Doubts of this kind can be more or less set at rest by invoking the idea of the force on a defect. If $E_{\text{int.}}$ changes by

$$\delta E_{\text{int.}} = -F_x\,\delta x + 0(\delta x^2) \tag{31}$$

when the defect is shifted by δx parallel to the x-axis, then F_x is, by definition, the x-component of the force on the defect. It is equal to the integral

$$F_x = \int_S \left(Wn_x - \frac{\partial \mathbf{u}}{\partial x} \cdot \mathbf{T}\right) dS \tag{32}$$

8   ESHELBY

Here S is a closed surface drawn in the undeformed material and enclosing the site of the defect, n_x is the normal (before deformation) to the element dS, $\mathbf{T}\,dS$ is the traction on dS and W is the energy density per unit undeformed volume. The writer has devised a number of proofs of (32) of varying generality and simplicity; for the best so far see ref. 11. Evaluation of (32) and similar expressions for F_y, F_z will verify any of the interaction energies we have exhibited, to within an additive constant, of course. However, they will do more. Equation (32) is valid for finite deformation and any non-Hookian behaviour, and, further, it is surface-independent in the sense that it is not affected by deforming S, provided S does not pass through some other singularity. For two defects a distance $2R$ apart evaluate F_x over a sphere S_R of radius R about one of them. If R is large enough linear elasticity will be adequate on S_R. Contract S_R into a small sphere about the defect, and evaluate the integral there. As the theory is linear we can pick out the cross term between the fields due to the two defects, and they turn out to be consistent with (3) and (5). This means that (3) and (5) are valid as long as we can separate the defects by some surface where linear elasticity is reasonably valid; the fact that that theory is hopeless near either defect is irrelevant. It also shows that in, say, (3) P is not the hydrostatic pressure that the second defect produces at the centre of the first (a more or less meaningless concept), but rather the pressure calculated at the site of the first defect from the elastic data on S_R and the linear theory. From its derivation[11] one can see that (31) is still valid when the specimen is a lattice of balls connected by springs; the second term in (32) then just comes from the elements of S which are cut through by springs. The only difficulty is that δx must be an integral number of lattice parameters, so that the term $0(\delta x^2)$ cannot be reduced indefinitely, but this does not matter if F_x varies only slowly with position.

JUMP FREQUENCIES

Figure 1 shows the conventional picture of a particle in a one-dimensional potential well of depth U, and the following equation,

$$f = \nu\, e^{-U/kT}, \qquad (33)$$

gives the usual expression for the jump frequency over the barrier B of height U. Here ν is the frequency of vibration of the particle at the bottom of the well A, and the usual rationalization of (33) is that ν gives the number of attempts in unit time, and the exponential the probability of one particle having enough energy to get to the top. The writer has never felt happy with this argument. The one-dimensional version of the treatment which follows was originally given in an article

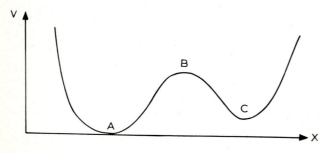

1 For explanation see text

written in 1966,[12] together with a statement that its extension to two or more dimensions was difficult. In the nine years which have elapsed between then and publication the writer has decided that this is not so, and the argument is here extended in outline to two, three or N dimensions.

It is a great help if we can imagine that there is a large number n of non-interacting particles in the well A, so that we may talk about the number which leave A via B in unit time rather than about the probability that a single particle leaves. To be able to do this we set up an ensemble in the sense of statistical mechanics, that is, a large number of replicas of Fig. 1 each containing only one particle and all in thermal equilibrium with one another and with a heat bath at temperature T. With the aid of a multiple-headed periscope we then superpose images of the members of the ensemble to get the required picture of a well apparently containing many particles. Or, begging some basic questions, we may take a very long film of the well with a single particle, cut it into many small lengths, print them on top of one another and run the resulting shorter film which will apparently show a well occupied by many, but mutually ignoring, particles.

To derive (33) for one dimension we only need two facts. First, if we write the number of particles between x and $x+dx$ as $\rho(x)\,dx$ then

$$\rho(x) = \text{const.}\; e^{-V(x)/kT}$$

where $V(x)$ is the potential at x (Boltzmann's barometric formula). Secondly the mean velocity in any given direction, $\bar{v}$ say, is independent of position (Maxwell distribution). We do not actually need to know that, for a particle of mass m, $\bar{v}$ is some multiple of $\sqrt{(kT/m)}$.

The flux of particles ϕ at a point in one direction, say left to right, is equal to the linear density of particles moving in the right direction, half the total, times their mean velocity. Evidently the jump frequency we are looking for may be written

$$f = \phi(\text{B})/n \qquad (34)$$

where n is the number of particles in the well A and $\phi(\text{B})$ is the left-to right flux at B. We have

$$\phi(\text{A}) = \tfrac{1}{2}\rho(\text{A})\bar{v}(\text{A}), \qquad \phi(\text{B}) = \tfrac{1}{2}\rho(\text{B})\bar{v}(\text{B})$$

which with $\bar{v}(\text{A}) = \bar{v}(\text{B})$ and

$$\rho(\text{B}) = \rho(\text{A})\, e^{-U/kT}, \qquad (35)$$

gives

$$\phi(\text{B})/\phi(\text{A}) = e^{-U/kT} \qquad (36)$$

We can get a value for $\phi(\text{A})$ very simply. If the particles oscillate with frequency ν in the well A each one crosses the bottom of the well from left to right once every $1/\nu$ seconds, so that $\phi(\text{A})$ is just $n\nu$, and so (36) becomes

$$\phi(\text{B}) = \phi(\text{A})\, e^{-U/kT} = n\nu\, e^{-U/kT} \qquad (37)$$

which in view of (34) is equivalent to (33). From the writer's point of view (37) gives the correct interpretation of (33): the $\exp(-U/kT)$ is there because the fluxes at B and A are obliged to be in this ratio, and the ν is there because the flux $\phi(\text{A})$ is proportional to ν as well as to n.

It is easy to see through the trick whereby we avoided the usual integrations of statistical mechanics. The

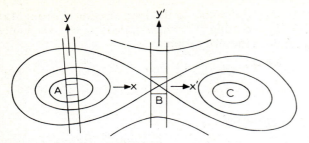

2 For explanation see text

ensemble can be regarded as a set of particles describing undisturbed orbits in phase space, with a certain distribution over energies. Of course particles are continually being thrown out of one orbit by thermal agitation, but as an equal number is being thrown into it this does not matter. Consequently we can combine purely mechanical results such as $\phi(A) = n\nu$ with statistical results such as (36) in a way which, with luck, will save us trouble.

Figure 2 shows the energy contours for a particle moving in two dimensions, and we want the flux of particles out of well A into well C by way of the saddle point B.

In the figure we have drawn a narrow strip embracing the y-axis. To avoid a large number of dx's we shall take the strip to be of unit width and choose our unit of length so small that nothing varies noticeably across the width of the strip. Also shown is a similar strip about the x-axis, or rather, merely the part of it which forms a square with the first strip, which is all that will concern us.

The x and y coordinates of a particle near the bottom of the well oscillate independently with frequencies ν_x, ν_y. The flux ϕ_x of particles across $x = 0$ can be written in two ways. Each of the n particles in the well crosses $x = 0$ in a given sense once every $1/\nu_x$ seconds. Alternatively ϕ_x is n(strip), the number of particles in the strip, times their mean velocity $\bar{v}$ times $\frac{1}{2}$, since half of them are moving the wrong way. So

$$(\phi_x =) \nu_x n = \tfrac{1}{2}\bar{v}n\,(\text{strip}) \tag{38}$$

Similarly the flux of particles across the unit length of the x-axis intercepted by the strip is either the number of particles n(square) in the square, times $\frac{1}{2}\bar{v}$, or it is the number of particles in the strip times the frequency ν_y with which they oscillate parallel to the y-axis. Of course a good many of them will drift sideways out of one side of the strip before they reach the x-axis, but as an equal number will drift in through the other side sideways drift is without effect. So we have

$$\nu_y n\,(\text{strip}) = \tfrac{1}{2}\bar{v}n\,(\text{square}) = \tfrac{1}{2}\bar{v}\rho\,(A) \tag{39}$$

where, because of our choice of unit of length n(square) is equal to the particle density ρ(A) near the bottom of the well. From (38) and (39) we can eliminate n(strip):

$$n = (\tfrac{1}{2}\bar{v})^2 \rho\,(A)/\nu_x\nu_y \tag{40}$$

If we construct similar strips about the y' and x' axes at the saddle-point we get by the same type of argument

$$\phi_{x'} = {}^* = \tfrac{1}{2}\bar{v}n'\,(\text{strip}) \tag{41}$$

in place of (38). The equations (38) and (41) are quite analogous, but, as the brackets indicate, we shall make

no use of the fact that (38) is equal to ϕ_x, and in (41) the $*$ indicates that there is no term $\nu_{x'}n'$ since there is no n' (a saddle-point will not hold water). The analogue of (39) is

$$\nu_{y'} n'\,(\text{strip}) = \tfrac{1}{2}\bar{v}n'\,(\text{square}) = \tfrac{1}{2}\bar{v}\rho\,(B) \tag{42}$$

From (41) and (42) we get

$$\phi_{x'} = (\tfrac{1}{2}\bar{v})^2 \rho\,(B)/\nu_{y'},$$

from which, using (40) and (35),

$$f = \frac{\phi_{x'}}{n} = \frac{\nu_x\nu_y}{\nu_{y'}} e^{-U/kT} \tag{43}$$

Next take Fig. 2 to represent a section through a three-dimensional potential diagram. Mark out planes $x = \pm\frac{1}{2}$ defining a slab whose trace is the strip about the y-axis, then planes $y = \pm\frac{1}{2}$ which intersect the slab in a square rod whose trace is the square at the centre of the well, and finally planes $z = \pm 1$ which cut out a unit cube from the rod. We can now find relations between n(slab), n(rod), n(cube), and analogous quantities for the saddle-point, just as we related n(strip), n(square) n'(strip), n'(square) in two dimensions. It is fairly obvious that in place of (33) or (43) we shall come up with

$$f = \frac{\nu_x\nu_y\nu_z}{\nu_{y'}\nu_{z'}} e^{-U/kT} \tag{44}$$

(For the details put $N = 3$ in (47), (48), (49) below.)

The model of, say, an interstitial particle moving in a fixed potential $V(x, y, z)$ is not realistic, as the lattice producing V deforms as the particle moves. In other cases, vacancy movement or the re-orientation of a split interstitial for example, it may not even be clear what the appropriate defect coordinate is. We then have to treat the whole lattice together. To conclude we shall outline a derivation of Vineyard's classic equation[13,12] (equation (49) below) by a generalization of the method used for (43).

The potential energy of a crystal containing $\frac{1}{3}N$ atoms may be plotted in N-dimensional space. Near a local minimum linear contributions q_i of the coordinates may be chosen so that the potential and kinetic energies take the forms

$$V + \text{const.} = \tfrac{1}{2}\Sigma k_i q_i^2, \qquad T = \tfrac{1}{2}\Sigma m\dot{q}_i^2 \tag{45}$$

with all the k_i positive, and near a saddle-point

$$V + \text{const.}' = \tfrac{1}{2}\Sigma k_{i'} q_{i'}^2, \qquad T = \tfrac{1}{2}\Sigma m\dot{q}_{i'}^2 \tag{46}$$

with, say, $k_{1'}$ negative and the rest positive. The state of the crystal is described by a representative point wandering about in q-space. To find the rate at which the system jumps out of the minimum via the indicated saddle-point we introduce an ensemble of representative points and carry on as in the derivation of (43). The number of representative points in the q-space volume element $dq = dq_1 dq_2 \ldots dq_N$ is proportional to $\exp(-V(q)/kT)dq$. Also the rate of change of any of the q_i has a constant mean value, $\bar{v}$ say (Maxwell distribution). This is a result of the fact that we have arranged for the 'mass' coefficients in (45) to be equal to each other and to those in (46). The only difficulty is one of N-dimensional visualization.

After choosing a sufficiently small unit of length in q-space (cf. the two-dimensional case) mark out successively slabs bounded by the planes $q_1 = \pm\frac{1}{2}$, $q_2 = \pm\frac{1}{2}\ldots$, $q_i = \pm\frac{1}{2}\ldots q_N = \pm\frac{1}{2}$ and let n_i denote the number of representative points in the box R_i formed by the intersection of the first i slabs. Then as an extension of (36) we have

$$(\phi_1 =)\nu_1 n = \tfrac{1}{2}\bar{v}n_1$$
$$\nu_i n_{i-1} = \tfrac{1}{2}\bar{v}n_i$$
$$\nu_N n_{N-1} = \tfrac{1}{2}\bar{v}n_N = \tfrac{1}{2}\bar{v}\rho\,(A);\qquad(47)$$

as in two dimensions sideways leakage has no effect. The middle equation says that the flux into R_i through one of its faces $q_i = \pm\frac{1}{2}$ is equal both to the number of points $\frac{1}{2}n_i$ in R_i itself which are moving in the right direction times $\bar{v}$ their mean velocity, and also to the number of points n_{i-1} in R_{i-1}, the previous region in the hierarchy, times the frequency ν_i with which they oscillate parallel to the q_i-axis. Likewise at the saddle-point we have

$$\phi_{1'} = * = \tfrac{1}{2}\bar{v}n_{1'}$$
$$\nu_{i'}n_{i'-1} = \tfrac{1}{2}\bar{v}n_{i'}$$
$$\nu_{N'}n_{N'-1} = \tfrac{1}{2}\bar{v}n_{N'} = \tfrac{1}{2}\bar{v}\rho\,(B)\qquad(48)$$

from which, eliminating the intermediate n_i from (45) and (46) and using (33) which holds here too, we finally reach the Vineyard formula

$$f = \frac{\phi_{1'}}{n} = \frac{\nu_1\nu_2\ldots\nu_N}{\nu_{2'}\nu_{3'}\ldots\nu_{N'}}e^{-U/kT}\qquad(49)$$

REFERENCES

1 J. D. ESHELBY: in 'Solid state physics' (Ed. F. SEITZ AND D. TURNBULL), 3, 79, 1956, New York, Academic Press
2 A. H. COTTRELL: Report Conf. on the 'Strength of solids', 30, 1948, London, The Physical Society
3 B. A. BILBY: Proc. Phys. Soc., 1950, A63, 191
4 G. ALBENGA: Atti Accad. Sci., Torino, Cl. Sci. fis. mat. nat., 1918/19, 54, 864
5 F. BITTER: Phys. Rev., 1931, 37, 1526
6 J. M. DEWEY: J. Appl. Phys., 1947, 18, 579
7 J. D. ESHELBY, Ann. Physik, 1958, 1, 116
8 J. R. HARDY AND R. BULLOUGH: Phil. Mag., 1967, 15, 1967
9 J. R. HARDY AND R. BULLOUGH: Phil. Mag., 1968, 17, 833
10 R. SIEMS: Physica Status Solidi, 1968, 30, 645
11 J. D. ESHELBY: J. Elasticity, 1975, 5, 321
12 J. D. ESHELBY: 'The physics of metals', 2, 1, 1975 (Ed. P. B. HIRSCH), Cambridge University Press
13 G. H. VINEYARD: J. Phys. Chem. Solids, 1957, 3, 121

Discrete lattice models of point defects

P. T. Heald

The calculation of point defect properties using discrete lattice models is reviewed. Particular attention is given to the distortion field around a point defect and the interaction between point defect pairs. Both real space and reciprocal space methods are considered. There are two direct space approaches commonly used: one involves the direct minimization, with respect to the lattice displacements, of the total energy of N atoms interacting via a pairwise potential $\phi(\mathbf{r})$; alternatively the crystal symmetry may be exploited by group theory procedures if the Defect Green's function method is used. This method is based on the Born–von Karman model of a lattice and uses the zero frequency limit of the phonon Green's function which gives the static response of the defect lattice to an applied force. The reciprocal space approach (the Kanzaki method) is based on the Fourier transformation of the real space equilibrium equations; this results in a set of decoupled equations which can be explicitly solved for the Fourier amplitudes of the displacement field which may then be found by Fourier inversion. The equivalence of the Defect Green's function and Kanzaki methods is demonstrated and it is shown that they reduce to the equivalent elasticity result in the asymptotic, small wave vector, limit. As one might expect, the lattice displacements in the immediate vicinity of a point defect are not adequately described by continuum theory. However, for neighbours more remote than about five atomic spacings it is shown that the displacement field is in agreement with elasticity theory provided that the volume relaxation around the defect is determined from a discrete lattice model. The interaction between near neighbour pairs of point defects is found to depend strongly on direction as well as spacing, but for more widely spaced defects the interaction energy agrees with the continuum result. Specific numerical results for the displacement fields and relaxation volumes of vacancies and interstitials and the interaction of point defect pairs in copper, aluminium, iron, and molybdenum are discussed.

The author is with the CEGB, Berkeley Nuclear Laboratories

If a point defect is modelled as a spherical inclusion of radius $(1+\epsilon)r_0$ placed into a hole of radius r_0, the surfaces being welded together and allowed to relax then, in isotropic elasticity, the inclusion is uniformly compressed to a final radius of $(1+\alpha\epsilon)r_0$ and the displacement field in the (infinite) matrix is[1]

$$\mathbf{u}^{\infty}(\mathbf{r}) = \alpha r_0^3 \epsilon \cdot \frac{\mathbf{r}}{r^3} \qquad (1)$$

where

$$\alpha = \frac{(1+\nu)}{3(1-\nu)} \qquad (2)$$

and ν denotes Poisson's ratio. In a finite crystal additional 'image' displacements $\mathbf{u}^{\mathrm{I}}$ are necessary to annul the surface tractions produced by $\mathbf{u}^{\infty}$ on the crystal surface. Consequently the total displacement field is[2] $\mathbf{u}^{\infty} + \mathbf{u}^{\mathrm{I}}$.

In general it is more convenient to work in terms of the fractional volume change produced by a point defect in a finite crystal:

$$\Delta V = 4\pi r_0^3 \epsilon \qquad (3)$$

With this notation the interaction energy between a system of internal or external stress σ_{jk} and a point defect is given by[3,4]

$$E'_{\mathrm{int}} = -\tfrac{1}{3}\sigma_{\mathrm{kk}} \Delta V \qquad (4)$$

provided that the elastic constants in the inclusion are the same as those in the matrix. However, if the shear modulus μ^* and bulk modulus K^* of the inclusion differ from those in the matrix (μ, K) there is an additional, induced, interaction with the stress field[5]

$$E''_{\mathrm{int}} = -\tfrac{1}{2}\int_{\mathrm{V}} \{Be_{\mathrm{mm}}^2 + 2M'e_{\mathrm{jk}}\,'e_{\mathrm{jk}}\}\,\mathrm{d}V \qquad (5)$$

where

V is the defect volume,

$$'e_{jk} = e_{jk} - \tfrac{1}{3}e_{mm}\delta_{jk},$$

$$B = \frac{K(K - K^*)}{K - \alpha(K - K^*)}, \tag{6}$$

$$M = \frac{\mu(\mu - \mu^*)}{\mu - \beta(\mu - \mu^*)} \tag{7}$$

and

$$\beta = \frac{2(4 - 5\nu)}{15(1 - \nu)} \tag{8}$$

The first-order interaction between two centres of spherical dilatation having relaxation volumes ΔV_a and ΔV_b is identically zero in isotropic media since, from equations (1), (3), and (4),

$$E'_{int} = \frac{\alpha K \Delta V_a \Delta V_b}{4\pi} \cdot \nabla^2\left(\frac{1}{R}\right) = -\alpha K \Delta V_a \Delta V_b \delta(R) \tag{9}$$

Provided that the inclusion is sufficiently small that the strains e_{jk} can be considered constant over the volume the induced interaction is, from equation (5),

$$E''_{int} = -\frac{6V\{M_a(\alpha\,\Delta V_b)^2 + M_b(\alpha\,\Delta V_a)^2\}}{(4\pi)^2 R^6} \tag{10}$$

If we had found the interaction by evaluating equation (5) over a surface surrounding defect (a) we might have expected to get only the first term in equation (10). The second term appears as the 'image' force on (a) due to the inhomogeneous sphere (b). If the defects are identical, a pair of vacancies for example, then

$$E''_{int} = -\frac{12V(\alpha\,\Delta V)^2}{(4\pi)^2 R^6} \cdot \frac{\mu(\mu - \mu^*)}{\mu - \beta(\mu - \mu^*)} \tag{11}$$

The first-order interaction is only zero for the spherically symmetric displacement field given by equation (1). In anisotropic media the displacement field around a vacancy is not spherically symmetric and the first-order interaction between two identical point defects is non-zero. For example, in weakly anisotropic cubic materials with elastic constants $c_{11}, c_{12},$ and c_{44} the displacement field around a vacancy is[5,6]

$$u_1(\mathbf{r}) = \frac{\alpha_A \Delta V}{4\pi} \cdot \frac{x_1}{r^3}\left\{1 - \left(\frac{3d}{2c_{11}}\right)\left[1 - \left(\frac{x_1}{r}\right)^2\right]\right\} \tag{12}$$

where

$$d = c_{11} - c_{12} - 2c_{44} \tag{13}$$

and

$$\alpha_A = \frac{c_{11} + 2c_{12}}{3c_{11}} \tag{14}$$

The first-order interaction between two similar defects is, in this approximation,[5]

$$E'_{int} = -\frac{15\alpha_A^2 \Delta V_a \Delta V_b}{8\pi} \cdot \frac{d}{R^3}\left\{\frac{3}{5} - \frac{(X_1^4 + X_2^4 + X_3^4)}{R^4}\right\} \tag{15}$$

We can calculate the defect formation energy in the continuum approximation as follows: the total *strain* energy in the matrix *and* inclusion due to the defect described by equation (1) is[7]

$$E_{elastic} = 8\pi\alpha\mu\epsilon^2 r_0^3 \tag{16}$$

If we ascribe a surface energy γ to the inclusion matrix interface, the increase in surface energy is

$$E_{surface} = 4\pi r_0^2(1 + \alpha\epsilon)^2\gamma \approx 4\pi r_0^2\gamma + 8\pi\alpha r_0^2\gamma\epsilon \tag{17}$$

Minimizing the sum $E_{elastic} + E_{surface}$ with respect to ϵ gives

$$\epsilon_{equilibrium} = -\gamma/2\mu r_0 \tag{18}$$

and the defect formation energy is found to be

$$E^f = 4\pi r_0^2\gamma - 2\pi\alpha\gamma^2 r_0/\mu \tag{19}$$

Unfortunately the continuum results are not quite as useful as they first appear: it is not possible to determine the parameters $\Delta V, B, M,$ and γ from continuum considerations alone. If numerical values of the displacement field or interaction energies are required it is necessary to estimate the parameters appearing in the continuum model.

These problems are not without considerable practical importance; for example, irradiation swelling occurs because of the difference in magnitude of the relaxation volume around an interstitial, ΔV_I, and that around a vacancy ΔV_V.[8] Irradiation creep has been attributed to the difference between the elastic properties of the defect region and the matrix, $\mu^* - \mu$.[9,10] The diffuse X-ray and thermal neutron scattering from point defects and the electrical resistivity due to point defects all depend on the lattice distortion in the neighbourhood of the defect. It is problems of this nature that discrete lattice theories attempt to describe more precisely.

Specifically lattice models have been used to determine point defect parameters such as the relaxation volume ΔV and the modulus change $\mu^* - \mu$. The displacement fields in the neighbourhood of point defects and the binding of small point defect clusters have been calculated. The interactions of point defects with extended defects have also been studied. In this article we shall concentrate on the intrinsic point defect properties rather than the interaction between point defects and extended defects. In addition the relationship between the lattice theories and the more familiar continuum theories will be discussed in some detail.

THE LATTICE MODEL

The perfect lattice

In second-order pseudopotential theory[11,12] the total energy of a perfect crystal is given by

$$U_0 = \tfrac{1}{2}\sum_{s\neq s'}\phi(\mathbf{s} - \mathbf{s}'; V) + F(V) \tag{20}$$

where $\mathbf{s}$ are the perfect lattice sites and V is the crystal volume. The pair potential $\phi(\mathbf{r}; V)$ is explicitly a function of the volume per atom and $F(V)$ represents a further large volume dependent term. In lattice calculations it is usual and often necessary to make certain simplifying assumptions with regard to equation (20). We shall assume that the pair potential $\phi(\mathbf{r})$ does not explicitly depend on the crystal volume and secondly we shall assume that the volume dependent part of the potential is linear in V, namely

$$U_0 = \tfrac{1}{2}\sum_{s\neq s'}\phi(\mathbf{s} - \mathbf{s}') + pV = \Phi_0 + pV \tag{21}$$

If equation (21) is minimized with respect to the lattice parameter we find that p, the pressure required to maintain the correct lattice parameter, is the Cauchy pressure. That is, for cubic crystals,

$$p = \tfrac{1}{2}(c_{12} - c_{44}) \tag{22}$$

If the atoms suffer displacements $\mathbf{u}(\mathbf{s})$ from their equilibrium positions then the pairwise term in equation (21) becomes

$$\Phi\{\mathbf{u}(\mathbf{s})\} = \tfrac{1}{2} \sum_{\mathbf{s} \neq \mathbf{s}'} \phi\{\mathbf{s} + \mathbf{u}(\mathbf{s}) - \mathbf{s}' - \mathbf{u}(\mathbf{s}')\} \tag{23}$$

which, for small displacements may be expanded in powers of $\mathbf{u}(\mathbf{s})$ to give

$$\Phi\{\mathbf{u}(\mathbf{s})\} = \Phi_0 + \sum_{\mathbf{s}} \left\{ \frac{\partial \Phi}{\partial u_j(\mathbf{s})} \right\}_0 u_j(\mathbf{s})$$
$$+ \tfrac{1}{2} \sum_{\mathbf{s} \neq \mathbf{s}'} \left\{ \frac{\partial^2 \Phi}{\partial u_j(\mathbf{s}) \, \partial u_k(\mathbf{s}')} \right\}_0 u_j(\mathbf{s}) u_k(\mathbf{s}') + \dots \tag{24}$$

where the subscript zero indicates that the derivatives are evaluated at the perfect lattice sites. Because of invariance requirements enforced by the lattice symmetry and stability[13] we have

(i) $\quad \left\{ \dfrac{\partial \Phi}{\partial u_j(\mathbf{s})} \right\}_0 = 0 \quad$ for all $\mathbf{s}$ \hfill (25)

(ii) $\quad \Phi_{jk}(\mathbf{s} - \mathbf{s}') = \left\{ \dfrac{\partial^2 \Phi}{\partial u_j(\mathbf{s}) \, \partial u_k(\mathbf{s}')} \right\}_0 \hfill$ (26)

depends on $\mathbf{s}$ and $\mathbf{s}'$ only through their difference and can therefore be labelled by the single index $\mathbf{s} - \mathbf{s}'$.

(iii) $\quad \sum_{\mathbf{s}'} \Phi_{jk}(\mathbf{s} - \mathbf{s}') = 0 \quad$ for all $\mathbf{s}$ \hfill (27)

In addition to the change in $\Phi\{\mathbf{u}(\mathbf{s})\}$ caused by the displacements the crystal volume will also change to a value $V + V^f$ where V^f is the defect formation volume. In the absence of relaxation V^f is one atomic volume for a vacancy. More generally $V_V^f = \Omega - \Delta V_V$ where ΔV_V is the relaxation volume around a vacancy. For an interstitial $V_I^f = \Delta V_I - \Omega$.

The defect lattice

The effect of introducing a point defect into an otherwise perfect lattice may be considered from two different but equivalent points of view. Since both are widely used it is advantageous to discuss both points of view. In the Kanzaki method developed by Kanzaki[14] and Hardy and his co-workers[15-18] the point defect is assumed to interact directly with the host atoms through a pairwise potential $\psi(\mathbf{r})$ and, as a consequence, the host atoms will be displaced from their perfect lattice sites. Due to this relaxation and because an atom will be added or removed from the crystal surface in order to create a vacancy or interstitial the crystal volume will be changed to $V + V^f$ where V^f is the defect formation volume. The total energy of the defect lattice may be written

$$U\{\mathbf{u}(\mathbf{s})\} = \sum_{\mathbf{s}} \psi\{\mathbf{s} + \mathbf{u}(\mathbf{s})\} + \tfrac{1}{2} \sum_{\mathbf{s} \neq \mathbf{s}'} \phi\{\mathbf{s} + \mathbf{u}(\mathbf{s}) - \mathbf{s}' - \mathbf{u}(\mathbf{s}')\}$$
$$+ p(V + V^f) \tag{28}$$

where the point defect is taken to be at the origin. If we expand the second term in equation (28) as far as the harmonic approximation we have

$$U\{\mathbf{u}(\mathbf{s})\} = \sum_{\mathbf{s}} \psi\{\mathbf{s} + \mathbf{u}(\mathbf{s})\} + \Phi_0$$
$$+ \tfrac{1}{2} \sum_{\mathbf{s} \neq \mathbf{s}'} \Phi_{jk}(\mathbf{s} - \mathbf{s}') u_j(\mathbf{s}) u_k(\mathbf{s}') + p(V + V^f) \tag{29}$$

and the defect formation energy is then

$$E^f\{\mathbf{u}(\mathbf{s})\} = U\{\mathbf{u}(\mathbf{s})\} - U_0 = \Psi\{\mathbf{u}(\mathbf{s})\}$$
$$+ \tfrac{1}{2} \sum_{\mathbf{s} \neq \mathbf{s}'} \Phi_{jk}(\mathbf{s} - \mathbf{s}') u_j(\mathbf{s}) u_k(\mathbf{s}') + p V^f \tag{30}$$

where

$$\Psi\{\mathbf{u}(\mathbf{s})\} = \sum_{\mathbf{s}} \psi\{\mathbf{s} + \mathbf{u}(\mathbf{s})\} \tag{31}$$

Minimizing $E^f\{\mathbf{u}(\mathbf{s})\}$ with respect to the lattice displacements gives

$$\frac{\partial E^f\{u(\mathbf{s})\}}{\partial u_j(\mathbf{s})} = 0 = \frac{\partial \Psi\{\mathbf{u}(\mathbf{s})\}}{\partial u_j(\mathbf{s})}$$
$$+ \sum_{\mathbf{s}'} \Phi_{jk}(\mathbf{s} - \mathbf{s}') u_k(\mathbf{s}') \tag{32}$$

The forces exerted by the point defect on the host atoms in their displaced positions may be defined as

$$F_j\{\mathbf{s} + \mathbf{u}(\mathbf{s})\} = -\frac{\partial \Psi\{\mathbf{u}(\mathbf{s})\}}{\partial u_j(\mathbf{s})} = \sum_{\mathbf{s}'} \Phi_{jk}(\mathbf{s} - \mathbf{s}') u_k(\mathbf{s}') \tag{33}$$

and the displacement field is given by

$$u_j(\mathbf{s}) = \sum_{\mathbf{s}'} U_{jk}(\mathbf{s} - \mathbf{s}') F_k\{\mathbf{s}' + \mathbf{u}(\mathbf{s}')\} \tag{34}$$

where $U_{jk}(\mathbf{s} - \mathbf{s}')$ is the inverse of $\Phi_{jk}(\mathbf{s} - \mathbf{s}')$ namely the perfect lattice Green's function. Equation (34) may be solved by an iterative procedure:

$$u_j(\mathbf{s}) = \sum_{\mathbf{s}'} U_{jk}(\mathbf{s} - \mathbf{s}') F_k(\mathbf{s}')$$
$$- \sum_{\mathbf{s}', \mathbf{s}''} U_{jk}(\mathbf{s} - \mathbf{s}') \Psi_{km}(\mathbf{s}', \mathbf{s}'') u_m(\mathbf{s}'') \tag{35}$$

where $F_k(\mathbf{s}')$ are the forces exerted by the defect on the host atoms in their perfect lattice sites and

$$\Psi_{jk}(\mathbf{s}', \mathbf{s}'') = \left\{ \frac{\partial^2 \Psi\{\mathbf{u}(\mathbf{s}')\}}{\partial u_j(\mathbf{s}') \, \partial u_k(\mathbf{s}'')} \right\}_0 \tag{36}$$

by virtue of equation (31) we see that

$$\Psi_{jk}(\mathbf{s}; \mathbf{s}') = 0 \quad \text{for } \mathbf{s} \neq \mathbf{s}' \tag{37}$$

Consequently $\Psi_{jk}(\mathbf{s}; \mathbf{s}')$ may be described by a single index.

The defect formation energy is given by equation (30) subject to the condition (32) and since

$$\Psi\{\mathbf{u}(\mathbf{s})\} = \Psi(\mathbf{s}) + \sum_{\mathbf{s}} \frac{\partial \Psi\{\mathbf{u}(\mathbf{s})\}}{\partial u_j(\mathbf{s})} \bigg|_{u=0} u_j(\mathbf{s}) + \dots$$
$$= \Psi(\mathbf{s}) - \sum_{\mathbf{s}} F_j(\mathbf{s}) \cdot u_j(\mathbf{s}) + \dots$$

then, to first order in the displacements, we have

$$E^f = \psi(\mathbf{s}) - \tfrac{1}{2} \sum_{\mathbf{s}} F_j(\mathbf{s}) \cdot u_j(\mathbf{s}) + p V^f \tag{38}$$

The first term in this expression is a geometric term which corresponds to the broken bond model; for example, we have for a vacancy in a bcc lattice

$$E_V^g = \Psi(\mathbf{s}) = -4\phi(r_1) - 3\phi(r_2) + \dots$$

The second term in equation (38) is the bond relaxation energy

$$E^r = \tfrac{1}{2}\sum_{\mathbf{s}} F_j(\mathbf{s}) \cdot u_j(\mathbf{s})$$

and pV^f is the volume-dependent contribution. Thus the defect formation energy may be visualized as being the sum of these contributions, namely

$$E^f = E^g - E^r + pV^f \tag{39}$$

The final, volume-dependent term is often omitted in calculations of defect formation energies despite the fact that it is comparable in magnitude with the other terms. However, since the displacement fields depend only on the gradient and curvature of the potential rather than its absolute value the calculation of the relaxation volumes may be correct despite any error in the energies.

If we had computed the vacancy formation energy at constant volume rather than constant pressure the pV^f term would be absent from equation (38); however the lattice vectors would be uniformly compressed so that $\mathbf{s} \to \mathbf{t} = \mathbf{s}(1 - V^f/3V)$ and, by virtue of the lattice equilibrium condition, the formation energy at constant volume is identical to that at constant pressure.

The alternative approach is to regard the point defect as having changed the host–host interatomic interactions from $\phi(\mathbf{r})$ to $\phi^*(\mathbf{r})$; this is the basis of the Defect Green's function method originally developed by Maradudin[13] for lattice dynamical problems and applied to static defect calculations by Grecu and Croitoru[20] and Tewary.[21] Due to the presence of the defect the host atoms are displaced from their perfect lattice sites and the crystal energy is, in the harmonic approximation

$$U\{\mathbf{u}(\mathbf{s})\} = \Phi_0^* + \sum_{\mathbf{s}} \left\{\frac{\partial \Phi^*}{\partial u_j(\mathbf{s})}\right\}_0 u_j(\mathbf{s})$$

$$+ \tfrac{1}{2}\sum_{\mathbf{s} \neq \mathbf{s}} \left\{\frac{\partial^2 \Phi^*}{\partial u_j(\mathbf{s})\,\partial u_k(\mathbf{s}')}\right\}_0 u_j(\mathbf{s})u_k(\mathbf{s}') + p(V + V^f)$$

Minimizing $U\{\mathbf{u}(\mathbf{s})\}$ with respect to the displacements gives

$$F_j^*(\mathbf{s}) = \sum_{\mathbf{s}'} \Phi_{jk}^*(\mathbf{s}; \mathbf{s}')u_k(\mathbf{s}') \tag{40}$$

where

$$F_j^*(\mathbf{s}) = -\left\{\frac{\partial \Phi^*}{\partial u_j(\mathbf{s})}\right\}_0 \tag{41}$$

and

$$\Phi_{jk}^*(\mathbf{s}, \mathbf{s}') = \left\{\frac{\partial^2 \Phi^*}{\partial u_j(\mathbf{s})\,\partial u_k(\mathbf{s}')}\right\}_0$$

since the defect has destroyed the translational invariance of the crystal $\Phi_{jk}^*(\mathbf{s}, \mathbf{s}')$ depends on $\mathbf{s}$ and $\mathbf{s}'$ independently. Inverting equation (40) gives the displacements

$$u_j(\mathbf{s}) = \sum_{\mathbf{s}'} U_{jk}^*(\mathbf{s}, \mathbf{s}')F_k^*(\mathbf{s}') \tag{42}$$

where $U_{jk}^*(\mathbf{s}, \mathbf{s}')$ is the defect lattice Green's function.

If $\Delta\Phi_{jk}^*(\mathbf{s}, \mathbf{s}')$ is the local change in the force constant matrix due to the defect we may write

$$\Phi_{jk}^*(\mathbf{s}, \mathbf{s}') = \Phi_{jk}(\mathbf{s} - \mathbf{s}') + \Delta\Phi_{jk}(\mathbf{s}, \mathbf{s}') \tag{43}$$

then

$$U_{jk}^*(\mathbf{s}, \mathbf{s}') = U_{jk}(\mathbf{s} - \mathbf{s}')$$

$$- \sum_{\mathbf{l}, \mathbf{l}'} U_{jm}(\mathbf{s} - \mathbf{l})\,\Delta\Phi_{mn}^*(\mathbf{l}, \mathbf{l}')U_{nk}^*(\mathbf{l}', \mathbf{s}') \tag{44}$$

Comparing these results with equation (35) we have

$$\left.\begin{array}{l} \mathbf{F}^*(\mathbf{s}) = \mathbf{F}(\mathbf{s}) \\[2mm] \text{and} \\[2mm] \Delta\Phi_{jk}^*(\mathbf{s}, \mathbf{s}') = \Psi_{jk}(\mathbf{s}, \mathbf{s}') \end{array}\right\} \tag{45}$$

Point defect interactions

If two point defects, labelled (a) and (b), are introduced individually into an otherwise perfect lattice the displacement field they would produce is given by equation (34) as

$$u_j^a(\mathbf{s}) = \sum_{\mathbf{s}'} U_{jk}(\mathbf{s} - \mathbf{s}')F_k^a\{\mathbf{s}' + \mathbf{u}(\mathbf{s}')\}$$

and a similar equation for defect (b). However, if defect (b) is introduced into a lattice already containing defect (a) the Green's function appropriate to defect (b) is the defect lattice Green's function $U_{jk}^a(\mathbf{s}, \mathbf{s}')$ not the perfect lattice Green's function. Consequently the total displacement field produced by both defects is

$$u_j(\mathbf{s}) = u_j^a(\mathbf{s}) + u_j^b(\mathbf{s}) + v_j(\mathbf{s})$$

$$= \sum_{\mathbf{s}'} U_{jk}(\mathbf{s} - \mathbf{s}')F^a\{\mathbf{s}' + \mathbf{u}(\mathbf{s}')\} \tag{46}$$

$$+ \sum_{\mathbf{s}'} U_{jk}^a(\mathbf{s}, \mathbf{s}')F_k^b\{\mathbf{s}' + \mathbf{u}(\mathbf{s}')\}$$

where $\mathbf{u}^{a,b}(\mathbf{s})$ is the displacement field that would be produced if either defect (a) or (b) was there alone and $\mathbf{v}(\mathbf{s})$ is the difference caused by the presence of the other defect. Thus

$$v_j(\mathbf{s}) = \sum_{\mathbf{s}'} \{U_{jk}^a(\mathbf{s}, \mathbf{s}') - U_{jk}(\mathbf{s} - \mathbf{s}')\}F_k^b\{\mathbf{s}' + \mathbf{u}(\mathbf{s}')\} \tag{47}$$

Since the force due to an isolated defect on the atoms in their displaced positions is $F^b\{\mathbf{s} + \mathbf{u}(\mathbf{s})\}$ the work done on introducing the second defect is

$$E = -\sum_{\mathbf{s}} F_j^b\{\mathbf{s} + \mathbf{u}(\mathbf{s})\} \cdot \{u_j^a(\mathbf{s}) + v_j(\mathbf{s})\} \tag{48}$$

and the first- and second-order interaction energies are

$$E'_{int} = -\sum_{\mathbf{s}} F_j^b\{\mathbf{s} + \mathbf{u}(\mathbf{s})\} \cdot u_j^a(\mathbf{s})$$

$$= -\sum_{\mathbf{s}, \mathbf{s}'} F_j^b\{\mathbf{s} + \mathbf{u}(\mathbf{s})\}U_{jk}(\mathbf{s} - \mathbf{s}')F^a\{\mathbf{s}' + \mathbf{u}(\mathbf{s}')\} \tag{49}$$

and

$$E''_{int} = -\sum_{\mathbf{s}} F_j^b\{\mathbf{s} + \mathbf{u}(\mathbf{s})\}v_j(\mathbf{s})$$

$$= \sum_{\substack{\mathbf{s}, \mathbf{s}' \\ \mathbf{l}, \mathbf{l}'}} F_j^b\{\mathbf{s} + \mathbf{u}(\mathbf{s})\}U_{jk}(\mathbf{s} - \mathbf{s}')\,\Delta\Phi_{km}^a(\mathbf{s}', \mathbf{l})$$

$$\times U_{mn}^a(\mathbf{l}, \mathbf{l}')F_n^b\{\mathbf{l}' + \mathbf{u}(\mathbf{l}')\} \tag{50}$$

respectively. To first order in $\Delta\Phi^a_{jk}$ equation (50) reduces to

$$E''_{int} = \sum_{s,s'} u^b_j(s)\,\Delta\Phi^a_{jk}(s,s')u^b_k(s') \qquad (51)$$

Since $\Delta\Phi^a_{jk}$ is only non-zero in the vicinity of defect (a) we may, for widely spaced defects, expand the displacement field $u^b(s)$ about defect (a):

$$u^b_j(s) = u^b_j(R+s^a) = u^b_j(R) + \frac{\partial u^b_j}{\partial x_k}s^a_k + \ldots \qquad (52)$$

where R is the defect separation and s^a are those lattice sites near defect (a) which have non-zero values of $\Delta\Phi^a_{jk}$. Since, by symmetry $\sum_{s'}\Delta\Phi^a_{jk}(s,s') = 0$ equation (52) reduces to

$$E''_{int} = \epsilon^b_{jm}(R)\Lambda_{jkmn}\epsilon^b_{kn}(R) \qquad (53)$$

where

$$\epsilon_{jm}(R) = \left(\frac{\partial u_j}{\partial x_k}\right)_R$$

and

$$\Lambda_{jkmn} = \sum_{s,s'} s^a_j\,\Delta\Phi^a_{km}(s,s')s'^a_n$$

$$\qquad (54)$$

$$= \sum_s s^a_j\Psi^a_{km}(s^a)s^a_n$$

In an isotropic continuum $\epsilon_{jk}(R)\alpha R^{-3}$ (equation (1)) and thus we have the result that $E''_{int}\alpha R^{-6}$. By equating the continuum result (equation (11)) to equation (53) we have a relationship between the defect modulus μ^* and the defect forces Ψ_{km}.

RELATIONSHIP BETWEEN THE LATTICE AND CONTINUUM MODELS

With real space discrete lattice models of the type pioneered by Tewordt[22] and Gibson *et al.*[23] the interatomic potential is assumed to act directly between the atoms in a discrete set near the defect and appropriate boundary conditions are applied to these atoms. The total energy of the configuration is minimized using numerical methods and consequently it is not possible to demonstrate analytically the asymptotic equivalence of these models and the continuum model. However, both the Green's function method and the Kanzaki method exploit the crystal symmetry by analytic methods and in both cases it is possible to show that the lattice equations reduce to the familiar elasticity results in the asymptotic limit. In this section we shall use the Kanzaki method to demonstrate this equivalence since the Fourier transform methods used are formally similar to the techniques used in lattice dynamics.

The Kanzaki method exploits the crystal symmetry by the use of the Fourier transforms of the displacements, the force constant matrix, and the forces, namely

$$Q_j(q) = \sum_s u_j(s)\exp(-i\,q\cdot s) \qquad (55)$$

$$\Lambda_{jk}(q) = \sum_s \Phi_{jk}(s)\exp(-i\,q\cdot s) \qquad (56)$$

and

$$\Gamma_j(q;u) = \sum_s F_j\{s+u(s)\}\exp(-i\,q\cdot s) \qquad (57)$$

where the sum extends over the N position vectors of the atoms. We shall regard the crystallite of N atoms as being one cell of an infinite superlattice, each supercell containing N unit cells and the appropriate defect. Accordingly we may impose periodic boundary conditions to determine the q vectors which are the N independent q vectors in the first Brillouin zone of the lattice. Thus we have

$$u_j(s) = \frac{1}{N}\sum_q Q_j(q)\exp(i\,q\cdot s) \qquad (58)$$

$$\Phi_{jk}(s) = \frac{1}{N}\sum_q \Lambda_{jk}(q)\exp(i\,q\cdot s) \qquad (59)$$

and

$$F_j\{s+u(s)\} = \frac{1}{N}\sum_q \Gamma_j(q,u)\exp(i\,q\cdot s) \qquad (60)$$

The notation $\Gamma_j(q,u)$ is used to indicate that the relaxed forces are used in equation (57) and $\Lambda_{jk}(q)$ is equivalent to the dynamical matrix used in lattice dynamics.

From equations (34) and (55)–(57) we may write

$$\Gamma_j(q,u) = \Lambda_{jk}(q)Q_k(q). \qquad (61)$$

Thus the $3N$ equations of the real space formulation are reduced to 3 reciprocal space equations which may be inverted analytically to give

$$Q_j(q) = \Omega_{jk}(q)\Gamma_k(q,u) \qquad (62)$$

where $\Omega_{jk}(q)$ is the inverse of $\Lambda_{jk}(q)$, namely

$$\Omega_{jk}(q)\Lambda_{km}(q) = \delta_{jm} \qquad (63)$$

The real space lattice displacements are now obtained from equations (58) and (62) as

$$u_j(s) = \frac{1}{N}\sum_q \Omega_{jk}(q)\Gamma_k(q,u)\exp(i\,q\cdot s) \qquad (64)$$

For a primitive lattice with one atom per unit cell Λ_{jk}, and consequently Ω_{jk} is real while the transformed forces Γ_k are imaginary with no real part. Under these conditions (64) reduces to

$$u_j(s) = \frac{i}{N}\sum_q \Omega_{jk}(q)\Gamma_k(q,u)\sin q\cdot s \qquad (65)$$

If the gradient and curvature of the potential functions $\phi(r)$ and $\psi(r)$ are known then the lattice displacements may be calculated from equation (65) by an iterative procedure. Specifically the unrelaxed forces are used to calculate a first approximation to the displacements which are then used to calculate an approximation to the relaxed forces etc.

The first-order strain field interaction between two point defects is, from equation (49),

$$E'_{int}(a,b) = -\frac{1}{N}\sum_q \Gamma^a_j(-q,u)\cdot Q^b_j(q)$$

$$\qquad (66)$$

$$= -\frac{1}{N}\sum_q \Gamma^a_j(-q,u)\Omega_{jk}(q)\Gamma^b_k(q,u)$$

For two identical defects Γ^a and Γ^b differ only by a 'phase factor' $\exp(i\,q\cdot R)$, where R is their separation. Thus, for identical defects,

$$E'_{int}(a,a) = -\frac{1}{N}\sum_q \Gamma_j(-q,u)\Omega_{jk}(q)\Gamma_j(q,u)\cos q\cdot R$$

$$\qquad (67)$$

Before discussing specific models it is worth noting a general result concerning the relaxation volume ΔV first derived by Hardy.[19] The elastic displacement field given by equation (1) may be regarded as being produced by the distribution of body force[24]

$$F_j(\mathbf{r}) = -G\frac{\partial}{\partial x_j}\{\delta(\mathbf{r})\}$$

or, more generally, for ellipsoidal inclusions

$$F_j(\mathbf{r}) = -G_{jk}\frac{\partial}{\partial x_k}\{\delta(\mathbf{r})\} \tag{68}$$

The Fourier transform of this distribution of body force is

$$\Gamma_j(\mathbf{q}) = \int_{\mathbf{r}} F_j(\mathbf{r})\exp(-i\mathbf{q}.\mathbf{r})\,d\mathbf{r} = -iG_{jk}q_k \tag{69}$$

while the equivalent lattice result is in the small wave vector limit ($\mathbf{q}\to 0$)

$$\Gamma_j(\mathbf{q}) = \sum_{\mathbf{s}} F_j\{\mathbf{s}+\mathbf{u}(\mathbf{s})\}\exp(-i\mathbf{q}.\mathbf{s})$$

$$= -iq_k\sum_{\mathbf{s}} s_k F_j\{\mathbf{s}+\mathbf{u}(\mathbf{s})\} + 0(q^2) \tag{70}$$

since $\sum_{\mathbf{s}}\mathbf{F}.(\mathbf{s}) = 0$ by symmetry. From equations (69) and (70) we have, if the displacement fields are to be compatible,

$$G_{jk} = \sum_{\mathbf{s}} s_k F_j\{\mathbf{s}+\mathbf{u}(\mathbf{s})\} \tag{71}$$

and the relaxation volume is[24]

$$\Delta V = \frac{1}{3K}\int_{\mathbf{r}}\mathbf{r}.\mathbf{F}\,d\mathbf{r} = \frac{G_{kk}}{3K} = \frac{1}{3K}\sum_{\mathbf{s}} s_k F_k\{\mathbf{s}+\mathbf{u}(\mathbf{s})\} \tag{72}$$

Thus the relaxation volume which determines the continuum displacement field is related to the defect–host atom interaction by[19]

$$\Delta V = -\frac{1}{3K}\sum_{\mathbf{s}} s_k \left.\frac{\partial\psi}{\partial x_k}\right|_{\mathbf{s}+\mathbf{u}(\mathbf{s})} \tag{73}$$

If a vacancy is modelled by taking $\psi(\mathbf{r}) = -\phi(\mathbf{r})$ then

$$\Delta V = \frac{1}{3K}\sum_{\mathbf{s}} s_k \left.\frac{\partial\phi}{\partial x_k}\right|_{\mathbf{s}+\mathbf{u}(\mathbf{s})}$$

$$\approx \frac{1}{3K}\left\{\sum_{\mathbf{s}} s_k \left.\frac{\partial\phi}{\partial x_k}\right|_{\mathbf{s}} + \sum_{\mathbf{s}} s_k \frac{\partial^2\phi}{\partial x_k\,\partial x_j}u_j(\mathbf{s}) + \ldots\right\} \tag{74}$$

Consequently the first-order approximation to the relaxation volume around a vacancy is[25]

$$\Delta V^0 = \frac{1}{3K}\sum_{\mathbf{s}} s_k \left.\frac{\partial\phi}{\partial x_k}\right|_{\mathbf{s}} = -\frac{2p\Omega}{K} \tag{75}$$

where p is the Cauchy pressure.

In order to obtain more detailed results it is necessary to consider specific lattice models. Initially we shall begin by considering one of the simplest models: a face-centred cubic lattice held together by first neighbour forces.

Face-centred cubic lattice: nearest neighbour generalized forces

This model is convenient because there are only three independent force constants[26] for example

$$\Phi_{jk}[\mathbf{s}=(1,1,0)a] = -\begin{pmatrix} \beta & \gamma & 0 \\ \gamma & \beta & 0 \\ 0 & 0 & \alpha \end{pmatrix} \tag{76}$$

The values for the other eleven nearest neighbour $\mathbf{s}$ vectors may be obtained by applying the symmetry operations of the fcc lattice to (76). The Fourier transform of the force constant matrix is (equation (56))

$$\Lambda_{jk}(\mathbf{q}) = \sum_{\mathbf{s}}\Phi_{jk}(\mathbf{s})\exp(-i\mathbf{q}.\mathbf{s})$$

$$= 4\begin{pmatrix} H_1 & \gamma s_1 s_2 & \gamma s_1 s_3 \\ \gamma s_2 s_1 & H_2 & \gamma s_2 s_3 \\ \gamma s_3 s_1 & \gamma s_3 s_2 & H_3 \end{pmatrix} \tag{77}$$

where the summation extends over nearest neighbours and

$$H_1 = \alpha(1-c_2 c_3) + \beta(2 - c_1 c_2 - c_1 c_3)\text{ etc.} \tag{78}$$

In equations (77) and (78)

$$s_\alpha = \sin q_\alpha a \quad\text{and}\quad c_\alpha = \cos q_\alpha a \tag{79}$$

For small wave vectors (the long wavelength limit) equation (77) reduces to

$$\Lambda_{jk}(\mathbf{q}\to 0) = 4\begin{pmatrix} \bar{H}_1 & \gamma q_1 q_2 a^2 & \gamma q_1 q_3 a^2 \\ \gamma q_2 q_1 a^2 & \bar{H}_2 & \gamma q_2 q_3 a^2 \\ \gamma q_3 q_1 a^2 & \gamma q_3 q_2 a^2 & \bar{H}_3 \end{pmatrix} \tag{80}$$

where

$$\bar{H}_1 = \frac{(\beta-\alpha)a^2}{2}\cdot q_1^2 + \frac{(\beta+\alpha)a^2}{2}(q_1^2+q_2^2+q_3^2)\text{ etc.} \tag{81}$$

By comparing equation (80) with the corresponding elements of the Navier–Stokes equations of continuum elasticity (i.e. with $\Omega c_{jklm}q_k q_m$) we find that

$$\alpha = a(2c_{44}-c_{11})/2, \qquad \beta = ac_{11}/2;$$

$$\gamma = a(c_{12}+c_{44})/2 \tag{82}$$

and consequently the force constant matrix in (77) is completely determined in terms of the elastic constants.

This model has been used to discuss point defect interactions in the fcc metals[27] and the following discussion is a more detailed account of that work. The real space forces $\mathbf{F}(\mathbf{s},\mathbf{u})$ may, for a spherical centre of dilatation, be written in the form

$$F_j(\mathbf{s},\mathbf{u}) = \frac{s_j}{|\mathbf{s}|}\mathscr{F}(\mathbf{u}) \tag{83}$$

where $\mathscr{F}(\mathbf{u})$ is independent of direction (i.e. a radial force). Consequently

$$\Gamma_j(\mathbf{s},\mathbf{u}) = \sum_{\mathbf{s}}\mathscr{F}(\mathbf{u})\frac{s_j}{|\mathbf{s}|}\exp-i\mathbf{q}.\mathbf{s}$$

$$= -i2\sqrt{2}\mathscr{F}(\mathbf{u})s_1(c_2+c_3) \tag{84}$$

and, from equation (72), we have

$$\Delta V = \frac{4\sqrt{2}a\mathscr{F}(\mathbf{u})}{K} \qquad (85)$$

for this model. The displacement field around a defect in the lattice described by equation (76) may be calculated by inserting the forces given by equation (84) and the inverse of $\Lambda_{jk}(\mathbf{q})$ from equation (77) into equation (65) and evaluating the sum. However rather than discuss detailed numerical results it is instructive to consider the correspondence between continuum elasticity and the lattice model. If we expand $\Gamma_j(\mathbf{q}, \mathbf{u})$ and $\Lambda_{jk}(\mathbf{q})$ as a power series in $\mathbf{q}$ we may write

$$\Gamma_j(\mathbf{q}, \mathbf{u}) = \Gamma_j^{(1)}(\mathbf{q}, \mathbf{u}) + \Gamma_j^{(3)}(\mathbf{q}, \mathbf{u}) + \dots \qquad (86)$$

and

$$\Lambda_{jk}(\mathbf{q}) = \Lambda_{jk}^{(2)}(\mathbf{q}) + \Lambda_{jk}^{(4)}(\mathbf{q}) + \dots \qquad (87)$$

where the superscript denotes the order in q of that term, for example

$$\mathbf{\Gamma}^{(1)}(\mathbf{q}; \mathbf{u}) = -i4\sqrt{2}\mathscr{F}(\mathbf{u})\mathbf{q}. \qquad (88)$$

Furthermore, if the deviation from isotropy is small (that is $d/c_{44} \ll 1$) then we may write $\Lambda_{jk}^{(2)}(\mathbf{q})$ in the form

$$\Lambda_{jk}^{(2)} = \Lambda_{jk}^I + \Lambda_{jk}^A + \text{terms of order } (d/c_{44})^2 \qquad (89)$$

where

$$\Lambda_{jk}^I(\mathbf{q}) = 2a^3 c_{44}\delta_{jk}q^2 + 2a^3(c_{12}+c_{44})q_j q_k \qquad (90)$$

and

$$\Lambda_{jk}^A(\mathbf{q}) = 2a^3 d \begin{pmatrix} q_1^2 & 0 & 0 \\ 0 & q_2^2 & 0 \\ 0 & 0 & q_3^2 \end{pmatrix} \qquad (91)$$

By inverting $\Lambda_{jk}^{(2)}$ we obtain $\Omega_{jk}^{(-2)}(\mathbf{q})$ to first order in (d/c_{44}) as

$$\Omega_{jk}^{(-2)} = \Omega_{jk}^I - \Omega_{jm}^I \Lambda_{mn}^A \Omega_{nk}^I \qquad (92)$$

where

$$\Omega_{jk}^I = \frac{1}{2a^3 c_{44}q^2}\left\{\delta_{jk} - \frac{(c_{12}+c_{44})}{(c_{12}+2c_{44})}q_j q_k\right\} \qquad (93)$$

In the long wavelength limit the sum in equation (65) may be replaced by an integral

$$\frac{1}{N}\sum_{\mathbf{q}} \to \frac{\Omega}{(2\pi)^3}\int_{\mathbf{q}} d\mathbf{q}$$

where Ω is the atomic volume. To lowest order in q equation (65) becomes

$$u_j(\mathbf{s}) = \frac{\Omega}{(2\pi)^3}\int_{\mathbf{q}} \Omega_{jk}^{(-2)}(\mathbf{q})\Gamma_k^{(1)}(\mathbf{q}, \mathbf{u})\sin \mathbf{q}.\mathbf{s}\, d\mathbf{q} \qquad (94)$$

For weakly anisotropic materials we have from equation (92)

$$u_j(\mathbf{s}) = \frac{\Omega}{(2\pi)^3}\int_{\mathbf{q}} [\Omega_{jk}^I - \Omega_{jm}^I \Lambda_{mn}^A \Omega_{nk}^I]$$
$$\times \Gamma_k^{(1)}(\mathbf{q}, \mathbf{u})\sin \mathbf{q}.\mathbf{s}\, d\mathbf{q} \qquad (95)$$

evaluating the integral leads to

$$u_1(\mathbf{s}) = \frac{4\sqrt{2}a\mathscr{F}(\mathbf{u})}{4\pi c_{11}} \cdot \frac{s_1}{|\mathbf{s}|^3}\left\{1 - \left(\frac{3d}{2c_{11}}\right)\left[1 - \left(\frac{s_1}{|\mathbf{s}|}\right)^2\right]\right\} \qquad (96)$$

which, with the aid of equation (85), reduces to the elasticity result given by equation (12). Thus the displacement field calculated from lattice theory is identical to the equivalent continuum result in the long wavelength limit. In a similar way the interaction energy E'_{int} may be written[27]

$$E'_{int} = -\frac{(2a^3)}{(2\pi)^3}\int_{\mathbf{q}} \left\{\Gamma_j^{(1)}(-\mathbf{q})\Omega_{jk}^{(-2)}(\mathbf{q})\Gamma_k^{(1)}(\mathbf{q})\right.$$
$$+ \Gamma_j^{(1)}(-\mathbf{q})\Omega_{jk}^{(-2)}(\mathbf{q})\Gamma_k^{(3)}(\mathbf{q})$$
$$+ \Gamma_j^{(3)}(-\mathbf{q})\Omega_{jk}^{(-2)}(\mathbf{q})\Gamma_k^{(1)}(\mathbf{q})$$
$$- \Gamma_j^{(1)}(-\mathbf{q})\Omega_{jm}^{(-2)}(\mathbf{q})\Lambda_{mn}^{(4)}(\mathbf{q})\Omega_{nk}^{(-2)}$$
$$\left. \times (\mathbf{q})\Gamma_k^{(1)}(\mathbf{q})\right\} \cos(\mathbf{q}.\mathbf{R})\, d\mathbf{q} \qquad (97)$$

The first term, of zero order in q, gives the delta function term of (7) together with the anisotropy term (13). The second and third term, of $0|q^2|$, arise because of the non local nature of the forces. The last term, also of $0|q^2|$ arises from the dispersion of the lattice. Since, for large values of R, $\cos \mathbf{q}.\mathbf{R}$ is rapidly oscillating, contributions to the integral come only from the vicinity of the pole at $\mathbf{q} = 0$ and we have[27]*

$$E'_{int} = -\frac{(4\sqrt{2}\mathscr{F}a)^2 a^2 \pi(1-2\nu)}{48(2\pi)^3 \mu(1-\nu)}$$
$$\times \left\{\frac{42(15-3\nu)}{(1-\nu)}\left[-\frac{3}{R^5} + \frac{5(X_1^4 + X_2^4 + X_3^4)}{R^9}\right]\right.$$
$$-\frac{405}{1-\nu}\left[-\frac{12}{R^5} + \frac{35(X_1^4 + X_2^4 + X_3^4)}{R^9}\right.$$
$$\left.\left. -\frac{21(X_1^6 + X_2^6 + X_3^6)}{R^{11}}\right]\right\} \qquad (98)$$

This interaction which is of order R^{-5} was first given by Hardy and Bullough[16] who gave a different angular variation due to the rather artificial lattice model which they used. For models which include only first neighbour interactions 4/3 of E_{int} comes from the second and third term in (97) and the remaining $-1/3$ comes from the final term. However, this is not a general result.

The induced interaction, E''_{int}, is in general given by equation 53. For vacancies we may take

$$\psi(\mathbf{r}) = -\phi(\mathbf{r}) \quad \text{(vacancy)} \qquad (99)$$

Since the operator $\partial/\partial s_j$ is equivalent to $-\partial/\partial s'_j$ we have from equations (54) and (99):

$$\Lambda_{jkmn} = \sum_{\mathbf{s}} s_j \Phi_{km}(\mathbf{s})s_n \qquad (100)$$

For the first neighbour generalized force model the coefficients are readily evaluated with the aid of equation (76) to give

$$\Lambda_{1111} = \sum_{\mathbf{s}} s_1^2 \Phi_{11}(\mathbf{s}) = -2\Omega c_{11}$$

$$\Lambda_{1112} = 0 = \Lambda_{1121} \quad \text{etc.}$$

$$\Lambda_{1212} = \sum_{\mathbf{s}} s_1 s_2 \Phi_{12}(\mathbf{s}) = \Lambda_{1122} = -\Omega(c_{12}+c_{44})$$

$$\Lambda_{1221} = \sum_{\mathbf{s}} s_1^2 \Phi_{22}(\mathbf{s}) = -2\Omega c_{44} = \Lambda_{2112}$$

$$(101)$$

* There are misprints in the expression given in this paper; the result given here is correct.

The remaining coefficients are obtainable from those above or directly from equation (100). In the isotropic case $c_{11} - c_{12} = 2c_{44}$ and the displacement field is of the form given in equation (1); consequently we have from equations (53) and (10)

$$E''_{int} = \epsilon_{jk} \Lambda_{jkmn} \epsilon_{mn} = -\frac{6\Omega(c_{12} + 3c_{44})}{R^6} \left(\frac{\alpha \Delta V}{4\pi}\right)^2 \quad (102)$$

By equating this result to the continuum result in equation (9) we obtain an expression for the defect modulus μ^*, namely

$$M = \frac{\mu(\mu - \mu^*)}{\mu - \beta(\mu - \mu^*)} = \frac{\Omega}{2V} \cdot \frac{\mu(3 - 4\nu)}{(1 - 2\nu)} \quad (103)$$

The elastic polarizability α^μ is given by

$$\alpha^\mu = -\Omega M = -\mu\Omega\left(\frac{\Omega}{2V}\right)\frac{(3 - 4\nu)}{(1 - 2\nu)}$$

For aluminium

$$\mu \sim 0.28 \times 10^{12} \text{ dynes/cm}^2$$

$$\Omega \sim 16 \times 10^{-24} \text{ cm}^3$$

$$\nu \sim \tfrac{1}{3}$$

thus

$$\alpha^\mu = -14\left(\frac{\Omega}{2V}\right) \quad \text{eV}$$

Taking the 'volume' of a vacancy as $V = \Omega - \Delta V \sim \frac{1}{2}\Omega$ this gives $\alpha^\mu = -14$ eV in good agreement with the experimental value of -16 eV quoted by Wenzl.[28]

While this simple model indicates the relationship between a discrete lattice approach to point defect properties and continuum theory, more sophisticated models have been used to calculate numerical values for the near neighbour defect interactions and the displacement field in the immediate vicinity of a point defect. These calculations are reviewed in the next section.

COMPUTER SIMULATION CALCULATIONS

This method involves the direct minimization of equation (28) with respect to the lattice displacements and consequently involves the use of advanced numerical procedures in conjunction with large computational facilities. It is apparent from the preceding discussion that the displacement field around a point defect may be determined from a knowledge of the gradient and curvature of the potential without knowing its absolute value. However, if the energies of the various defect configurations are to be determined the interaction potential must be specified. Since the form of the potential function influences the results of computer simulation calculations it is appropriate to describe the types of potential that have been used.

The pioneering calculations of Tewordt[22] and Gibson et al.[23] made use of the purely repulsive Born–Mayer potential and Girifalco and Streetman[29] have used a series of different Lennard-Jones potentials to study the relaxation around a single vacancy in a bcc lattice. Extensive use has been made of Morse potentials.[30–32] The major advantage of these potential functions is that they have a simple analytic form, their disadvantages include truncation problems and their inability to reproduce the harmonic data of the crystals, in particular the

elastic constants. In 1964 Johnson[33] proposed a simple method of overcoming the problems associated with these traditional potentials. He represented the potential by a series of polynomials:

$$\phi_\alpha(r) = \sum_{\beta=0}^{N} A_{\alpha\beta} r^\beta, \qquad r_\alpha \leqslant r \leqslant r_{\alpha+1}$$

where N is typically 3, 4 or 5. Thus the potential has a simple analytic form and the range of the potential is restricted to the near neighbours. Typically the potential is made to approach zero smoothly between second and third nearest neighbours. The coefficients $A_{\alpha\beta}$ are determined from the conditions that the various segments of the potential join smoothly and from experimental data. For cubic materials there are three conditions relating $\phi(r)$ to the elastic constants, namely for a body-centred cubic lattice with interactions extending out to second nearest neighbours we have

$$3ac_{11} = \phi''(r_1) + \frac{2}{r_1}\phi'(r_1) + 2\phi''(r_2)$$

$$3ac_{12} = \phi''(r_1) - \frac{4}{r_1}\phi'(r_1) - \frac{2}{r_2}\phi'(r_2)$$

$$3ac_{44} = \phi''(r_1) + \frac{2}{r_1}\phi'(r_1) + \frac{2}{r_2}\phi'(r_2)$$

where, as before, $2a$ is the cube cell side, r_1 is the nearest neighbour separation, $\sqrt{3}a$, and $r_2 = 2a$ is the second nearest neighbour separation. The vacancy formation energy, E_V^f, is frequently used to determine one of the coefficients $A_{\alpha\beta}$ through the relationship

$$4\phi(r_1) + 3\phi(r_2) = -E_V^f - E_V^r + pV_V^f.$$

Since E_V^r and V_V^f are not known *a priori* this relationship has to be satisfied by an iterative procedure. At small values of r the potential is smoothly matched to the Born–Mayer repulsion determined by the radiation damage threshold or by calculation.[34,35] Potentials constructed in this way have many advantages, they are short range and analytic in form thus facilitating computer calculations. They reproduce the correct harmonic response and short range repulsion. The disadvantages are that the method of fitting to the perfect lattice properties is not unique and there is no theoretical justification for truncating the potential. A series of potentials constructed by this method are given in a paper by Johnson and Wilson.[36]

More recently, following the work of Harrison,[11] several authors have applied pseudopotential theory to the calculation of point defect properties,[36–40] and these methods will be the subject of a subsequent paper in this volume by Dr Evans. However, for completeness, a brief description of the formalism adopted by Ho[39] will be given here. Essentially there are three steps involved in the calculation for vacancies:

1 The vacancy is introduced by removing an ion from the centre of a finite solid containing N atoms and placing it on the surface. In this stage there is no relaxation of the individual atoms from their lattice positions, but the electron density decreases by a factor $1/(1 + N)$ due to the additional lattice site. The structural energy required for changing the atomic arrangement in order to create the vacancy is analogous to E_V^g.

2 A lattice statics method is used to calculate the relaxation around the vacancy. The energy gained in this process is E_V^r.

3 Finally the defect lattice is allowed to dilate uniformly to the equilibrium configuration. This produces a term analogous to pV_V^f.

It is usual in this type of calculation for the relaxation procedure (2) to be carried out with a truncated form of the potential or even a short range potential different from that used in step (1). Thus the principal difference between this method and that of Johnson[33] is that while Johnson uses an approximate potential function the minimization procedure is carried out self-consistently, the pseudopotential method makes use of the correct perfect lattice potential but often the minimization of the total energy is achieved by approximate methods

RESULTS

In this section some results for the face-centred cubic metals copper and aluminium and the body-centred cubic metals iron and molybdenum are presented. This is by no means intended to be a comprehensive list; for example Flocken and Hardy[41] have studied vacancies in the alkali metals and their results are not included in this discussion. Nor is it meant to imply that only cubic materials have been considered; Popović and Carbotte[42] and Doneghan and Heald[43] have calculated the relaxation around a vacancy in magnesium and Nicholson and Bacon[44] have considered vacancies in graphite. However, copper and iron have been the subject of extensive studies by various authors and the work on aluminium and molybdenum is fairly complete and has been included for comparison.

Copper

The nearest neighbours of a vacancy in copper have been found to relax inwards by about 1% of the cube cell side,[45,31,46,25] while the second nearest neighbours relax outwards by a fraction of 1%. The displacements of more distant neighbours are small (less than $0.01 \times 2a$) and consequently the use of the harmonic approximation in the lattice statics reciprocal space method would appear to be justified. The relaxation volume around a vacancy in copper has been found to be $\Delta V_V = -(0.46 \pm 0.02)\Omega$.[47,46,25] The numerical procedures used by the various authors were different and in most cases there were differences in the interatomic potential used. In view of this the results are encouragingly consistent. It would appear that provided the interatomic potential used is consistent with the harmonic properties of the crystal then the lattice calculations of displacement fields and relaxation volumes for vacancies are quite reproducible. Indeed one would have anticipated this from equations (74) and (75).

The nearest neighbour divacancy configuration has been found to be the most stable,[31,46] although the absolute values of the binding energy depend on the value of the interatomic potential at the first nearest neighbour separation distance. Specifically Doyama and Cotterill[31] give $E_{2V}^B = 0.53$ eV while Englert et al.[48] obtain $E_{2V}^B = 0.31$ eV.

Johnson and Brown[47] have calculated the activation energy for vacancy migration to be $E_V^m = 0.43$ eV.

With interstitial calculations the situation is rather more complicated since the nearest neighbour separations are of the order of a, that is, much less than the equilibrium spacing. In this region the interatomic potential is rapidly changing and consequently the defect energies are very sensitive to the near neighbour displacements. Typically $E_I^f \sim 4E_V^f$,[49-51] and the $\langle 100 \rangle$ split configuration is found to be the most stable. However, the energy differences between the various possible configurations are relatively small ($\sim \frac{1}{10}E_I^f$). The nearest neighbours of an interstitial move fairly large distances ($\sim 10\%$ of $2a$).[49,52,53] The relaxation volume around the octahedral interstitial is $\Delta V_I(0) = (1.25 \pm 0.16)\Omega$,[49,51-53,31] while for the $\langle 100 \rangle$ split interstitial $\Delta V_I(H_0) = (1.29 \pm 0.16)$.[50,51,31,53]

Doyama and Cotterill[31] have studied the stability of Frenkel pairs; they found that if the vacancy and interstitial had a common nearest neighbour they spontaneously recombine. While this common nearest neighbour condition is sufficient to cause spontaneous recombination it is not, in some cases, necessary and even more distant vacancy–interstitial pairs were found to be unstable.

Aluminium

All the atoms around a vacancy in aluminium relax inwards.[46,25] The first and second nearest neighbours move by about 1% of $2a$ and more distant neighbours by a small fraction of this. Aluminium is fairly isotropic with $(c_{11} - c_{12} - 2c_{44})/c_{11} = 0.093$; consequently the continuum displacement field is given to a good approximation by equation (1) and

$$r^2 u_r = \frac{\alpha \, \Delta V}{4\pi}$$

Figure 1 shows the displacement field calculated from the lattice statics method by means of equation (65). For large r the values of $r^2 u_r$ converge to a constant value which is the continuum limit; from Fig. 1 it is clear that for distances greater than $8a$ from the vacancy the displacement field is adequately described by equation (1). The relaxation volume has been calculated to be $\Delta V_V = -(0.49 \pm 0.02)\Omega$.[46,25] Bullough and Hardy[46] found the nearest neighbour divacancy pair to be the most stable with $E_{2V}^B \sim 0.1 \rightarrow 0.2$ eV.

The displacements around the octahedral interstitial were found to be large ($\sim 10\%$ of $2a$) and the corresponding relaxation volume was calculated to be $\Delta V_I(0) = (1.3 \pm 0.07)\Omega$[53] while that for the $\langle 100 \rangle$ split is $\Delta V_I(H_0) = (2.06 \pm 0.03)\Omega$.[53]

Iron

Lattice calculations indicate that the nearest neighbours of a vacancy in iron collapse inwards by 1–2% of $2a$ and

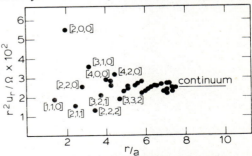

1 Displacement field around a vacancy in aluminium (Miller and Heald, 1975; Bullough and Hardy, 1968)

the second nearest neighbours move outwards by a similar distance.[54,55] All other displacements are a fraction of 1% of $2a$. Flocken[54] has attempted to compare the lattice displacement field with the isotropic elastic continuum result; however, since iron is quite anisotropic the continuum displacement field is not adequately described by equation (1) and in this respect his comparison and conclusions are incorrect. The relaxation volume around a vacancy in iron has been calculated as $\Delta V_V = -0.23\Omega$.[55] Beeler[56] and Flocken[54] have shown that the second nearest neighbour divacancy configuration is the most stable with a binding energy of $E_{2V}^B(2, 0, 0) = 0.2 \rightarrow 0.3$ eV; the second most stable configuration is the nearest neighbour pair with $E_{2V}^B(1, 1, 1) \sim 0.14$ eV, and the only other divacancy with appreciable binding is the fourth nearest neighbour pair with $E_{2V}^B(3, 1, 1) \sim 0.05$ eV. Johnson[33] has calculated the activation energy for vacancy motion to be $E_V^m = 0.68$ eV.

The relaxation volume around the $\langle 110 \rangle$ split and the octahedral interstitial configurations have been reported[33] as $\Delta V_I(H_0) = 1.6\Omega$ and $\Delta V_I(0) = 1.4\Omega$ respectively. The $\langle 110 \rangle$ split configuration was found to be the most stable and the activation energy for its migration is given as 0.33 eV while that for the unstable crowdion configuration is 0.04 eV.[33]

Gibson et al.[23] have studied the stability of near neighbour Frenkel pairs and found that all pairs with a nearest neighbour atom in common spontaneously recombined. In addition some more distant pairs were found to be unstable.

Molybdenum

The nearest neighbours of a vacancy in molybdenum collapse in by about 3% of $2a$ and the second nearest neighbours move away from the vacant site by a similar amount. The relaxation volume around the vacancy was found to be $\Delta V_V = -0.29\Omega$.[55] Since molybdenum is only weakly anisotropic with $(c_{11} - c_{12} - 2c_{44})/c_{11} = 0.13$ the displacement field remote from the vacancy may be compared with the continuum displacement field given by equation (1). In Fig. 2 the lattice displacements computed from equation (65) are shown as a function of r. For values of $r > 10a$ the displacements derived from lattice theory converge to the continuum approximation.

The second nearest neighbour divacancy configuration was found to be the most stable with a binding

TABLE 1   A summary of the defect properties discussed in the results section

Metal	ΔV_I 0	H_0	ΔV_V	E_{2V}^B Nearest neighbour	2nd nearest	E_V^m
Copper	1.25	1.29	-0.46	$0.3 \rightarrow 0.5$		0.43
Aluminium	1.30	2.06	-0.49	$0.1 \rightarrow 0.2$		
Iron	1.4	1.6	-0.23	0.14	$0.2 \rightarrow 0.3$	0.68
Molybdenum			-0.29	0.02	0.44	

Volumes are in units of the atomic volume and energies are in eV.

energy $E_{2V}^B(2, 0, 0) = 0.44$ eV.[36] The only other configuration with appreciable binding is the fourth nearest neighbour pair with $E_{2V}^B(3, 1, 1) = 0.1 \rightarrow 0.2$ eV.[55] This suggests that in molybdenum divacancy diffusion takes place by a succession of $(2, 0, 0) \rightarrow (3, 1, 1) \rightarrow (2, 0, 0)$ configurations.

The defect properties discussed in this section are summarized in Table 1.

DISCUSSION

In this article we have attempted to present the formulation of the lattice theory of point defects in general terms. It is shown in the second section that defect formation energy and displacement field is determined by the minimization of the energy difference between the defect lattice $U\{\mathbf{u}(\mathbf{s})\}$ and the perfect lattice U_0 with respect to the lattice displacements $\mathbf{u}(\mathbf{s})$. Thus the formation energy is

$$E^f(\mathbf{u}(\mathbf{s})) = \sum_{\mathbf{s}} \psi\{\mathbf{s} + \mathbf{u}(\mathbf{s})\}$$

$$+ \tfrac{1}{2} \sum_{\mathbf{s} + \mathbf{s}'} \phi\{\mathbf{s} + \mathbf{u}(\mathbf{s}) - \mathbf{s}' - \mathbf{u}(\mathbf{s}')\} - \Phi_0 + p V^f \quad (104)$$

and the displacement field is determined by

$$\frac{\partial E^f\{u(\mathbf{s})\}}{\partial u_j(\mathbf{s})} = 0 \quad (105)$$

The computer simulation methods minimize equation (104) directly while the lattice statics procedures (either the Kanzaki or Greens function methods) solve equation (105) in the harmonic approximation and use the displacement field so determined to evaluate the defect formation energies from the harmonic approximation to equation (104), namely equation (30). For vacancies the near neighbour displacements are of the order of 1 or 2% of the interatomic spacing and the use of the harmonic approximation is justified. However, for interstitials the near neighbour displacements are in excess of 10% of the lattice spacing and the use of the harmonic approximation should be viewed with some caution.

In the third section we have demonstrated that the displacement field calculated by lattice methods is, in regions remote from the defect, identical to the continuum elasticity result provided that the elastic strength of the defect is determined from the lattice model. Furthermore, detailed numerical calculations enable us to determine the range for which the continuum calculations are valid and Figs. 1 and 2 show that for values of r greater than $10a$ the defect displacement field is adequately described by continuum theory. The numerical values also show that the relaxation volume around

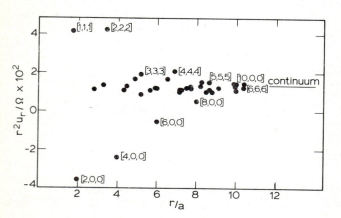

Displacement field around a vacancy in molybdenum (Kenny et al., 1973)

an interstitial is much larger than that around a vacancy as one might expect. An important consequence of this is that the interaction between a dislocation and an interstitial is much stronger than that between a dislocation and a vacancy and this provides the basis for the theory of irradiation swelling.

Similar results are obtained for the interaction of point defect pairs where, for widely spaced defects, the interaction is of the form

$$E_{int}(R) = \frac{Af_1(\theta, \phi)}{R^3} + \frac{Bf_2(\theta, \phi)}{R^5} + \frac{C}{R^6} \qquad (106)$$

The first term in equation (106) is identical to Eshelby's[5] elasticity result. A term similar to the second in equation (106) may be obtained from continuum theory if one considers force multipoles;[57] however, it is really a lattice term representing the non locality of the defect forces and the dispersion of the medium. The final term is analogous to Eshelby's[5] induced interaction term and the relationship between the constant C and the modulus M given by equation (7) enables the defect modulus μ^* to be determined as in equation (103).

Finally we should remark that more extensive calculations have been carried out. Although a discussion of these results is beyond the scope of this article it is worth pointing out that discrete lattice models have been used to study the cascade damage produced by energetic particles.[58-61] The interaction of vacancies with extended defects has received much attention, for example the vacancy–dislocation interaction has been studied by Bullough and Perrin,[62] Perrin et al.[63] and Boyer and Hardy[64] and the interaction of vacancies and divacancies with twin boundaries has been discussed by Ingle et al.[65]

ACKNOWLEDGMENT

This paper is published by permission of the Central Electricity Generating Board.

REFERENCES

1 N. F. MOTT AND F. R. N. NABARRO: *Proc. Phys. Soc.*, 1940, **52**, 86
2 J. D. ESHELBY: *Phil. Trans. Roy. Soc.*, 1951, **A244**, 87
3 A. H. COTTRELL: 'Progress in metal physics' (Ed. B. CHALMERS), Vol. I (Ch. II), 1949, London, Butterworths
4 B. A. BILBY: *Proc. Phys. Soc.*, 1950, **A63**, 191
5 J. D. ESHELBY: *Acta Met.*, 1955, **3**, 487
6 G. LEIBFRIED: *Z. Physik.*, 1953, **135**, 23
7 F. R. N. NABARRO: *Proc. Roy. Soc.*, 1940, **A175**, 519
8 G. W. GREENWOOD et al.: *J. Nucl. Mat.*, 1959, **4**, 305
9 P. T. HEALD AND M. V. SPEIGHT: *Phil. Mag.*, 1974, **29**, 1075
10 W. G. WOLFER et al.: ASTM, STP 570, 1976, p. 233
11 W. A. HARRISON: 'Pseudopotentials in the theory of metals', 1966, Massachusetts, L. A. Benjamin Inc.
12 V. HEINE AND D. WEAIRE: *Solid State Physics*, 1970, **24**
13 A. A. MARADUDIN: *Rep. Progress Physics*, 1965, **28**, 331
14 H. KANZAKI: *J. Physics Chem. Solids*, 1957, **2**, 24
15 J. R. HARDY: *J. Physics Chem. Solids*, 1960, **15**, 39
16 J. R. HARDY AND R. BULLOUGH: *Phil. Mag.*, 1967, **15**, 237
17 J. R. HARDY AND R. BULLOUGH: *Phil. Mag.*, 1967, **16**, 405
18 J. R. HARDY AND A. B. LIDIARD: *Phil. Mag.*, 1967, **15**, 825
19 J. R. HARDY: *J. Physics Chem. Solids*, 1968, **28**, 331
20 D. GRECU AND M. CROITORU: Int. Conf. on 'Vacancies and interstitials in metals', Vol. II, 574, 1968, Jülich, Kernforschungsanlage
21 V. K. TEWARY: *Advances Physics*, 1973, **22**, 757
22 L. TEWORDT: *Phys. Rev.*, 1958, **109**, 61
23 J. B. GIBSON et al.: *Phys. Rev.*, 1960, **120**, 1229
24 J. D. ESHELBY: *Solid State Physics*, 1956, **3**, 79
25 K. M. MILLER AND P. T. HEALD: *Physica Status Solidi* (b), 1975, **67**, 569
26 G. H. BEGBIE: *Proc. Roy. Soc.*, 1947, **A188**, 189
27 P. T. HEALD: *Phil. Mag.*, 1970, **22**, 751
28 H. WENZL: 'Vacancies and interstitials in metals' (Eds. A. SEEGER et al.), 363, 1970, Amsterdam, North-Holland
29 L. A. GIRIFALCO AND J. R. STREETMAN: *J. Physics Chem. Solids*, 1958, **4**, 182
30 A. C. DEMASK et al.: *Phys. Rev.*, 1959, **113**, 781
31 M. DOYAMA AND R. M. J. COTTERILL: 'Lattice defects and their interactions (Ed. R. R. HASIQUTI), 79, 1967, New York, Gordon and Breach
32 A. DE PINO et al.: *Radiat. Effects*, 1970, **3**, 23
33 R. A. JOHNSON: *Phys. Rev.*, 1964, **134**, 1329
34 O. B. FIRSOV: *Soviet Physics*, JETP, 1958, **6**, 534
35 P. T. WEDERPOHL: *Proc. Phys. Soc.*, 1967, **92**, 79
36 R. A. JOHNSON AND W. D. WILSON: 'Interatomic potentials and simulation of lattice defects' (Eds. P. C. GEHLEN et al.), 301, 1972, New York, Plenum
37 R. BENDEDEK AND A. BARATOFF: *J. Physics Chem. Solids*, 1971, **32**, 1015
38 R. CHANG AND L. M. FALICOV: *J. Physics Chem. Solids*, 1971, **32**, 465
39 P. S. HO: 'Interatomic potentials and simulation of lattice defects' (Eds. P. C. GEHLEN et al.), 321, 1972, New York, Plenum
40 R. EVANS AND M. W. FINNIS: *J. Physics F.*, 1976, **6**, 483
41 J. W. FLOCKEN AND J. R. HARDY: *Phys. Rev.*, 1968, **175**, 919
42 Z. D. POPOVIĆ AND J. P. CARBOTTE: *J. Physics F.*, 1974, **4**, 1599
43 M. DONEGHAN AND P. T. HEALD: *Physica Status Solidi* (a), 1975, **30**, 403
44 A. P. P. NICHOLSON AND D. J. BACON: *Physica Status Solidi* (a), 1975, **28**, 613
45 G. SCHOTTKY et al.: *Physica Status Solidi*, 1964, **4**, 419
46 R. BULLOUGH AND J. R. HARDY: *Phil. Mag.*, 1968, **17**, 833
47 R. A. JOHNSON AND E. BROWN: *Phys. Rev.*, 1962, **127**, 446
48 A. ENGLERT et al.: 'Fundamental aspects of dislocation theory', NBS Spec. Publ. 317, p. 273
49 P. HOCKSTRA AND D. R. BEHRENDT: *Phys. Rev.*, 1962, **128**, 560
50 K. H. BENNEMANN: *Phys. Rev.*, 1961, **124**, 669
51 A. SEEGER et al.: *J. Physics Chem. Solids*, 1962, **23**, 639
52 J. W. FLOCKEN AND J. R. HARDY: *Phys. Rev.*, 1968, **175**, 919
53 K. M. MILLER AND P. T. HEALD: *Physica Status Solidi* (b), 1976, **78**, 341
54 J. W. FLOCKEN: *Phys. Rev.*, 1970, **2**, 1743
55 P. N. KENNY et al.: *J. Physics F.*, 1973, **3**, 513
56 J. R. BEELER: Symp. on the 'Nature of small defect clusters', Vol. I, 173, 1966, AERE Harwell Report 5269
57 R. SIEMS: *Physica Status Solidi*, 1968, **30**, 645
58 I. M. TORRENS AND M. T. ROBINSON: 'Interatomic potentials and simulation of lattice defects' (Eds. P. C. GEHLEN et al.), 423, 1972, New York, Plenum
59 D. G. DORAN AND R. A. BURNETT: 'Interatomic potentials and simulation of lattice defects' (Eds. P. C. GEHLEN et al.), 403, 1972, New York, Plenum
60 J. R. BEELER: *Phys. Rev.*, 1966, **150**, 470
61 M. J. NORGETT: 'The physics of irradiation produced voids', 44, 1974, AERE Harwell Report R7924
62 R. BULLOUGH AND R. C. PERRIN: Radiation damage in reactor materials, Vol. II, 233, 1969, Vienna, IAEA
63 R. C. PERRIN et al.: 'Interatomic potentials and simulation of lattice defects' (Eds. P. C. GEHLEN et al.), 507, 1972, New York, Plenum
64 L. L. BOYER AND J. R. HARDY: *Phil. Mag.*, 1972, **26**, 225
65 K. W. INGLE et al.: *Phil. Mag.*, 1976, **33**, 663

Dynamical properties of interstitials in metals

P. H. Dederichs

Recent developments in the understanding of the dynamical properties of self-interstitials are reviewed. Computer simulations have shown that interstitials in metals can vibrate with low-frequency resonance modes in addition to the well known high-frequency localized modes. The low frequencies are due to the strong negative 'leaf springs' which are typical for the strongly compressed lattice around the interstitial atom and introduce a tendency towards instability into the lattice. Analytical model calculations of the local-frequency-spectra of the interstitial show that it vibrates practically only with the frequencies of the localized and resonant modes, but not with the 'normal' lattice frequencies. The tendency of the interstitial towards instability also influences its static behaviour. The interstitial configuration is strongly polarizable by external forces. The experimentally observed large and anisotropic changes of the elastic shear constants due to irradiation can be traced back to the symmetry of the 100-dumbbell. The structure and dynamical behaviour of di- and tri-interstitials is also briefly discussed. They are characterized on the one hand by very high binding energies and on the other hand by high mobilities similar to that of the single interstitial.

The author is at the Institut für Festkörperforschung der KFA, Jülich, Germany

In the last few years our knowledge of the properties of self-interstitials has increased considerably. Up to about five years ago the only experimental information stemmed from resistivity and lattice parameter measurements of the annealing behaviour, the interpretation of which is always somewhat questionable. By the development of X-ray diffraction techniques a breakthrough has been achieved which allows a rigorous determination of the interstitial position: the 100-dumbbell has been identified as the interstitial configuration in Al and Cu.[1-3] Further very useful microscopic information has been obtained, for example by elastic relaxation measurements[4-6] and by channelling studies of interstitial-impurity processes.[7]

The theoretical side of this problem has already been tackled many years ago by computer simulation. The weak point of this approach is the uncertainty in the empirical two-body potentials used. However, the results of the above measurements are essentially in agreement with the very early predictions of computer simulation and this gives us more confidence in such studies.

Whereas most simulations are concerned either with high energy-radiation damage or with the structure of interstitials and vacancies, Lehmann and Scholz[9] made a detailed study of the vibrations of interstitials. Surprisingly they found that in addition to high-frequency localized modes, interstitials can also vibrate with low-frequency resonant vibrations. These resonance vibrations are due to the strong compression of the lattice around the interstitial atom and are thus typical for interstitials. We think that they are important for an understanding of many physical properties of the interstitial.[10-11]

In the next section we give a qualitative explanation of the unusual dynamical behaviour of interstitials, i.e. the simultaneous occurrence of localized *and* resonant modes. In the following two sections we consider in detail the vibrational behaviour of an octahedral interstitial and of a 100-dumbbell. The resonances also influence the static behaviour of interstitials: by external forces interstitials can become strongly polarized, leading to relatively large changes of the elastic constants (fifth section). In the final section we discuss the dynamical behaviour of di- and tri-interstitials.

QUALITATIVE EXPLANATION OF THE DYNAMICS OF SELF INTERSTITIALS

Defects can locally change the vibrational behaviour of an ideal lattice. Two kinds of characteristic defect modes can occur: (*a*) localized modes with frequencies above the maximum frequency ω_{max} of the ideal lattice, representing the real undamped eigenmode of the lattice, localized near the defect, or (*b*) resonance modes with very low frequencies $\omega \ll \omega_{max}$. The latter are only weakly damped since in this frequency range the ideal lattice has only very few phonons which can dissipate the energy. Usually one argues that the characteristic frequency associated with the defect is $\omega^2 \approx f^d/M^d$ where f^d is the Einstein-force constant of the defect and M^d is the defect mass. Thus localized modes are expected for very light or for strongly bound atoms such as interstitials, whereas resonances ought to occur for heavy and/or weakly bound atoms.

By studying the dynamics of interstitials, Scholz and Lehmann[9] found in a computer simulation that the 100-dumbbell interstitial vibrates, in addition to several high-frequency localized modes, also with low-frequency resonance modes which one would not expect from the Einstein model. The simultaneous occurrence of both kinds of modes can be explained by the strongly compressed lattice around the interstitial.[11]

For a central potential $V(R)$ the coupling constants between two atoms at distance $\mathbf{R} = (R, 0, 0)$ are given by

$$\phi_{ij}(R) = -\partial_{R_i} \partial_{R_j} V(R) = \begin{Bmatrix} f_\parallel & 0 & 0 \\ 0 & f_\perp & 0 \\ 0 & 0 & f_\perp \end{Bmatrix}$$

with

$$\begin{aligned} f_\parallel &= V''(R) \\ f_\perp &= V'(R)/R \end{aligned} \qquad (1)$$

The 'longitudinal' force constant $f_\parallel$ enters for displacements parallel to $\mathbf{R}$ ('spiral spring'), whereas the 'transversal' force constants $f_\perp$ enters for displacements perpendicular to $\mathbf{R}$ ('leaf' or 'bending spring'). For an ideal lattice, the nearest neighbour constant $f_\parallel = f_\parallel^0$ is the dominating force constant—for example, in fcc lattices it is usually a factor of 10 larger than all the other springs—and $f_\perp$ can be neglected in a first approximation.

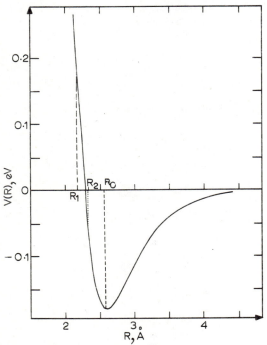

1   Morse potential for Cu:[9]
   $V(R) = D\{\exp -2\alpha(R-R_0) - 2 \exp - \alpha(R-R_0)\}$;
   $D = 0.18$ eV, $\alpha = 2.33$ Å, $R_0 = 2.57$ Å;
   R_1 = separation of the dumbbell atoms,
   R_2 = separation of a dumbbell atom from its
   nearest neighbours

However in the compressed region of an interstitial both force constants are much larger and $f_\perp$ cannot be neglected. A typical interaction potential (Morse potential) is shown in Fig. 1. The parameters of the potential are fitted to Cu.[9] The separation R_1 of the two dumbbell-atoms and the separation R_2 of a dumbbell atom with its nearest neighbours is appreciably smaller than the nearest neighbour distance R_0 in the ideal lattice. For such small distances the atoms are strongly repelling each other.

With the Morse potential one obtains, for example for the interaction of the two dumbbell atoms $f_\parallel \cong 7f_\parallel^0$ and $f_\perp \cong -0.6f_\parallel^0$; and for the interaction between one dumbbell atom and its nearest neighbours: $f_\parallel \cong 4f_\parallel^0$, $f_\perp \cong -0.3f_\parallel^0$. Thus, because of the strong interaction, the longitudinal force constant $f_\parallel$ is very large, and because of the repulsion of the atoms $f_\perp$ is always negative and becomes comparable to $f_\parallel^0$ which represents the restoring forces of the ideal lattice. Such a negative bending spring can be visualized as being produced by a compressed spiral spring which is unstable against displacements perpendicular to the spring axis.

Due to the two different force constants $f_\parallel$ and $f_\perp$ two different types of characteristic mode can occur:

   (i) vibrations which considerably stress the strong spiral springs in the interstitial region giving rise to high-frequency localized modes; in this case the relatively weak leaf springs are not important
   (ii) vibrations which stress the strong springs only slightly, in which case the negative leaf springs become important; they lower the frequency and hence give rise to resonant modes.

The negative leaf springs arising from the repulsion of the atoms at close distances introduce a tendency towards instability into the lattice since for some configurations they may even lower the frequency so much that $\omega^2 < 0$; then this interstitial configuration is unstable. Thus the negative bending springs determine the stability of a configuration and lead to resonance modes for the stable configuration.

The strong repulsion of the atoms near the interstitial also has more far-reaching consequences: as is well known from computer simulation, different interstitial configurations differ only slightly in energy. Migration energies especially are very small, typically about 0.1 eV for many metals. Thus as a function of the interstitial coordinate, the entire energy contour is rather flat. The resonance modes are a local reflection of this overall flatness of the energy contour and of the high mobility of the interstitial. In this way the repulsion of the atoms at short distances determines in a very characteristic way the dynamical behaviour of the interstitial.

OCTAHEDRAL INTERSTITIAL

In a fcc lattice the configuration with the highest symmetry, i.e. cubic symmetry, is the octahedral interstitial, shown in Fig. 2. The high symmetry facilitates the calculation of the dynamical behaviour. As the simplest model we consider an interaction to nearest neighbours only with a force constant $f_\parallel^0$ between the atoms of the ideal lattice. The interstitial is coupled to its nearest neighbours, i.e. the six face-centered atoms in Fig. 2, by a spiral spring $f_\parallel$ and a bending spring $f_\perp$. The calculations are strongly simplified by standard group theoretical methods which we will not discuss here. The intersti-

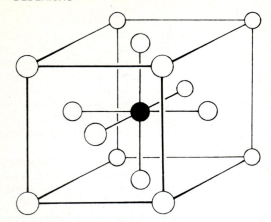

2 Octahedral interstitial in fcc crystal

tial moves only in a three-fold degenerate F_{1u}-representation.

Most naturally the vibrational behaviour of the defect is described by its Green's function $G_{xx}^{dd}(\omega)$, which is the response of the defect, i.e. its displacement in the x-direction, due to a unit force of frequency ω on the defect (in the x-direction). The calculations[12] yield

$$G_{xx}^{dd}(\omega) = \left\{ \frac{2f_{\parallel} + 4f_{\perp} + f_{\parallel}f_{\perp}(4a + 2c - 4\sqrt{2}b)}{1 + af_{\parallel} + cf_{\perp} + (ac - b^2)f_{\parallel}f_{\perp}} - M\omega^2 \right\}^{-1} \quad (2)$$

with

$$a = \overset{\circ}{G}_{xx}^{(0)} + \overset{\circ}{G}_{xx}^{(200)}, \qquad b = 2\sqrt{2}\overset{\circ}{G}_{xx}^{(110)},$$
$$c = \overset{\circ}{G}_{xx}^{(0)} + \overset{\circ}{G}_{zz}^{(200)} + 2\overset{\circ}{G}_{zz}^{(110)}$$

Here $\overset{\circ}{G}_{ij}^{mn} = \overset{\circ}{G}_{ij}^{(m-n)}(\omega)$ is the Green's function of the ideal lattice. The local vibration spectrum of the defect

$$z_x^d(\omega) = \frac{2M\omega}{\pi} \operatorname{Im} G_{xx}^{dd}(\omega) \quad (3)$$

describes with which frequencies the defect can vibrate.

By choosing, as a typical example, a strong longitudinal force constant $f_{\parallel} = 5f_{\parallel}^0$ between the defect and its neighbours the resulting local spectrum of the defect is shown in Fig. 3 for various values of the leaf spring $f_{\perp}$. The spectrum always contains a localized mode, with a frequency almost independent of $f_{\perp}$. Contrary to this,

for increasing negative values of $f_{\perp}$ a pronounced resonant mode at low frequencies appears. The defect hardly vibrates with the eigenfrequencies of the ideal lattice (dotted line) and the local spectrum can be composed in a first approximation from the resonant and localized modes alone. The motion of the defect and its nearest neighbours in the direction of the vibration is indicated in Fig. 3 by the arrows below the modes. In the localized mode the defect and the neighbours move with opposite phase ('optical vibration'), thus compressing the strong springs, whereas they move in phase in the resonant mode ('acoustical vibration').

For $f_{\perp} \cong -0.96f_{\parallel}^0$ the resonance frequency approaches zero and the static Green's function $G_{xx}^{dd}(0)$ diverges: the octahedral configuration becomes unstable. The stability limit can be determined directly from the condition $G_{xx}^{dd}(0) = \infty$. Equation (2) yields:

$$2f_{\parallel} + 4f_{\perp} + 1.235\frac{f_{\parallel}f_{\perp}}{f_{\parallel}^0} = 0 \quad (4)$$

which for $f_{\parallel} = 5f_{\parallel}^0$ gives $f_{\perp} = -0.96f_{\parallel}^0$. Even for an infinitely strong spring $f_{\parallel}$ the configuration can become unstable, if $f_{\perp} \leq -1.544f_{\parallel}^0$. According to most computer simulations, the octahedral configuration is indeed unstable, which is in agreement with the recent experimental findings.

100-DUMBBELL

X-ray scattering experiments show the 100-split interstitial (dumbbell) as the stable configuration in Al[1-3] and Cu.[3] This configuration (shown in Fig. 4) was also obtained by numerous computer calculations.[13] To get insight into the vibrational behaviour of this 100-dumbbell we consider a simple model with nearest neighbour interaction only.

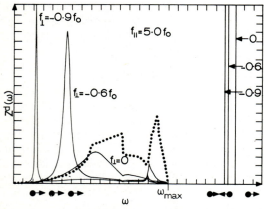

3 Local frequency spectrum of octahedral interstitial for different values of the leaf spring $f_{\perp}$

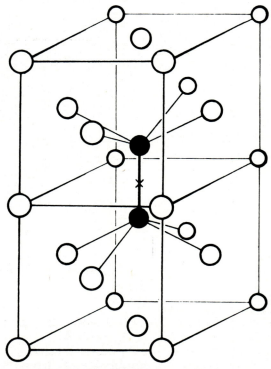

4 100-dumbbell configuration in fcc crystal

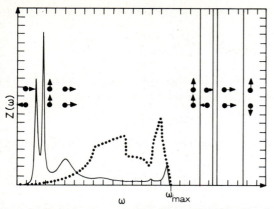

5 Local frequency spectrum of 100-dumbbell atoms

For symmetry reasons we describe the defect by a vacancy at position $(0, 0, 0)$ and two interstitials at $(0, 0, \pm d)$. The coupling constants are given by ideal springs $f_\parallel^0$ between the host atoms, by a spiral spring $f_\parallel$ and a leaf spring $f_\perp$ between the dumbbell atoms, and by spiral springs $F_\parallel$ and leaf springs $F_\perp$ between an interstitial and its four nearest neighbours (Fig. 4). The vacancy is conveniently described by zero-coupling constants to its neighbours. The following force constants were chosen by fitting some of the resulting frequencies of the resonant and localized modes to those obtained by computer simulation for Cu using the Morse potential.[9]

$$f_\parallel = 6 \cdot 4 f_\parallel^0, \quad f_\perp = -0 \cdot 62 f_\parallel^0, \quad F_\parallel = 4 \cdot 8 f_\parallel^0,$$

$$F_\perp = -0 \cdot 36 f_\parallel^0$$

The resulting vibrational spectrum of the dumbbell atoms[12] is shown in Fig. 5. Due to the lower symmetry of the 100-dumbbell, the spectrum is more complicated than the one of the octahedral interstitial: it contains four different localized modes and three pronounced resonant modes, but nearly no intensity at the normal lattice frequencies. The motion of the two dumbbell atoms is indicated for each mode by the arrows. Figure 6 shows the amplitude distribution of the interstitial atoms and the nearest neighbours for the localized mode A_{1g} with the highest frequency and the resonant modes A_{2u} and E_g. The localized 'dumbbell-stretching' mode A_{1g} is a typical 'optical' mode where all neighbours move in opposite direction to each other yielding a frequency of about $1 \cdot 5\omega_{max}$. In contrast to this the neighbours move in phase with the dumbbell atoms in the resonant modes.

TABLE 1 Characteristic frequencies of the 100-dumbbell for Morse potential and Born–Mayer potential

Potential	$\omega_{max}, 10^{12} \text{ s}^{-1}$	$\dfrac{\omega_{res}^{E_g}}{\omega_{max}}$	$\dfrac{\omega_{res}^{A_{2u}}}{\omega_{max}}$	$\dfrac{\omega_{loc}^{A_{1g}}}{\omega_{max}}$
Morse	8·4	0·12	0·21	1·65
Born–Mayer	6·3	0·13	0·18	1·52

The resonant libration mode E_g is important for the elastic behaviour of the interstitial (next section).

We have calculated the spectra also for different potentials. The resulting frequencies for the Morse potential and for the Born–Mayer potential are listed in Table 1 for the librational resonant mode E_g, the translational resonant mode A_{2u}, and the localized mode A_{1g}. A modified Morse potential, fitted to the measured volume change $\Delta V_I = 1 \cdot 6 V_c$ of the interstitial in Cu, gives practically the same values as the Morse potential. The close agreement between the different potentials might be fortuitous.

Elementary jumps

The elementary jump of the 100-dumbbell with the lowest activation energy is shown in Fig. 7. The centre of the dumbbell moves to a nearest neighbour site, e.g. $\frac{1}{2}(110)$, whereas the axis rotates in the (001)-plane by $90°$. For the Morse potential (Born–Mayer potential) for Cu, the activation energy for this process is $0 \cdot 08 \text{ eV}$

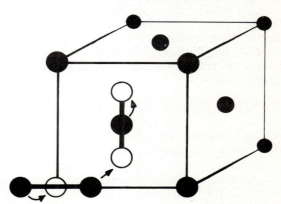

● positions before jump; ○ changed positions after jump

7 Elementary jump of 100-dumbbell

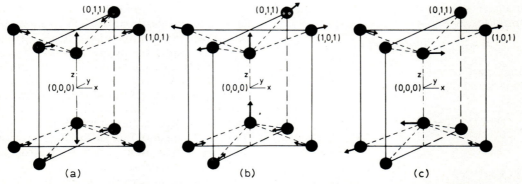

a localized mode A_{1g}; *b* translational resonant mode A_{2u}; *c* librational resonant mode E_g

6 Amplitude distribution of localized and resonant modes of 100-dumbbell

(0·04 eV). All other changes of the configuration are associated with an appreciably higher activation energy. In particular a simple rotation of the dumbbell around its centre by 90° needs about 0·20 eV (0·14 eV) and therefore can not be activated at low temperatures. Experimentally the above migration process has been confirmed by magnetic aftereffect measurements in Ni [14] and recently by elastic aftereffect measurements of Spiric *et al.*[6] in Al.

ELASTIC POLARIZABILITY OF INTERSTITIALS

The tendency of the interstitial towards instability also influences its static behaviour. For certain static displacements the restoring forces around the interstitial are especially weak, or for given external forces on the interstitial the static displacements are especially large and can diverge, if ω_{res} approaches zero (*see* for example equation (2) for the defect Green's function). The same is true if we apply homogeneous surface forces on a macroscopic crystal; in addition to the homogeneous displacements we obtain, localized near the interstitials, large induced displacements which are the larger the lower the resonance frequency. For a large crystal with many interstitials this polarization of the interstitials by a homogeneous external field leads to a change of the elastic constants. Such changes of the elastic data due to irradiation have been known experimentally for more than ten years already. For instance, after neutron irradiation König *et al.*[15] and Wenzl *et al.*[15] obtained very large decreases of the shear modulus of polycrystals of order of 50% per at.% Frenkel pairs.

The key point for a theoretical treatment[16] is the change of the coupling parameters, $\phi = \mathring{\phi} + \varphi$, near the interstitial. By applying homogeneous forces on the surface of the crystal, the corresponding displacements are determined by

$$\mathbf{s} = \mathring{\mathbf{s}} - \mathring{G}\varphi\mathbf{s} \quad \text{with} \quad \mathring{s}_i^m = \sum_{j=1}^{3} \epsilon_{ij} X_j^m \tag{5}$$

Here $\mathring{G} = 1/\mathring{\phi}$ is the static Green's function of the ideal crystal; $\mathring{\mathbf{s}} = \epsilon\mathbf{X}$ is the homogeneous deformation in the ideal crystal (ϵ_{ij} = strain tensor, X_j^m = equilibrium positions). By introducing the t-matrix, equation (5) can also be written in the form

$$\mathbf{s} = \mathring{\mathbf{s}} - \mathring{G}t\mathring{\mathbf{s}} \quad \text{with} \quad t = \varphi - \varphi\frac{1}{\mathring{\phi} + \varphi}\varphi \tag{6}$$

Thus, in addition to the homogeneous displacements $\mathring{\mathbf{s}}$, near the defect additional displacements $\mathbf{s}_{ind} = -\mathring{G}\varphi\mathbf{s} = -Gt\mathring{\mathbf{s}}$ are induced which can be formally related to induced Kanzaki forces $\mathbf{K}_{ind}(\epsilon)$

$$\mathbf{K}_{ind}(\epsilon) = -\varphi\mathbf{s} = -t\mathring{\mathbf{s}}, \qquad \mathbf{s}_{ind}(\epsilon) = \mathring{G}\mathbf{K}_{ind} \tag{7}$$

The long-range part of these displacements is determined by the induced dipole tensor:

$$P_{ij}^{ind}(\epsilon) = \sum_m K_{i\,ind}^m X_j^m = \sum_{k,l=1}^{3} \alpha_{ij,kl}\epsilon_{kl}$$

with

$$\alpha_{ij,kl} = -\sum_{mn} X_j^m t_{ik}^{mn} X_l^n \tag{8}$$

where α is the diaelastic polarizability of the single defect. For a crystal with a statistical defect distribution (concentration c) we obtain on the average an induced stress

$$\sigma_{ij}^{ind} = \frac{c}{V_c}P_{ij}^{ind} = \frac{c}{V_c}\sum_{kl} \alpha_{ij,kl}\epsilon_{kl} \tag{9}$$

where c/V_c is the volume density. Inserting this into Hooke's law which may be written as

$$\sigma + \sigma^{ind} = \mathring{C}\epsilon \quad \text{or} \quad \sigma = (\mathring{C} + \Delta C)\epsilon \tag{10}$$

we obtain for the change ΔC of the elastic constants $\mathring{C}$

$$\Delta C_{ijkl} = -\frac{c}{V_c}\alpha_{ij,kl} = \frac{c}{V_c}\sum_{mn} X_j^m t_{ik}^{mn} X_l^n \tag{11}$$

For a weakly disturbing defect, for example a substitutional defect, the first Born approximation $t \cong \varphi$ holds and ΔC is directly given by the changed coupling parameters φ. However, for a nearly unstable interstitial the second term $t \cong -\varphi(1/(\mathring{\phi} + \varphi))\varphi$ is the dominating one, since the local response $1/(\mathring{\phi} + \varphi)$ is very large. Thus we expect large and negative changes of the elastic constants for interstitials. The change is always negative because in such cases the induced stress enhances the external stress $\sigma: \sigma^{ind} \sim +\sigma$.

The influence of the resonance vibrations on the different elastic moduli depends to a large extent on *selection rules* based on symmetry. This may be seen from the equation for the t-matrix, if we expand the Green's function $1/(\mathring{\phi} + \varphi)$ in the perturbed subspace, i.e. where $\varphi \neq 0$, into eigenfunctions $|\alpha\rangle$

$$t \cong -\varphi\frac{1}{\mathring{\phi} + \varphi}\varphi = -\varphi\sum_\alpha\frac{|\alpha\rangle\langle\alpha|}{f_\alpha^{eff}}\varphi \tag{12}$$

In the case of a resonance the eigenvalue f_α^{eff}, representing an effective restoring force for the resonance mode $|\alpha\rangle$, is very small. In calculating ΔC we have to form matrix elements of t with homogeneous deformations ($\sum_j \epsilon_{ij}X_j$ or X_j). Since these have even symmetry ($X_j^{\mathbf{n}} = -X_j^{-\mathbf{n}}$), only resonance modes with even ('gerade') symmetry ($s_i^{\mathbf{n}}(\alpha) = -s_i^{-\mathbf{n}}(\alpha)$) can contribute to the elastic constants. For the 100-dumbbell this means (*see* Fig. 4) that the resonance modes A_{2u} and E_u do not influence the elastic behaviour. Only the resonant libration mode E_g (Fig. 6b) remains. However for symmetry reasons this mode can only couple to 100-shear deformations (Fig. 8a) which determine the shear modulus c_{44}, but not to a homogeneous compression (Fig. 8b) determining the compression modulus $c_{11} + 2c_{12}$ nor to 110-shear deformations (Fig. 8c) determining the shear modulus $c_{11} - c_{12}$.

For the 100-dumbbell we have calculated[16] the changes of the elastic constants by computer simulation using a Morse potential fitted to Cu. The actual calculations are complicated by the facts that in addition to the t-matrix term of equation (11) one obtains a second contribution due to the volume expansion and that one has to average over all three 100-directions of the dumbbells. For details we refer to the original paper.[16] The largest effect is shown by the two dumbbell atoms. For the 100-shear of Fig. 8a the angular change ϵ_d of the dumbbell axis is about a factor of 20 larger than the applied shear angle ϵ.

Table 2 summarizes the results of the theory for the 100-dumbbell and for the vacancy in Cu. Included are also some recent results of single crystal measurements of Cu after thermal neutron irradiation by Rehn *et al.*[5]

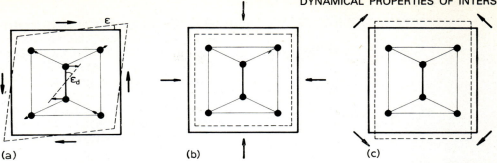

a 100-shear (modulus c_{44}); *b* compression (modulus $1/3(c_{11}+2c_{12})$); *c* 110-shear (modulus $1/2(c_{11}-_{12})$s

8 Homogeneous deformation of crystal with 100-dumbbell

TABLE 2 Change of elastic constants

	$\dfrac{\Delta c_{44}}{cc_{44}}$	$\dfrac{\Delta(c_{11}-c_{12})}{c(c_{11}-c_{12})}$	$\dfrac{\Delta(c_{11}+2c_{12})}{c(c_{11}+2c_{12})}$
Theory[16] (Morse potential for Cu)			
100-dumbbell	−37	−15·3	−7·9
vacancy	−3·1	−4·4	−2·7
Frenkelpair	−40·1	−19·7	−10·7
Experiment			
Frenkelpair Cu[5]	−31	−15	0±1
Frenkelpair Al[4]	−27	−15	—

and on Al after electron irradiation by Robrock *et al.*[6] In general one sees that the theoretical values indeed give large and negative changes as have been observed experimentally. In particular, the anisotropy of the shear moduli $[|\Delta c_{44}|>|\Delta(c_{11}-c_{12})/2|]$ which is a direct consequence of the large resonance contribution to c_{44} is in agreement with the experiments. Since the symmetry (E_g) of this resonant mode is a direct consequence of the 100-dumbbell symmetry, this is a further indication of the existence of the 100-dumbbell in Cu and Al and in agreement with the recent X-ray results.[1−3]

The relative large changes of the elastic constants suggest similar large changes in the phonon dispersion curves. This effect has been calculated by Wood and Mostoller[17] and Schober *et al.*[18] The first experimental results are not quite conclusive.[19,20] The polarizability is also important for the interaction of interstitials with each other. Due to the polarization one obtains in addition to the usual $1/R^3$ interaction an attractive van der Waals type $1/R^6$ interaction.[21] The polarization by an external field seems to be decisive for an understanding of irradiation creep as has been noted by Speight and Heald[22] and Bullough and Willis.[23]

By analogy to electrostatics one can obtain in addition to the diaelastic polarizability a paraelastic polarizibility which is due to the reorientation of anisotropic dipole tensors in an external strain field ϵ_{ij}. The paraelastic change of the dipole tensor is temperature dependent and is given by:

$$\Delta P_{ij}^{\text{para}}=\sum_{kl}\alpha_{ij,kl}^{\text{para}}\,\epsilon_{kl}$$

with

$$\alpha_{ij,kl}=\frac{1}{kT}\{\langle P_{ij}P_{kl}\rangle-\langle P_{ij}\rangle\langle P_{kl}\rangle\} \qquad (13)$$

where $\langle\ \rangle$ means the average overall equivalent orientations of the defect. However, in order to observe this effect, the defects must be able to reorient in an external field so that for interstitials the effect can only be observed[6] during migration. It occurs only for anisotropy defects, where it can be an order of magnitude larger than the dielastic polarizability.

DI- AND TRI-INTERSTITIALS

Because of their strong interaction interstitials will always, when given a chance, cluster together and form di-interstitials, tri-interstitials, etc and finally large dislocation loops. The structure of the small aggregates has been studied by Johnson[24] and recently by Schober[25] using computer simulation.

In both cases di-interstitials with very high binding energies of about 1 eV were found. The most stable pair I_2 consists of two 100-dumbbells aligned parallel to each other on nearest neighbour sites (Fig. 9*a*). The second most tightly bound pair I_2', shown in Fig. 9*b* has a somewhat smaller binding energy than the configuration I_2. This metastable configuration represents an intermediate step for the elementary migration process of the di-interstitial I_2. One dumbbell performs the usual

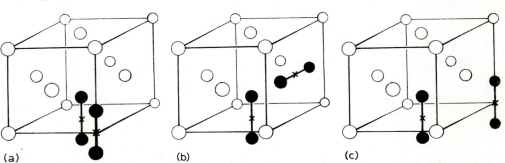

a and *c* stable configuration I_2; *b* metastable configuration I_2'

9 Di-interstitial in fcc

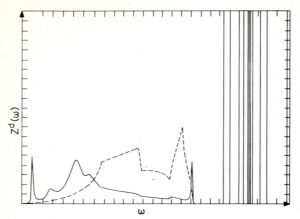

10   Local frequency spectrum of di-interstitial atoms (broken line = spectrum of the ideal crystal)

elementary jump as shown in Fig. 7, leading from the stable configuration I_2 to the metastable configuration I_2' (Fig. 9b). From there the dumbbell can jump further to the stable configuration I_2 of Fig. 9c. The activation energy for migration is $E_{I_2}^M = 0.13$ eV for the Morse potential (0·29 eV for Johnson's Ni potential), slightly higher than the migration energy of $E_I^M = 0.08$ eV (0·15 eV) of the single dumbbell.

The vibrational behaviour of the di-interstitial is similar to that of the single interstitial.[26] Its local vibrational spectrum, as calculated for the Morse potential, is shown in Fig. 10 and again consists of localized and resonant modes. However, owing to the lower symmetry, all degeneracies are removed leading to a large number of different localized and resonant modes. The latter partially overlap.

The binding energy of the tri-interstitial[25] is even higher. For the Morse potential the dissociation of the tri-interstitial into a single and a di-interstitial costs about 1·5 eV. The stable configuration consists of three orthogonal 100-dumbbells on nearest neighbour positions (see Fig. 11a). Contrary to the single- and di-interstitial, the tri-interstitial has a lower activation energy for reorientation than for migration. The reorientation process leads over a metastable configuration I_3' (Fig. 11b) to a tri-interstitial I_3 (Fig. 11c) which is centred at the same octahedral position as the original one. The activation energy for this process is about the same as the one for the jump of the single interstitial.

Experimentally di-interstitials have first been identified in Al by Ehrhart and Schilling,[1] who measured the

Huang-diffuse scattering of electron-irradiated Al after stage I_D. Haubold and Schilling[3] confirmed for Al the configuration I_2 of Fig. 9a. Elastic constant measurements by Robrock and Schilling[4] indicate that its polarizability is even larger than the one of two single interstitials. A recovery stage II_1, just above stage I_D, is attributed to the free migration of di-interstitials. Recent elastic-after-effect measurements by Spiric et al.[6] are in agreement with the diffusion process shown in Fig. 9a–c. They also observe a process which can be explained by a reorientation of the tri-interstitial as shown in Fig. 11a–c.

In Cu, the situation seems to be quite different. After electron irradiation, directly after stage I_D large interstitial clusters are formed containing about 5–10 interstitials.[27] Since di-interstitials are thought to be a necessary intermediate step for the formation of these clusters, one concludes that very small interstitial agglomerates such as di- or tri-interstitials must have a higher mobility in Cu than the single interstitial, so that once formed they immediately migrate and form bigger and more stable clusters.

REFERENCES

1. P. EHRHART AND W. SCHILLING: *Phys. Rev.* 1973, **B8**, 2604
2. P. EHRHART *et al.*: in 'Advances in solid state physics', 1974 XIV, 87
3. H. G. HAUBOLD in Proc. Int. Conf. on 'Fundamental aspects of radiation damage in metals', 268, Gatlinburg, USA, Oct. 1976
4. K. H. ROBROCK AND W. SCHILLING: *J. Physics F (Metal Physics)* 1976, **5**, 303
5. L. E. REHN *et al.*: *Phys. Rev.*, 1974, **B10**, 349; J. HOLDER *et al.*: *Phys. Rev.*, 1974, **B10**, 363; *Phys. Rev. Lett.*, 1974, **32**, 1054
6. V. SPIRIC *et al.*: *Phys. Rev.*, in press
7. M. L. SWANSON: *Can. J. Phys.*, 1975, **53**, 1117
8. M. L. SWANSON: ref. 3, p. 316
9. A. SCHOLZ AND C. LEHMANN: *Phys. Rev.*, 1972, **B6**, 813
10. P. H. DEDERICHS *et al.*: *Phys. Rev. Lett.*, 1973, **31**, 1130
11. P. H. DEDERICHS: ref. 3, p. 187
12. R. ZELLER AND P. H. DEDERICHS: *Z. Physik*, 1976, **B25**, 139
13. R. A. JOHNSON, *J. Physics.*, F (*Metal Physics*), 1973, **3**, 295
14. K. FORSCH *et al.*: *Physica Status Solidi (a)*, 1974, **23**, 223
15. H. WENZL: in 'Vacancies and interstitials in metals', (Ed. A. SEEGER *et al.*), 1969, Amsterdam, North-Holland
16. P. H. DEDERICHS *et al.*: *Z. Physik*, 1975, **B20**, 155
17. R. F. WOOD AND M. MOSTOLLER: *Phys. Rev. Lett.*, 1975, **35**, 45
18. H. R. SCHOBER *et al.*: *Z. Physik*, 1975, **B21**, 255

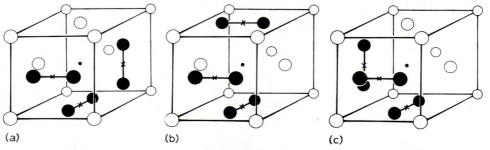

(a)   (b)   (c)

a and *c* stable configuration I_3; *b* metastable configuration I_3'; the dot (●) indicates the octahedral position at which both configurations are centred

11   Tri-interstitial in fcc

19 R. M. NICKLOW et al.: *Phys. Rev. Lett.*, 1975, **35**, 1444

20 K. BÖNING: private communication

21 H. TRINKAUS: ref. 3, p. 254

22 P. T. HEALD AND M. V. SPEIGHT, *Phil. Mag.*, 1974, **29**, 1075; *Acta Met.*, 1975, **23**, 1389

23 R. BULLOUGH AND J. R. WILLIS: *Phil. Mag.*, 1975, **31**, 855

24 R. A. JOHNSON: *Phys. Rev.*, 1966, **152**, 629

25 H. R. SCHOBER: *J. Physics F (Metal Physics)*, to be published

26 R. ZELLER AND H. R. SCHOBER: to be published

27 P. EHRHART AND U. SCHLAGHECK: *J. Physics F (Metal Physics)*, 1974, **4**, 1575; 1974, **4**, 1589

Calculation of point defect formation energies in metals

R. Evans

Current theoretical models for calculating point defect formation energies in metals are critically reviewed. We concentrate on the monovacancy formation energy since this is the problem which has received the most theoretical and experimental attention and discuss in detail the various approaches which have been used to calculate this quantity for simple (non-transition) metals. We conclude that none of the existing theories give a satisfactory account of the vacancy formation energy except, perhaps, for some of the alkali metals. We comment on the use of empirical pairwise potentials and show how such potentials play a different role in the theory of the formation energy and the formation volume from the effective pairwise potentials derived in the pseudopotential formalism. Finally we briefly mention some alternative approaches to the energetics of point defects. These aim to provide an accurate description of the electron density around the defect but have not yet been applied to the calculation of formation energies.

The author is at the University of Bristol

At the Jülich conference Friedel[1] commented on the diversity of approaches used to study the properties of point defects. Eight years on, one is again struck by the number of different theoretical techniques which are currently employed in point defect studies. In this volume Eshelby has shown how much one can learn from continuum models and Heald has reviewed some results from calculations based on discrete lattice models in which the atoms are assumed to interact via empirical pairwise interatomic potentials. In each case the relative simplicity of the approach permits the study of rather complicated interactions between point defects. However, in a real metal the presence of a point defect causes a significant redistribution of the electron density and this makes any proper theory of the energetics of the system necessarily complex. While solid-state theory and computers have developed to an extent that it is now possible to make meaningful calculations of the binding energy, equilibrium lattice constant and compressibility of a perfect metallic crystal at $T = 0$ K from essentially first principles (i.e. given the atomic number and the lattice type) it is still a challenging problem to describe the energetics away from $T = 0$ K. When one turns to defected crystals or crystals with surfaces one is forced to make many more approximations and assumptions. This is necessary even for the simplest point defect situations, i.e. a monovacancy and a single interstitial.

'Electron' theorists have usually restricted themselves to this class of problem and asked 'What is the energy required to create a single vacancy or a single interstitial in a metal at $T = 0$ K and how does it arise?' Here we will attempt to review the progress that has been made in providing an answer to this question. Experimentally it is easier to measure the concentration of monovacancies and hence the vacancy formation energy and there now exist fairly reliable experimental estimates of this quantity for many metals (for example Seeger[2] and Franklin[3]). Consequently most theoretical attention has also been focused on the monovacancy problem. In this article we will concentrate mainly on the calculation of the vacancy formation energy for simple (non-transition) metals. For such systems it is usually assumed that the core electrons are rigidly attached to the nuclei and that all redistribution of charge is confined to the conduction electrons which are then treated in some appropriate theory. Such an assumption leads to considerable simplifications. In transition metals the outermost d electrons do not behave like atomic d states nor are they easily amenable to theories developed for the 'nearly free electron like' conduction electrons of the simple metals.

Even after restricting the discussion to vacancies in simple metals one finds that two separate and what appear to be unrelated approaches have developed. The first was pioneered by Fumi[4] who calculated formation energies assuming that the effect of the vacancy on the electron distribution could be represented by a single repulsive impurity potential in an otherwise free-electron gas. Later Stott et al.[5] tried to extend Fumi's calculation and recently Nieminen et al.[6] and Robinson and de Chatel[7] have carried out properly self-consistent calculations for a jellium model of the type introduced by Fumi. The discrete nature of the lattice is not usually considered in this approach.

The second scheme for calculating the formation energy was suggested by Harrison.[8] This is based on the pseudopotential treatment of the electron-ion interaction and many calculations have been performed using this approach, for example Ho,[9] Chang and Falicov,[10] DuCharme and Weaver,[11] Popović et al.[12] and Minchin et al.[13] Since this theory explicitly takes into account discrete lattice effects one might expect it to be superior to the first approach. It assumes, however, that the perturbation in the conduction electron distribution caused by the creation of a vacancy is sufficiently weak to be treated in a simple linear response formalism. In the jellium models the electron distribution around the vacancy is calculated using more sophisticated non-linear theories.

Since the practitioners of both methods have frequently obtained results for the formation energy which agree well with the estimates obtained from experiment, and in order to judge what the current state of the art really is, one must look critically at the different theories and the various calculations. In the second and third sections such a critique is presented.* In the fourth section we briefly discuss the use of empirical pairwise potentials in the study of the energetics of a single vacancy. We show how these empirical potentials differ from the effective pairwise potentials calculated from pseudopotential theory and how they lead to different results for the vacancy formation volume. We also comment on the validity of published 'derivations' of Mukherjees' empirical correlation. In the final section, we discuss other possible schemes for calculating the electron density and formation energy associated with a vacancy.

JELLIUM MODELS OF THE VACANCY

Review of some earlier calculations

The first jellium models started from the premise that a vacancy in a simple metal could be considered as a single impurity with a valence equal to the negative of the valence of the metal. The 'host' metal is then replaced by a uniform electron gas with electron density equal to the average value of the conduction electron density in the metal. The existence of some compensating uniform positive background of charge is also implied. If the valence of the metal is Z the formation of a vacancy is concerned with a redistribution of Z electrons around a Fermi energy of E_F. Thus one might naively expect the

formation energies E^f to scale roughly with ZE_F in a jellium model. We shall discuss how well this argument stands up to detailed investigation.

Fumi[4] represented the metal as a large spherical box of volume Ω in which the positive charge of N_i ions was assumed to be uniformly distributed. He then calculated the change in energy of the system of N electrons, $N = N_i Z$, which occurs when an amount Z of positive charge is taken from the centre of the sphere and spread uniformly over its surface. The contribution to E^f which arises from the change in the one-electron eigenvalues, ΔE_{eigen}, can be expressed in terms of the phase shifts $\eta_l(k)$ which describe the scattering of the electrons by the impurity potential associated with the vacancy:

$$\Delta E_{eigen} = -\frac{2}{\pi} \int_0^{K_F} dk \sum_{l=0}^{\infty} (2l+1)k\eta_l(k) \qquad (1)$$

where K_F is the free-electron Fermi momentum, i.e. $E_F = K_F^2$. If the scattering is assumed to be weak enough $\eta_l(k)$ can be evaluated in the first Born approximation and it follows that ΔE_{eigen} is proportional to the average value of the scattering potential. The latter can be obtained from the Friedel sum rule

$$\frac{2}{\pi} \sum_{l=0}^{\infty} (2l+1)\eta_l(K_F) = -Z \qquad (2)$$

if the Born approximation is used again. With these approximations ΔE_{eigen} is independent of the form assumed for the scattering potential and is given by

$$\Delta E_{eigen} = \tfrac{2}{3}ZE_F \qquad (3)$$

Fumi then assumed that on creating the vacancy the spherical box expands by a volume $\delta\Omega$. The change in kinetic energy of a system of free electrons due to such an expansion is $-\tfrac{2}{5}ZE_F(\delta\Omega/\Omega_0)$ where $\Omega_0 = \Omega/N_i$ is the atomic volume.

Taking $\delta\Omega = \Omega_0$ the formation energy is

$$E^f = \tfrac{2}{3}ZE_F - \tfrac{2}{5}ZE_F = \tfrac{4}{15}ZE_F \qquad (4)$$

Arguing that the use of the Born approximation overestimates the magnitude of the phase shifts for a repulsive potential, Fumi estimated that a more rigorous calculation of $\eta_l(k)$ could reduce E^f to about $ZE_F/6$. Although this latter result is in good qualitative agreement with experiment for the alkali and the noble metals (see Table 1) it leads to gross overestimates for metals with $Z > 1$ and this led Stott et al.[5] (see also March[14]) to re-examine Fumi's approach.

These authors recognized that Fumi's calculation: (a) did not correct for the double counting of electron–electron interactions in the eigenvalue sum of equation (1) and (b) did not consider electron exchange and correlation contributions to the formation energy. In their own calculations they attempted to include these effects and did not make use of the Born approximation. For the impurity potentials Stott et al. chose the negative of screened Herman–Skillman free-ion potentials but claimed that the displaced charge density around the vacancy did not depend sensitively on the particular choice of potential. Their resultant charge densities are very small in the interior of the vacancy indicating that the electrons are very strongly expelled by the repulsive potential. Their calculated formation energies are sensitive to the form chosen for the impurity potential. Whereas ΔE_{eigen} and the exchange contribution to E^f

* Throughout the paper we will use atomic units so that $\hbar = 2m = 1$ and the unit of length is the Bohr radius. We will use E^f to denote the vacancy formation energy. E_V^f denotes the formation energy calculated under constant volume conditions. The subscripts 0, 1, and 2 are used to label the order to which the electron-ion pseudopotential is included in the calculation, for example E_0^f is the result of a jellium calculation while E_{V2}^f denotes a second-order pseudopotential calculation carried out at constant volume. Ω_0 is the atomic volume. Ω^f denotes the formation volume and its accompanying subscripts again label the order of perturbation theory. Similarly p_0 denotes the pressure of the jellium model, p_1 the pressure calculated in first-order pseudopotential theory, and p_2 the pressure to second order in the pseudopotential. Other symbols are defined as they occur.

TABLE 1   Vacancy formation energies E^f calculated from jellium models and comparison with experiment;* all energies are given in eV

	r_s, a.u.	ZE_F	$\frac{1}{6}ZE_F$	E^f (Stott et al.[5])	E^f (Manninen et al.[15])	E^f (expt)
Cu	2·67	7·03	1·17	0·77		1·17[G]
Ag	3·01	5·53	0·92			1·01[G]
Au	3·02	5·49	0·92			1·01[G]
Li	3·25	4·74	0·79		0·26	0·34[M]
Na	3·93	3·24	0·54		0·18	0·42[M]
K	4·86	2·12	0·35		0·17	0·39[M]
Rb	5·23	1·83	0·31			0·27[G]
Cs	5·63	1·58	0·26		0·18	0·28[M]
Be	1·88	28·34	4·72		−2·17	
Mg	2·65	14·27	2·38	1·28	−0·22	0·81[M]
Al	2·07	35·07	5·85	1·05	−1·20	0·66[M]
Tl	2·48	24·43	4·07		−0·05	
Pb	2·30	37·88	6·31	1·52		0·53[G]

* The experimental results are taken from the compilations of Górecki[45] (G) or Manninen et al.[15] (M).

are not strongly dependent on the potential, Stott et al. state that the double counting term (a) is, and has the opposite sign (i.e. it is negative) to these earlier contributions. Nevertheless, Stott et al.'s results for the formation energies of several metals (see Table 1) are in fair agreement with experiment. It appears that the double counting correction is sufficiently large to reduce significantly the sum of all the other contributions.

Stott et al. conclude that the essential physics of the vacancy problem are contained in this jellium approach. We find it disturbing, however, that E^f should be so sensitive to the choice of impurity potential and since the latter must be somewhat arbitrary this would seem to throw doubt on Stott et al.'s conclusion. Furthermore it is not clear from their paper whether or not their model can be made completely self-consistent.

Recently, Nieminen et al.[6]* have queried Stott et al.'s conclusions. Nieminen et al. performed a fully self-consistent, non-linear calculation of the charge density and formation energy associated with a jellium vacancy. They model the vacancy by a spherical hole of radius R, centred at the origin, in an otherwise uniform background of positive charge density ρ_u. The appropriate positive charge distribution is

$$\rho_0^+(r) = \rho_u(1 - \Theta(R - r)) \qquad (5)$$

where Θ is the usual step function. Nieminen et al. assume R is equal to the Wigner–Seitz radius, i.e. $4\pi R^3/3 = \Omega_0$ and, following Fumi, allow the unperturbed metallic sphere which is of volume Ω to expand by Ω_0 when the vacancy is created. Charge neutrality requires that the total positive charge is equal to the number of electrons so

$$\int d\mathbf{r}\, \rho_0^+(r) = N = N_i Z$$

where the integral is over the volume $\Omega + \Omega_0$ and hence

$$\rho_u = \frac{N}{\Omega} = Z\Omega_0^{-1}.$$

Although one could invent other shapes and sizes for the hole this specification would seem to be eminently plausible. From now on we will refer to this particular

* A fuller account of this work is given by Manninen et al.[15]

model as the canonical jellium vacancy. Because of its simplicity it could be viewed as the model on which proposed electron theories should be tested. It would then have a status equivalent to that of the planar uniform background model which is frequently used in the study of metal surfaces (for example Lang[16]).

The formation energies calculated by Manninen et al. for several jellium metals are listed in Table 1. The striking feature of these results is that E^f is negative for metals with small values of the electron radius, r_s.† This is in contradiction to the results of Stott et al. The calculated electron densities also differ from those obtained by Stott et al. since, for each metal, these densities tend to a finite value of about $0·2\rho_u$ in the centre of each vacancy (see Figs. 1 and 2). The spherical hole is not nearly so repulsive as the impurity potential assumed by Stott et al.

† This is defined by $\frac{4\pi}{3}r_s^3 = \rho_u^{-1}$.

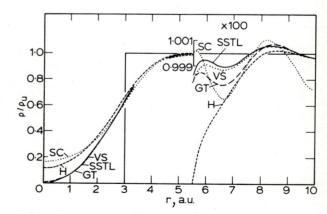

1   Electron density profiles at the canonical jellium model of a vacancy in Al($r_s = 2·07$ a.u.); SC denotes the self-consistent, non-linear result of Nieminen et al.;[6] the other profiles are linear screening results with various approximations for the exchange and correlation correction as discussed in the text

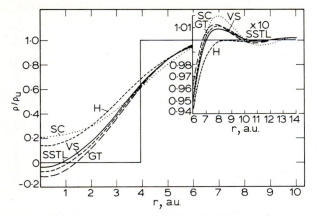

2 Electron density profiles at the canonical jellium model of a vacancy in Na ($r_s = 3 \cdot 93$ a.u.); the labelling follows that of Fig. 1

As Manninen *et al.*'s work constitutes the most complete treatment of the jellium model it is important to examine the density functional formalism on which their calculations are based and look in detail at the consequences of their results. In particular we should understand why E^f is negative for some values of r_s. This is not obvious from Manninen *et al.*'s presentation.

Density functional formalism

The Hohenberg–Kohn–Sham density functional theory is well documented and has been ably reviewed in the recent article by Lang.[16] Here we will merely list the main points of the theory. Hohenberg and Kohn[17] proved that the total ground state energy U (at $T = 0\,K$) of any system of electrons and positive charge could be written as

$$U = U_{es} + G[\rho] \qquad (6)$$

where U_{es} is the total electrostatic energy and $G[\rho]$ is an unknown but unique functional of the electron density $\rho(\mathbf{r})$. Later Kohn and Sham[18] used this result to transform the many-electron problem into an effective one-electron problem. In their theory the exact electron density is given by

$$\rho(\mathbf{r}) = \sum_i |\psi_i(\mathbf{r})|^2 \qquad (7a)$$

where the ψ_i are the eigenfunctions of an effective one-electron Schrödinger equation

$$(-\nabla^2 + V_{eff}[\rho\,;\mathbf{r}])\psi_i(\mathbf{r}) = \epsilon_i\psi_i(\mathbf{r}) \qquad (7b)$$

The summation in equation ($7a$) extends over the N lowest-lying orthonormal solutions of equation ($7b$). N is the total number of electrons. The effective potential has the form

$$V_{eff}[\rho\,;\mathbf{r}] = \phi(\mathbf{r}) + \frac{\delta E_{xc}[\rho]}{\delta\rho(\mathbf{r})} \qquad (8)$$

where $\phi(\mathbf{r})$ is the total electrostatic (Hartree) potential and the functional $E_{xc}[\rho]$ is defined by

$$E_{xc}[\rho] \equiv G[\rho] - T[\rho] \qquad (9)$$

The functional $T[\rho]$ is just the kinetic energy of the non-interacting system defined by equations (7) and it is

straightforward to show

$$T[\rho] = \sum_i \epsilon_i - \int d\mathbf{r}\,\rho(\mathbf{r})V_{eff}[\rho\,;\mathbf{r}] \qquad (10)$$

The second term in equation (10) corrects for the double counting involved in the eigenvalue summation. $E_{xc}[\rho]$ is the exchange and correlation energy which is again a unique functional of the electron density. All the difficulties associated with the many-electron aspects of the problem are concealed in this object. Leaving such difficulties aside one is left to solve what is essentially a Hartree self-consistent field problem. One proceeds by selecting a trial V_{eff} and solves equations (7) self-consistently to obtain $\rho(\mathbf{r})$. The total ground state energy is easily obtained from equations (6), (9), and (10) as

$$U = U_{es} + E_{xc}[\rho] + \sum_i \epsilon_i - \int d\mathbf{r}\,\rho(\mathbf{r})V_{eff}[\rho\,;\mathbf{r}] \qquad (11)$$

If the electron density is expected to vary slowly with $\mathbf{r}$ it is natural to expand E_{xc} as a series of density gradients, i.e.

$$E_{xc}[\rho] = \int d\mathbf{r}[\epsilon_{xc}(\rho(\mathbf{r}))\rho(\mathbf{r}) + \epsilon_{xc}^{(2)}(\rho(\mathbf{r}))|\nabla\rho(\mathbf{r})|^2 + \ldots] \qquad (12)$$

where $\epsilon_{xc}(\rho)$ is the average exchange and correlation energy per particle of the *uniform* electron gas. In almost all applications of the theory it is assumed that all gradient terms can be omitted in equation (12).* This constitutes the 'local density approximation', the validity of which depends on the particular application (*see* for example Lang[16]). Within this approximation the effective potential of equation (8) reduces to

$$V_{eff}[\rho\,;\mathbf{r}] = \phi(\mathbf{r}) + \mu_{xc}(\rho(\mathbf{r})) \qquad (13)$$

where

$$\mu_{xc}(\rho) = \frac{d}{d\rho}(\rho\epsilon_{xc}(\rho))$$

and equation (11) becomes

$$U = U_{es} + \sum_i \epsilon_i - \int d\mathbf{r}\,\rho(\mathbf{r})\left[\phi(\mathbf{r}) + \rho(\mathbf{r})\frac{d\epsilon_{xc}(\rho(\mathbf{r}))}{d\rho(\mathbf{r})}\right] \qquad (14)$$

The electrostatic potential is

$$\phi(\mathbf{r}) = \int d\mathbf{r}'\frac{(\rho(\mathbf{r}') - \rho^+(\mathbf{r}'))}{|\mathbf{r} - \mathbf{r}'|} \qquad (15)$$

where $\rho^+(\mathbf{r})$ is the appropriate positive charge distribution and the electrostatic energy is

$$U_{es} = \tfrac{1}{2}\int d\mathbf{r}(\rho(\mathbf{r}) - \rho^+(\mathbf{r}))\phi(\mathbf{r}) \qquad (16)$$

Manninen *et al.*[15] used the positive background of equation (5) and solved equations (7), using the local density approximation of equation (13), for the electron density. The energy of the spherical metal with a vacancy can then be calculated from equation (14). The formation energy E_0^f in this model is obtained by subtracting the energy of the uniform jellium metal consisting of N electrons in a volume Ω in the presence of a uniform

---

* We should note that the Hartree approximation sets $E_{xc}[\rho] \equiv 0$, i.e. it ignores all exchange and correlation effects. Thomas–Fermi theory sets $E_{xc}[\rho] \equiv 0$ *and* approximates the kinetic energy functional by the local density expression $T[\rho] = \int d\mathbf{r}\,t(\rho(\mathbf{r}))\rho(\mathbf{r})$ where $t(\rho)$ is the average kinetic energy per particle of the uniform electron gas.

positive background of charge:

$$E_0^f = U - G[\rho_u]$$

where

$$G[\rho_u] = N(t(\rho_u) + \epsilon_{xc}(\rho_u)) = N u_{eg}(\rho_u) \qquad (17)$$

Since $\phi(\mathbf{r}) = 0$ for the uniform system it follows from equation (10) that

$$G[\rho_u] = \sum_k \epsilon_k^{\text{uniform}} - N\mu_{xc}(\rho_u) + N\epsilon_{xc}(\rho_u)$$

where the eigenvalues are simply $\epsilon_k^{\text{uniform}} = k^2 + \mu_{xc}(\rho_u)$. After some manipulation we find

$$E_0^f = \sum_i \epsilon_i - \sum_k \epsilon_k^{\text{uniform}} + N\rho_u \frac{d\epsilon_{xc}(\rho_u)}{d\rho_u}$$

$$- \int d\mathbf{r}\, \rho(\mathbf{r})\left(\phi(\mathbf{r}) + \rho(\mathbf{r}) \frac{d\epsilon_{xc}(\rho(\mathbf{r}))}{d\rho(\mathbf{r})}\right) + U_{es} \qquad (18)$$

The first term in equation (18) refers to the eigenvalue summation for the defect metal of volume $\Omega + \Omega_0$ while the second refers to the uniform metal of volume Ω. Consequently the sum of these is just the quantity derived by Fumi, i.e. $\Delta E_{eigen} - \frac{2}{5}ZE_F$, where the phase shifts in equation (1) now refer to scattering from the potential $V_{eff}[\rho; \mathbf{r}]$. The integration in equation (18) extends over the volume $\Omega + \Omega_0$. After further rearrangement the formation energy becomes

$$E_0^f = \Delta E_{eigen} - \int d\mathbf{r}\, \rho(\mathbf{r})\phi(\mathbf{r})$$

$$- \int d\mathbf{r}\left[\rho^2(\mathbf{r}) \frac{d\epsilon_{xc}(\rho(\mathbf{r}))}{d\rho(\mathbf{r})} - \rho_u^2 \frac{d\epsilon_{xc}(\rho_u)}{d\rho_u}\right]$$

$$+ U_{es} - \frac{2}{5}ZE_F - \rho_u Z \frac{d\epsilon_{xc}(\rho_u)}{d\rho_u} \qquad (19)$$

The integrals, including that for U_{es}, refer to the metal with a vacancy and all converge as $r \to \infty$, since $\phi(\mathbf{r}) \to 0$ in this limit. It is easy to show that this result for E_0^f is equivalent to that obtained by Manninen *et al.*[15] Noting that the sum of the last two terms in equation (19) is simply $-p_0\Omega_0$ where p_0 is the pressure of the uniform electron gas

$$p_0 = -\frac{dG[\rho_u]}{d\Omega} = \rho_u^2 \frac{du_{eg}(\rho_u)}{d\rho_u} \qquad (20)$$

the formation energy can be written in the form

$$E_0^f = E_{V0}^f - p_0\Omega_0 \qquad (21)$$

E_{V0}^f is the formation energy which would be calculated under constant volume conditions.

Equation (21) is merely a special case of the general result (*see* for example Evans and Finnis[19])

$$E^f = E_{V0}^f - p\Omega^f \qquad (22)$$

where E_V^f is the constant volume formation energy, p is the pressure of the system under consideration, and Ω^f is the expansion in volume associated with the formation process. If the system is in equilibrium, $p = 0$, then E^f can be calculated at constant volume.

In many *models*, however, the pressure is finite and for the jellium the magnitude and sign of p_0 depend on the value of r_s.

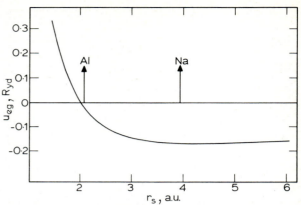

3 Total energy per electron of the uniform electron gas, u_{eg}, plotted v. the electron radius r_s. The arrows label the values of r_s corresponding to Al and Na; note 1 Ryd = 13·6 eV

In Fig. 3 the total energy per particle of the uniform electron gas, $u_{eg}(\rho_u)$, is plotted as a function of r_s. This has been calculated from the standard results (in Ryd units)

$$t(\rho_u) = 2 \cdot 21/r_s^2$$

$$\epsilon_{xc}(\rho_u) = -0 \cdot 916/r_s - 0 \cdot 88/(r_s + 7 \cdot 8) \qquad (23)$$

where

$$\frac{4\pi}{3} r_s^3 = \rho_u^{-1}$$

and the Wigner interpolation formula has been used for the correlation energy. From the graph it can be seen that for high electron densities $-p_0\Omega_0$ is large and negative while for $r_s \sim 4$ a.u. (corresponding roughly to Na) the system is almost in equilibrium and for large values of $r_s - p_0\Omega_0$ is small and positive. Negative values of E_0^f arise in Manninen *et al.*'s calculations for small r_s because their assumed expansion of the jellium by one atomic volume brings about a reduction in energy which is greater than the increase in energy caused by the distortion of the electron density around the hole.

There is, of course, no reason *a priori* for assuming that the volume of formation associated with the vacancy is Ω_0. From general thermodynamic considerations (for example Franklin[3]) it can be shown that the formation volume is given by

$$\Omega^f = \left(\frac{\partial\Delta G}{\partial p}\right)_T \qquad (24)$$

where ΔG is the change in Gibbs free energy associated with the formation of the defect. At $T = 0$ K

$$\Delta G = \Delta U + p\Omega^f$$

but the total energy required to create the defect is from equation (22)

$$\Delta U = E^f = E_V^f - p\Omega^f$$

and so

$$\Omega^f = \left(\frac{\partial E_V^f}{\partial p}\right)_T = -\frac{\Omega}{B}\left(\frac{\partial E_V^f}{\partial\Omega}\right)_T \qquad (25)$$

where B is the bulk modulus. Thus Ω^f can, in general, be calculated from the constant volume formation energy.

It would be useful to evaluate the formation volumes in this way for the canonical jellium model and investigate whether or not E_0^f remains negative for metals with small values of r_s. Such a study is now under way (M. W. Finnis, private communication).

The fact that $p_0 \neq 0$ is a serious shortcoming of the jellium approach. Indeed one might argue that the constant volume formation energy E_{V0}^f has more physical significance than E_0^f. Since a jellium metal is not, in general, in equilibrium one could arbitrarily impose an external pressure which exactly cancels p_0. E_0^f then reduces to E_{V0}^f, the constant volume result.*

As mentioned earlier the important qualitative difference between the canonical jellium model and the model employed by Stott et al.[5] lies in the relative strengths of the repulsive potentials associated with the vacancy. Stott et al. argued that their impurity potentials were too strong to permit a realistic calculation of the displaced electron densities using linear screening theory (first-order perturbation theory). For the canonical model, on the other hand, one might imagine that linear theory would not be completely useless and so it is instructive to solve for the charge density and formation energy in this approximation. The formalism which we develop will prove useful when we later discuss the pseudopotential formulation of the vacancy problem.

Linear screening of a jellium vacancy

In the linear screening approximation the change in electrostatic potential caused by the introduction of the vacancy is treated as a perturbation on the uniform electron gas of density ρ_u. If the vacancy is represented by some positive charge distribution $\rho_0^+(\mathbf{r})$ the perturbation potential is simply

$$\delta V^+(\mathbf{r}) = -\int d\mathbf{r}' \frac{\rho_0^+(r')}{|\mathbf{r} - \mathbf{r}'|} + \int \frac{d\mathbf{r}'\rho_u}{|\mathbf{r} - \mathbf{r}'|} \quad (26)$$

and the perturbation in electron density is

$$\delta\rho(\mathbf{r}) = \rho_{lin}(\mathbf{r}) - \rho_u = \int d\mathbf{r}' X(|\mathbf{r} - \mathbf{r}'|)\delta V^+(\mathbf{r}') \quad (27)$$

where $X(|\mathbf{r} - \mathbf{r}'|)$ is the interacting linear response function of the uniform electron gas of density ρ_u. Clearly the electron density in the perturbed metal is

$$\rho_{lin}(\mathbf{r}) = \rho_u + (2\pi)^{-3}\int d\mathbf{q}\exp(-i\mathbf{q}\cdot\mathbf{r})X(q)\delta V^+(\mathbf{q}) \quad (28)$$

where

$$\delta V^+(\mathbf{q}) = -\frac{4\pi}{q^2}\rho_0^+(\mathbf{q}) \quad (q \neq 0)$$

The interacting response function can be expressed in terms of the non-interacting (Lindhard) function $X_0(q)$ and the Coulomb interaction $4\pi/q^2$:

$$X(q) = X_0(q)[1 - X_0(q)(1 - f_{ex}(q))4\pi/q^2]^{-1} \quad (29)$$

where $f_{ex}(q)$ corrects for exchange and correlation effects. The Hartree approximation sets $f_{ex} \equiv 0$. The

Lindhard function has the form:

$$X_0(q) = -\frac{K_F}{(2\pi^2)}\left[1 + \frac{1 - \eta^2}{2\eta}\ln\left(\left|\frac{1 + \eta}{1 - \eta}\right|\right)\right] \quad (30)$$

where $\eta = q/2K_F$. For the canonical jellium vacancy described by equation (5)

$$\delta V^+(q) = 16\pi^2 R^2\rho_u j_1(qR)/q^3 \quad (q \neq 0) \quad (31)$$

and

$$\rho_{lin}(r) = \rho_u\left(1 + 8R^2\int_0^\infty dq\, X(q)j_0(qr)j_1(qR)/q\right) \quad (32)$$

where j_0 and j_1 are spherical Bessel functions.

In Figs. 1 and 2 the charge densities calculated from equation (32) are compared with those obtained by Nieminen et al.[6] from their non-linear Kohn–Sham calculations. Several different prescriptions for $f_{ex}(q)$ have been used in the linear theory. SSTL refers to the f_{ex} of Singwi et al.,[21] VS to Vashista and Singwi,[22] and GT to Geldart and Taylor.[23] The latter theories are to be preferred over SSTL since they are based on self-consistent treatments of the uniform electron gas which satisfy, almost exactly, the electron compressibility sum rule. For jellium with a value of r_s corresponding to Al (Fig. 1) the three prescriptions yield very similar density profiles and $\rho_{lin}(r)$ is close to zero in the centre of the vacancy. The Hartree (H) calculation produces a profile which, for small r, is closer to the Kohn–Sham result. For large r all the theories exhibit Friedel oscillations of very small amplitude. These oscillations arise in the usual fashion from the logarithmic singularity in $X_0(q)$ (see equation (30)). In Fig. 2 we plot the corresponding results for the jellium model of Na. The above prescriptions for f_{ex} lead to negative values of the charge density inside the hole while the Hartree profile is again closer to the Kohn–Sham result. The amplitude of the Friedel oscillations is larger in this case.

Clearly the linear theory is not too successful for the charge densities. In the calculation of the formation energies it is convenient to work under constant volume conditions.

Expanding the functional G about the uniform density ρ_u we have

$$G[\rho] = G[\rho_u] + \int d\mathbf{r}\,\frac{\delta G[\rho]}{\delta\rho(\mathbf{r})}\bigg|_{\rho_u}\delta\rho(\mathbf{r})$$
$$+ \frac{1}{2}\int\int d\mathbf{r}\,d\mathbf{r}'\frac{\delta^2 G[\rho]}{\delta\rho(\mathbf{r})\delta\rho(\mathbf{r}')}\bigg|_{\rho_u}\delta\rho(\mathbf{r})\delta\rho(\mathbf{r}') + \dots$$

At constant volume the term linear in $\delta\rho(\mathbf{r})$ vanishes since

$$\int d\mathbf{r}\,\delta\rho(\mathbf{r}) = 0 \quad (33)$$

and on Fourier transforming we find

$$G[\rho] = G[\rho_u] + \frac{1}{\Omega}\sum_\mathbf{q} K(q)\delta\rho(\mathbf{q})\delta\rho(-\mathbf{q}) \quad (34)$$

where

$$K(|\mathbf{r} - \mathbf{r}'|) = \frac{1}{2}\frac{\delta^2 G[\rho]}{\delta\rho(\mathbf{r})\delta\rho(\mathbf{r}')}\bigg|_{\rho_u}$$

This function is related to the interacting response function of the uniform electron gas (Hohenberg and

---

* An analogous situation occurs in the uniform planar background treatment of the surface energy. In their Kohn–Sham calculations for this model of the surface, Lang and Kohn[20] find that for $r_s < 2\cdot4$ a.u. the surface energy is negative. However, as pointed out by Lang,[16] their calculation of the energy required to cleave the metal ignores the work done against the arbitrary forces which are postulated to hold the jellium together.

Kohn[17]):

$$K(q) = -\frac{1}{2}\left(\frac{4\pi}{q^2} + \frac{1}{X(q)}\right) \tag{35}$$

The $q = 0$ term in equation (34) is zero from equations (33) and (35) and after substituting for $\delta\rho$ from equation (27) we find

$$G[\rho] = G[\rho_u] - \frac{1}{2\Omega}\sum_{\mathbf{q}\neq 0}\rho_{\mathrm{lin}}(-\mathbf{q})\phi(\mathbf{q}) \tag{36}$$

where $\phi(\mathbf{q})$ is the Fourier transform of the electrostatic potential (equation (15)) and in the linear approximation is

$$\phi(\mathbf{q}) = \left(1 + \frac{4\pi}{q^2}X(q)\right)\delta V^+(\mathbf{q}) \quad (q \neq 0) \tag{37}$$

The electrostatic energy U_{es} (equation (16)) can be written as

$$U_{\mathrm{es}} = \frac{1}{2\Omega}\sum_{\mathbf{q}}\phi(\mathbf{q})(\rho_{\mathrm{lin}}(-\mathbf{q}) - \rho_0^+(-\mathbf{q})) \tag{38}$$

where the $q = 0$ term again gives zero contribution. Since the energy of the unperturbed jellium metal is just $G[\rho_u]$ the formation energy is

$$E_{V0}^{\mathrm{f}} = -\frac{1}{2\Omega}\sum_{\mathbf{q}\neq 0}\rho_0^+(-\mathbf{q})\phi(\mathbf{q})$$

or in terms of δV^+

$$E_{V0}^{\mathrm{f}} = \frac{1}{2\Omega}\sum_{\mathbf{q}\neq 0}\left(\frac{q^2}{4\pi} + X(q)\right)|\delta V^+(\mathbf{q})|^2 \tag{39}$$

For the canonical jellium model

$$E_{V0}^{\mathrm{f}} = 16\pi\rho_u^2 R^4 \int_0^\infty dq\left(1 + \frac{4\pi}{q^2}X(q)\right)\frac{j_1^2(qR)}{q^2} \tag{40}$$

In Fig. 4 we have plotted the values of the formation energy calculated from equation (40) and compared them with those of Manninen et al. The GT prescription was used for $f_{\mathrm{ex}}(q)$ but the results obtained with the VS form lie within about 2% of these. The linear screening theory gives a good qualitative account of the variation

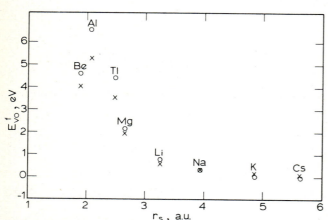

4   Constant volume vacancy formation energies E_{V0}^{f} calculated for the canonical jellium model; (0) labels the self-consistent, non-linear results obtained from Manninen et al.'s[15] calculations; (X) labels the linear-screening results as discussed in the text

of E_{V0}^{f} between different metals. There are, of course, important quantitative discrepancies especially for the polyvalent metals where the results from the linear theory are typically 1 eV smaller than those of Manninen et al. For the alkali metals the magnitude of the discrepancy is typically ~0·2 eV and for the particular case of Na the linear screening result is fortuitously close to that of Manninen et al. The Hartree ($f_{\mathrm{ex}} = 0$) results for E_{V0}^{f} are consistently larger, by 0·3–0·5 eV, than those obtained in the Kohn–Sham treatment but they also exhibit the same trend with Z and r_s.

Equation (39) indicates that E_{V0}^{f} might be sensitive to the form of the potential δV^+ and it is feasible that other choices of δV^+ (corresponding to alternative prescriptions for the positive background ρ_0^+) might predict a different behaviour of E_{V0}^{f} and hence E_0^{f} as a function of Z and r_s. This remains to be investigated.

DISCRETE LATTICE THEORIES

Jellium models are, of course, far removed from real metals. They would be useful for describing the formation energy if the replacement of the deformed positive background by the appropriate discrete lattice of ions constitutes, in some sense, a weak perturbation on the system. Only under these circumstances could jellium be thought of as a reasonable starting approximation for a discussion of the energetics of point defects. This point has been emphasized recently by Evans and Finnis[19] and we will return to it on p. 40.

Almost all the existing calculations of the vacancy formation energy for simple metals which include the discrete nature of the lattice are based on the pseudo or model potential formalism. In this theory it is assumed that the strongly attractive potential that an ion exerts on the conduction electrons can be replaced by a much weaker effective potential—the pseudopotential. The latter should be chosen so that the eigenvalue spectrum of the conduction electron states is the same as that which would be produced by the true electron-ion potential and should be sufficiently weak for use in perturbation theory. Several schemes exist for constructing pseudopotentials (see for example Cohen and Heine[24]) but, of these, only the 'screened-ion' variety is readily applicable to the calculation of total energies. As we have seen earlier, electron–electron interactions must be included self-consistently in any treatment of energetics and this is possible if one defines a 'bare-ion' pseudopotential and properly calculates the screening of this object by the interacting electrons. Harrison[8] seems to have been the first to apply such a pseudopotential model to the study of formation energies. Here we present a derivation of Harrison's result for the vacancy formation energy which illustrates the linear nature of the approximations implicit in this method. This presentation should clarify some of the confusion expressed at the Sixth Battelle Colloquium (see the discussion after the papers by Ho,[25] Chang[26] and the contributions by Hirth[27] and Duesbery[28]).

The Harrison model: linear screening theory
We consider the uniform jellium of the last section. This consists of N electrons interacting in a volume Ω in the presence of a uniform compensating positive background. The latter is then replaced by an array of N_i pseudo-ions located at ionic positions $\mathbf{R}_i$. If the 'bare' pseudopotential associated with each pseudo-ion is v_{ps}

then the change in potential due to such a replacement is

$$\delta V(\mathbf{r}) = \sum_{i=1}^{N_i} v_{ps}(\mathbf{r}-\mathbf{R}_i) + \int d\mathbf{r}' \frac{\rho_u}{|\mathbf{r}-\mathbf{r}'|} \quad (41)$$

where as before $\rho_u = N/\Omega$ and we have assumed the pseudopotential is a local function of position. The electron density $\rho_{ps}(\mathbf{r})$ can be calculated in the linear screening approximation as in equations (27) and (28) and we find

$$\rho_{ps}(\mathbf{q}) = X(q)\,\delta V(\mathbf{q}) \qquad q \neq 0$$
$$= N \qquad\qquad\qquad q = 0 \qquad (42)$$

If we introduce the structure factor $S(\mathbf{q})$

$$S(\mathbf{q}) = \sum_{i=1}^{N_i} \exp(i\mathbf{q}.\mathbf{R}_i) \qquad (43)$$

where the summation is over all *ion* sites. $\delta V(\mathbf{q})$ can be expressed as

$$\delta V(\mathbf{q}) = S(\mathbf{q})v_{ps}(q) \qquad q \neq 0$$
$$= N_i w_c(q=0) \qquad q = 0 \qquad (44)$$

The non-Coulombic part of the pseudopotential, w_c, is defined by

$$v_{ps}(r) = w_c(r) - \frac{Z}{r}.$$

We can now follow the analysis leading up to equation (36) and write

$$G[\rho] = G[\rho_u] - \frac{1}{2\Omega} \sum_{\mathbf{q}\neq 0} \rho_{ps}(-\mathbf{q})\left(1 + \frac{4\pi}{q^2}X(q)\right)\delta V(\mathbf{q}) \qquad (45)$$

The electrostatic energy in this model is

$$U_{es} = \frac{1}{2\Omega}\sum_{\mathbf{q}} \frac{4\pi}{q^2}\rho_{ps}(-\mathbf{q})\rho_{ps}(\mathbf{q})$$

$$+ \frac{1}{\Omega}\sum_{\mathbf{q}} \rho_{ps}(-\mathbf{q})\left(\delta V(\mathbf{q}) - \frac{4\pi}{q^2}N\delta_{q,0}\right)$$

$$+ \frac{1}{2\Omega}\sum_{\mathbf{q}} \frac{4\pi Z^2}{q^2} \sum_{i\neq j} \exp(i\mathbf{q}.(\mathbf{R}_i - \mathbf{R}_j)) \qquad (46)$$

where the first term is the self-energy of interaction of the electrons, the second is the electron-pseudo-ion contribution and the third is the self-energy of interaction of the pseudo-ions. The latter are assumed to interact solely via Coulomb repulsion. The total energy U of the system of pseudo-ions and conduction electrons in a volume Ω is obtained by adding equations (45) and (46). After some rearrangement and cancellation of the $q = 0$ terms we find

$$U = G[\rho_u] + \rho_u N_i w_c\,(q=0)$$

$$+ \frac{1}{2\Omega}\sum_{\mathbf{q}\neq 0} \frac{4\pi Z^2}{q^2}(|S(\mathbf{q})|^2 - N_i)$$

$$+ \frac{1}{2\Omega}\sum_{\mathbf{q}\neq 0} X(q)|v_{ps}(q)|^2|S(\mathbf{q})|^2 \qquad (47)$$

The result is valid for *any* arrangement of the pseudo-ions provided Ω is kept constant and so it is convenient to formulate the defect problem under such a constraint.

Harrison[8] considered the constant volume formation energy of a single vacancy located at the origin. Ignoring

any relaxation of the ions around the vacancy, the structure factor of the appropriate defect lattice is

$$S_v(\mathbf{q}) = S_p(\mathbf{q}) - 1 \qquad (48)$$

where

$$S_p(\mathbf{q}) = \sum_{i=1}^{N_i+1} \exp(i\mathbf{q}.\mathbf{R}'_i)$$

is the structure factor of a perfect lattice of $N_i + 1$ sites in which the lattice spacing is reduced by a factor μ^{-1} (where $\mu^3 = 1 + 1/N_i$) over that in the original perfect lattice of N_i sites. Harrison proved that the Madelung energy (represented by the third term of equation (47)) changes by $-\frac{2}{3}E_M$ where

$$E_M = \frac{\alpha Z^{\frac{5}{3}}}{2r_s} \qquad (49)$$

is the Madelung energy per ion of the original perfect crystal and α is the appropriate Ewald constant, on creating a vacancy under such conditions. The change in 'band structure energy' given by the last term of equation (47) is also readily computed. If we write

$$F(q) = X(q)|v_{ps}(q)|^2$$

and $\qquad\qquad\qquad\qquad\qquad\qquad\qquad\qquad (50)$

$$U_{bs} = \frac{1}{2\Omega}\sum_{\mathbf{q}\neq 0} F(q)|S_v(\mathbf{q})|^2$$

and note that $S_p(\mathbf{q})$ is only non-zero at reciprocal lattice vectors $\mathbf{g}'$ we find*

$$U_{bs} = \frac{(N_i^2 - 1)}{2\Omega}\sum_{\mathbf{g}'\neq 0} F(g') + \frac{1}{4\pi^2}\int_0^\infty dq\, q^2 F(q)$$

with

$$g' \equiv |\mathbf{g}'| = \mu|\mathbf{g}|$$

where $\mathbf{g}$ is a reciprocal lattice vector of the original lattice. Since

$$F(g') \approx F(g) + (\mu - 1)g\frac{\partial F(g)}{\partial g}$$

$$= F(g) + \frac{g}{3N_i}\frac{\partial F(g)}{\partial g}$$

the constant volume formation energy in this model, E^f_{V2}, reduces to

$$E^f_{V2} = -\frac{2}{3}E_M + \frac{1}{4\pi^2}\int_0^\infty dq\, q^2 F(q) + \frac{1}{2\Omega_0}\sum_{\mathbf{g}\neq 0}\frac{g}{3}\frac{\partial F(g)}{\partial g} \qquad (51)$$

where terms $0(1/N_i)$ have been ignored.

This result for the formation energy is in a form which is suitable for computation since both the integral and the summation are rapidly convergent. All that is required is the bare-ion pseudopotential form factor $v_{ps}(q)$.† The calculations of DuCharme and Weaver[11]

---

* We restrict ourselves to primitive cubic lattices.
† This result for E^f_{V2} is also valid for non-local pseudopotentials. $F(q)$ is simply $2\Omega_0\Phi_{bs}(q)$ where $\Phi_{bs}(q)$ is the usual energy-wave number characteristic.

and Minchin *et al.*[13] are based on this formula. Chang and Falicov[10] also work under constant volume conditions but they include relaxation effects.

If we allow for an expansion in volume then the corresponding formation energy is

$$E_2^f = E_{V2}^f - p_2\Omega_2^f \qquad (52)$$

which follows from equation (22); p_2 is the pressure given by differentiating equation (47) with respect to volume:

$$\Omega_0 p_2 = \Omega_0 p_0 + \rho_u w_c(q=0) + \frac{E_M}{3}$$
$$+ \frac{1}{2\Omega_0} \sum_{g \neq 0} \left(F(g) + \frac{K_F}{3} \frac{\partial F(g)}{\partial K_F} + \frac{g}{3} \frac{\partial F(g)}{\partial g} \right) \qquad (53)$$

The formation volume in this model is simply obtained from the general result of equation (25):

$$\Omega_2^f = -\frac{\Omega_0}{B_2} \left(\frac{\partial E_{V2}^f}{\partial \Omega_0} \right)_T \qquad (54)$$

where $B_2 = -\Omega(\partial p_2/\partial\Omega)_T$.

Although it is not obvious from their paper, Popović *et al.*[12] used equations (51) and (54) in their calculations of vacancy formation energies. Their derivation does not invoke constant volume conditions but it is possible to show algebraically that their final expression for the formation energy (their equation (3.4)) is exactly equivalent to

$$E^f(\text{Popović}) = E_R + E_{V2}^f - p_2\Omega_0$$

where E_R is the relaxation energy (defined at constant volume) and Popović *et al.* deliberately chose their pseudopotential so that $p_2 = 0$, i.e. their lattice was in equilibrium. It can also be shown that their result for the formation volume is just

$$\Omega^f(\text{Popović}) = \Omega_2^f - \frac{\Omega_0}{B_2} \left(\frac{\partial E_R}{\partial \Omega_0} \right)$$

For completeness we should also refer to other derivations of the vacancy formation energy in this scheme. Ho's[9] result includes a term which explicitly involves putting an atom at the surface and hence requires some specification of the local configuration of atoms at the surface. As Popović *et al.* have pointed out, this cannot be correct as vacancy formation is certainly a *bulk* phenomenon. August[29] has also produced an erroneous derivation for E_2^f which involves the surface configuration. Very recently Tanigawa and Doyama[30] have re-derived our above results by treating the monovacancy in the framework of alloy theory. Their final result* is exactly of the form

$$E^f(TD) = E_{V2}^f - p_2\Omega^f$$

but they do not derive the formation volume.

Real space version of the Harrison model

The linear screening treatment of vacancy formation can be reformulated in terms of effective pairwise potentials.

This real space representation is appealing because it casts the problem into a form which is *similar* to that which has been studied at length by those workers who use empirical pairwise potentials.

Equation (47) for the total energy of a metallic array of pseudo-ions and conduction electrons can be transformed to

$$U = N_i u(\rho_u) + \frac{1}{2} \sum_{i \neq j} \phi_{\text{eff}}(|\mathbf{R}_i - \mathbf{R}_j|) \qquad (55)$$

where the first term is structure independent and the second involves the summation of an effective pairwise potential ϕ_{eff} between all the pseudo-ions. For our present discussion it is not important to know the detailed form of $u(\rho_u)$ but an explicit expression for this quantity is easily derived provided the pseudopotential is local (for example Evans[31] or Finnis[32]). If the pseudopotential is non-local the analysis is much more complicated. The pair potential includes the direct Coulomb repulsion between the ions as well as a band structure contribution

$$\phi_{\text{eff}}(R) = \frac{Z^2}{R} + (2\pi)^{-3} \int d\mathbf{q} \exp(-i\mathbf{q} \cdot \mathbf{R}) F(q) \qquad (56)$$

Since $F(q)$ depends on the average electron density ρ_u, the pair potential is volume dependent.

If a vacancy is created at the origin, under constant volume conditions, the formation energy is easily shown to be

$$E_{V2}^f = -\frac{1}{2} \sum_{i \neq 0} \left(\phi_{\text{eff}}(R) + \frac{R}{3} \frac{\partial \phi_{\text{eff}}(R)}{\partial R} \right)_{R=|\mathbf{R}_i|} \qquad (57)$$

where the summation is over all the *ion* positions.

The first term in equation (57) is due to the breaking and re-making of pairwise bonds in the formation process while the second takes into account the shortening of the interatomic distances which occurs under these constant volume conditions. If we ignore relaxation effects, i.e. we take $\mathbf{R}_i$ to be perfect lattice sites, the first term is equal to the pairwise contribution to the binding energy while the second is proportional to the virial term in the pressure since

$$\Omega_0 p_2 = \rho_u \frac{du}{d\rho_u} - \frac{1}{2} \sum_{i \neq 0} \left(\frac{R}{3} \frac{\partial \phi_{\text{eff}}(R)}{\partial R} - \rho_u \frac{\partial \phi_{\text{eff}}(R)}{\partial \rho_u} \right)_{R=|\mathbf{R}_i|} \qquad (58)$$

It is clear that the virial term will be, in general, non-zero and calculations indicate this contribution can be as large as or larger than the first contribution to E_{V2}^f (Minchin *et al.*[13]). Equation (57) is not as convenient for computational purposes as its reciprocal space equivalent since the effective pairwise potentials have to be calculated by Fourier transforming and exhibit long-range oscillatory behaviour which means many shells of neighbours must be included in the summation.

Calculations based on the Harrison model

Although it would be inappropriate to present a comprehensive review of all the calculations based on Harrison's pseudopotential model it is useful to summarize the main results. In Table 2 we list the formation energies calculated by Minchin *et al.*[13] and by Popović *et al.*[12]

Minchin *et al.* used the Ashcroft empty-core model potential and the Geldart and Vosko[33] form of exchange

---

*The second term of their equation (47) seems to have the wrong sign (*see* their equation (44)).

TABLE 2 Vacancy formation energies E^f calculated from Harrison's pseudopotential model and comparison with experiment;* all energies are given in eV; Popović et al.[12] include relaxation in their calculations while Minchin et al.[13] do not

	E^f (Minchin et al.[13])	E^f (Popović et al.[12])	E^f (expt)
Li	0·41	0·29	0·34[M]
Na	0·31	0·41	0·42[M]
K	0·21	0·35	0·39[M]
Rb		0·31	0·27[G]
Cs		0·29	0·28[M]
Mg	0·90		0·81[M]
Zn	0·81		0·50[G]
Cd	0·59		0·44[G]
Al	0·39	0·86	0·66[M]
In	−0·56		0·45, 0.57
Pb	−1·92		0·53[G]

* The experimental results are taken from the compilations of Górecki[45] (G) and Manninen et al.[15] (M) except for In where we have quoted the values listed by Minchin et al.[13]

and correlation correction $f_{ex}(q)$. Their calculations do not include relaxation or expansion, i.e. they implicitly assume $p_2 = 0$. We see from the table that their results for the alkali and divalent metals are in rough agreement with experiment. For In and Pb, however, their calculated values are negative. Popović et al. used a local version of the Heine–Abarenkov model potential and the Singwi et al.[21] f_{ex}. They adjusted the parameters of the pseudopotential so that $p_2 = 0$ at the observed lattice spacing and B_2 was equal to the observed compressibility. Relaxation was included using a lattice statics technique. Their results for the alkali metals and Al are in reasonable agreement with experiment. They also calculate formation volumes for these metals.

Recently Rao[34] and Das et al.[35] have carried out detailed calculations of the relaxation energy associated with vacancy formation using force constants derived from pseudopotential theory. These authors review earlier work on this problem. Since it seems to be fairly well accepted that relaxation reduces the formation energy by $<0·1$ eV for the simple metals (with the possible exception of Li) we can focus our attention on the unrelaxed energy of formation. DuCharme and Weaver[11] and Chang and Falicov[10] found that E_{V2}^f was sensitive to the choice of pseudopotential for the divalent metals Mg and Cd and trivalent Al which they investigated. This sensitivity can be traced to the cancellation of terms of differing sign which occur in equation (51). For a polyvalent metal each term is large in magnitude. In real space, the effective pairwise potential, which is a combination of the Coulomb repulsion and the band structure contribution, shows considerable variation when one pseudopotential is substituted for another and as one might expect from equation (57) this leads to drastic changes in the formation energy. In fact the reasonable agreement with experiment obtained by Popović et al.[12] results to a large extent from their particular prescription for the pseudopotential. Pairwise potentials calculated from their parameters are very different from those obtained by other authors and lead to poor phonon spectra and liquid structure factors (see Evans and Finnis[19] and Kumaravadivel and Evans[36]).

The effective pairwise potential is also sensitive to the choice of $f_{ex}(q)$ particularly in the neighbourhood of the first minimum. Again the polyvalent metals are most strongly affected because the cancellation of the Coulomb repulsion by the band structure energy is critical in these systems. To indicate how important this can be for the formation energy we calculated E_{V2}^f for Al using the same Ashcroft empty core radius as that employed by Minchin et al.[13] but taking different forms for $f_{ex}(q)$. Our resultant values (see Table 3) were −0·41 eV (SSTL) and −0·23 eV (VS) compared with Minchin et al.'s result of +0·39 eV and the Hartree (H) result of 2·2 eV. Clearly exchange and correlation is important!

In view of these considerations it would be useful to have the results of a systematic investigation of E_{V2}^f and

TABLE 3 Vacancy formation energies calculated in various schemes;* these results are taken from Evans and Finnis;[19] all energies are given in eV

		E_{V0}^f	E_{V0}^f (M)	$E_{V1}^f(\rho_{lin})$	$E_{V1}^f(\rho)$	E_{V2}^f	E_{V2}^f (SLS)
Na	H	0·66	0·31	0·63 (0·78)	(0·38)	0·53 (0·66)	(0·39)
	SSTL	0·39		0·36 (0·54)		0·26 (0·41)	(0·38)
	VS	0·34		0·31 (0·50)		0·20 (0·36)	(0·38)
Al	H	7·05	6·56	2·78 (3·04)	(1·56)	2·20 (2·23)	(1·67)
	SSTL	5·11		0·42 (0·72)		−0·41 (−0·39)	(1·37)
	VS	5·24		0·58 (0·88)		−0·23 (−0·21)	(1·39)

* E_{V0}^f is the constant volume formation energy of the canonical jellium model treated in linear screening theory. E_{V0}^f(M) is the corresponding quantity obtained from Manninen et al.'s[15] non-linear treatment. The other formation energies are defined in the text. The numbers in brackets were calculated using the Ashcroft empty core model potential with the radius parameter $R_c = 1·758$ a.u. for Na and $R_c = 1·131$ a.u. for Al. The same values were used by Manninen et al. in their first-order pseudopotential calculations. The corresponding unbracketed energies were calculated with $R_c = 1·67$ a.u. for Na and $R_c = 1·12$ a.u. for Al.

Ω_2^f (ignoring the complications of relaxation) based on the best available pseudopotentials and realistic $f_{ex}(q)$ for a wide range of metals. At the present we can tentatively conclude that the Harrison prescription leads to sensible estimates of the formation energy for the alkali metals but is unreliable for other metals in that it often predicts negative values of E_{V2}^f.

The formation volume has only been properly calculated* by Popović et al.[12] and recently by Finnis and Sachdev.[37] Although their results agree well with experiment for those few cases where appropriate experimental information is available, all of this work is based on the Popović et al. prescription for the pseudopotential and thus is subject to the criticisms mentioned above. In the case of Na, relaxation effects constitute a large fraction of the calculated formation volume while in Al these contribute only a few percent of this quantity.

So far we have only referred to theories which work to second order in the pseudopotential. One can, of course, include higher order terms in the expansion of the total energy. In real space, the equation corresponding to equation (55) then contains 3 body and higher interactions. Yoshioki and Mori[38] have recently attempted to include third-order terms in the calculation of the formation energy of Na and K and they claim that these additional contributions are very large and hence important. This is surprising because the third-order contributions to the binding energy and phonon spectra of these metals are very small. For the polyvalent metals, on the other hand, we might expect higher order contributions to be significant in the calculation of the formation energy since such terms play an important role in the calculation of phonon spectra in some of these metals (Bertoni et al.[39]). It is feasible that even if the binding energy and phonon spectrum of a given metal can be well-accounted for in low-order perturbation theory, the corresponding expansion of the vacancy formation energy in powers of the pseudopotential may not be rapidly convergent. This would be the case if the distortion of the electron density around the defect was too strong to be treated in low-order perturbation theory. Under such circumstances one might abandon the uniform positive background starting point and begin with a deformed positive background (hole) of the kind discussed in the second section. Given the electron density corresponding to the hole one could then replace the background by the appropriate defect lattice of pseudo-ions and calculate the accompanying change in energy using perturbation theory. If this replacement constitutes a weaker perturbation on the electron density around the hole than that exerted on the *uniform* electron gas by the introduction of a lattice with a vacancy, then one might expect perturbation theory to be more appropriate. Such a philosophy was adopted by Manninen et al.[15] and discussed by Evans and Finnis.[19] This work is of importance because it links the Harrison type of approach with the jellium models and indicates the limitations of both techniques.

Replacement of a positive background by a discrete lattice

We consider the jellium model of the second section in which the positive background is described by a charge distribution $\rho_0^+(\mathbf{r})$. Replacing this background by the

* See p. 42.

discrete lattice of pseudo-ions introduces a perturbation potential:

$$\delta V'(\mathbf{r}) = \sum_i v_{ps}(\mathbf{r}-\mathbf{R}_i) + \int d\mathbf{r}' \frac{\rho_0^+(\mathbf{r}')}{|\mathbf{r}-\mathbf{r}'|} \qquad (59)$$

To first order in $\delta V'$ the energy of the system is changed by

$$\Delta U_1^d = \int d\mathbf{r}\, \rho(\mathbf{r})\, \delta V'(\mathbf{r}) + \frac{1}{2} \sum_{i \neq j} \frac{Z^2}{|\mathbf{R}_i - \mathbf{R}_j|}$$
$$- \frac{1}{2} \int \int \frac{d\mathbf{r}\, d\mathbf{r}'\, \rho_0^+(\mathbf{r})\rho_0^+(\mathbf{r}')}{|\mathbf{r}-\mathbf{r}'|} \qquad (60)$$

where $\rho(\mathbf{r})$ is the electron density corresponding to the original positive background $\rho_0^+(\mathbf{r})$. The second term of equation (60) is the direct ion–ion interaction in the defect lattice while the third term is the self-interaction energy of the positive background. By subtracting the corresponding quantity ΔU_1^p for the perfect crystal the change in formation energy can easily be evaluated and at constant volume the formation energy has the form

$$E_{V1}^f = E_{V0}^f + \Delta U_1^d - \Delta U_1^p \qquad (61)$$

Explicit expressions for E_{V1}^f are given by Evans and Finnis[19] for the case of the canonical jellium model and no relaxation of the ions about the vacant site. If we allow for an expansion of Ω_1^f then

$$E_1^f = E_{V1}^f - p_1 \Omega_1^f \qquad (62)$$

where

$$\Omega_0 p_1 = \Omega_0 p_0 + \rho_u w_c\,(q=0) + E_M/3$$

which follows from equation (53). In Table 3 we list the values of E_{V1}^f calculated by Evans and Finnis for Na and Al and compare these with the jellium results, i.e. E_{V0}^f.

For Na the formation energy increases or decreases, depending on the choice of pseudopotential, on adding the first-order contribution. In all cases the magnitude of the change in energy is <0.2 eV. In Al the first-order terms reduce the formation energies by ~ 4.4 eV and hence cancel most of the large jellium value. The electron–ion interaction is clearly of importance in this metal. The first-order corrections do not depend strongly on whether the jellium is treated in a linear approximation or in the Kohn–Sham formalism.

The perturbation potential $\delta V'$ causes a change $\delta \rho'(\mathbf{r})$ in the electron density distribution and this in turn gives a contribution (in second order in $\delta V'$) of

$$\Delta U_2^d = \frac{1}{2} \int d\mathbf{r}\, \delta\rho'(\mathbf{r})\, \delta V'(\mathbf{r})$$

to the energy of the system. In general a calculation of $\delta\rho'(\mathbf{r})$ requires knowledge of the response function of the *inhomogeneous* electron distribution around the hole and this is not easily calculated. If we assume that this response function can be approximated by that of the homogeneous electron gas, i.e.

$$\delta\rho'(\mathbf{r}) \approx \int d\mathbf{r}'\, X(|\mathbf{r}-\mathbf{r}'|)\, \delta V'(\mathbf{r}')$$

then the second-order calculation can be carried through.

One can show that if the electron density and energy of the jellium model are calculated in linear screening theory as on p. 35 then the total energy of the defected lattice, which is $G[\rho_u] + E_{V0}^f + \Delta U_1^d + \Delta U_2^d$, is independent of $\rho_0^+(\mathbf{r})$, i.e. the original positive background cancels out in this order and the total energy reduces to that of equation (47). This must happen, of course, since the theory is linear. The formation energy in this case is just E_{V2}^f of equation (51).

If, on the other hand, the original jellium vacancy is treated in a non-linear theory, as we have argued it should be, the approximate second-order formation energy E_{V2}^f(SLS). where SLS stands for semi-linear screening, is given by

$$E_{V2}^f(\text{SLS}) = E_{V1}^f(\rho) + E_{V2}^f - E_{V1}^f(\rho_{lin}) \qquad (63)$$

where $E_{V1}^f(\rho)$ is the first-order formation energy calculated from the non-linear charge density ρ and $E_{V1}^f(\rho_{lin})$ is the corresponding energy calculated from the linear screening treatment of the jellium vacancy. The pressure in this model is simply p_2.

The second-order correction reduces E_{V1}^f by <0.15 eV for Na while in Al this contribution decreases the formation energy by typically 1 eV (see Table 3). The energies E_{V2}^f calculated in the Harrison scheme for Na are realistic but, as we discussed earlier, are quite sensitive to the choice of pseudopotential parameter and to $f_{ex}(q)$. In Al, the values of E_{V2}^f calculated with both forms of f_{ex} and both choices of pseudopotential parameter are negative. The semi-linear results for Na are close to the values of E_{V2}^f and this is because the first-order formation energies $E_{V1}^f(\rho)$ are close to $E_{V1}^f(\rho_{lin})$. The second-order term reduces $E_{V1}^f(\rho)$ in Al from 2.48 eV to about 1.39 eV, but this is still far too large.

It should be noted (Evans and Finnis[19]) that the pressures p_1 and p_2 are non-zero for both Na and Al when these are calculated using the Ashcroft model potentials of Table 3. Consequently attempts to allow for a change in volume on vacancy formation introduce large energies $-p_1\Omega_1^f$ and $-p_2\Omega_2^f$ in the appropriate order of perturbation theory. As was the case with the jellium problem of the section on 'Density functional formalism', it is not clear which formation energy is more physically relevant under these circumstances, but we are inclined to emphasize the constant volume results since these do not depend on the long-wavelength limit of the pseudopotential, $w_c(q=0)$. It is this parameter which is usually adjusted to satisfy $p_2 = 0$ in second-order pseudopotential calculations (Ashcroft and Langreth[40]).

We can summarize the results of Evans and Finnis as follows. For Na it appears that low order perturbation theory will converge and the semi-linear results seem to give a reasonable account of the formation energy. (The energies E_{V2}^f(SLS) are probably still too low since relaxation will reduce them further.) Although the effects of the discrete lattice (electron–ion interaction) are non-negligible it is clear that a significant contribution to the formation energy is due to a redistribution of electrons which can, to some extent, be modelled in a jellium calculation. The Harrison formulation can be expected to be reasonably adequate in Na because linear screening gives a good account of the energetics of this metal. For the polyvalent metal Al it seems as though all existing low-order perturbation schemes are unsuccessful. The semi-linear results might be improved by a better approximation to the response function but we think it unlikely that this would reduce the formation

energy by the required 0.6 or 0.7 eV. The introduction of discrete pseudo-ions brings about gross perturbations and the expansion of the formation energy in powers of the pseudopotential does not appear to be rapidly convergent.*

EMPIRICAL PAIRWISE POTENTIALS

Since Heald has already reviewed calculations of the energetics of point defects which are based on empirical pairwise potentials, in the present section we will only point out the similarities and differences between this approach and that discussed in the previous section. The studies which employ empirical pair potentials rarely attempt to calculate the formation energy of a monovacancy. They are primarily concerned with relaxation effects and interactions between defects and often use the vacancy formation energy as input data for calculations. Furthermore they are frequently directed at transition and noble metals for which a pseudopotential description of the electron-ion interaction is inappropriate and, consequently, for which there is no simple ab initio theory of the total energy of an arbitrary configuration of atoms.

The standard model assumes that the total energy of the metal can be expressed as

$$U = \tfrac{1}{2} \sum_{i\neq j} \phi(|\mathbf{R}_i - \mathbf{R}_j|) + F(\Omega) \qquad (64)$$

where $\phi(R)$ is a density independent or 'rigid' pairwise potential and $F(\Omega)$ is a structure independent contribution which is assumed to depend on the volume Ω. The pressure in this model is

$$p = -\frac{1}{2\Omega_0} \sum_{i\neq 0} \left(\frac{R}{3}\phi'(R)\right)_{R=|\mathbf{R}_i|} - \frac{dF(\Omega)}{d\Omega}$$

where $\phi'(R) = d\phi/dR$. For the lattice to be in equilibrium at the observed atomic volume Ω_0 we require

$$\left.\frac{dF(\Omega)}{d\Omega}\right|_{\Omega_0} = -\frac{1}{2\Omega_0} \sum_{i\neq 0} \left(\frac{R}{3}\phi'(R)\right)_{R=|\mathbf{R}_i|} \qquad (65)$$

where the $\mathbf{R}_i$ refer to the observed perfect lattice sites. The constant volume vacancy formation energy in this model has the same form as that in equation (57), i.e.

$$E_V^f = -\tfrac{1}{2} \sum_{i\neq 0} \left(\phi(R) + \frac{R}{3}\phi'(R)\right)_{R=|\mathbf{R}_i|} \qquad (66)$$

If the volume term $F(\Omega)$ is ignored and the lattice is assumed to be in equilibrium under pairwise forces alone then the virial term vanishes and the formation energy reduces to the cohesive energy per atom—providing relaxation is ignored. Experimentally the formation energies of metals are typically about one-third

---

* It is significant that Finnis[41] reached essentially equivalent conclusions regarding the calculation of the surface energy U^s. As mentioned earlier a jellium calculation yields a large negative value for U_0^s in Al. The replacement of the uniform planar background by a discrete lattice of pseudo-ions yields a large positive value for U_1^s, which is close to the experimental estimate, in first-order perturbation theory but the second-order correction of Finnis is large and negative leading to a value for U_2^s(SLS) which is about one third of the experimental estimate. The analogue of the Harrison model, i.e. linear screening of pseudopotentials, gives even smaller surface energies U_2^s. For Na semi-linear theory seems to be quite adequate and leads to a reasonable value of the surface energy. The calculated values of U_2^s are only about one-third of the experimental estimate in Na.

of the cohesive energies so relaxation would have to bring about enormous reductions of energy if this model were to be appropriate. (For the rare-gas solids the model may be better founded since E^f is roughly 0·7 to 0·8 times the cohesive energy in these systems.)

In the general case the unrelaxed formation energy is

$$E^f = E^f_V = -\frac{1}{2}\sum_{i\neq 0}\phi(|\mathbf{R}_i|) + \Omega_0\frac{dF(\Omega)}{d\Omega}\bigg|_{\Omega_0} \qquad (67)$$

where we have used the fact that $p = 0$. One can then argue that if most of the binding energy of the metal resides in $F(\Omega)$ the pairwise contribution to E^f may not necessarily be too large. The situation would then be similar to that pertaining in the section on the 'Real space version of the Harrison model'. In most calculations, however, the derivatives of $\phi(R)$ have been obtained by direct fitting to measured elastic constants or phonon spectra, $F(\Omega)$ has been set equal to zero, and the magnitude of $\phi(R)$ determined from equation (67) by taking a selection of values for the unrelaxed formation energy (for example Johnston and Wilson[42]). Not surprisingly the pair potentials obtained in this fashion are very different from those $\phi_{eff}(R)$ calculated from pseudopotential theory. The latter have first minima which are typically 0·02–0·05 eV in depth while the empirical potentials are usually one order of magnitude deeper. The relaxation contribution to the formation energy also comes out to be large for some of the empirical potentials, that is to say much larger than the $-0·1$ eV quoted earlier from the pseudopotential calculations.

The empirical pairwise potentials are often used to calculate vacancy formation volumes so it is useful to compare the calculation of Ω^f in this approach with our earlier result from the Harrison model. From equations (25) and (66) we find

$$\Omega^f = \frac{1}{9B}\sum_{i\neq 0}\left(2R\phi'(R) + \frac{R^2}{2}\phi''(R)\right)_{R=|\mathbf{R}_i|}$$

and we again ignore relaxation.

The bulk modulus is given by

$$B = \Omega\frac{d^2F(\Omega)}{d\Omega^2}\bigg|_{\Omega_0} + \frac{1}{9\Omega_0}\sum_{i\neq 0}\left(-R\phi'(R)\right.$$
$$\left. + \frac{R^2}{2}\phi''(R)\right)_{R=|\mathbf{R}_i|}$$

where $\phi''(R) = d^2\phi/dR^2$ and so the formation volume can be written as

$$\frac{\Omega^f}{\Omega_0} = 1 + \frac{1}{B}\left(\frac{1}{3\Omega_0}\sum_{i\neq 0}(R\phi'(R))\bigg|_{R=|\mathbf{R}_i|} - \Omega\frac{d^2F(\Omega)}{d\Omega^2}\bigg|_{\Omega_0}\right) \qquad (68)$$

This unrelaxed Ω^f reduces to Ω_0 in the case where $F(\Omega)$ is zero and the lattice is in equilibrium under pairwise forces alone. More generally we can rewrite equation (68) as

$$\frac{\Omega^f}{\Omega_0} = 1 - \frac{2}{B}p^{Cauchy} \qquad (69)$$

since the bracketed quantity in equation (68) is proportional to the Cauchy pressure p^{Cauchy} in this model. p^{Cauchy} is a measure of the difference between the elastic

constants c_{12} and c_{44}, i.e.

$$p^{Cauchy} = \frac{1}{2}(c_{12} - c_{44}) = \frac{dF(\Omega)}{d\Omega}\bigg|_{\Omega_0} + \frac{\Omega}{2}\frac{d^2F(\Omega)}{d\Omega^2}\bigg|_{\Omega_0} \qquad (70)$$

where the last step follows from equation (65).

Most authors have followed Hardy's[43] scheme for calculating the formation volume. This is based on the harmonic lattice statics formalism. It is straightforward to show (e.g. Finnis and Sachdev[37]) that the unrelaxed contribution to Ω^f in this method is just that given in equation (68) provided $d^2F(\Omega)/d\Omega^2$ is set equal to zero. Rao[34] and Das et al.[35] calculated effective pairwise potentials from pseudopotential theory and then used these in the Hardy scheme to calculate Ω^f. As Finnis and Sachdev have recently pointed out this corresponds to neglecting the density dependence of ϕ_{eff}, i.e. the pair potential is assumed to be 'rigid'.

The proper result for the unrelaxed formation volume in the Harrison pseudopotential approach should be obtained from equation (54) and this *cannot* be written in the simple form of equation (69). Terms such as $\partial^2\phi_{eff}/\partial^2 K_F$, $\partial\phi_{eff}/\partial K_F$ and $\partial^2\phi_{eff}/\partial K_F\,\partial R$ arise from the differentiation with respect to volume which are absent in the 'rigid' pair potential description. The calculations of Finnis and Sachdev suggest that these missing terms can be very important in some metals. For example in Na the full calculation leads to a realistic estimate $\Omega^f = 0·54\Omega_0$, but if ϕ_{eff} is assumed to be rigid and the Hardy prescription is employed, the calculated value of Ω^f is negative. Das et al.[35] obtained a negative formation volume for Li presumably for the same reasons. For Al, on the other hand, Finnis and Sachdev find that both prescriptions yield values of about $0·7\Omega_0$.

Many authors have attempted to correlate the measured vacancy formation energies of metals with other physical properties such as the surface energy, melting temperature, cohesive energy, etc. (*see* for example Grimvall and Sjödin[44] and Górecki[45]). Although such correlations are useful for obtaining rough order of magnitude estimates they are of little use in understanding the basic physics of the formation problem. In fact, the attempts at explanation of empirical correlations have often proved to be extremely misleading. We illustrate this point by reference to the successful empirical relation of Mukherjee[46] which relates the formation energy to the Debye temperature θ_D

$$E^f = K\theta_D^2 M\Omega_0^{\frac{2}{3}} \qquad (71)$$

where M is the atomic mass and K is a constant. Tewary[47] set $F(\Omega) = 0$ in the model of equation (64) and used a simple Debye approximation to the phonon spectrum to relate θ_D to the empirical pairwise potential. He claims

$$\frac{1}{2}\sum_{i\neq 0}\phi(|\mathbf{R}_i|) = K'\theta_D^2 M\Omega_0^{\frac{2}{3}} \qquad (72)$$

where the summation is over perfect lattice sites and K' is a constant which is very close to the empirical value K obtained by Mukherjee[46]. Tewary identifies the left-hand side of equation (72) with the unrelaxed vacancy formation energy. However from equation (67) we see that this quantity is really the negative of the cohesive energy per atom. March's[14] derivation of Mukherjee's relation ignores the discrete nature of the lattice. Using the Bohm–Staver formula for the sound velocity of a free-electron jellium and the Debye theory he argues that $\theta_D^2 M \Omega_0^{\frac{2}{3}} \propto Z E_F$ and, following Fumi, should therefore be proportional to the formation energy. Such an argument might convince the unsuspecting reader that the jellium description is a realistic one. We feel, however, it is merely an example of how two poor theories (those of Bohm–Staver and Fumi) can be brought together to produce what appears to be a very reasonable result.

CONCLUSIONS AND PERSPECTIVES

In the second section we concluded that jellium models were unsatisfactory for any quantitative description of the formation energy of a vacancy, except perhaps for Na, the 'theorist's metal'. In the third section we came to essentially the same conclusion regarding models in which the positive background of the jellium is replaced by a discrete lattice of pseudo-ions and the accompanying change in energy is calculated in low-order perturbation theory. It seems that one is forced to abandon all models which use jellium, in one form or another, as a starting point. Clearly a more 'local', non-perturbative scheme, which takes into account the true nature of the electronic wavefunctions in the neighbourhood of the vacancy, is required. For transition metals where the d electrons which contribute to the binding are fairly well localized around the nuclei it is self-evident that a local, atomic-like description is appropriate.* Several possible schemes can be envisaged which incorporate these features.

For a perfect lattice one can calculate the total crystalline electron density using the density functional formalism. The electrostatic potential due to the full nuclear charge is employed in equation (8) and the eigenfunctions of equation (7) separate into core states and conduction states. The latter can be calculated using standard band structure techniques. For a given structure and lattice spacing the total energy can then be calculated and by varying the lattice constant the pressure and bulk modulus can be evaluated. Slater and co-workers and Williams and his co-workers have carried through such calculations for several metals and the results are very encouraging. A programme of this kind is made possible because the imposition of periodic boundary conditions permits the solution of the effective one-electron Schrödinger equation. In a defect problem Bloch's theorem does not apply and there is no possibility of solving the Schrödinger equation for an arbitrary, *infinite* array of nuclei. If one neglects relaxation of atoms around a single vacancy then the eigenvalue problem reduces to the calculation of the scattering of Bloch waves by the missing atom. Beeby[49] formulated the appropriate multiple scattering theory but to our knowledge no calculation along these lines has yet been performed since the computation is necessarily heavy. Although one might hope to learn about the electron density around a vacancy from such a calculation one would probably be restricted to a first pass estimate. After the first iteration of equations (7) the atoms in the neighbourhood of the vacancy would not be equivalent to those far away and this causes complications in the solution of the Schrödinger equation. A realistic calcula-

tion of the formation energy is hardly feasible in this framework.

One can avoid these particular difficulties by constructing a perfect crystal with a large unit cell in which the central atom is missing. The eigenvalue problem then presents, in principle, no difficulties. Recently Louie *et al.*[50] have performed a calculation along these lines for Si using 54 atoms per unit cell. They discuss the electron density around the missing atom. Provided the unit cell is large enough the latter should mimic the electron density near a vacancy and such a scheme may also be useful for estimating unrelaxed formation energies.

A third approach which would strongly emphasize the 'local' aspects of the defect problem is the self-consistent cluster treatment. In this model the Kohn–Sham equations are solved for a finite number of atoms which are either located in free space or embedded in a medium of constant potential. For the vacancy problem one could consider a cluster consisting of a central missing atom surrounded by say two or three shells of nearest neighbours. Again one should be able to learn about the redistribution of electrons near a vacancy from such a model. By varying the positions of the nuclei it may even be possible to learn something about relaxation energies. Of course, by considering only a finite number of atoms one is neglecting all long-range effects. These may or may not prove to be important in the calculation of the formation energy.

Clearly any shift in emphasis towards a local description constitutes a significant departure from the discrete lattice models described in the third and fourth sections. The latter have been widely employed since they involve explicit forms for the total energy as a function of the positions of the atoms and hence properly include long range effects. We are arguing that the failures of such models can be attributed to an improper treatment of the electron distribution near the vacancy and that we must now concentrate our efforts in this direction. In this context more attention should be paid to the interpretation of positron annihilation experiments since these provide some information about the electron distribution. Such sentiments have been expressed before! (for example March[14]).

The difficulties encountered with the vacancy are bound to occur with other point defects and will probably be more pronounced since other defects cause larger perturbations on the lattice and the electronic structure. It is also worth noting that our understanding of the surface energy of metals is very limited (Evans and Kumaravadivel[51]). We are not short of problems to solve!

ACKNOWLEDGMENTS

The author would like to thank Drs D. A. Greenwood, B. L. Gyorffy, J. W. Martin, F. Nizzoli, and G. M. Stocks for many useful discussions concerning various aspects of the vacancy problem. He is particularly grateful to Dr M. W. Finnis for stimulating correspondence and for sending details of his work on the formation volume prior to publication.

---

* Allan and Lannoo[48] have made some tight-binding calculations of the vacancy formation energy and formation volume in transition metals but we feel their model is too crude to be realistic.

REFERENCES

1 J. FRIEDEL: 'Vacancies and interstitials in metals' (Ed. A. SEEGER *et al.*), 787, 1970, Amsterdam, North-Holland
2 A. SEEGER: *J. Physics F (Metal Physics)*, 1973, **3**, 248

3 A. D. FRANKLIN: 'Point defects in solids', Vol. 1: 'General and ionic crystals' (Ed. J. H. CRAWFORD AND L. M. SLIFKIN), 1, 1972, New York, Plenum
4 F. G. FUMI: *Phil. Mag.*, 1955, **46**, 1007
5 M. J. STOTT *et al.*: *Proc. Roy. Soc.*, 1970, **A316**, 201
6 R. NIEMINEN *et al.*: *Solid-State Commun.*, 1975, **16**, 831
7 G. G. ROBINSON AND P. F. DE CHATEL: *J. Physics F (Metal Physics)*, 1975, **5**, 1502
8 W. A. HARRISON: 'Pseudopotentials in the theory of metals', 1966, New York, Benjamin
9 P. S. HO: *Phys. Rev.*, 1971, **B3**, 4035
10 R. CHANG AND L. M. FALICOV: *J. Physics Chem. Solids*, 1971, **32**, 465
11 A. R. DU CHARME AND H. T. WEAVER: *Phys. Rev.*, 1972, **B5**, 330
12 Z. D. POPOVIĆ *et al.*: *J. Physics F (Metal Physics)*, 1974, **4**, 351
13 P. MINCHIN *et al.*: *J. Physics F (Metal Physics)*, 1974, **4**, 2117
14 N. H. MARCH: *J. Physics F (Metal Physics)*, 1973, **3**, 233
15 M. MANNINEN *et al.*: *Phys. Rev.*, 1975, **B12**, 4012
16 N. D. LANG: *Solid State Physics*, 1973, **28**, 224 (Eds. F. SEITZ *et al.*), New York, Academic Press
17 P. HOHENBERG AND W. KOHN: *Phys. Rev.*, 1964, **136**, B864
18 W. KOHN AND L. J. SHAM: *Phys. Rev.*, 1965, **140**, A1133
19 R. EVANS AND M. W. FINNIS: *J. Physics F (Metal Physics)*, 1976, **6**, 483
20 N. D. LANG AND W. KOHN: *Phys. Rev. B*, 1970, **1**, 4555
21 K. S. SINGWI *et al.*: *Phys. Rev. B*, 1970, **1**, 1044
22 P. VASHISTA AND K. S. SINGWI: *Phys. Rev. B*, 1972, **6**, 875; and *Errata Phys. Rev.*, **B6**, 4883
23 D. J. W. GELDART AND R. TAYLOR: *Can. J. Phys.*, 1970, **48**, 167
24 M. L. COHEN AND V. HEINE: *Solid State Phys.*, 1970, **24**, 37 (Ed. F. SEITZ *et al.*), New York, Academic Press
25 P. S. HO: 'Interatomic potentials and simulation of lattice defects' (Eds. P. C. GEHLEN *et al.*), 321, 1972, New York, Plenum
26 R. CHANG: 'Interatomic potentials and simulation of lattice defects' (Eds. P. C. GEHLEN *et al.*), 391, 1972, New York, Plenum
27 J. P. HIRTH: 'Interatomic potentials and simulation of lattice defects' (Eds. P. C. GEHLEN *et al.*), 456, 1972, New York, Plenum
28 M. S. DUESBERY: 'Interatomic potentials and simulation of lattice defects' (Ed. P. C. GEHLEN *et al.*), 458, 1972, New York, Plenum
29 G. R. AUGST: *Physica Status Solidi (b)*, 1973, **60**, 491
30 S. TANIGAWA AND M. DOYAMA: *Physica Status Solidi (b)*, 1976, **73**, 517
31 R. EVANS: *J. Physics C (Solid State Physics)*, 1974, **7**, 2808
32 M. W. FINNIS: *J. Physics F (Metal Physics)*, 1974, **4**, 1645
33 D. J. W. GELDART AND S. H. VOSKO: *Can. J. Phys.*, 1966, **44**, 2137
34 P. V. S. RAO: *J. Physics F (Metal Physics)*, 1975, **5**, 843
35 S. G. DAS *et al.*; *J. Physics F (Metal Physics)*, 1975, **5**, L35
36 R. KUMARAVADIVEL AND R. EVANS: *J. Physics C (Solid State Physics)*, 1976, **9**, 3877
37 M. W. FINNIS AND M. SACHDEV: *J. Physics F (Metal Physics)*, 1976, **6**, 965
38 S. YOSHIOKI AND G. MORI: *J. Physics F (Metal Physics)*, 1976, **6**, 1743
39 C. M. BERTONI *et al.*: *J. Physics F (Metal Physics)*, 1974, **4**, 19
40 N. W. ASHCROFT AND D. C. LANGRETH: *Phys. Rev.*, 1967, **155**, 682
41 M. W. FINNIS: *J. Physics F (Metal Physics)*, 1975, **5**, 2227
42 R. A. JOHNSTON AND W. D. WILSON: 'Interatomic potentials and simulation of lattice defects' (Ed. P. C. GEHLEN *et al.*), 301, 1972, New York, Plenum
43 J. R. HARDY: *J. Physics Chem. Solids*, 1968, **28**, 331
44 G. GRIMVALL AND S. SJÖDIN: *Physica Scripta*, 1974, **10**, 340
45 T. GÓRECKI: *Z. Metallkunde*, 1974, **65**, 426
46 K. MUKHERJEE: *Phil. Mag.*, 1965, **12**, 915
47 V. K. TEWARY: *J. Physics F (Metal Physics)*, 1973, **3**, 704
48 G. ALLAN AND M. LANNOO: *J. Physics Chem. Solids*, 1976, **37**, 699
49 J. L. BEEBY: *Proc. Roy. Soc.*, 1967, **A302**, 113
50 S. G. LOUIE *et al.*: *Phys. Rev. B*, 1976, **13**, 1654
51 R. EVANS AND R. KUMARAVADIVEL: *J. Physics C (Solid State Physics)*, 1976, **9**, 1891

Point defects in non-metals

A. M. Stoneham

The aims of this survey are to contrast aspects of defects and diffusion in metals and non-metals and to summarize some of the important features of defects in non-metals. The ultimate source of the difference between metals and non-metals is the presence of conduction electrons. This difference has three main effects. The first is purely practical because it affects the physical properties and hence the uses to which these materials are put. In metals one is primarily interested in mechanical properties, and hence in those defects (such as dislocations, grain boundaries, and certain specific impurities, e.g. H, C, O or N) which help to determine mechanical behaviour. However, it is the optical and semiconductor properties of non-metals which are usually of interest, and here it is the roles of point defects and of selected impurities (e.g. P in Si) which are of prime importance. Even in studies of intrinsic point defects there are major differences, for spectroscopic methods (optical and spin resonance) can be used to give detailed information about defects, whereas metal physicists must be content with techniques which merely label defects (annealing temperatures or internal friction peaks) or methods which do not even discriminate between different defects present (such as resistivity or specific heat). The second main influence of the conduction electrons is theoretical, and concerns the methods one uses. In insulators and semiconductors it is both useful and practical to calculate the detailed electronic structure of defects, whereas in metals one usually reduces the electronic features to a combination of volume-dependent and repulsive forces, with aspects such as screening limited to a free electron model. The third effect of a large number of conduction electrons is that one is never faced with questions of the charge-state of a defect. This can be crucial in insulators and semiconductors, notably in diffusion.

The author is at AERE, Harwell

In many respects the study of defects in metals has proceeded completely separately from the developments in the study of non-metals. I want to outline the reasons for this, and I shall point to some of the areas where contact remains. Clearly some topics involve both metals and non-metals, notably those concerned with interfaces, as in the oxidation of metals or in metal-insulator device systems. However, other points of contact exist in analogies between apparently entirely different systems, and in the methods of analysing defect properties.

The major difference between metals and non-metals lies in the presence of conduction electrons. This difference has a wide range of effects. One concerns the practical uses of these materials, and hence decides which defects are of most importance. Another is that theoretical approaches to defects are usually different when conduction electrons are present, with an emphasis on a free-electron gas in metals. The absence of an electron gas in insulators leads to one other important difference, for the long-range Coulomb fields of point defects are not screened out. Moreover, defects in non-metals can exist in several charge states, often with very different properties.

This survey will be far too brief to cover any but a few special cases of defects in insulators; for more general reviews, there are recent books by Flynn[1] and myself.[2] I have concentrated on those defects whose properties can be contrasted with those discussed in the other papers in this volume.

DEFECTS IN METALS AND NONMETALS

The defects in metals discussed in this volume fall into four broad groups:

(i) intrinsic point defects such as the interstitials and vacancies produced in radiation damage

(ii) alloying elements, usually similar in character in some sense to the host, e.g. Cr in Fe or K in Na

(iii) impurities, especially the light interstitial elements H, C, O, and N which are so important in the technological properties of metals

(iv) dislocations and other extended defects.

In non-metals, dislocations play a similar role in the

mechanical properties and as a sink for point defects. They lie outside the topic of point defects so, apart from observing that dislocations can be charged, with consequent effects[3] they will not be discussed further. All the other classes of defects have their counterparts in insulators and semiconductors. However, the detail with which they have been studied is much greater, both experimentally and theoretically. Experimentally the reason is that accurate spectroscopic measurements can be made without difficulty. The nature of the ground and excited states can be probed, with results of a quality to make rather accurate calculations worthwhile. Defects in metals are usually observed by techniques which merely label the defect (e.g. annealing temperatures), if indeed they discriminate between the various defect species present. Fortunately, X-ray methods have produced some precise geometrical information, but the lack of a reliable spectroscopic technique is acutely felt. Theoretically too, one has a much better understanding of defects in non-metals, and accurate predictions are more common. Partly this is a consequence of the experimental achievement, but partly (if I may be controversial) it is because theories of defects in metals have concentrated so much on electron gas or jellium models. These are excellent zero-order models, but not ones which allow extension to accurate and systematic quantitative predictions.

ELECTRONIC STRUCTURE OF POINT DEFECTS IN NON-METALS

The commonest experimental methods used in studying point defects in non-metals are optical absorption and electron spin resonance. Since the hyperfine structure seen in spin resonance gives a measure of the defect wavefunction amplitude at various neighbouring nuclei, and since the optical bands measure the energy differences between various states, a very complete understanding can be achieved. This can be supplemented by other approaches: the effects of stress and of electric or magnetic fields on sharp lines, or the observation of perturbed lattice vibrations all have their value.

The types of experiment have determined the kind of theory which is useful, since one needs to know both the energy levels and wavefunctions of electrons localized near the defect. Given these it is possible (if not always easy) to calculate the things which are observed (*see* for example part III of ref. 2). The theoretical methods used fall into four broad classes:

(i) effective mass theory, e.g. for shallow donors (Si : P) or acceptors (Si : Al)
(ii) Green's function methods, e.g. for isoelectronic defects in semiconductors, such as GaP : N;
(iii) variational approaches, including pseudopotential methods, as for F-centres (electron trapped at anion vacancy)
(iv) molecular approaches, potentially useful for any compact defect

Effective mass theory is often the easiest and most robust method to use. In it one concentrates on a small number of electrons in a conduction band (or holes in a valence band), representing the effect of the other electrons by a dielectric constant and an effective mass different from the free electron mass. The method has had some remarkable quantitative successes. The Green's function methods have a closer resemblance to

some of the scattering methods used in metals theory, but have proved primarily of pedagogical value in insulators and semiconductors. I shall comment in more detail on the other two approaches, for these pseudopotential and molecular methods show the areas in common between metals and non-metals. However, I also feel that the emphasis on the free-electron gas in metals has been greatly overstressed, and that there are important problems for which molecular methods are essential.

Pseudopotential theory in non-metals has been used mainly as a means for correcting a simple Hamiltonian. For instance, both the simple donors in semiconductors and the F-centres in alkali halides involve a single defect electron trapped by a potential which is Coulomb far from the defect. A simple model, giving rather good first-order results, is to scale the energy levels and wavefunctions of a hydrogen atom. However, this is not always satisfactory. The F-centre is observed in some 17 crystals analogous to NaCl, and a more accurate approach is needed to understand trends. In this case there are two sources of the trends. First, the Coulomb fields of the individual ions (e.g. Na^+, Cl^-) must be put in properly. These point-ion terms are the source of the main dependence on lattice parameter. Secondly, one must recognize that the ions are not merely point ions, but have structure. To lowest order one may use an effective potential for each ion γ

$$V_\gamma = V^{\text{point ion}} + C_\gamma \delta(r - r_\gamma) \qquad (1)$$

with s-wave scatter only. However, the energy dependence of the pseudopotential term in C_γ is important and so, in addition, there is an indirect consequence of the point-ion contribution: C_γ depends on the point-ion potential at site γ as well as on properties of the ion cores alone. Equations of the type (1) and their generalizations can now give transition energies of colour centres to about 5–10%, plus reliable trends along a series of related materials; the wavefunctions also agree well with experiment.

The use of molecular methods has been enhanced recently now that fast, convenient, Hartree–Fock programs have become available. Molecular methods in the most general sense—regarding a crystal as a large molecule—are old, of course, band theory being a limiting case of molecular orbital theory. However, it is not generally realized how small a group of atoms behaves sensibly like bulk solid. In an ionic crystal, one must include Madelung terms separately; in covalent systems there are problems of various types which lead to slower convergence than one would expect; in metals, the problems seem few. Indeed, there are some important cases where molecular approaches appear to give the only simple quantitative approaches. One example is that of the interstitials H, C, O, and N in metals.[4] Here one cannot use conventional pseudopotential theory: basically, the division into valence and core electrons is not simple, and one has to compromise between the inaccuracy of regarding all the interstitial electrons as a rigid core or the other problems of dealing with several valence electrons on each site. Further, conventional pseudopotential methods are only self-consistent to linear order, and this causes further difficulty (one can, of course, treat a proton in an electron gas,[5] in essence, one then handles the screening better at the expense of omitting some of the geometric features determined by the lattice structure). Conventional molecular methods prove convenient both in the full Hartree–Fock sense and in various approximate forms, such as $X\alpha$ or the semi-empirical schemes pioneered by organic chemists. It is worth mentioning that the semi-empirical schemes need not be used as approximations to Hartree–Fock theory, and they are best used as a means of extrapolation from simpler systems. For example one fruitful application has been to calculate potential energy surfaces for H in metals after choosing parameters to fit suitable diatomic molecule or pure metal data. It should be stressed, incidentally, that the muffin-tin approximation used in most $X\alpha$ approaches makes the calculation of potential energy surfaces most unreliable.

LATTICE DISTORTION AND RELATED TOPICS

Here I want to mention only a few special items which illustrate the differences between metals and non-metals.

One simple example concerns the interatomic forces. In ionic crystals one must handle long-range Coulomb interactions. In semiconductors one must include bond-bending and similar many-body forces. Yet in most metal problems, one can get acceptable results with short-range central forces plus a volume-dependent term.

Another example concerns the Kanzaki method. In non-metals one must recognize that the wavefunction of any trapped electron may vary rapidly with the local geometry. If so, then a force-constant picture becomes inappropriate, and a generalization is needed.[2] In the simplest terms, the Kanzaki theory minimizes the sum of two energies: an elastic term $\frac{1}{2}\mathbf{x} . \mathbf{A} . \mathbf{x}$ (where $\mathbf{A}^{-1}$ is the force-constant matrix, and $\mathbf{x}$ describes all atomic displacements) and a term $V(\mathbf{x})$ from the presence of the defect. This leads to a formal condition for equilibrium:

$$\frac{\partial}{\partial \mathbf{x}}(\tfrac{1}{2}\mathbf{x} . \mathbf{A} . \mathbf{x} + V(x)) = 0 \qquad (2)$$

If V simply contains constant terms and those from defect forces (i.e. $V(\mathbf{x}) = V_0 - \mathbf{F} . \mathbf{x}$) then the equilibrium distortion is given by:

$$\mathbf{x}_{eq} = \mathbf{A}^{-1} . \mathbf{F} \qquad (3)$$

and there is a lattice relaxation energy $-\tfrac{1}{2}\mathbf{F} . \mathbf{A}^{-1} . \mathbf{F}$. Here $\mathbf{A}^{-1}$ is the lattice analogue of the continuous Green's function, and there are many methods (computer simulation, Fourier transform methods, Ewald–Kellermann approaches, etc) of estimating it. The defect term V is more complicated in non-metals when there may be trapped electrons whose wavefunction varies with the distortion. Suppose λ is a parameter describing these wavefunctions, and let λ_0 be its value when $\mathbf{x} = 0$. Now λ satisfies a variational principle. If the Born–Oppenheimer approximation holds, then $\partial V(\lambda, x)/\partial\lambda|_x = 0$. We must expand V in terms of $(\lambda - \lambda_0)$ as well as $\mathbf{x}$, but the variational principle allows us to eliminate explicit dependence on λ. In its place there is a more complex dependence on $\mathbf{x}$, and it may be necessary to handle some of the non-linear terms. In essence, the defect forces $\mathbf{F}$ are functions of $\mathbf{x}$ when the electronic structure must be handled separately. They can have further complications, the Jahn–Yeller effect being an

example. If the defect has orbital degeneracy, in general there will be a distortion which lowers both the energy and the symmetry. Here one has a many-valued potential energy surface. Fortunately, this can be handled by an appropriate generalization of the linear response approach, and detailed quantitative calculations have been made.

One final example which emphasizes some of the differences between metals and non-metals concerns donors in silicon and germanium. Shallow donors have energy levels roughly analogous to a hydrogen atom, but scaled because of the altered effective mass and large dielectric constant. Further differences include chemical effects on s-state energies (the 'central cell correction') and the splitting of p-states when the conduction band minima have low symmetry. This splitting is important in Si and Ge, where separate transitions $1s \rightarrow 2p_0$ and $1s \rightarrow 2p_{\pm 1}$ are seen. Now the splitting of these transitions contains no chemical component, for the 2p states have a node at the impurity site. But there is a contribution from the long-range strain field of the donor. So it is possible (at least in a few cases) to estimate the dilatation the donor produces by purely spectroscopic means.[2]

DISCUSSION

I now turn from the methods used for defect calculations to the type of studies that are worthwhile. Roughly, three types of information are wanted:

(i) what defects are present?
(ii) what are the defect properties?
(iii) what processes occur involving defects?

Much progress has been made in all areas, and I shall illustrate this with some recent examples. However, it would be wrong to suppose that success in these areas is new. Many remarkable insights and predictions, mainly valid to this day, can be found in the classic 1940 text of Mott and Gurney,[6] for example.

Deciding which defects are present is often difficult, and many sensible suggestions have proved wrong. A good example is in TiO_{2-x}, where one might (and many did) naturally assume that oxygen vacancies and/or titanium interstitials would be present. In fact neither are present: defects thought to be interstitial titanium proved to be hydrogen impurity, and the missing oxygens were found to be assimilated into shear planes. So first thoughts can be wrong, and this is especially so when there are chemical impurities too: most systems of interest technologically are dirty.

Defect properties in non-metals include a whole range of spectroscopic parameters. Moreover, these parameters are often known in a series of related crystals, such as the set of 17 alkali halides with the NaCl structure. Thus, these are very detailed checks in one's understanding. It is fair to say that there is a detailed understanding of most (not all) of the well-characterized simple defects, and there is quantitative agreement with experiment, sometimes (as for donors in silicon) of very high accuracy. The role of theory has been to unravel some of the more complicated systems. This is especially important when the timescale is very fast. An example where theory has played an important part is the self-trapped exciton in ionic crystals. In essence, when a (perfect) alkali halide is excited optically, an electron moves in the conduction band while the hole forms a localized halogen molecular ion. The electron remains bound in the vicinity of the hole, and its energy levels can be studied despite the short lifetime. The system is important partly because it can decay non-radiatively to give an interstitial and a vacancy: radiation damage can be induced by optical excitation only. The damage mechanism is an example of the non-radiative transitions which are important in areas ranging from diffusion to electrode processes and the efficiency limits of semiconductor devices.

Finally, *defect processes* are the usual point of contact between pure science and practical application. Corrosion is a good example. Here the rate of corrosion involves, *inter alia*, the motion of ions (either oxygen or metal) and of electrons through an oxide layer. If one is to influence this rate by appropriate dopants, then one must know precisely which species move and how. This is not really possible by experiment alone for there are indirect effects of unwanted impurities which are hard to allow for. However, the theory has now reached the stage where one can sort out the mechanisms involved.

A few years ago it was probably true to say that the only contribution fundamental studies could make to corrosion was a framework within which a range of empirical data could be placed. It is now getting close to the time where atomistic studies, both experimental and theoretical, can guide developments in this and similar areas.

REFERENCES

1 C. P. FLYNN: 'Point defects and diffusion,' 1972, Oxford University Press
2 A. M. STONEHAM: 'Theory of defects in solids', 1975, Oxford University Press (this book contains references to most of the work referred to in the present paper)
3 R. W. WHITWORTH: *Adv. Phys.*, 1975, **24**, 205
4 A. MAINWOOD AND A. M. STONEHAM: *J. Less Common Metals*, 1976, in press
5 Z. D. POPOVIC *et al.*: *Phys. Rev.*, 1976, **B13**, 590
6 N. F. MOTT AND R. W. GURNEY: 1940, 'Electronic processes in ionic materials', 1940 and 1964, Oxford University Press

Session II

Point defect aggregates and their effects on properties

Chairman (morning): R. E. Smallman (University of Birmingham)
Technical Secretary (morning): K. H. G. Ashbee (University of Bristol)

Chairman (afternoon): G. W. Greenwood (University of Sheffield)
Technical Secretary (afternoon): B. L. Eyre (AERE)

Point defect interaction and early stages of clustering

K. Schroeder

The existing evidence for the formation and properties of small clusters of self-interstitials and vacancies are reviewed. First, the long-range elastic interaction and its implication for cluster formation is discussed. Irrespective of the nature of the defects one always finds an attractive interaction in some directions. This means there is always a tendency for clustering. It turns out that the interaction of the sink with the mobile defect in the saddle point configuration determines the agglomeration rate. Computer simulation is the only method to obtain theoretical information about structure, stability, and mobility of small clusters. One finds that small interstitial clusters should be very stable and have a high mobility comparable to that of single interstitials. Small vacancy clusters, on the other hand, are less stable and have a lower mobility comparable to that of single vacancies. The experimental evidence for small clusters is very limited. It comes mostly from diffuse X-ray scattering and positron annihilation. We discuss briefly the background of the two methods and applications to irradiation defects, e.g. in Al, Cu, and Mo. The results generally confirm the theoretical picture.

The author is with the Institut für Festkörperforschung der KFA Jülich, Germany

Small clusters of point defects are of special interest for irradiation studies because they are a necessary intermediate step for the production of large defects agglomerates. The understanding of their structure and properties such as stability and mobility and their interaction might eventually yield some idea of how to prevent or minimize the growth of voids. However, very little is known about small clusters, mainly because they are not accessible to electron microscopy study. Theoretical investigations are also very difficult because small clusters can neither be treated by analytical methods using continuum elasticity theory nor can they easily be simulated on the computer because they have many more degrees of freedom than single point defects.

Most experimental evidence for the existence of small aggregates of self-interstitials and vacancies is rather indirect. Small clusters must exist at an intermediate stage because larger clusters have been found in the electron microscope also after electron irradiation which produces only single defects. The existence of di-vacancies is deduced from the deviation of the self-diffusion coefficient from a simple Arrhenius law at temperatures close to the melting point in many metals.[1,2] Small binding energies (0.1–0.3 eV) and mobilities larger than for single vacancies are generally found. The existence of di-interstitials or small intersti-tial agglomerates has been deduced from the incomplete recovery of electrical resistivity after long-range interstitial migration in electron-irradiated metals.[3,4] It was concluded that the interaction of two interstitials is at least as strong as the one of an interstitial with a vacancy.

Some more direct evidence is due to recent application of new experimental methods such as X-ray diffuse scattering and positron annihilation. Before presenting some experimental results (fifth section) I would like to discuss briefly the interaction of point defects, which leads to clustering, and computer simulation of small clusters (second section), further the rate theory used to describe the kinetics of defect agglomeration (third section) and the connection between rate constant and long-range interaction (fourth section).

INTERACTION BETWEEN POINT DEFECTS

When a point defect is introduced into a crystal, forces are exerted on the neighbouring atoms which lead to long-range displacements. In the vicinity of the defect the displacements from regular lattice sites can be so large that the coupling between the atoms of the crystal also is changed. For the long-range displacement field, however, a defect can be described by so-called Kanzaki forces,[5] i.e. effective forces to be applied to the atoms of the ideal harmonic crystal which yield the same displacements as the defect produces in the real crystal. In metals these forces are very localized and extend only to a few shells of close neighbours.

If one introduces a second defect into the crystal the forces due to this second defect have to do work against the displacements produced by the first one and *vice versa*. This is the cause of the elastic interaction between defects.

Long-range interaction

For distances R which are large compared to the range of the forces ($R \geqslant$ a few lattice constants) the leading term of the interaction is the dipole–dipole interaction:

$$W_d(\mathbf{R}) = \sum_{ik} \overset{1}{P}_{is} \left(\frac{\partial}{\partial X_s} \frac{\partial}{\partial X_t} G_{ik}(\mathbf{R}) \right) \overset{2}{P}_{kt} \qquad (1)$$

Here $G_{ik}(\mathbf{R})$ is the static elastic Green's function which is determined by the elastic constants.[6] P_{ij} is the dipole force tensor of a defect, i.e. the first moment of the Kanzaki forces $K_i^{\mathbf{n}}$:

$$P_{ij} = \sum_n K_i^{\mathbf{n}} X_j^{\mathbf{n}} \qquad (2)$$

($\mathbf{R^n} = \{X_j^{\mathbf{n}}\}$ = equilibrium position of atom n).

The dipole force tensor fully characterizes the long-range displacement field of a defect, e.g. Tr P is proportional to the volume change due to the defect. The Green's function and thus the dipole interaction cannot be given in closed form for anisotropic crystals. We will discuss briefly the main features:

1 $W_d(\mathbf{R})$ falls off with distance $\sim 1/R^3$ in analogy to the electrical dipole interaction.

2 For isotropic defects ($\overset{1,2}{P}_{ij} = \overset{1,2}{P} \delta_{ij}$) in an isotropic medium the dipole interaction vanishes. Then higher order multipoles determine the interaction which then drops off faster with distance. The dipole interaction is non-zero only if the medium and/or at least one of the defects are anisotropic.

3 For cubic crystals we obtain to lowest order in the anisotropy parameter $d = c_{11} - c_{12} - 2c_{44}$ (c_{ij} = elastic constants in Voigt's notation) for isotropic defects ($\overset{1,2}{P}_{ij} = \overset{1,2}{P} \delta_{ij}$, i.e. pure dilatation or compression centres) the well known Eshelby formula[7]

$$W_d(\mathbf{R}) = -\frac{15}{8\pi} d \left(\frac{5}{3c_{11} + 2c_{12} + 4c_{44}} \right)^2 \frac{\overset{1}{P}\overset{2}{P}}{R^3} \left(\frac{3}{5} - \sum_i \left(\frac{X_i}{R} \right)^4 \right) \qquad (3)$$

For metals ($d < 0$) and defects of the same kind (e.g. interstitials both $P^{1,2} > 0$) the interaction is attractive in $\langle 100 \rangle$- and repulsive in $\langle 100 \rangle$- and $\langle 111 \rangle$ directions.

4 The dipole interaction averaged over the solid angle at fixed distance is zero. The same is true for higher multipole interactions. This implies that there are always attractive and repulsive directions irrespective of the strength and symmetry of the defects. Thus the long-range part of the interaction always causes a tendency for clustering: when the defects become mobile they will preferentially jump to sites with low potential energy (i.e. with attractive interaction) and eventually form agglomerates. The same argument holds for the interaction of point defects with small clusters and dislocation loops whose long-range displacement fields can also be characterized by dipole force tensors.

5 The strength of the interaction is determined by the strength of the defects (dipole force tensor). Thus self-interstitials which have a much stronger displacement field than vacancies will interact more strongly with any inhomogeneity.

6 Corrections to the dipole interaction, equation (1), include higher multipole contributions,[8] dispersion corrections to the static Green's function [9] and contributions by induced dipole moments due to the elastic polarizability of the defects[10] (for a detailed discussion see the paper by Heald,[11] this volume).

Computer simulation of small clusters

The structure and stability of defect agglomerates is determined by the interaction at short distances. Then, however, the simple description of the interacting defects by Kanzaki forces breaks down since the strongly distorted regions around the defects overlap. Thus analytical calculations are impossible and computer simulations have been made to obtain information on defect clusters (for a recent review, *see* ref. 12). Usually defects are introduced into a small crystallite containing a few thousand atoms which interact via a given potential. The stable configurations of the defects are found by minimizing the total energy of the crystallite. In principle, one could map the entire energy contour as a function of the defect coordinates by minimization under geometrical constraints. However, this procedure is very tedious and time-consuming especially for higher agglomeration states. Thus, in practice, only a few configurations are calculated. The parameters obtained from such calculations include structure and binding energy of the most stable configuration and a few metastable configurations of single defects and small clusters, activation energies for migration and reorientation (where applicable), and structure of the respective saddle point configurations.[12]

The necessary input for the calculations is the interatomic potential. Usually two-body interaction potentials are used fitted to equilibrium data. Thus the potential is determined at atomic distances of the ideal crystal. However in defect calculations other distances are involved where the potential function is less reliable. In the case of interstitials in the close-packed fcc structure the repulsive part is most important[13] whose exponential variation known from electron theory is usually incorporated in the potential. Thus, in this case the results are least sensitive to the potential. If the attractive parts becomes more important, as is the case for vacancies in fcc and particularly for defects in the more open bcc structure, the results become more sensitive to the potential chosen, in particular to the range,[14] and the results are less reliable. Nevertheless, some general features (discussed below) can be obtained from computer simulation work which can serve as valuable guidelines for the interpretation of experiments.

Most calculations reported are with potentials chosen to fit Cu and Fe as examples for fcc and bcc structure, respectively. Small agglomerates containing up to four single defects were investigated rather systematically but also larger planar defect aggregates were treated in order to study the collapse of a cluster into a dislocation loop.[15,16]

Interstitials and interstitial clusters

The stable configuration of the single interstitial turns out to be the ⟨100⟩-split in Cu[13,17] and ⟨110⟩-split in Fe.[18]

Dederichs[19] has discussed the properties of single and multiple interstitials in fcc metals. Figure 1 shows the most stable configurations of small clusters containing up to four interstitials as found by Schober[13] with a 'Cu'-potential. The smallest clusters are close-packed arrangements of ⟨100⟩-dumbbell interstitials, for the tetra-interstitial the fourth dumbbell decays into the octahedral position. All multi-interstitials have a large binding energy (of the order of 1 eV per interstitial). Due to the strong displacements the interaction extends rather far and there are a large number of metastable configurations with only slightly higher energy. Thus the low migration energies (of the order of 0·1–0·3 eV) of the small clusters are plausible. For the single and di-interstitials the jumps with the lowest activation energy lead to migration with reorientation whereas the tri-interstitials can reorient with lower activation energy than for migration.

Measurements of the mechanical relaxation[20,21] after electron irradiation show that in Al the migration energies of small interstitial clusters increase with size in accord with the calculation. This is supported by the results of diffuse X-ray scattering in the same metal,[22] whereas the results on electron irradiated Cu[23] and the absence of mechanical relaxation peaks in this metal suggest that small interstitial clusters are more mobile than single interstitials in Cu and thus cannot be observed. For a detailed discussion see p. 58.

Small clusters of interstitials in bcc metals ('Fe') consist of parallel ⟨110⟩ dumbbell interstitials aggregated on close neighbour sites in the {110}-plane (see Fig. 2). At

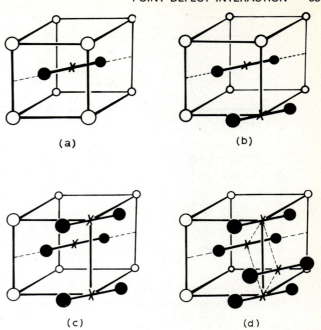

a single interstitial in the ⟨110⟩ dumbbell configuration; *b* di-interstitial; *c* tri-interstitial; *d* tetra-interstitial

2 Stable configurations of small interstitial clusters in bcc crystals (Fe) as found by computer simulation[15,18]

larger sizes (n ≥ 4) a tendency for shear into loops on {111} planes is observed.[15] The binding energy of the di-interstitial is large (about 1 eV and the migration energy turns out to be lower than that of the single interstitial (0·18 eV compared to 0·33 eV).[18] No values for the migration or reorientation energy for larger interstitial clusters are reported.

Vacancy clusters

The results for small vacancy clusters in fcc metals depend on the potential used. Figure 3 shows the structures. Close-packed clusters are found for short-range potentials and the interaction between vacancies is essentially limited to nearest neighbour distances.[24] The binding energy can be obtained by bond counting (0·25 eV per vacancy bond for 'Ni'[24]). The migration energy of di-vacancies is smaller than that of a single vacancy (0·9 eV compared to 1·32 eV for 'Ni'[24]). Because of the necessity of partial dissociation during migration the migration energies increase with cluster size. Longer range potentials yield vacancy platelets on {111} planes as the stable configuration[25] which collapse rather early (n ≥ 6) to form loops.[16] The migration energy for the di-vacancy turns out to be unrealistically small for such a potential (0·03 eV compared to 0·69 eV for single vacancies for 'Cu'[25]).

For bcc metals close-packed clusters were found to be most stable with the exception of the di-vacancy which consists of two vacancies on second nearest neighbour sites[26] (Fig. 4). Single, di- and tri-vacancies have essentially the same migration energy (0·68 eV for 'Fe'[26]) whereas the tetra-vacancy can only migrate by dissociating with an energy of 1·12 eV (for 'Fe') and thus seems to be the first stable nucleus for further clustering.[26]

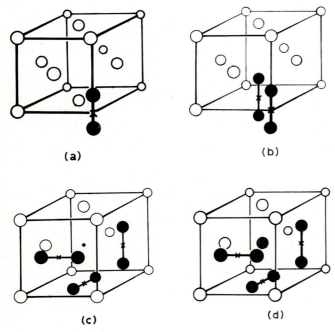

a single interstitial in the ⟨100⟩ dumbbell configuration; *b* di-interstitial; *c* tri-interstitial; *d* tetra-interstitial

1 Stable configurations of small interstitial clusters in fcc crystals (Cu) as found by computer simulation[13]

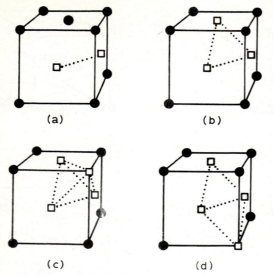

a di-vacancy; *b* tri-vacancy; *c* tetra-vacancy for short-range potential;[24] *d* tetra-vacancy for long-range potential[25]

3   Stable configurations of small vacancy clusters in fcc crystals (Ni, Cu) as found by computer simulation[24,25]

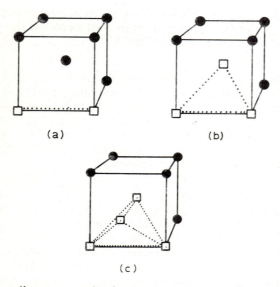

a di-vacancy; *b* tri-vacancy; *c* tetra-vacancy

4   Stable configuration of small vacancy clusters in bcc crystals (Fe) as found by computer simulation[26]

There is no experimental result which has been directly related to multi-vacancy migration or reorientation.

KINETICS OF CLUSTER GROWTH

The growth of clusters is determined by the mobility and interaction of the various reaction partners and by the stability of the complexes formed. Under irradiation the following reactions are conceivable:

(i)   recombination of interstitials and vacancies
(ii) formation and growth of clusters (interstitial and vacancy type) by agglomeration of single defects and mobile small clusters

(iii) shrinkage of clusters by absorption of single or multiple defects of the opposite kind
(iv) Dissociation of clusters by emission of single defects or small clusters.

In order to give a realistic description of clustering phenomena one has to consider the mobility of small defect clusters also. This has not been done in calculations up to now.

At low temperatures ($T \leqslant 200$ K) only interstitials and interstitial clusters are mobile and thus no vacancy clusters need to be considered. At higher temperatures also vacancies become mobile and thus all reactions are possible.

Usually the kinetics are described in terms of rate equations. For small defect densities each reaction can be considered independently and the reaction rate is proportional to the densities of the reaction partners times an appropriate rate constant. For example, the rate of loss of single interstitials due to recombination with single vacancies is given by

$$\frac{\mathrm{d}}{\mathrm{d}t}c_{i\,\mathrm{rec}} = -K_0 c_i c_v \qquad (4)$$

Here c_i, c_v are the average densities of single interstitials and vacancies, respectively. The rate constant K_0 can easily be calculated using continuum theory of diffusion and assuming that recombination takes place instantaneously whenever the defects approach each other closer than the 'recombination radius' R_0.[27,28] Considering mobile interstitials (diffusion constant D_i) and immobile vacancies (low temperatures) K_0 is given by the steady-state diffusion flux of interstitials into the recombination sphere of a vacancy with the interstitial density maintained at unity at large distances

$$K_0 = 4\pi R_0^2 D_i \left(\frac{\partial\rho}{\partial r}\right)_{R_0} = 4\pi D_i R_0 \qquad (5)$$

Here $\rho(\mathbf{r})$ is the correlation function of interstitial–vacancy pairs. $\rho(\mathbf{r})$ obeys the stationary diffusion equation $\Delta\rho(r) = 0$ and the boundary conditions $\rho(R_0) = 0$, $\rho(r \to \infty) = 1$ with the solution $\rho(r) = 1 - R_0/r$. Thus the recombination rate is proportional to the radius R_0 and not to the cross-section πR_0^2 of the interaction sphere as a direct consequence of correlations ($\rho(\mathbf{r}) \neq 1$) between interstitials and vacancies built up because of diffusion. Similar expressions are valid for other reactions like formation, growth or shrinkage of clusters which involve the diffusion of defects. The reaction radii are phenomenological parameters which incorporate two effects.

1   The actual instable sites around a vacancy in the case of recombination or trapping sites in the case of clustering processes where the interaction is much larger than the thermal energy. It can be shown that the replacement of these discrete reaction sites by a spherical reaction volume and the use of continuum theory instead of jump diffusion on a discrete lattice is justified even for very few reaction sites.[29] The radius R_0 of the appropriate reaction volume turns out to be approximately equal to the distance of the outermost reaction sites from the defect centre.

2   The long-range interaction of the reaction partners giving rise to a drift of mobile defects towards the regions of lower potential energy. This leads to a temperature dependence of the effective reaction radius.[30,31]

For strong attractive potentials the effective reaction radius is given by the distance at which the decrease of the potential energy in the saddle point configuration due to interaction equals the thermal energy of the defect: $-\varphi_s(R_{eff}) = kT$. We shall discuss this effect in more detail in the next section.

Dissociation processes are described by a Boltzmann factor containing the dissociation energy which is the sum of the binding energy of the complex and the migration energy of the defect jumping away.

There are a number of papers recently reviewed by Wiedersich[32] which calculate the growth and annealing of large clusters by numerically solving a large set of rate equations. This seems to be the only way to obtain theoretical information about the density and size distribution of clusters under various temperature and irradiation conditions.

Of course, the rate equations describe the development of the true physical system only approximately. There are a number of complications not considered so far. Balluffi[33] has discussed the situation for large clusters such as voids and dislocation loops. For example, the incorporation of an arriving defect into a cluster is a multistep process which involves more than one activated process. It is not clear at the moment whether this can be described by an effective reaction constant. Other complications arise since the defects involved are non-cubic. Johnson[24] and Schober[13] have discussed the 'diffusion channels' which are caused by the particular geometry of the elementary jump of the $\langle 100 \rangle$-dumbbell. As a consequence interstitials might not always find the most stable configuration. The same is true for vacancy clustering if multi-vacancies become mobile. Thus several different configurations of clusters containing the same number of single defects might be formed which have to be distinguished. The same effect can be introduced by the orientation dependence of the interaction of non-cubic defects. This can inhibit certain diffusion paths for the $\langle 100 \rangle$-dumbbells because they change orientation with every jump. Thus interstitials might become trapped in metastable configurations.

RATE CONSTANT FOR DIFFUSION IN LONG-RANGE POTENTIALS

Within the classical picture of diffusion the jump rate of a mobile defect is determined by a Boltzmann factor containing the energy difference between saddle point configuration and stable configuration. Due to interaction with other defects the potential energies in both configurations and thus the jump rates become dependent on position (see Fig. 5) and for non-cubic defects also on orientation and jump direction.[34,35] We shall discuss only cubic defects.

The leading term of the long-range interaction is the dipole–dipole interaction discussed on p. 52. The strength is proportional to the dipole force tensor. During a jump of the mobile defect in general the dipole force tensor changes. The associated volume change, the migration volume, is the cause of the well known pressure dependence of diffusivities. Thus we have to distinguish two interaction potentials, $\varphi(\mathbf{r})$ for the stable configuration and $\varphi_s(\mathbf{r})$ for the saddle point configuration.

The particle current is obtained by considering the net jump rate across the energy barrier between two adjacent sites (see Fig. 5). With the indicated energies we

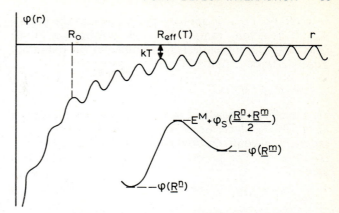

5   Sketch of the interaction potential of a mobile particle with a sink at $r = 0$; the unstable sites are incorporated in the microscopic reaction radius R_0, the long-range interaction yields a temperature dependent effective reaction radius $R_{eff}(T)$; in the insert details are shown for two adjacent sites m and n; for jumps from n to m a larger activation energy is required than for the reverse jump

obtain for the net flux from **n** to **m** across the barrier at $\mathbf{R^n} + \mathbf{R^m}/2$

$$j((\mathbf{R^n} + \mathbf{R^m})/2) = \nu_0 \exp(-\beta[E^M + \varphi_s((\mathbf{R^n} + \mathbf{R^m})/2)])$$
$$\times [w(\mathbf{R^n})\exp(\beta\varphi(\mathbf{R^n})) - w(\mathbf{R^m})\exp(\beta\varphi(\mathbf{R^m}))] \quad (6)$$

with ν_0 the attempt frequency, E^M the activation energy for diffusion in the ideal lattice, $w(\mathbf{R^n})$ the occupation probability of site **n**, and $\beta = 1/kT$.

For slowly varying functions we obtain the continuum expression for the current density by expanding to lowest order in the potential gradient[31,36]

$$\mathbf{j}(\mathbf{r}) = -D \exp(-\beta\varphi_s(\mathbf{r})) \, \text{grad}[\rho(\mathbf{r})\exp(\beta\varphi(\mathbf{r}))]$$
$$= -D \exp(-\beta(\varphi_s(\mathbf{r}) - \varphi(\mathbf{r}))$$
$$\times [\text{grad } \rho(\mathbf{r}) + \rho(\mathbf{r}) \, \text{grad } \beta\varphi(\mathbf{r})] \quad (7)$$

Here D is the ideal diffusion constant and $\rho(\mathbf{r})$ the particle density. The first term of the last line of equation (7) is the usual diffusion current with the local diffusion constant $D(\mathbf{r}) = D \exp(-\beta(\varphi_s(\mathbf{r}) - \varphi(\mathbf{r}))$, the second term is the drift current with the mobility related to the local diffusion constant by Einstein's relation: $B(\mathbf{r}) = \beta D(\mathbf{r})$.

With this current density we can evaluate the rate constant K_{eff}, i.e. the total flux of mobile defects into the reaction sphere of a sink. The density $\rho(r)$ has to obey the boundary conditions $\rho(R_0) = 0$ and $\rho(r \to \infty) = 1$. As a result of the interaction the rate constant K_{eff} is temperature dependent. It can be interpreted in terms of an effective reaction radius[30,31]

$$K_{eff} = 4\pi D R_{eff}(T) \quad (8)$$

It turns out that R_{eff} is determined by the interaction in the saddle point configuration alone. The temperature dependence of R_{eff} is related to the radial dependence of the interaction, the absolute value depends on the strength and the angular dependence.

The result is particularly simple in the case of spherically symmetric potentials. For weak potentials

$(\beta|\varphi_s(R_0)| \ll 1) R_{eff}$ is approximately equal to the microscopic reaction radius R_0, for strong potentials $(\beta|\varphi_s(R_0)| \gg 1)$ it is entirely determined by the potential. Purely attractive potentials yield approximately

$$-\varphi_s(R_{eff}(T)) = kT \tag{9}$$

i.e. the effective reaction radius is given by the distance at which the decrease of the potential energy in the saddle point configuration is equal to the thermal energy (see Fig. 5). For power potentials $\varphi_s(r) = -\alpha/r^n$ this yields

$$R_{eff}(T) = (\alpha/kT)^{1/n} \tag{10}$$

For strong repulsive power potentials R_{eff} is given by

$$R_{eff} \cong R_0 n\beta\varphi_s(R_0) \exp(-\beta\varphi_s(R_0)) \tag{11}$$

i.e. the interaction creates an effective activation barrier $\varphi(R_0)$ which lowers the reaction rate substantially at low temperatures.

Realistic potentials, however, are angular dependent. For example, the elastic dipole–dipole interaction has attractive and repulsive parts of equal weight. From the above results we would expect that the repulsive parts of the solid angle do not contribute to the total flux and that the attractive directions alone determine the rate constant. This is indeed the case. The absolute value of R_{eff} can only be calculated approximately either analytically using a variational principle[31] or numerically.[37] The results agree quite well. They yield a reduction of R_{eff} for the dipole–dipole interaction, equation (3), by a factor of about 0·7 compared to R_{eff} for a purely attractive potential with the same strength as in the most attractive direction.

The temperature dependence of the recombination radius of interstitial–vacancy pairs in Cu has been determined from production rate measurements[38] to be proportional to $T^{-1/3}$ over the range $50\,K < T < 200\,K$. This is in accord with the elastic dipole interaction, equation (3).

The dependence of the effective reaction radius on the strength of the potential can also be obtained from equations (9) and (10). In the case of elastic interaction, equation (3), α is proportional to the volume change of the mobile particle. Because of the larger volume change of interstitials compared to vacancies this leads to a preferential absorption of interstitials at dislocation loops.[77]

EXPERIMENTAL RESULTS FOR SMALL DEFECT AGGLOMERATES

There is only limited direct information about small defect clusters. One reason is that rather favourable conditions have to be met in order to detect small clusters. They must be stable and must not be mobile at the temperature at which they are formed because otherwise they will dissociate or migrate and immediately form larger clusters. Such favourable conditions seem to be fulfilled for interstitial clusters in Al.

The second reason is the lack of experimental methods sensitive enough to monitor cluster formation and understood well enough to allow quantitative interpretation. One notable exception is X-ray diffuse scattering. In particular the so-called Huang scattering close to the Bragg peaks has proved to be a valuable method to investigate the size and symmetry of interstitial clusters. We shall discuss the main features of the method shortly below. The method has been used in Al,[22] Cu,[23,39,40] Au,[41] Mo,[42] Zn.[42] Additional and more detailed information on specific multi-interstitials in Al has been obtained using mechanical relaxation methods.[20,21] The formation of interstitial–impurity complexes in Al has been investigated using Mössbauer spectroscopy[43] and backchannelling,[44] as well as production rate measurements.[38,45]

Small vacancy agglomerates have been studied by other methods. Di-vacancies have been found in field ion microscopy studies of quenched Pt.[46] In Cu small vacancy platelets have been observed by diffuse X-ray scattering close to the forward direction.[47] Recently positron annihilation has received much attention because of its specific sensitivity to vacancies and vacancy agglomerates. We shall shortly discuss the method below. Unfortunately a quantitative understanding of the effect of vacancy clustering on the positron annihilation characteristics like lifetime, angular correlation, and Doppler-broadening is still lacking. Recently, first attempts in this direction have been reported. For example, the method has been applied to the study of vacancy agglomeration in Mo,[48] Cu,[49] and Au.[49]

Theoretical background of experimental methods

Diffuse X-ray scattering

There are a number of excellent review papers available on both the theoretical[50–52] and experimental[53–56] aspects of the method. We shall shortly discuss the main features. Details can be found in the review by Dederichs[50] for example.

The diffuse intensity is caused by the destruction of the perfect order of the crystal by defects. This makes the destructive interference between the Bragg peaks incomplete. The intensity (per atom) of X-ray diffuse scattering for small concentrations c of randomly distributed defects is given by

$$S_{diff}(\mathbf{K}) = c|F(\mathbf{K})|^2 \tag{12}$$

Here $\mathbf{K}$ is the scattering vector, i.e. the difference between the wave vector of the incident and scattered waves. The intensity is proportional to the concentration since waves scattered by different defects add incoherently. The 'defect structure factor'

$$F(\mathbf{K}) = f_{\mathbf{K}}^D + f_{\mathbf{K}} \sum_{\mathbf{n}} \exp(i\mathbf{K}\mathbf{R}^n)[\exp(i\mathbf{K}\mathbf{t}^n) - 1] \tag{13}$$

consists of the atomic scattering amplitude, $f_{\mathbf{K}}^D$, of the defect itself and a contribution due to the displacements $\mathbf{t}^n$ of the crystal atoms $\mathbf{n}$ from their regular lattice positions $\mathbf{R}^n$. Depending on the scattering vector $\mathbf{K}$ different parts of the displacement field determine $F(\mathbf{K})$.

Huang scattering

The intensity is especially large in the so-called Huang region close to the Bragg peaks: $\mathbf{K} = \mathbf{h} + q$, $q \ll h$. Here the main contribution to $F(\mathbf{K})$ comes from the long-range displacement field which falls off as $1/R^2$. The strength and symmetry is determined by the dipole force tensor P_{ij} of the defect. The following important features are found:

(i) the intensity increases sharply as q^{-2} as one approaches the Bragg peak; this q^{-2} dependence is observed in a limited region of q values

(ii) the intensity is proportional to $c(\Delta V^{rel})^2$ with ΔV^{rel} the relaxation volume of a defect, by a combined measurement with the change of the lattice parameter $\Delta a/a \sim c \cdot \Delta V^{rel}$ the defect density and relaxation volume can be determined separately; because of the quadratic dependence on ΔV^{rel} Huang scattering is particularly sensitive to interstitials whose relaxation volume is usually much larger than that of vacancies.

(iii) the shape of the intensity distribution around the Bragg reflections is solely determined by the symmetry of the long-range displacement field of the defects; characteristic planes or lines of zero intensity show up for highly symmetric defects[51] which allow a unique determination of the defect symmetry.

Forward scattering

Near the forward direction $\mathbf{K} = q \approx 0$, the main contributions to $F(\mathbf{K})$ come from the atomic scattering and the long-range displacement field. The magnitude of the intensity is determined by the change of the *local* atomic density at the defects. For interstitials this is very small since the change due to the additional atom is largely cancelled by the large relaxations. For vacancies, however, the relaxations are small and the density change is appreciable because of the missing atom. Forward scattering is thus primarily sensitive to vacancies and uncollapsed vacancy clusters.

Scattering between Bragg reflections

The diffuse intensity between Bragg peaks yields information about the position of interstitial atoms and the strongly distorted neighborhood. Since the displacements are large close to interstitials they have to be calculated by lattice statics. Haubold[55] has identified the structure of single interstitials in Al and Cu and of di-interstitials in Al by this method. In order to interpret the experiments he used model calculations characterizing the defects by a set of Kanzaki forces and compared the resulting intensity distributions with the experimental results.

Scattering by defect clusters

Whenever the asymptotic displacement field is most important, the scattering from clusters can be calculated by considering a cluster as a new entity with a larger defect strength. For example, the influence of clustering of defects on the Huang scattering can easily be understood assuming a linear superposition of displacements due to the single defects forming a cluster. Then at large distances the apparent dipole force tensor of a cluster containing n defects is $P^{cl} = nP^{pd}$ (disregarding possible relaxation effects). Since the Huang intensity is proportional to the square of the dipole force tensor and the density of clusters is smaller by a factor of $1/n$ compared to the density of single defects, in total the Huang intensity increases linearly with the size of the clusters. For the case of dislocation loops Dederichs[50] obtained the same result. Thus Huang scattering is a very sensitive tool for monitoring clustering of defects. The analysis of the intensity distributions gives information about the symmetry of the clusters.

The most characteristic differences of the scattering due to clusters as compared to point defects are:

(i) a large increase of the Huang intensity per single defect linear with the size of the clusters

(ii) due to the larger distortions near the clusters the q^{-2} dependence characteristic of the long-range displacement field is observed in a smaller region $q \leqslant 1/R$ where R is approximately the linear dimension of a cluster.

(iii) For larger q a steeper decrease of the scattering intensity $\sim q^{-4}$ is found (Stokes–Wilson region). Here the intensity is only linearly dependent on the defect strength and in certain directions of reciprocal space shows radial oscillations characteristic of the size of the clusters.[57]

Mixture of defects

Complications result from a mixture of different defects and in particular from a size distribution of clusters. Vacancies and interstitials can be distinguished by their different contributions in different regions of reciprocal space (as discussed above). For size distributions of clusters several different moments can be determined in different regions of reciprocal space.[51] In practice, exponential-like size distributions are found[52] consistent with the results of electron microscopy studies[58] and used to determine the average cluster size.[54,22]

Positron annihilation

A detailed theory of the behaviour of positrons in a defect lattice is still missing and to a large extent phenomenological arguments are used for the interpretation of experiments. Nevertheless, the observations have given information about the presence of vacancy type defects and about changes of the agglomeration state. We shall shortly discuss the background and refer to some recent review articles by Goland,[59] Triftshäuser,[60] Seeger,[61] and in particular by West[62] for details.

Positrons which are injected from a source such as Na[22] into a metallic sample with high energy (0·5 MeV) are quickly thermalized. Due to their positive change they are repelled by the positive ion cores. Thus in a homogeneous metal they preferentially stay in the interstitial region. Eventually they annihilate with a metal electron, mostly by emission of two photons. The lifetime of a positron (typically about 10^{-10} s) is determined by the local electron density. Since energy and momentum are conserved during the annihilation process and the positron can be considered at rest, the analysis of the photon pair momentum distribution as measured by angular correlation or Doppler broadening yields information about the momentum distribution of the metal electrons. Both the conduction electrons with momentum smaller than the Fermi momentum and the tightly bound core electrons with a much broader momentum distribution contribute. The weight of the two components is determined by the spatial variation of the positron density over the lattice. Angular correlation measurements have successfully been used to determine Fermi surfaces.[63]

Positron traps

If a metal contains defects some of the positrons can be trapped. In the trap they see a different electronic environment and thus the annihilation characteristics are changed.

Theoretical calculations show that vacancies[64,65] are favourable traps for positrons in many metals (but not in alkali metals) with rather high binding energies which make detrapping during the lifetime of a positron improbable. Also voids were found to trap positrons[66] whereas single interstitials were found to repel positrons.[67] Generally one expects positrons to be trapped in regions of reduced nuclear density.

Change of lifetime

The positrons trapped in a vacancy see a reduced electron density and thus their lifetime is larger than in the perfect lattice. The weight of this long lifetime component in a measured lifetime spectrum is determined by the density of vacancies. If the vacancies agglomerate into voids or other uncollapsed clusters a further reduction of the electron density and thus an increase of the lifetime is expected. First calculations which correlate the lifetime of a trapped positron with the size of a void have recently been reported by Hautojärvi et al.[68]

Change of angular correlation and Doppler broadening

Because of the missing ion core in a vacancy the trapped positrons have less overlap with core electrons and relatively more annihilations will occur with conduction electrons having small momentum. Thus the angular correlation curve and the Doppler broadened annihilation line sharpen. The details of the changed curves are determined by the positron and electron density distribution in the trap and are thus characteristic of the nature of the trap. The measured curves are a sum of the contributions of trapped and free positrons. To obtain information about the traps an analysis is often made with the trapping model.[69-71]

Trapping model

In its simplest version the model supposes that a positron can annihilate from two distinct states; a free state in the perfect lattice and a trapped state at a defect. Escape from traps is neglected. The experimental value of a quantity F characteristic of the annihilation process which is a linear function of the positron state is given by an average over the annihilation probabilities in the different states:

$$F = F^f(1-\eta_t) + F^t\eta^t \quad \text{with} \quad \eta_t = \frac{\mu c}{\lambda_f + \mu c} \quad (14)$$

Here $F^{f,t}$ is the specific value of F in the free and trapped state, respectively. η_t is the fraction of positrons annihilating in the trapped state, λ_f the annihilation rate in the free lattice, and μc is the trapping rate of positrons for a concentration c of traps. For F one can insert for example the positron lifetime or a shape parameter of the angular correlation curve or the Doppler broadened photon line. For example, the peak count rate I_v around zero angle or zero energy shift which is characteristic of the valence or conduction electron contribution and the count rate I_c at the wings of the distribution which is characteristic of the core electron contribution have been used.

Relating the changes $\Delta F = F - F^f$ of two such quantities, e.g. ΔI_v and ΔI_c, one obtains a defect specific parameter

$$R = \left| \frac{I_v^t - I_v^f}{I_c^t - I_c^f} \right| \quad (15)$$

Since R does not depend on the concentration of traps any change of R indicates a change of the nature of the traps. This is, however, only true for the simple trapping model with only one type of traps present or predominating.

Interstitial clusters

Results for Al

Ehrhart and Schilling[22] have investigated the structure of defects in electron-irradiated Al and the annealing behaviour using Huang scattering. Immediately after irradiation at 4 K they found that most probably the stable configuration of single interstitials is the $\langle 100 \rangle$ dumbbell configuration. This was confirmed by Haubold[55] who measured X-ray diffuse scattering over a wide range in reciprocal space. The change of Huang scattering upon annealing indicates the change of defect density and configuration. Figure 6 (taken from ref. 53) shows the Huang intensity in electron-irradiated Al close to the (400)-reflection for a q parallel $\langle 100 \rangle$ and

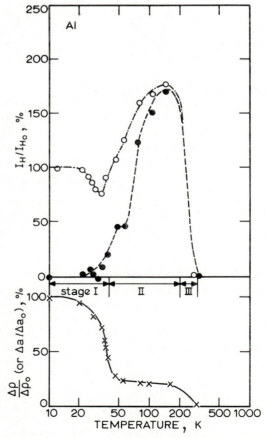

6   Isochronal recovery of electrical resistivity (points ×, lower figure) and of Huang intensity of electron-irradiated Al;[22] measurements are shown close to the (400)-reflection for $I_H^{\parallel}(q\|100$, points $\bigcirc$) and $I_H^{\perp}(q\|011$, points ●)

⟨011⟩ as a function of annealing temperature. For comparison also the electrical resistivity is shown. In the early part of stage I (4–28 K) the Huang intensity $I_H^\parallel$ for **q**∥**h** essentially decreases proportional to the electrical resistivity. The intensity $I_H^\perp$ for **q**⊥**h** remains zero as it is immediately after irradiation. This supports the interpretation of close pair recovery in this temperature range. At the end of stage I (28–40 K) the angular distribution and the magnitude of the intensity change significantly. $I_H^\parallel$ first decreases more slowly than the resistivity and then increases again. In the same temperature range $I_H^\perp$ starts to deviate from zero which indicates a change of symmetry of the predominant defect. A detailed analysis suggests the formation of predominantly di-interstitials with ⟨110⟩ preferred orientation. This result has been confirmed by measurements over a large range of reciprocal space as shown in Fig. 7 (from ref. 53). The experimental results are compared to model calculations for two possible di-interstitial configurations. As can be seen the best agreement is obtained for the configuration of two parallel ⟨100⟩ dumbbell interstitials on nearest neighbour positions which has been found to be the most stable one by computer simulation.[13,24] Upon further annealing the interstitial clusters grow to sizes of about 10 interstitials (assuming all clusters to be of the same size). At this stage the characteristic q^{-4} dependence of the intensity is found for larger q. The angular distribution in the Huang region is consistent with the formation of loops on {111} or {110} planes.

Huang scattering gives information about the average size and symmetry of clusters. In contrast, mechanical relaxation experiments can yield information about specific types of defects as characterized by relaxation times of reorientation processes under external stress.[73] Figure 8a shows the spectrum of reorientation processes appearing in electron-irradiated Al as measured by Robrock et al.[20,21] The processes are indicated at the temperature at which the characteristic relaxation times are 100 s and are labelled according to convention. Figure 8b shows the annealing behaviour of the relaxation strength of some of the processes which is a measure of the density of the participating defects. Process 1 has been identified as the reorientation of Frenkel pairs in a close pair configuration and process A as the combined reorientation and migration of the ⟨100⟩ dumbbell.[22] All

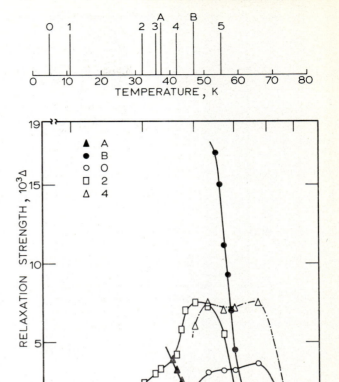

8 *a* Relaxation spectrum of reorientation processes in electron-irradiated Al;[20,21] the processes are indicated at the temperature at which the respective relaxation time is 100 s.; *b* annealing behaviour of the relaxation strength of some processes shown in *a*; process A disappears at the end of stage I where the other processes appear

other processes (0, 2–5, B) appear upon annealing through stage I where process A disappears (*see* Fig. 8b). Particularly interesting are the processes B and 2. From the symmetry and annealing behaviour it was concluded that process B is due to simultaneous reorientation and migration of the di-interstitials formed by two parallel ⟨100⟩ dumbbells on nearest neighbour sites. The process disappears at about 50K in stage II₁ indicating that di-interstitials are only slightly less mobile than single interstitials in Al. All other processes are due to reorientation without long-range migration. Process 2 has been tentatively assigned to tri-interstitials. It also disappears in stage II₁ whereas the other processes are stable to higher annealing temperatures.

Results for Cu

A different picture seems to apply to Cu.[23] In Fig. 9 (taken from ref. 53) the Huang intensity is shown for the (220)-reflection in Cu for q parallel to ⟨110⟩ and to ⟨1̄10⟩ in the same plot as for Al (note the difference in scale!). The steep rise of the intensity at the end of stage I yielding clusters about 5–10 interstitials can be understood if the smaller clusters are at least as mobile as single interstitials. Then after being formed they migrate and immediately form larger clusters. This would also

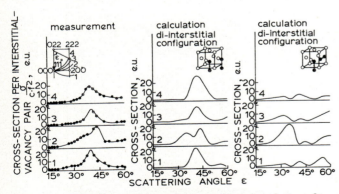

7 Comparison of the diffuse scattering intensity from Al electron irradiated at 4·5 K($\Delta\rho_0$ = 190 nΩcm) and annealed to 40 K with model calculations for two di-interstitial models;[53] intensities are shown along four Ewald circles indicated in the insert

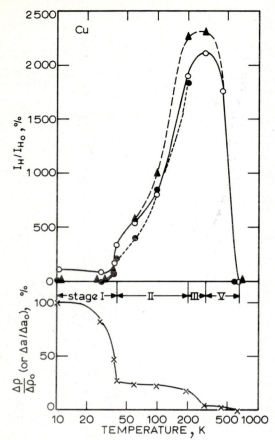

9   Isochronal recovery of electrical resistivity $\Delta\rho_0$ (points ×, lower figure) and of Huang intensity of electron-irradiated Cu;[23] measurements are shown close to the (220) reflection for $I_H^{\parallel}$ ($q \parallel 110$, points ○) and $I_H^{\perp}$ for two directions ($q \parallel 110$, points ● and $q \parallel 001$, points ▲)

explain the absence of mechanical relaxation peaks after stage I annealing in Cu. Upon further annealing the clusters grow to very large sizes in Cu. They contain about 500 interstitials at 300 K (assuming no distribution of sizes). These large clusters are most probably dislocation loops on {111} planes[23] which have also been found by electron microscopy.[58]

Other metals

Huang scattering has also been measured in Au,[41] Mo,[42] and Zn.[42] Clusters were found in Au immediately after low-temperature irradiation and in Mo and Zn after annealing through stage I. No detailed analysis was made.

Vacancy clusters

Di-Vacancies in Pt

The only experiments in which di-vacancies have been positively identified are the field ion microscopy studies of Berger et al.[46] on quenched platinum. They have identified a few di-vacancies on nearest neighbour sites after quenching from 1 700°C. From the ratio of the di-vacancy concentration to mono-vacancy concentration they calculated a binding free energy of 0·23 eV. The preferential loss of di-vacancies in samples exposed

to temperatures around 350°C after quenching was attributed to a larger mobility of di-vacancies compared to mono-vacancies. No quantitative result could be obtained for the mobilities.

Vacancy clusters in Mo

Eldrup et al.[48] investigated vacancy type defects in electron-irradiated (at 50°C) Mo by positron lifetime measurements. They found a long lifetime component attributed to positrons trapped at vacancies. Figure 10 shows the change of this long lifetime upon annealing through stage III (200°–400°C) and for higher temperatures compared to the recovery of radiation-induced resistivity. The large increase of the lifetime in stage III where the resistivity drops to about 20% of its original value indicates a drastic change of the type of traps. These new traps must have a much reduced electron density as is expected for small voids or other uncollapsed vacancy clusters. No size estimate was given by the authors but recently Hautojärvi et al.[68] have correlated the lifetime with the void size. Their analysis of the data of Eldrup et al. yields about 15 vacancies per void after stage III. Such a large number can only be obtained if the smaller vacancy clusters are also mobile; it can be shown[74] that homogeneous nucleation assuming only mono-vacancies to be mobile results in a size distribution which is strongly peaked at small sizes even if the capture radius of the larger clusters increases linearly with size. The formation of vacancy clusters is also seen by angular correlation which shows a considerable narrowing during stage III annealing.[48] After stage III no clusters could be seen in the electron microscope but after further annealing to 900°C which resulted in further increase of the lifetime (see Fig. 10) voids with diameters of about 3 nm were observed.

Vacancy clusters in Cu

In Cu vacancy cluster formation was investigated by Haubold[47,55] measuring X-ray diffuse scattering in forward direction and Mantl[49] measuring Doppler broadening of the positron annihilation line. Figure 11a (after ref. 55) shows the scattering intensity per defect close to the forward direction in a so-called Guinier plot

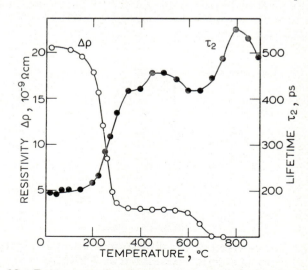

10   Recovery of radiation-induced electrical resistivity (points ○) and behaviour of long positron lifetime, τ_2, (points ●) of electron-irradiated Mo[48]

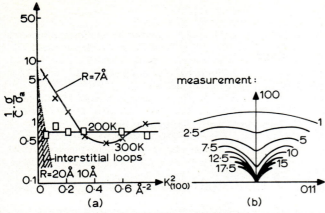

11 Small-angle scattering of electron-irradiated Cu in a Guinier plot; shown are measurements[55] before stage III (200 K) and afterwards (300 K); for comparison the expected cross-sections for interstitial loops detected by Huang scattering are also shown

(log of intensity v. square of scattering vector) for different annealing temperatures. Before stage III, at 200 K, the scattering cross-section is constant and has a value of about 0·7 a.u. as one would expect for a relaxed vacancy. After stage III, at 300 K a strong increase at small angles is observed indicating the formation of small vacancy clusters. The analysis shows that the clusters have a radius of gyration of 0·7 nm and contain about 10–20 vacancies. The anisotropy of the intensity distribution is shown in Fig. 11*b* (after ref. 47). It is consistent with agglomeration of vacancies on {111} planes. The absolute intensity at larger angles suggests that these agglomerates are not loops but some kind of uncollapsed platelets or possibly stacking-fault tetrahedra.[76] Figure 12 (from ref. 60) shows the annealing behaviour of the defect specific parameter R (defined on p. 58) as measured by Doppler broadening[49] in electron-irradiated Cu($\Delta\rho_0 = 65$ nΩ cm). Before stage III the value of R agrees well with that obtained for single vacancies in thermal equilibrium.[75] During stage III, R *increases* indicating a change in the nature of traps. In the new traps the positron has less overlap with core electrons as one would expect for small uncollapsed vacancy agglomerates.

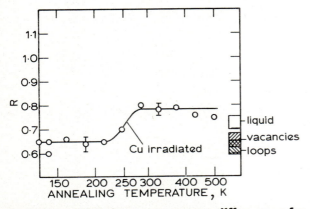

12 Ratio of lineshape parameter differences for electron-irradiated Cu versus annealing temperature;[49] indicated are values of R for dislocation loops, vacancies in thermal equilibrium, and trapping sites in liquid Cu

CONCLUSION

We have shown that there is a sound basis for the discussion of structure, stability, and mobility of small clusters as well as for the kinetics of cluster growth. Theoretical and experimental evidence suggest that small aggregates of interstitials and vacancies have to be considered mobile with jump frequencies comparable to the respective single defects. This has to be considered in the interpretation of high-temperature irradiation studies.

In a few fortunate cases the identity of clusters has been disclosed and all parameters could be determined, e.g. for the di-interstitial in Al. The question of the first immobile nuclei for the formation of larger clusters is still unresolved.

ACKNOWLEDGMENTS

The author has benefitted from many discussions with colleagues. In particular he wishes to thank P. H. Dederichs, P. Ehrhart, H.-G. Haubold, S. Mantl, K.-H. Robrock, H. Schober, and H. Trinkaus for discussions on their work and for making unpublished results available to him. Many thanks are due to W. Schilling and F. W. Young, Jr. for their continuous encouragement.

REFERENCES

1 A. SEEGER AND H. MEHRER: 'Vacancies and interstitials in metals' (Eds. A. SEEGER *et al.*), 1970, Amsterdam, North-Holland
2 N. L. PETERSON: 'Diffusion in solids, recent developments' (Eds. A. S. NOWICK AND J. J. BURTON), 115, 1975, New York, Academic Press
3 J. W. CORBETT *et. al.*: *Phys. Rev.*, 1959, **114**, 1452, 1460
4 W. SCHILLING *et al.*: ref. 1, p. 255
5 H. KANZAKI: *J. Physics Chem. Solids*, 1957, **2**, 107
6 H. R. SCHOBER *et al.*: *Physica Status Solidi (b)*, 1974, **64**, 173
7 J. D. ESHELBY: *Solid State Phys.*, 1956, **3**, 79
8 R. SIEMS: *Physica Status Solidi*, 1968, **30**, 645
9 J. R. HARDY AND R. BULLOUGH: *Phil Mag.*, 1967, **15**, 237
10 H. TRINKAUS: 'Fundamental aspects of radiation damage in metals', Vol. 1, 254, 1975, ERDA CONF-75 1006
11 P. T. HEALD: this volume
12 R. A. JOHNSON: *J. Physics F (Metal Physics)*, 1973, **3**, 295
13 H. R. SCHOBER: *J. Physics F*, to be published
14 A. DE PINO, JR *et al.*: *Radiat. Effects*, 1970, **3**, 23
15 R. BULLOUGH AND R. C. PERRIN: *Proc. Roy. Soc.*, 1968, **A305**, 541
16 E. J. SAVINO AND R. C. PERRIN: *J. Physics F (Metal Physics)*, 1974, **4**, 188
17 R. A. JOHNSON AND E. BROWN: *Phys. Rev.* 1962, **127**, 446
18 R. A. JOHNSON: *Phys. Rev.*, 1964, **134**, A1329
19 P. H. DEDERICHS: this volume
20 V. SPIRIC *et al.*: *Phys. Rev. B*, 1977, **B15**, 672
21 K. H. ROBROCK *et al.*, *Phys. Rev. B*, 1977, **B15**, 680
22 P. EHRHART AND W. SCHILLING: *Phys. Rev. B* 1973, **8**, 2604
23 P. EHRHART AND K. SCHLAGHECK: *J. Physics F (Metal Physics)*, 1974, **4**, 1589
24 R. A. JOHNSON: *Phys. Rev.*, 1966, **152**, 629
25 M. DOYAMA AND R. M. J. COTTERILL: 'Lattice defects and their interaction', (Ed. R. R. HASIGUTI), 79, 1967 New York, Gordon and Breach
26 J. R. BEELER, JR AND R. A. JOHNSON: *Phys. Rev.*, 1967, **156**, 677
27 M. V. SMOLUCHOWSKI: *Z. physikal Chem. (Leipzig)*, 1917, **92**, 129

28 T. R. WAITE: *Phys. Rev.*, 1957, **107**, 463, 471
29 K. SCHROEDER AND E. EBERLEIN: *Z. Physik B*, 1975, **22**, 181
30 T. R. WAITE: *J. Chem. Phys.*, 1958, **28**, 103
31 K. SCHROEDER AND K. DETTMANN: *Z. Physik B*, 1975, **22**, 343
32 H. WIEDERSICH: ASM Materials Science Seminar: Radiation Damage in Metals, Proc. Seminar held Nov. 9–10, 1975, Cincinnati, Ohio, to be published
33 R. W. BALLUFFI: in ref. 10, Vol. 2, 852
34 J. S. KOEHLER: *Phys. Rev.*, 1969, **181**, 1015
35 C. P. FLYNN: 'Point defects and diffusion', 411 ff, 1972, Oxford, Clarendon Press
36 H. JOHNSON: *Scripta Met.*, 1970, **4**, 771
37 M. PROFANT AND H. WOLLENBERGER: *Physica Status Solidi (b)*, 1975, **71**, 515
38 D. BECKER *et al.*: *Physica Status Solidi (b)*, 1972, **54**, 455
39 P. EHRHART AND U. SCHLAGHECK: in ref. 10, Vol. 2, 839
40 B. VON GUERARD AND J. PEISL: in ref. 10, Vol. 1, 287, 309
41 P. EHRHART AND F. SEGURA, in ref. 10. Vol. 1, 295
42 P. EHRHART: in ref. 10, Vol. 1, 302
43 G. VOGL AND W. MANSEL: in ref. 10, Vol. 1, 349
44 M. L. SWANSON *et al.*: in ref. 10, Vol. 1, 316
45 H. WOLLENBERGER: in ref. 10, Vol. 1, 582
46 A. S. BERGER *et al.*: *Acta Met.*, 1973, **21**, 123, 137; *see also* D. N. SEIDMAN; *J. Physics F (Metal Physics)*, 1973, **3**, 393
47 H. G. HAUBOLD: ERDA Symposium on 'X-ray and gamma-ray sources and applications', May 19–21, 1976, Ann Arbor, Mich. USA, to be published
48 M. ELDRUP, *ET al.*: *J. Physics F (Metal Physics)*, 1976, **6**, 499; *see also* ref. 10, Vol. 2, 1127
49 S. MANTL AND W. TRIFTSHÄUSER; *Phys. Rev. Lett.*, 1975, **34**, 1554; *see also* ref. 10, Vol. 2, 1122
50 P. H. DEDERICHS: *J. Physics F (Metal Physics)*, 1973, **3**, 471
51 H. TRINKAUS: *Physica Status Solidi (b)*, 1972, **54**, 209
52 B. C. LARSON: *J. Appl. Cryst.*, 1975, **8**, 150
53 P. EHRHART, *et al.*: 'Festkörperprobleme' (Advances in solid state physics), (Ed. H.-J. QUEISSER), Vol. XIV, 87, 1974, Braunschweig, Pergamon/Vieweg
54 B. C. LARSON: in ref. 10, 820
55 H. G. HAUBOLD: in ref. 10, 268
56 W. SCHMATZ: 'Treatise of materials science and technology', Vol. 2, 105, 1973, New York, Academic Press
57 H. TRINKAUS: *Z. angew. Physik*, 1971, **31**, 229
58 B. L. EYRE: *J. Physics F (Metal Physics)*, 1973, **3**, 422
59 A. N. GOLAND: in ref. 10, Vol. 2, 1107
60 W. TRIFTSHÄUSER: 'Festkörperprobleme' (Advances in solid state physics), Vol. XV, 381 (Ed. H. QUEISSER), 1975, Braunschweig, Pergamon/Vieweg
61 A. SEEGER: *J. Physics F (Metal Physics)*, 1973, **3**, 248–94
62 R. N. WEST: 'Positron studies in condensed matter', 1974, London, Taylor and Francis Ltd., a reprint of *Adv. Phys.*, 1973, **22**, 263
63 K. FUJIWARA AND O. SUEOKA: *J. Phys. Soc. Japan*, 1966, **29**, 1479
64 C. H. HODGES: *Phys. Rev. Lett.*, 1970, **28**, 284
65 J. ARPONEN *et al*: *Solid-State Commun.*, 1973, **12**, 143
66 C. H. HODGES AND M. J. STOTT: *Solid-State Commun.*, 1973, **12**, 1154
67 M. J. STOTT AND P. KUBICA: *Phys. Rev.*, 1975, **B11**, 1
68 P. HAUTOJÄRVI *et al.*: 4th Int. Conf. Positron annihilation, Helsingør, Denmark, 23–26. Aug. 1976, Vol. 2, 88
69 W. BRAND: Proc. Int. Conf. on 'Positron annihilation', (Eds. A. T. STEWART AND L. O. ROELLIG) 80, 1967, New York, Academic Press
70 B. BERGERSEN AND M. J. STOTT: Solid-State Commun., 1969, **7**, 1203
71 D. C. CONNORS AND R. N. WEST: *Phys. Lett.*, 1969, **30A**, 24
72 W. TRIFTSHÄUSER: *Phys. Rev.*, 1975, **B12**, 4634
73 A. S. NOWICK AND B. S. BERRY, 'Anelastic relaxation in crystalline solids', 1972, New York, Academic Press
74 K. SCHROEDER: *Radiat. Effects*, 1973, **17**, 103
75 W. TRIFTSHÄUSER AND J. D. MCGERVEY: *Appl. Phys.*, 1975, **6**, 177
76 H.-G. HAUBOLD AND D. MARTINSEN: to be published in Proc. Int. Conf. on 'Properties of atomic defects in metals', Argonne Nat. Lab., Argonne, Ill. USA 18–22 Oct. 1976
77 W. G. WOLFER AND M. ASHKIN: *J. Appl. Phys.*, 1975, **46**, 1975

Electron microscopy studies of point defect clusters in metals

B. L. Eyre, M. H. Loretto, and R. E. Smallman

Transmission electron microscopy (TEM) has been widely used to study point defect clusters in metals following quenching and irradiation. As a result of this work considerable progress has been made in our understanding of the nucleation, morphology, and growth of clusters and this paper reviews the current state of this understanding. The cubic metals (fcc and bcc) have been studied most widely. Cluster geometry and how it is affected by a wide range of variables such as the temperature at which clustering occurs has been well characterized. Detailed quantitative studies have also been made of cluster nucleation and growth kinetics and a self-consistent picture has been developed of the physical processes involved which describes the behaviour of quenched-in and radiation-induced defects. It has been shown that fundamental differences exist between fcc and bcc metals particularly with regard to the formation of vacancy clusters. In contrast, comparatively little work has been carried out on hcp metals and a clear picture has not yet emerged of the factors influencing cluster geometry. The results indicate that considerable differences exist on going from one metal to another within the general class of hexagonal metals. Moreover, virtually no quantitative results have been reported on the nucleation and growth of clusters in hcp metals.

B. L. Eyre is at AERE, Harwell, and M. H. Loretto and R. E. Smallman are at the University of Birmingham

The physical processes involved in the clustering of vacancy and interstitial point defects in metals and alloys have received widespread attention over the last 20–30 years, both because of their importance in achieving a fundamental understanding of crystal defects and because of their practical relevance to the problem of irradiation damage in reactor materials. The technique of transmission microscopy has played a prominent role in this work (*see* refs. 1–4 for earlier reviews) and a major factor underlying the interpretation of observations has been the establishment of a sound theoretical understanding of diffraction contrast images from a wide range of defects, ranging down to just resolvable clusters of 10–15 Å diameter. Thus, it is now possible under favourable conditions to determine with precision, number densities, size spectra, and geometries of cluster populations in thin foil specimens. The coupling of this capability with well designed systematic experiments has enabled considerable advances to be made in our understanding of point defect clustering processes. The purpose of this paper is to summarize the current state of this understanding which has come mostly from studies of quenched or irradiated metals.

In the first part of the paper we discuss the clustering of interstitials in the main classes of metals as defined by crystal structure, i.e. fcc, bcc and hcp, before going on to discuss the clustering of vacancies in these same groups of metals. The studies of interstitial clusters have mostly been made on irradiated metals and the picture regarding their behaviour is comparatively simple in fcc and bcc metals. The interstitials aggregate to form faulted loops and the main variations in these two classes of metals are related to the scale on which the loops are nucleated and the stage at which they unfault. Important factors influencing these aspects of interstitial behaviour are impurity content and stacking fault energy. In irradiated hcp metals interstitial point defects again aggregate to form loops, but their geometry is considerably more complex than in the cubic metals and the greatest emphasis has been placed on the study of this aspect. Considerably more work has been carried out on the · clustering of vacancies in both quenched and irradiated metals. However, in contrast to interstitial clustering, the picture is complex with regard to both vacancy cluster geometry and the parameters influencing cluster nucleation and growth.

INTERSTITIAL CLUSTERS
Before discussing the experimental observations of interstitial clusters in metals and alloys, it is worth

summarizing the important physical factors governing cluster nucleation and growth. First, self interstitials cause considerable distortion of the surrounding lattice and calculated formation energies are very high–about 4–5 eV for many metals. Secondly, there is a body of experimental evidence to show that the migration energy for self interstitials is low, i.e. typically 0·1– 0·2 eV, and they are, therefore, mobile at temperatures down to < 100 K in most metals. Virtually all of the firm evidence concerning the behaviour of self interstitials has come from irradiated metals. Moreover, most electron microscopy studies have been carried out on specimens irradiated at temperatures $\geqslant 20°C$, and at such temperatures the interstitial point defects surviving recombination aggregate extremely rapidly to form clusters. Again, because of the large associated distortion, three-dimensional clusters are energetically unfavourable and the evidence shows that generally interstitials aggregate on close-packed planes to form dislocation loops. In hcp metals dislocation loops may also form on high index planes.

With regard to the nucleation of such loops, theoretical treatments have, in general, assumed that the minimum stable loop nucleus is a di-interstitial.[5-7] In pure crystals these are assumed to form homogeneously and on completion of the nucleation stage the concentration of loops, N_L^i, as a function of temperature, T K, and defect production rate, K, is given by:

$$N_L^i = N_0^i K^{\frac{1}{2}} \exp(E_m^i / 2KT) \qquad (1)$$

where E_m^i is the activation energy for interstitial migration and N_0^i is a constant containing the necessary geometrical factors representing the capture cross-sections for loss of interstitials by recombination at fixed sinks such as grain boundaries and dislocation lines and at the loop embryos themselves. In the case of crystals containing a concentration C_i of impurities which trap interstitials with a binding energy E_b, equation (1) must be modified to take account of the reduced interstitial mobility:[6]

$$N_L^i = N_0^i K^{\frac{1}{2}} C_i^{\frac{1}{2}} \exp\frac{E_m^i + E_b}{2kT} \qquad (2)$$

where the exponential term represents the release rate of interstitials from impurity traps.

With regard to the change in interstitial loop radii during irradiation, two factors stemming from the properties of interstitial point defects have an important influence. First, the elastic size effect interaction causes dislocations to attract interstitials more strongly than vacancies.[8,9] Secondly, because the formation energy of self interstitials is greater than that for vacancies, i.e. $E_f^i \gg E_f^v$, the dominant process at elevated temperatures is vacancy emission, and in the case of interstitial loops, this leads to growth. The importance of these factors to loop stability during irradiation is demonstrated by the rate equation defining the point-defect diffusion-controlled changes in loop radius:

$$\left[\frac{dr_L}{dt}\right]_i = \frac{1}{\mathbf{b}}\left\{Z_i D_i c_i - D_v c_v + D_v c_v^e \exp - \left[\frac{(F_{el}+\gamma)\mathbf{b}^2}{kT}\right]\right\} \qquad (3)$$

where c_i and c_v are the interstitial and vacancy concentrations respectively, D_i and D_v are their diffusivities, F_{el} is the elastic energy of the loop, $\mathbf{b}$ its Burgers vector, γ is its stacking fault energy if faulted, and z_i is a bias term

defining the preferred attraction of the loops for interstitials. Values for c_i and c_v are obtained by solving rate equations containing the point defect generation rate, loss rates at sinks, and recombination rates.[10] The important point demonstrated by equation (3) is that interstitial loops are intrinsically stable defects which, in the presence of neutral sinks, can undergo continuous growth during irradiation. At low temperatures, this growth is bias-driven, i.e. $Z_i D_i c_i > D_v c_v$, whereas at higher temperatures, when $D_v c_v^e$ becomes significant, growth by vacancy emission makes an increasingly important contribution.

Interstitial loops in fcc metals

Interstitial loops have been observed in a range of irradiated fcc metals and alloys including Cu, Ag, Au, Ni, Al, and various austenitic alloys. Whereas the geometry of the loops has been well characterized, comparatively little systematic work has been carried out on their nucleation and growth. In all of the cases studied, the interstitials aggregate initially on close-packed planes to form faulted Frank loops with $\mathbf{b} = \frac{a}{3}\langle 111 \rangle$.

Interstitial loop nucleation has been studied in a number of fcc metals using a high-voltage electron microscope (HVEM) to both irradiate specimens and to continuously observe loop formation during irradiation.[11,12] By fitting the temperature dependence of N_L for Al and Au to an Arrhenius plot, values have been obtained for the effective activation energy for interstitial migration.[11] The following interesting points emerge from these results:

(i) for low-temperature irradiation, i.e. $T < 300$ K, at which vacancies are comparatively immobile, N_L quickly saturates with dose

(ii) values for activation energies obtained from Arrhenius plots of the data over this temperature range are summarized in column 1 of Table 1

(iii) over this temperature range, the absolute values of N_L^i are sensitive to small changes in impurity level, but the temperature dependence is not changed; thus, it is deduced[11] that the activation energies given in column 1 of Table 1 represent $(E_m^i + E_b)$

(iv) at very low temperatures, < 100 K, the slopes of the Arrhenius plots change to lower values and it is considered that the interstitials mostly combine to form di-interstitials without encountering impurity traps; thus, the activation energies derived from the plots (presented in column 2 of Table 1) are probably closer to that for interstitial migration

TABLE 1   Activation energies for interstitial loop nucleation

Metal	Intermediate temperature $(E_m^i + E_b)$, eV	Low temperature E_m^i, eV
Au	0·19	0·04
Al	0·08	0·03
Fe	0·26	—
Mo	0·18	0·05
W	0·21	—

(v) at elevated temperatures, >300 K, when the vacancies are very mobile, interstitial loops were observed to nucleate continuously during irradiation.

Very little quantitative evidence has been obtained regarding the temperature dependence of interstitial loop numbers in commercial alloys of the type used for reactor core components. The few results that have been reported for ion- and neutron-irradiated stainless steels at temperatures in the void swelling regime (400°–600°C) show a much steeper dependence of loop numbers on temperature[13–16] than was found for the comparatively pure fcc metals electron irradiated at low temperatures. Arrhenius plots of the data (plotted assuming a bimolecular nucleation model) yield activation energies well in excess of 1.0 eV, implying strong binding to solute atoms.[17] The scale on which interstitial loops nucleate has an important influence on void formation and growth[18] and thus there is a strong practical motivation for more systematic studies of loop nucleation in commercial alloys.

A large body of evidence has been obtained to show that interstitial loops continue to grow in fcc metals during irradiation and this is consistent with them having a preferred interaction for interstitial point defects. In most fcc metals the loops are observed to remain faulted until they intersect with other loops and unfault to $\mathbf{b} = \frac{a}{2}\langle 110 \rangle$. An exception to this general behaviour was found in aluminium foils irradiated with α-particles in which a majority of the loops are perfect at an early stage in their development.[19] Once the loops unfault, they are frequently observed to be rhombohedral in shape. With increasing growth and intersections, the loops are eventually incorporated into a network.

Direct experimental studies have been made of interstitial loop growth in Al, Au, and Ni by carrying out *in situ* irradiation in an HVEM.[11,20,21] In the interpretation of the results from such experiments account has to be taken of the possible influence the free surfaces have on point defect survival and therefore on loop growth. The simplest situuation is when the vacancies are comparatively immobile, i.e. at low temperatures, when loop radius increases with $(\text{time})^{1/3}$. At high temperatures when both vacancies and interstitials are very mobile, loop radius increases linearly with time. In thick foils, the loop growth is predicted to be dependent on the square root of vacancy diffusivity, i.e. $dr_L/dt \propto (D_v)^{\frac{1}{2}}$ and thus the rate is expected to be temperature-dependent. This is observed experimentally over a restricted temperature range. For example, the observed growth rates in Ni agree with the prediction for a thick foil geometry up to 750 K.[22] For the situation when the foil surfaces act as a dominant sink for the point defects (the so-called thin foil case) loop growth, although still linearly dependent on time, is now independent of vacancy mobility and therefore of temperature. The foil thickness for which the thin foil result applies increases with increasing irradiation temperature.[11] In all of the cases studied, loop growth was bias-driven (vacancy emission negligible). Using loop growth data from Ni,[21] a quantitative estimate has been made of the preference the loops have for interstitials compared to vacancies.[23] This is expressed as a ratio of the point defect capture radius, r_c, of a dislocation for interstitial and vacancies, i.e. $r_c^i/r_c^v = 5.5 \pm 1.0$.

Interstitial loops in bcc metals

Direct studies have been made of interstitial loop nucleation in bcc metals using the same approach as for fcc metals.[11] Estimates of $E_m^i + E_b$ and E_m^i obtained from Arrhenius plots of the data for Fe, Mo, and W are included in Table 1 and it can be seen that the values are similar to those for fcc metals.

Loop geometry is an important feature distinguishing bcc and fcc metals. It has been predicted[23] that interstitials aggregate initially to form faulted $\frac{a}{2}\langle 110 \rangle$ loops on $\{110\}$ in bcc metals. Because of the high stacking fault energy, such loops unfault at an early stage to $\frac{a}{2}\langle 111 \rangle$ or $a\langle 100 \rangle$. Of the refractory metals, molybdenum has been studied the most widely and it is observed that when the loops have grown to a visible size (~ 20 Å diameter) they are predominantly perfect with $\mathbf{b} = \frac{a}{2}\langle 111 \rangle$. In α-iron, on the other hand, large rectilinear $a\langle 100 \rangle$ loops have been observed following irradiation with electrons or self ions at elevated temperatures.[24,25]

The glissile nature of interstitial loops in bcc metals enables them to aggregate and coalesce by a process of glide and self-climb. The driving force for this process is the elastic interaction between adjacent loops,[26] which is a function of irradiation dose and temperature. An important material variable is the impurity content because of its effect on loop glide stress.[27] If the critical conditions for loop coalescence by glide and self-climb are not satisfied, the loops grow by point defect diffusion controlled processes (bias-driven interstitial flow or thermal vacancy emission) in the same way as for fcc metals. However, if the conditions for loop coalescence are satisfied, it is the dominant loop growth process and it can lead to rapid development of a coarse dislocation network. An important feature of the loop coalescence mechanism is that the rate-limiting process is pipe diffusion around the loop perimeters and it can therefore occur at temperatures appreciably below that for self-diffusion. Results from neutron-irradiated molybdenum show that for low irradiation temperatures, e.g. $< 0.15 T_m$, loops having identical values of $\mathbf{b}$ glide together to form closely spaced loops commonly referred to as 'rafts'.[28,29] The critical irradiation temperature for loops to migrate sufficiently for coalescence to occur by self-climb appears to be $\sim 0.2 T_m$ and at temperatures in the range $0.2–0.35 T_m$, coarse dislocation networks develop at low doses, i.e. $\leqslant 10^{19}$ n/cm^{-2}.[30] In contrast, the development of a dislocation network in fcc metals containing a distribution of interstitial loops growing by point defect diffusion controlled process is much slower. Moreover, the network spacing is more directly related to initial loop spacing in fcc metals and it is therefore in general considerably less than in bcc metals at temperatures $< 0.35 T_m$. This fundamental difference in behaviour between fcc and bcc metals can have important consequences on void formation and growth.

Interstitial loops in hcp metals

The experimental study of interstitial loops in hcp metals has been dominated by the question of geometry, which is considerably more complicated than for fcc and bcc metals. There are no published quantitative data elucidating loop nucleation and growth kinetics during irradiation. The main features of the experimental observations that have been obtained from a range of hexagonal metals can be summarized as follows.

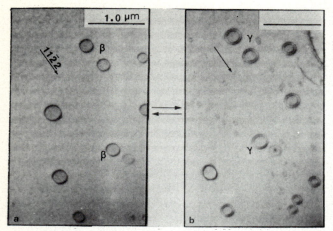

1   Micrographs of interstitial loops of b = [0001] in zinc imaged: *a* below the threshold voltage at 200 keV and *b* above the threshold voltage at 400 kV; the loops (e.g. those labelled β) split into concentric loops (e.g. labelled γ) as they grow (S. Karim, unpublished work)

1 Zirconium has been studied following electron,[31,32] heavy ion, and neutron irradiation.[34] Only interstitial loops were identified in specimens irradiated with Kr^+ followed by an anneal at about 600°C. On the other hand, neutron-irradiated specimens post-irradiation annealed at 500°C contained a 2:1 mixed population of vacancy and interstitial loops which all had Burgers vectors lying in the basal plane of the type $\mathbf{b} = \frac{1}{3}\langle 11\bar{2}0 \rangle$. To complicate the situation further other workers have observed loops (of unspecified nature) having Burgers vectors not contained in the basal plane in Zr following electron irradiation at 300°C and ion irradiation.

2 Only vacancy loops have been positively identified[35] in titanium following neutron irradiation and we shall return to discuss these observations further in the section on p. 76.

3 Limited studies of pure magnesium have shown that loops of the type $\mathbf{b} = \frac{1}{3}\langle 11\bar{2}0 \rangle^*$ are formed on non-basal planes during electron irradiation. However, the behaviour in electron-irradiated Mg–0·2% Mn alloy is considerably more complex with interstitial loops having $\mathbf{b} = \frac{1}{2}[0001]$ and $\frac{1}{3}\langle 11\bar{2}0 \rangle$ forming on the basal and $\{10\bar{1}1\}$ planes at low doses. At high doses these two families of loops are apparently replaced by vacancy loops having $\mathbf{b} = [0001]$ and $\frac{1}{6}\langle 20\bar{2}3 \rangle$.

4 Only interstitial loops have been observed to form in Zn and Cd during electron irradiation.[36,38,39] At low doses these loops are all observed to have $\mathbf{b} = \frac{1}{2}[0001]$ and subsequently transform to loops having $\mathbf{b} = [0001]$ and $\frac{1}{6}\langle 20\bar{2}3 \rangle$ at high doses. Detailed analysis of the growth behaviour of the [0001] loops shows that they are of two types: those which split into two concentric $\frac{1}{2}[0001]$ loops during high flux irradiation and those which remain apparently as [0001] loops as they grow. Micrographs illustrating examples of these two types of behaviour are presented in Figs. 1*a* and *b*. At low dose

* It is emphasized that loops having $\mathbf{b} = \frac{1}{3}\langle 11\bar{2}0 \rangle$ cannot be nucleated by the condensation of vacancies or interstitials on the most densely packed planes, i.e. (0001), and thus the observation of such loops represents a fundamental difference between hcp metals on the one hand and fcc and bcc metals on the other.

rates and low temperatures all the [0001] loops facet along $\langle 11\bar{2}0 \rangle$ directions.[36,38]

It is clear from this brief summary that the behaviour of hcp metals during irradiation is complex and that there is no obvious correlation between the types of dislocation loop observed and the dimensional instability during irradiation of Zr, Zn, Cd, and Ti; no dimensional changes are observed in Be and Mg. Equally there is no obvious correlation with c/a ratio although the two metals Cd and Zn which have very large c/a appear to behave identically. It is worth emphasizing that the purity of the materials studied has not been closely controlled and it is clear from the observations on fcc and bcc metals that this is an extremely important variable.

VACANCY CLUSTERS

Studies have been made of the formation and growth of vacancy clusters in metals following quenching or irradiation. In contrast to interstitials, vacancies can aggregate to form planar defects or voids. Before discussing the experimental observations, it is again worth summarizing the general physical framework that has been developed for describing the formation and growth of such clusters.

Considering first the case of vacancy loops or voids formed from a quenched-in supersaturation of vacancies, defined by c_v/c_v^e, where c_v is the actual vacancy concentration and c_v^e is the chemical equilibrium concentration, the rate equations defining the dependent change in loop or void radii can be written as follows:

$$\frac{dr_L}{dt} = \frac{D_s^v}{b}\left[\frac{c_v}{c_v^e} - \exp\frac{(F_{el}+\gamma)b^2}{kT}\right] \quad (4a)$$

$$\frac{dr_c}{dt} = \frac{D_s^v[1+(\rho_d r_c)^{\frac{1}{2}}]}{r_c}\left[\frac{c_v}{c_v^e}\right.$$

$$\left. - \exp\frac{[(2\gamma_s/r_c)-p_g]b^3}{kT}\right] \quad (4b)$$

where r_1 and r_c are the loop and void radii respectively, D_s^v is the bulk diffusibity, ρ_d is dislocation density, γ_s is surface energy, and p_g is the gas pressure in a void. In both cases loop or void growth continues until the vacancy concentration decays to that in equilibrium with the loop perimeter or void surface. Thereafter growth of large clusters can continue at the expense of small ones.

In the case of loop or void growth during irradiation, vacancies and interstitial point defects are generated continuously and the rate equations defining the time-dependent change in loop or void radii can be written as follows:[10]

$$\frac{dr_L}{dt} = \frac{1}{b}\left[D_v c_v - Z_i D_i c_i - D_v c_v \exp\frac{(F_{el}+\gamma)b^2}{kT}\right] \quad (5a)$$

$$\frac{dr_c}{dt} = \frac{1}{r_c}\left\{[1+(\rho_d r_c)^{\frac{1}{2}}]D_v c_v - [1+(Z_i\rho_d r_c)^{\frac{1}{2}}]D_i c_i\right.$$

$$\left. - [1+(\rho_d r_c)^{\frac{1}{2}}]D_v c_v^e \exp\left[\frac{[(2\gamma_s/r_c)-p_g]b^3}{kT}\right]\right\} \quad (5b)$$

An important aspect exposed by these equations is that planar defects bounded by dislocations attract interstitials preferentially and thus, at low temperatures, loops undergo bias-driven shrinkage. Moreover, at high temperatures thermal vacancy emission also leads to loop shrinkage. Therefore, vacancy loops are basically unstable defects during irradiation and they are unlikely to form when point defect generation is uniform such as during electron irradiation. Voids, on the other hand, undergo biased-driven growth in the presence of biased sinks at low temperatures. At higher temperatures when the thermal emission term becomes dominant, whether voids grow or shrink depends on the sign of $[(2\gamma_s/r_c) - p_g]$. During neutron irradiation when He gas is being continuously created a flux of gas atoms can arrive at the voids causing gas-driven growth.[18]

Vacancy clusters in quenched metals

Since the first observations of dislocation loops in quenched Al[40] there has been a great deal of work on fcc and hcp metals. Interpretation of the results is complicated by the role played by impurities and gases. Impurities will inevitably be present even in nominally pure metals at about the same order of concentration as the concentration of quenched-in vacancies. Moreover gases, as well as being present as an impurity in most metals, can also be (inadvertently) introduced during heat treatment before quenching. Other complicating factors are the variation in quenching rate, the influence of different aging treatments after quenching, and the influence of stacking fault energy. All of these can play an important role in determining the type of vacancy clusters which are formed and it is necessary to account for their influence in constructing a mechanistic framework to explain the observed complexities in behaviour reported in the literature. Rather than describe all the (frequently conflicting) data an attempt will be made to summarize the present understanding of the observations in quenched fcc and hcp metals. No discussion will be attempted for bcc metals since, as already indicated, there appears to be no clear evidence for the formation of defect clusters in quenched bcc metals.

Influence of stacking fault energy

The early observation of prismatic dislocation loops of $\mathbf{b} = \frac{1}{2}\langle 110 \rangle$ in quenched Al[40] and of stacking fault tetrahedra in quenched gold,[41] suggested that stacking fault energy plays a dominant role in determining the type of defect formed in quenched metals. Thus, on the one hand, it seemed reasonable to conclude that high stacking fault energy of Al led either to the unfaulting of any $\frac{1}{3}\langle 111 \rangle$ loops by a reaction of the type:

$$\tfrac{1}{3}\langle 111 \rangle + \tfrac{1}{6}\langle 112 \rangle \to \tfrac{1}{2}\langle 110 \rangle$$

or simply that $\frac{1}{2}\langle 110 \rangle$ loops were formed directly. On the other hand, the low stacking fault energy of gold would allow the formation of faulted loops of $\mathbf{b} = \frac{1}{3}\langle 111 \rangle$ which could further dissociate to stacking fault tetrahedra by reactions of the type:[41]

$$\tfrac{1}{3}\langle 111 \rangle \to \tfrac{1}{6}\langle 110 \rangle + \tfrac{1}{6}\langle 112 \rangle$$

along the $\langle 110 \rangle$ edges of the $\frac{1}{3}\langle 111 \rangle$ loop, followed by reactions of the type

$$\tfrac{1}{6}\langle 112 \rangle + \tfrac{1}{6}\langle 121 \rangle \to \tfrac{1}{6}\langle 110 \rangle$$

along the $\langle 110 \rangle$ direction inclined to the plane of the original dislocation loop of $\mathbf{b} = \frac{1}{3}\langle 111 \rangle$, so that a complete

tetrahedron is formed if the fault energy is not too high. However, subsequent work[42-44] has shown that the most common defect in quenched aluminium is a (single or multilayer) loop of dislocation of $\mathbf{b} = \frac{1}{3}\langle 111 \rangle$ and that many of the small loops which were inferred to be of the type $\mathbf{b} = \frac{1}{2}\langle 110 \rangle$ were in fact $\frac{1}{3}\langle 111 \rangle$ loops. Local stresses due either to misfitting solute atoms[45] or simply to quenching or handling stresses can cause unfaulting of the $\frac{1}{3}\langle 111 \rangle$ dislocation to $\frac{1}{2}\langle 110 \rangle$ dislocations, and such unfaulting is commonly observed during examination in the electron microscope. This unfaulting can arise by any one of the three Shockley dislocations of $\mathbf{b} = \frac{1}{6}\langle 112 \rangle$ contained in the plane of the fault leading to the formation of non-edge perfect loops having one of the three possible $\frac{1}{2}\langle 110 \rangle$ dislocations. It has been shown that the mechanism for forming stacking fault tetrahedra in quenched metals is one of continuous growth from a nucleus which may be a hexavacancy cluster.[46-48] Thus, the observation of stacking fault tetrahedra does not necessarily imply that the stacking fault energy is low, but as discussed later reflects the stability of the tetrahedron nucleus at the post-quench aging temperature. Stacking fault tetrahedra have now been observed in quenched gold,[41] silver,[49] copper,[49] and nickel[50] as well as in some alloys quenched and aged under appropriate conditions.

In hcp metals the influence of stacking fault energy again appears to be small. Thus, in Cd,[51] Zn,[52] and Mg,[53-55] quenched under appropriate conditions, faulted loops of $\mathbf{b} = \frac{1}{3}\langle 20\bar{2}3 \rangle$ and $\frac{1}{2}[0001]$ are found, but small amounts of impurity can change the type of loop which is observed in quenched magnesium as discussed later.[55] Again, it seems reasonable to conclude that the influence of other variables outweighs any small influence stacking fault energy may have on the nature of secondary defects in quenched metals. Nevertheless the degree of dissociation of dislocation loops (and hence their shapes) and their subsequent stability against climb is clearly dependent on stacking fault energy and this aspect is discussed in a later section.

Influence of quench rate and post-quench aging

Quench rate has a direct effect on vacancy supersaturation and it is also associated with the generation of thermal stress. It has been observed that these two factors influence cluster distributions and morphologies in quenched Al and Au. Specifically a higher quench rate both reduces the temperature for the onset of cluster nucleation during quenching and increases the vacancy supersaturation. In both metals the higher supersaturation leads to a higher nucleation rate and therefor higher cluster density. We might also expect the interactions of vacancies and impurity atoms to be a direct function of supersaturation of both species, but this aspect has not been studied systematically. Moreover, in Au the temperature at which vacancy clusters nucleate has a dominant influence on their morphology (this aspect is discussed further later) and the lowering of the threshold temperature for cluster nucleation during a quench due to a higher cooling rate promotes the formation of stacking fault tetrahedra over $\frac{a}{3}\langle 111 \rangle$ Frank loops. As expected, the lower quench stresses associated with slower quenching rates leads to a higher proportion of faulted loops in quenched Al.[44] The influence of quench stresses on cluster morphology is somewhat different in

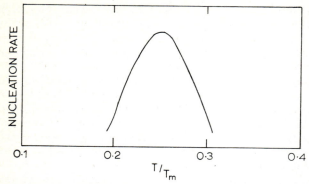

2   Schematic diagram representing nucleation rate of dislocation loops in quenched aluminium as a function of aging temperature[56]

gold with higher stress favouring the formation of Frank loops over stacking fault tetrahedra at a given nucleation temperature. However, it is emphasized that the influence of quench stress is secondary to that of nucleation temperature on vacancy cluster morphology in Au.

The important influence that nucleation temperature has on the nature of clustered defects is best illustrated by the studies of post-quench aging on aluminium[56,57] and gold.[48] The experiment consists of quenching from temperature T_p and then carrying out a two-stage aging sequence at an initial temperature T_{AL} and final temperature T_{AF}. In aluminium a systematic study has been made of the effect of varying T_{AL} and T_{AF} and the following conclusions emerge from this work.[56,57] By varying T_{AL} between 183 K and 303 K the nucleation rate was shown to reach a maximum at about 253 K with the loop concentration achieving a saturation value in the very short time of about 1 s (*see* Fig. 2). However, the final concentration of loops is not simply a function of time at T_{AL} but also depends on T_{AF} (in the range 273–373 K) which implies that the stability of a nucleus of a given size decreases as the temperature increases. As discussed below this contrasts with observation of stacking fault tetrahedra nucleation in gold.

The formation of double (and multilayer) loops is favoured by aging treatments which inhibit nucleation of new loops. For example, the proportion of double layer loops is favoured by slow quenching rates[44] and by double aging treatments in which T_{AL} is higher than the T_{AF}. Such treatments ensure a significant concentration of singly faulted loops and a low (but sufficient) supersaturation of vacancies to promote the nucleation of a second loop within an existing loop rather than further growth of pre-existing loops.

It has been argued[44] that the occurrence of multilayer loops in Al and in Mg suggests that impurity nucleation plays a signficant role in the formation of faulted loops since in an fcc crystal the three-layer loop on which a four-layer loop nucleates is, to a first approximation, a region of perfect crystal. The recent observation of a subsidiary displacement associated with stacking faults in silicon[58] may be relevant (if a similar displacement occurs in other materials) in giving rise to both multilayer loops in Al and the observed formation of complex loops in quenched Ag and Cu–Al alloys.[49,59] These loops in Ag and Cu–Al are formed by climb of the planar fault in a single-layer loop and this fault climb requires the formation of jog lines in the fault. No systematic

work has been carried out to determine the influence of aging temperature on the occurrence of these complex loops, but it seems reasonable to assume that similar aging treatments to those giving rise to multilayer loops in Al would favour the nucleation of jog lines.

The importance of aging temperature is particularly important in quenched gold. Thus, it has been shown by various workers[47,60] that if aging of quenched gold is done at temperatures above about 400 K the proportion of stacking fault tetrahedra which forms is very small and Frank loops are the dominant defect. However, progressively lowering the aging temperature down to as low as 250 K results in the nucleation of stacking fault tetrahedra, with the peak in the nucleation rate occurring at a temperature between the two extremes. In gold this maximum occurs at about 335 K. These observations are consistent with the concept that the hexavacancy is the smallest stable nucleus giving rise to stacking fault tetrahedra. On this basis the formation of tetrahedra is favoured by divacancy interactions and the formation of planar loops by interactions involving single vacancies. The stability of di-vacancies decreases with increasing aging temperature and hence the probability of forming tetrahedra nuclei decreases. This type of model, based as it is on a continuous growth mechanism, implies that the formation of tetrahedra is not dependent on the stacking fault energy but depends instead on the stability of divacancies at the aging temperature. It would seem in principle that tetrahedra could be formed in fcc metals such as platinum and even in aluminium if the correct aging procedure were used. If we take $0.25T_m$ (where T_m is the melting point in degrees K) as the approximate upper limit for stable tetrahedra nuclei in gold, then an equivalent aging temperature is about 240 K for Al and 500 K for Pt. Aging at about these temperatures has been carried out[56,61] and the resulting defects are predominantly faulted loops. On this basis it could be inferred that stacking fault tetrahedra cannot be formed in Pt and Al, but such a conclusion assumes that the stabilities of divacancies and higher-order vacancy clusters are equivalent (in terms of T/T_m) in these metals to gold and further experimental work is clearly needed before it can be concluded that it is not possible to form tetrahedra in Pt and Al and indeed in all fcc metals.

Influence of impurities

It has been observed that vacancy clustering occurs on a finer scale with increasing impurity content and it is widely accepted that nucleation is always heterogeneous on impurity atoms. For example, in gold the addition of impurities results in a decrease in size and increase in density of tetrahedra.[62] It appears that the valency of the impurity plays a part in determining the magnitude of its effect on nucleation rates of tetrahedra. Thus, divalent metallic impurities (Cd, Zn, and Mg) are more effective than monovalent metals such as Cu and Ag. It has been suggested[63] that the potency of divalent atoms arises from their stronger electrostatic interaction with divacancies. Similarly, the importance of hydrogen in apparently promoting nucleation has been attributed to its role in trapping divacancies.[63] Other nucleation models involve trivacancy–impurity atom interactions in promoting tetrahedra nucleation, but the precise nucleation mechanism is still open to argument in the absence of direct observation. It has also been argued that the increased nucleation rate in gold preannealed in, or quenched from, a CO atmosphere can be attributed to the presence of H_2 in the CO (rather than the reduction of impurities by the CO).[63] However, other experiments[64] have failed to demonstrate any direct influence of H_2 and thus it has been suggested more recently[61] that the main effect of heating in CO or H_2 is to allow C (and perhaps other impurities), in interstitial solution to act as a potent nucleating agent and experiments carried out on Pt of various C contents appear to support the view that C is a very efficient nucleating agent. Other experiments[65] have indicated that oxygen is also a powerful nucleating agent and residual oxygen in copper, nickel, and silver is considered to be responsible for the difficulty experienced in observing vacancy clusters in these metals following quenching,[50] the vacancy clusters being nucleated on too fine a scale to be visible. Support for this view comes from the ease with which a high concentration of visible dislocation loops and stacking fault tetrahedra can be quenched into Cu–Al and Ni–1% Al alloys.[65] In these cases the Al solute acts as a getter for dissolved oxygen and is thus effective in reducing its level in solution.

There is a range of evidence to show that impurities can have an important influence on the geometry of vacancy clusters in quenched metals. In Al it has been shown that highly misfitting impurities promote unfaulting of $\frac{a}{3}\langle 111\rangle$ Frank loops.[45] In Mg impurities have been observed to have a more fundamental effect on loop geometry. Specifically, in relatively impure Mg the predominant quenched-in defects are loops of $\mathbf{b}=\frac{1}{2}[0001]$, $\frac{1}{3}\langle 20\bar{2}3\rangle$ etc. on the basal plane.[54,55] However, recent work has shown that quenching very pure Mg leads to the formation of highly elongated loops of $\mathbf{b}=\frac{1}{3}\langle 11\bar{2}0\rangle$ lying on high index planes.[55] This observation cannot be explained in terms of unfaulting in conjunction with slip on the appropriate glide cylinder since $\langle 11\bar{2}0\rangle$ is contained in the basal plane and loops of $\mathbf{b}=\frac{1}{2}\langle 11\bar{2}0\rangle$ on the basal plane would be shear loops. Thus, the vacancies must have condensed on planes other than the basal plane in the pure Mg. It is by no means obvious why this should happen, but we have already seen that electron irradiation of Mg can also result in interstitial loops of $\mathbf{b}=\frac{1}{3}\langle 11\bar{2}0\rangle$ being formed on high index planes. Perhaps nucleation of clusters is difficult and the point defects cause climb of $\langle 11\bar{2}0\rangle$ dislocations with the climbed segment pinching off to form loops. This is somewhat analogous to helix formation in Al and Al alloys which can give rise to loops having $\mathbf{b}=\frac{1}{2}\langle 110\rangle$.

As well as acting as nucleation sites for faulted loops and tetrahedra, gaseous impurities can act in a totally different manner by promoting the nucleation of voids.[65,66] Void formation is also sensitive to quenching conditions and is favoured by low supersaturations, but it appears that gas is essential as a nucleating agent for void formation and Al, Ag, Cu, and Pt quenched from hydrogen particularly develop high void densities. In the case of copper and magnesium[54,67] the voids (or gas bubbles) give rise to punching stresses which are large enough to produce many prismatic (presumably interstitial) loops. The ease with which voids form in platinum and aluminium may be an important factor in explaining the absence of stacking fault tetrahedra in these metals—the growth and nucleation conditions appear to be somewhat similar for voids and tetrahedra and the prolific nucleation of voids will presumably reduce the possibility of the formation of stacking fault tetrahedra.

Stability of vacancy clusters

The direct observation of clustered interstitial and vacancy defects, discussed briefly in the previous sections, has contributed a great deal to the understanding of point defect behaviour in crystals. Further information has also been obtained by studying the thermal stability of the clustered defects. Most of this work has been carried out on quenched metals and hence the annealing of vacancy defects will be discussed in this section. The clustered vacancy defects shrink by vacancy emission when they are heated to a high enough temperature and direct observation of this shrinkage has been used to calculate self-diffusion energies and stacking fault energies.

The first direct observations of loop climb were carried out on quenched Al specimens which were subject to *in situ* annealing in an electron microscope.[68] Following these early experiments, this approach has been adopted for measuring self-diffusion parameters and stacking fault energy by comparing the shrinkage rates of faulted and perfect loops.[44] For large loops the line tension driving force for shrinkage is negligible compared with the force provided by the stacking fault, and it can be shown that equation (4a) can be rewritten to describe the shrinkage rates of faulted and perfect loops during post-quench annealing $(C_v/C_v^e l)$ of a thin foil of half thickness L:

$$\left(\frac{dr_L}{dt}\right)_F = -\frac{2\pi D_s}{b \ln L/b}\left[\exp\frac{\gamma b^2}{kT}-1\right] \quad (6a)$$

$$\left(\frac{dr_L}{dt}\right)_p = -\frac{2\pi D_s}{b \ln L/b}\cdot\frac{\alpha b}{r_L} \quad (6b)$$

where the subscripts F and p refer to faulted and perfect loops, and α is a constant given by the approximation that $(\alpha b/r)\sim\exp(F_{el}b^2/kT)-1$ which is valid for large loops. From (6b) we can write

$$\ln\frac{(\dot{r}_L^2)_p}{2\alpha b}=A\frac{Q}{kT} \quad (7)$$

where $(\dot{r}_L)_p=(dr_L/dt)_p$, $A=2\pi D_0/(b \ln L/b)$, and Q is the activation energy for self-diffusion. Thus, Q can be obtained from the slope of an Arrhenius plot of $(\dot{r}_L^2)_p/2\alpha b$ v. $1/kT$ for perfect loops. Moreover, γ can be determined by comparing similar plots for perfect and imperfect loops using the following equation obtained from (6a) and (6b):

$$\exp\frac{\gamma b^2}{kT}-1=\frac{2\alpha b(\dot{r}_L)_F}{(\dot{r}_L^2)_p} \quad (8)$$

In the case of Al the values obtained over the temperature range 130°–150°C for Q and the extrinsic fault energy using this approach are $(1\cdot26\pm0\cdot04)$ eV and (135 ± 20) erg cm^{-2} respectively.[69]

This method of determining stacking fault energy is the only direct method for materials of high stacking fault energy. It cannot be applied to materials of low fault energy because the dissociation of the $\frac{1}{3}\langle 111\rangle$ dislocation leads to jog nucleation difficulty and it is found that jog nucleation/propagation becomes rate-controlling.[70] In fact, jog nucleation activation energies have been obtained from measurements of the temperature dependence of loop annealing and jog velocity.[71,72] Similarly, the energy of a jog line in Au has been determined from the temperature and size dependence of the annealing of stacking fault tetrahedra.[73]

The annealing of voids has been studied using the same approach and hence a value of surface energy obtained.[74] Work has also been carried out[75] in which it was shown that voids in aluminium, which were connected to the foil surface by dislocations, annealed more quickly than isolated voids. This was attributed to pipe diffusion along dislocations although in similar experiments on copper[76] no difference was observed between the shrinkage rate of isolated voids and voids threaded by dislocations. It has been suggested[77] that this somewhat unexpected result may be due to the fact that the voids in copper contain significant amounts of gas and that the removal of this gas is rate-controlling.

As indicated above, annealing studies on quenched-in loops can be used to determine diffusion data. Comparison of the annealing rates of pure Al and a dilute Al alloy can then be used to determine the apparent activation energy for loop annealing in the alloy. This has been done for an Al–Si alloy and a value of 0·18 eV was deduced for the binding energy between a vacancy and a silicon atom.[78]

Similar loop annealing studies have been carried out on hcp metals, but these are complicated by growth of loops due to oxidation-induced vacancies as discussed in the paper by Smallman and Dobson at this conference.

Vacancy clusters in irradiated metals

fcc metals

We showed in the introduction to this section that vacancy loops are basically unstable defects during irradiation when point defect generation is uniform, such as for electron irradiation. Results from *in situ* studies in HVEMs are, in general, consistent with this predicted behaviour. The preferred attraction of vacancy loops for interstitials has been confirmed by observing shrinkage of vacancy loops induced by electron irradiation.[79] Exceptions to the predicted behaviour have been observed in the vicinity of dislocations where the preferred flow of interstitials to the side of the dislocation in dilatation creates an excess vacancy flux on the compressive side, thus allowing vacancy loops to nucleate and grow.[21] Moreover, vacancy loops can form during electron irradiation in thin regions of a foil where the surfaces act as a dominant sink for the faster moving interstitials.[11] Stacking fault tetrahedra have also been observed to form in thin regions of copper and dilute copper alloy foils during electron irradiation.[80] Lastly, vacancy loops can, of course, form immediately following irradiation because the interstitial point defect concentration decays much faster, leaving a vacancy supersaturation.

It is well known that vacancies also aggregate to form voids in metals and alloys under a wide range of irradiation conditions including electron irradiation. Because of its practical importance to the dimensional stability of reactor core components, the phenomenon of void swelling has been widely studied and *in situ* electron irradiation in HVEMs is one of the approaches that have been used. In this paper we will not attempt to review the very large complex range of data on void swelling, but we wish to highlight some aspects which illustrate how the formation and growth of voids fits into the overall picture of point defect clustering in metals.

In situ HVEM irradiation experiments have shown that if the nucleation conditions are satisfied, voids form in fcc metals over a wide temperature range. As expected, from studies of quenched metals, the lower temperature threshold approaches that for long-range single vacancy migration. The upper cut-off is given by the temperature at which thermal vacancy emission from voids exceeds the net vacancy flow into them. Curves showing the variation in total void volume, expressed as ($\Delta V/V \times 100$), as a function of irradiation temperature are shown for copper[81] and an austenitic stainless steel,[82] FV548, in Figs. 3a and b.

With regard to the factors influencing void nucleation, gases are known to play an important role. They may be divided into two main classes: first, surface-active gases such as O, N, and H, which are frequently present as residual impurities in metals and alloys; second, inert gases such as He which may be preinjected or generated

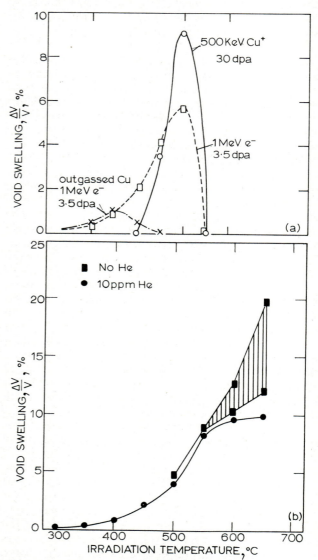

3 **Plots of void swelling *v.* irradiation temperature for *a* annealed and outgassed copper irradiated with 1 MeV electrons to 3·5 dpa and annealed copper irradiated with 500 KeV copper ions to 30 dpa; *b* FV548 stainless steel pre-injected with 10 appm He and He-free, irradiated with 1 MeV electrons to 30 dpa**

continuously during irradiation due to transmutation reactions. The influence of residual gas on the volume swelling of electron-irradiated copper is shown[81] in Fig. 3a. The annealed samples contained appreciably more residual gases than the outgassed specimens. It can be seen that the presence of the gases both increases very considerably the void swelling levels and shifts the void swelling peak to higher temperatures. Results from heavy ion irradiation experiments showed that the presence of residual gases such as O enhance void nucleation. All of these observations are consistent with the gas migrating to embryo vacancy clusters and reducing surface energy. Similar results were obtained for nickel.[81] It is well established that He preinjection enhances void nucleation, but its effect on swelling can be complex causing an increase or decrease depending on the neutral and biased sink distributions. The results presented in Fig. 3b for FV548 illustrate a case where the preinjection causes a decrease in void swelling. Moreover, an important feature of the He preinjected specimens is the weak temperature dependence of the void numbers compared with the He-free specimen (see Fig. 4). A detailed discussion of these observations has been presented elsewhere.[82] To summarize, it was proposed that the residual gas content in the steel was sufficient to promote void nucleation in the He-free specimens. However, in the He preinjected specimens, the He atoms migrate during room temperature preinjection to form a finely distributed population of gas bubbles which then act as the dominant nucleation sites for voids over the lower temperature range, thus explaining the insensitivity of void numbers to temperature. Above 600°C, thermal vacancy emission becomes increasingly important and the critical sized bubble to nucleate a void increases. Thus, a decreasing fraction of the bubble population can grow into voids and the void numbers decrease. It is emphasized that in situations where the residual gas content is too low to promote void nucleation, He preinjection will increase void swelling.

The rate theory of void swelling[83] predicts that for moderate dislocation densities, swelling should increase linearly with dose while the dislocation structure is evolving and that this rate should increase as $(\text{dose})^{\frac{3}{2}}$ when the dislocation density reaches a quasi-steady state. For extremely high dislocation densities, i.e. $k_d^2 \gg k_v^2$, where k_d^2 and k_v^2 are the dislocation and void sink strengths respectively, the dose dependence of void swelling is expected to follow a cube law, whereas for extremely high void densities, together with low dislocation densities, i.e. $k_v^2 \gg k_d^2$ swelling should increase as $(\text{dose})^{\frac{3}{2}}$. The evidence regarding the dose dependence of swelling in electron-irradiated metals has come mostly from experiments on austenitic stainless steels. Results from type 316 and FV548 steels[82,84] are consistent in showing that following an initial transient period of a few dpa, the swelling increase is linear with dose up to 30–40 dpa. There is no tendency for the swelling rate to increase towards a $(\text{dose})^{\frac{3}{2}}$ law which is consistent with the dislocation structure continuing to evolve over the dose and temperature range in question. This is in agreement with Kiritani's observation[11] of continuing interstitial loop nucleation during electron irradiation at elevated temperatures, i.e. $> 0 \cdot 3 T_m$.

Going from electron irradiation to irradiation with light charged particles such as protons, heavy ions and neutrons results in a progressive increase in the mean primary recoil energy. This results in an increasingly non-uniform point defect generation due to the production of displacement cascades by primary knock-ons. During the creation of cascades, the interstitials are transported outwards, most probably by focused collision sequences, leaving a vacancy-rich region at the cascade centre. There is now considerable evidence from TEM studies to show that such vacancy-rich regions can collapse to form vacancy loops in a wide range of metals. As the irradiation temperature increases vacancies can also aggregate to form voids, although whether or not this occurs in situ in cascades is an unresolved question. In the remainder of this section we will summarize the evidence from fcc metals and alloys on factors affecting vacancy loop formation in cascades and their influence on void formation and growth.

Most of the evidence regarding vacancy loop formation during irradiation of fcc metals has been obtained from Cu, Ag, and Au. It has been established[34,35] that small vacancy loops form in neutron-irradiated copper at temperatures up to 300°C. The loop sizes are consistent with the collapse of cascade centres and it can be argued, following the predictions regarding vacancy loop stability (see equation (5a)), that only in such regions is vacancy loop formation possible during irradiation. A series of experiments has also been carried out on Cu, Ag, Au, Ni, and stainless steel using heavy ion irradiation at energies <100 keV to simulate primary knock-ons generated during neutron irradiation. In these experiments, the displacement cascades are deposited close (within ~100 Å) to the incident surface and thus it is believed that a large fraction of the self interstitials are lost at the surface. The conditions are therefore particularly favourable for collapse of the vacancy-rich regions and studies have been made of defect geometries, numbers, and sizes. The influence of ion mass, addition of alloy elements, and irradiation temperature have also been explored. The main features of the results can be summarized as follows.

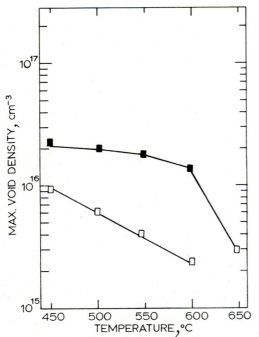

4 Plots C_v v. T for electron-irradiated FV548 with and without He pre-injection to 10 appm

1 All of the results from Cu, Ag, and Au show that the cascades collapse to form faulted $\frac{a}{3}\langle 111\rangle$ loops initially. It has been established[36] for Cu that many of the loops undergo some secondary dissociation on intersecting {111} planes towards stacking fault tetrahedra by the Silcox–Hirsch mechanism.[41]

2 The fraction of cascades collapsing to form visible loops, defined as defect yield,[84] is high in all three metals ranging from ~0·5 in Cu to 1·0 in Au irradiated with self ions >30 keV. Moreover, the fraction of vacancies taking part in the collapse process, expressed as cascade efficiency,[88] is also high ranging from 0·3 to 0·5 in copper to $\gtrsim$ in Au. These values are more uncertain because of the difficulties in measuring small defect sizes and in calculating the total number of vacancies generated in each cascade.

3 On increasing ion mass, the defect yield and cascade efficiency is increased. This is illustrated qualitatively for Cu in Figs. 5a and b, which are micrographs of specimens irradiated with 30 keV Cu$^+$ and W$^+$ respectively.[89] The result of increasing ion mass while keeping energy constant is to generate the same number of defects in a smaller volume of crystal. The vacancy supersaturation at the cascade centre and hence the driving force for loop nucleation is therefore increased.

4 Little work has been reported regarding the influence of alloy elements on cascade collapse. Experiments on a series of binary copper alloys containing Al, Si, Zn, and Be at the 1–15 wt.-% level have shown that in these cases the additives have two effects.[89] First, the tendency for faulted loops to undergo secondary dissociation is increased and this is expected since all of these elements reduce the stacking fault energy markedly. Secondly, a preliminary quantitative analysis shows that defect numbers are higher but mean defect sizes are reduced. This is consistent with the alloy elements promoting loop nucleation in the cascades on the one hand and reducing the fraction of vacancies surviving there on the other. Examination of a more complex alloy, type 316 stainless steel, irradiated with 80 keV Cr$^+$, has shown that the number of visible vacancy loops and their mean size is considerably less than in pure Cu.[14] However, increasing the ion mass again has the effect of increasing both defect yield and cascade efficiency as shown in Fig. 6. The vacancy loops undergo more secondary dissociation compared to Cu and this is consistent

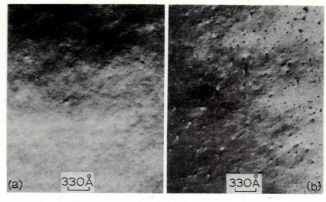

6 **Dark field micrographs of 1 050°C solution-treated type 316 stainless steel irradiated at room temperature with *a* 80 KeV Cr$^+$ and *b* 80 KeV W$^+$ to a dose of 5×10^{11} ions cm^{-2}; g = (200)**

with the low stacking fault energy of stainless steel (~15 erg cm^{-2}) compared to Cu (~40 erg cm^{-2}). It is clear from these observations that the influence of alloy elements on cascade collapse is complex. In general, they appear to reduce vacancy survival in cascades which implies enhanced recombination with interstitials. This could be a direct consequence of defocusing collision sequences, thus reducing the separation of interstitials from the cascade centres. The copper alloy results also show that alloy additions can promote vacancy loop formation in the cascades and thus they must reduce the energy barrier to nucleation. One direct effect is to reduce the stacking fault energy although this is not considered to be a limiting factor to loop nucleation.

5 Limited evidence has been obtained regarding the influence of irradiation temperature on the formation of visible ($\gtrsim$15 Å diameter) vacancy loops. A comparatively simple picture emerges from self ion irradiation experiments on copper.[91] The number of loops, N_L, remains constant up to 250°C and then decreases sharply between 300° and 400°C. It has been shown[90,91] that these results fit a model in which loop generation by cascade collapse is essentially athermal. On the other hand, as irradiation temperature increases, loop shrinkage by thermal emission becomes increasingly dominant leading to a decrease in N_L. Calculations show that the temperature range over which loop loss by thermal shrinkage increases sharply is $0·4$–$0·5 T_m$ (300°–400°) which is in good agreement with the observations. Results from neutron-irradiated copper show that visible vacancy loops are present in specimens irradiated at temperatures up to 300°C which is consistent with the ion irradiation results.[86] Moreover, vacancy loops continue to be observed with little change in N_L in stainless steel irradiated with Cr$^+$ and W$^+$ ions at temperatures up to 450°C.[14] In this case, the thermal emission shrinkage model predicts that N_L should sharply decrease above ~500°C.

As in the case of electron irradiation, increasing the irradiation temperature results in vacancies aggregating to form voids in ion- and neutron-irradiated fcc metals. However, increasing the recoil energy has a direct effect on the shape of the swelling v. temperature curve. This is illustrated for copper[81] and type 316 stainless steel[92,93] in Figs. 3a and 7 in which the swelling curves for ion and electron irradiation are compared. It can be seen that the

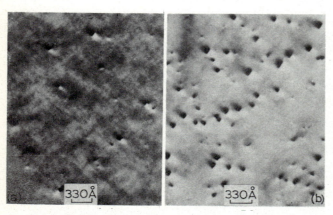

5 **Dark field micrographs of copper irradiated at room temperature with: *a* 30 KeV Cu$^+$ and *b* 30 KeV W$^+$ to a dose of 2×10^{11} ions cm^{-2}; g = (200)**

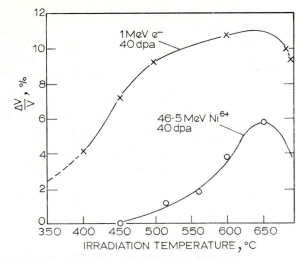

7 Plots of void swelling v. irradiation temperature for 1 050°C solution-treated type 316 irradiated with 1 MeV electrons and 46 MeV Ni⁶⁺ to a dose of 40 dpa

lower threshold temperature for void swelling is increased markedly in the ion-irradiated specimens. It is considered[18] that this is a direct consequence of cascade collapse to form vacancy loops, thus effectively removing a fraction of the vacancy point defects from the system. The lower threshold temperature is then governed by the feeding back of vacancies to the system by thermal emission controlled shrinkage of loops. The threshold temperatures for void swelling in both ion-irradiated copper and stainless steel are consistent with the observed thermal stability of vacancy loops in the low-energy self ion irradiation experiments. At higher temperatures, the loss of vacancies to loops is temporary since it is eventually balanced by re-emission from the loops. This results in an incubation period within which void nucleation and growth is suppressed. However, beyond the incubation period the quasi-steady state vacancy loop population also contributes to the dislocation sink strength and calculations show that this can reduce the void growth rate up to the peak swelling temperature.[18] Again, the swelling curves for stainless steel are consistent with this explanation. The effect is masked in Fig. 3a for copper because of the order of magnitude difference in dose level between the electron- and ion-irradiated specimens. Positive evidence for the presence of vacancy loops within the void swelling temperature range has come from TEM studies of low-dose neutron-irradiated copper. Specimens irradiated to doses in the range 10^{17}–10^{18} n/cm^{-2} at 250° and 350°C contain a population of small vacancy loops as well as voids.[82]

A second important effect of increasing recoil energy is re-solution which can affect both void nucleation and distributions of second phases. This aspect is considered in more detail in another paper in these proceedings.[93] In the present paper, we wish only to draw attention to the effect recoil energy has on void distribution in specimens preinjected with He. We have already seen that in electron-irradiated FV548, room-temperature preinjection with He results in a weak temperature dependence of C_v up to 600°C (Fig. 4). Irradiation of the same material with heavy ions, on the other hand, results in a much more marked temperature dependence in C_v (also

illustrated in Fig. 2). It has been proposed[30] that the heavy ion irradiation results in resolution of the He bubble population established during preinjection and that the bubbles which grow into voids are renucleated on a scale that is a function of temperature. It can be seen that the shape of the ln C_v v. $1/kT$ plot is similar to that for the gas-free electron-irradiated specimens. This suggests that in ion irradiation, the preinjected He affects the absolute number of voids, but it is the residual gases, through their effect on surface energy, that control the temperature dependence of C_v.

bcc metals

There is no reported evidence for vacancy loop formation in electron-irradiated bcc metals. We shall see that the available evidence suggests that the stacking fault energy, at least for molybdenum, is extremely high, i.e. $\gg 1000$ erg. cm^{-2}, and thus the vacancy supersaturation will never be high enough during electron irradiation to nucleate faulted loops. However, vacancy-induced climb of dislocations could, in principle, lead directly to loops or dipoles having $\mathbf{b} = \frac{1}{2}\langle 111 \rangle$. It has been shown that electron irradiation at elevated temperatures can lead to void formation in α-iron and ferritic steels,[95] but there is no published evidence regarding the factors affecting void nucleation and growth. Moreover, there are no reported observations of voids in other electron-irradiated bcc metals.

Small vacancy loops have been observed in neutron-irradiated Mo[96,97] and Nb[2] and, as in the case of fcc metals, their size is consistent with formation being a consequence of cascade collapse. Limited evidence from molybdenum shows that the numbers of vacancy loops decrease continuously with temperature over the range 200°–550°C (Fig. 9).[30]

Heavy ion irradiation has again been used to simulate cascades generated by primary knock-ons in Mo,[98,99] W,[100] and α-Fe.[101] The results have exposed a number of interesting aspects regarding defect geometries and the effect of ion mass, impurities, and irradiation temperature. The main highlights can be summarized as follows.

1 Irradiation at room temperature with self-ions or ions of comparable mass results in the formation of visible vacancy loops in Mo and W,[98–100] but not in α-Fe.[101] In the Mo and W the loops are mostly found to lie on {110} with $\mathbf{b} = \frac{a}{2}\langle 111 \rangle$.[99,102] Thus, the results are

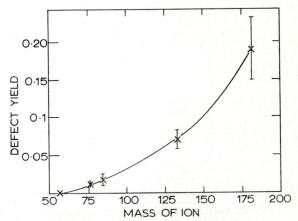

8 Plot of defect yield v. ion mass for α-iron irradiated with 80 KeV ions at room temperature to a dose of 5×10^{12} ions cm^{-2}

consistent with vacancies following the same pattern of behaviour as interstitials regarding loop nucleation.

2 Because the loops have glissile Burgers vectors, those having an appreciable component normal to the specimen surface slip out owing to the image forces. Moreover, the surface is also believed to influence the unfaulting of loops during nucleation so that, where possible, they shear to $\mathbf{b} = \frac{a}{2}\langle 111 \rangle$ having the maximum component normal to the surface.[103] Thus, the fraction of loops that glide out to the surface is a function of foil orientation. For these reasons no accurate estimates have yet been made for defect yield in Mo and W, but results for high-purity molybdenum suggest that it is at least comparable to Cu for room-temperature irradiation. The cascade efficiency also appears to be high in pure molybdenum, i.e. ~70%, which is comparable to Au.

3 Increasing the mass of the incident ion increases both defect yield and cascade efficiency. This is exhibited most dramatically in α-iron in which no visible loops are formed by self-ion irradiation. The effect of increasing ion mass (energy 80 keV) on defect yield is shown[101] in Fig. 8 and it can be seen that it reaches a value of ~0.2 for an ion mass of 184 (W). The effect of increasing ion mass is to maintain the defect number approximately constant while reducing the cascade volume.[104] Thus, the vacancy supersaturation is increased and the results suggest that it achieves the necessary critical value to nucleate loops for an ion mass of ~75. A particularly interesting feature of the iron results is that vacancy loops lying on {110} with $\mathbf{b} = a\langle 100 \rangle$ and $\frac{a}{2}\langle 111 \rangle$ were both identified. These results again show that the vacancy loops follow the same nucleation process as for interstitial loops.[23]

4 Little evidence has been reported regarding the influence of impurities or alloy elements on vacancy loop formation in bcc metals. Results from molybdenum[99] have shown that both substitutional and interstitial impurities reduce defect yield and cascade efficiency. Interstitial impurities appear to have the larger effect and this is consistent with the associated elastic distortion around the impurity atoms being more effective in defocusing collision sequences, thus enhancing recombi-

nation in the cascades. However, the interstitial solutes have very low solubility in molybdenum and they are probably highly surface active. They may, therefore, have a marked effect in reducing the surface energy leading to cavity rather than loop nucleation in the cascade.

5 It has already been shown that the vacancy loop numbers decrease between 200° and 500°C in neutron-irradiated Mo (Fig. 9). We also plot on this figure the observed decrease in vacancy loop numbers in self-ion irradiated Mo as a function of irradiation temperature.[99] In this latter case, we see that N_L remains constant up to ~200°C and then decreases sharply at higher temperatures. These results are not consistent with the simple thermal emission model that fitted the copper observations. The calculated temperature for the shrinkage of the perfect loops observed in the room-temperature irradiated specimens is >700°C. This has been confirmed by post-irradiation annealing experiments. From equation (3a) the criterion for a loop to nucleate and grow in a cascade is $c_v/c_v^e > \exp(F_{el} + \gamma)b^2/kT$. In copper, because γ is low, there is no difficulty in satisfying this criterion and loops are formed over a wide temperature range. In molybdenum, on the other hand, γ is much higher and the observations are consistent with the supersaturation and shrinkage terms being more nearly in balance during the critical nucleation stage. Thus, as the irradiation temperature exceeds that for long-range vacancy migration and vacancies can diffuse out of the cascades reducing c_v/c_v^e, the balance is tipped so that the shrinkage term remains dominant and loop nucleation is impossible. A quantitative treatment of the cascade collapse problem[105] suggests that the stacking fault energy in Mo is ~4 000 erg. cm^{-2}. We shall see that this high barrier to loop nucleation has important consequences for the void swelling behaviour.

Ion or neutron irradiation at elevated temperatures results in vacancies aggregating to form voids in bcc metals. Quite extensive studies have been made of Mo following both ion and neutron irradiation, the other bcc metals having been studied less extensively. Considering first the results from molybdenum, the following aspects can be highlighted.

1 Neutron irradiation experiments[30] have demonstrated that the lower threshold temperature for void formation is much lower, in terms of homologous temperature, than for fcc metals, i.e. $0.2T_m$ compared to ~$0.3T_m$ in most fcc metals. This low threshold temperature is thought to be a direct consequence of the sharp fall in cascade collapse to form vacancy loops at temperatures >200°C (see Fig. 6). Following irradiation at 330°C ($0.2T_m$), a population of vacancy loops was identified as well as voids, but at a much lower density and thus their influence on void formation and growth is expected to be small. These results suggest that, in contrast to fcc metals, recoil energy has little direct effect on void swelling in Mo.

2 With regard to void nucleation, the two outstanding features exhibited by both ion- and neutron-irradiated molybdenum are first, the high void concentration, i.e. ~10^{17} cm^{-3}, and second, the weak temperature dependence of C_v over a wide temperature range. The variation in C_v with temperature is shown[30,112] for Mo neutron irradiated to different doses in Fig. 10. It can be seen that C_v saturates at a low dose (<10^{20} n cm^{-2}) at low temperatures and that the saturation dose increases

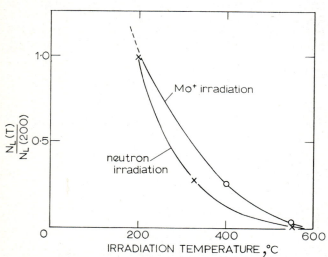

9 Plots of vacancy loop concentrations, normalized with respect to the value at 200°C, v. irradiation temperature in Mo irradiated with 60 KeV Mo+ and neutrons

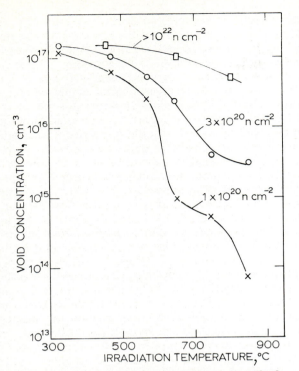

10   Plots of void concentration v. irradiation temperature for Mo neutron irradiated to different doses

which produces a high density of gas bubbles, show that the bubbles first exhibit short-range order before developing long-range order. Once formed the void lattice is stable being very resistant to post-irradiation annealing. A surprising feature of the annealing behaviour is that there is no growth of the larger voids at the expense of the small ones. What happens is that voids disappear from some areas leaving other areas where the void lattice is unchanged, giving the structure an overall patchy appearance. These observations are consistent with there being a strong interaction between voids in agreement with theoretical predictions.[109]

4   Evidence from both neutron- and ion-irradiated molybdenum shows that once C_v has saturated and the void lattice has developed, void growth is slow, the dose dependence being less than linear. This is a consequence of having a high void density together with a low dislocation density which results in the voids acting as the predominant sink for both interstitials and vacancies. A particularly interesting feature of the low-dose neutron-irradiation results is that during the void nucleation period when C_v is increasing, a fraction of the void population actually shrinks, contributing to an overall decrease in mean void radius. An example of this behaviour is shown for pure molybdenum, irradiated at 850°C to 1×10^{20} n cm^{-2} and 3×10^{20} n cm^{-2}, by the two size spectra in Fig. 11. Similar results were[30] obtained for high-purity and commercial-purity molybdenum and TZM alloy irradiated at temperatures down to 650°C. These results suggest that the voids formed initially grow rapidly, but as C_v builds up the newer voids grow much more slowly giving rise to two maxima in the void size spectra. Then, as the dose increases from 10^{20} to 3×10^{20} n cm^{-2}, the larger voids apparently become unstable and shrink. It is emphasized that this is completely contrary to the usual behaviour in a system containing a size distribution of defects or precipitates in which the largest component is the most stable. The driving force for the shrinkage of the largest voids is not

with temperature. Figure 10 also illustrates the weak temperature dependence of C_v which is in contrast to the observed behaviour in fcc metals and group VA bcc metals. This is consistent with there being a low activation barrier to void nucleation. An important metallurgical property of the group VIA bcc metals distinguishing them from many other metals, including the group VA bcc metals, is their extremely low solubility for O and N at temperatures $<0.5T_m$. Thus, these gases are expected to be surface active and they could be having a major effect on void nucleation by lowering the surface energy. In this context it is interesting to note that C_v is not changed markedly on going from high purity to commercial purity Mo[30,106] suggesting that only a low gas content is necessary to promote void nucleation. Moreover, the three or four orders of magnitude increase in dose rate on going from neutron to ion irradiation does not have a large effect on C_v at intermediate temperatures (700°–800°) which is a reflection of the weak temperature dependence in C_v.

3   A distinctive feature of the void distributions observed in many of the bcc metals (although it is not unique to this class of crystal structure) is the alignment of the voids to form a three-dimensional lattice having the same symmetry properties as the host lattice. This was first observed in ion irradiated molybdenum[107] and has since been shown to be a general feature of molybdenum irradiated to high doses with neutrons or ions. An important property of the void lattice is the ratio a_v/r_v, of the void lattice parameter, a_v, to void radius r_v, which remains approximately constant in Mo at ~ 8. Comparing results from low and high dose neutron-irradiated molybdenum[30,106] suggests that the voids are initially nucleated randomly and that the order develops once C_v has saturated. Results from molybdenum irradiated with α-particles at room temperature,[108]

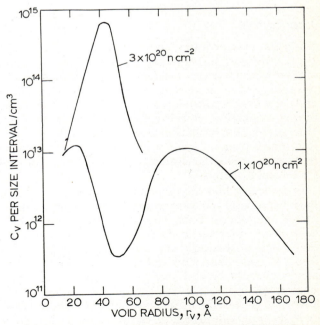

11   Plots of the void size spectra for Mo neutron irradiated at 850°C to 1×10^{20} n cm^{-2} and 3×10^{20} n cm^{-2} (fission)

understood. The greater stability of the smaller voids could be due to a larger inert gas pressure resulting from the accumulation of He, or a markedly lower surface energy because of a higher concentration of residual gases at the void surface. The larger voids could then shrink preferentially by thermal vacancy emission as the total neutral sink strength increases leading to a reduction in the vacancy and interstitial fluxes. Alternatively, the need for the voids to order into a bcc lattice as C_v increases may provide the necessary driving force since the larger voids will increasingly violate the a_v/r_v ratio that is a basic property of the lattice.

Considering briefly the results from the group VA bcc metals, voids have also been observed in these metals following ion or neutron irradiation at elevated temperatures. The lower threshold temperature has not been determined systematically, but voids have been observed[110] in Nb and V neutron irradiated to low doses (6×10^{19} n cm^{-2}) down to 420°C. Voids have also been observed to order to form a bcc lattice in Nb and Ta irradiated to high doses and it has been shown for Nb that ordering is promoted by the presence of oxygen in solution.[111] Two features distinguish Nb and V from Mo. First, C_v exhibits much greater temperature dependence in the group VA metals, indicating that the activation energy for void nucleation is higher. This is consistent with their considerably greater solubility for oxygen and nitrogen. Thus, the tendency for these gases to separate out at void surfaces will be less and their effect on surface energy is also expected to be considerably smaller. The second feature is that the addition of alloy elements such as Ti and Cr in V and Zr in Nb drastically reduce the void swelling, whereas the addition of Ti and Zr to Mo has little effect on the swelling levels. It has been shown that the alloy addition to Nb and V results in a considerable increase in the irradiation-induced dislocation density together with a considerably lower void density.[113] However, no systematic studies have been carried out to determine the role played by the alloy additions in changing the damage structure in these metals.

hcp metals

As in the case of interstitial clusters (p. 65) comparatively little work has been carried out on the formation and growth of vacancy clusters in hcp metals. Much of the published evidence is concerned with void formation and there has been little work on vacancy loop formation during irradiation. A self-consistent picture does not emerge from the results but there is again a clear indication that hcp metals do not conform with the general pattern regarding the clustering of vacancies developed for the cubic metals.

Considering first the question of vacancy loops that have been observed to form in hcp metals under three sets of conditions. First, large loops with $b = \frac{1}{3}\langle 11\bar{2}0 \rangle$ have been observed in Zr following neutron irradiation[34] and post-irradiation annealing at 490°C. This result is consistent with similar observations on bcc metals and can be accounted for by the formation of small loops by cascade collapse during irradiation followed by growth during the subsequent anneal. However, the observation of large vacancy loops in Ti following neutron irradiation[35] at elevated temperatures cannot be explained in this way. The apparent growth of the loops during irradiation would appear to be inconsistent

with the expected preferred drift of interstitials to the loops. Even more puzzling is the apparent replacement of interstitial loops by vacancy loops as dose increases in an electron-irradiated Mg–0·2% Mn alloy.[37]

As in cubic metals, vacancies can aggregate to form voids in hcp metals during irradiation at elevated temperatures and gas again plays an important role in void nucleation. However, comparison of the void formation and growth behaviour for different types of irradiation (electrons, ions, and neutrons) has exposed more dramatic differences than is generally found for the cubic metals. For example, electron irradiation of 99·9% pure Ti in an HVEM produces voids[43] whereas neutron irradiation results in vacancy loop formation. Similarly, electron irradiation of 99·9% pure Zr, Zr–0·1% Fe, and Zr–0·1% Ni results in void formation whereas voids have not been observed in these materials following ion or neutron irradiation.[35,114] These results are consistent with vacancy loop formation by cascade collapse occurring more readily in ion- and neutron-irradiated Ti and Zr than in fcc and bcc metals. It can be shown that when cascade collapse efficiency (the fraction of vacancies retained in vacancy loops) exceeds a critical value, void growth is completely suppressed. However, it is important to recognize that in the case of hexagonal metals significant differences in behaviour can occur on going from one hcp metal to another. For example,[114] voids have been observed in both electron and C^{2+} ion irradiated Co. In this case direct observations in an HVEM suggested that dislocation glide as well as climb plays an important role in void growth.

Little work has been carried out on the influence exerted by impurities and alloy elements on void formation in hcp metals. It has already been pointed out that small additions of Fe and Ni to Zr do not inhibit void formation during electron irradiation but voids are not observed in Zr–0·1% Sn and Zircaloy irradiated under identical conditions. In the case of electron-irradiated Zn voids are not observed in high-purity (99·999%) material but they form in less pure material.[114] Voids are also observed to form readily in Mg.

SUMMARY AND CONCLUSIONS

Results from TEM studies have provided considerable physical insight into the clustering of point defects in quenched and irradiated metals, although some important aspects are still not well understood. Considerable progress has been made in elucidating cluster geometries and the main factors influencing their nucleation and growth in pure fcc and bcc metals. The main area requiring further systematic work on these classes of metals is on the role played by impurities and alloy elements in governing the geometry, distribution, and stability of clusters. Trapping of point defects at solute atoms modifies their mobility, and results from quenching and irradiation experiments have shown that this can have important effects on cluster nucleation. Moreover, basic physical parameters such as stacking fault energy can be a sensitive function of composition and this, in turn, can have important effects on cluster geometry. It is also important to recognize that the influence of solutes can be greatly enhanced by segregation to clusters during the early stages of their formation. Clearly the role played by impurities and alloy elements in governing cluster formation is of considerable technological importance because of its relevance to the behaviour of reactor core materials which are mostly complex alloys.

In the present paper, emphasis has been placed on highlighting the differences in behaviour between fcc and bcc metals. They are distinguished by a large difference in stacking fault energy and this accounts for many of the observed differences, particularly in the geometry and growth behaviour of planar defects. The main conclusions to be drawn from the experimental observations can be summarized as follows:

1   The critical vacancy supersaturation required to form vacancy loops appears to be much higher in bcc metals than in fcc metals. Thus, there are no unambiguous observations of vacancy loops in quenched bcc metals. Even in the case of room-temperature irradiation so as to generate displacement cascades containing regions of extremely high vacancy supersaturation, vacancy loops are not always formed in bcc metals, e.g. α-iron irradiated with self ions whereas they have been observed in a wide range of fcc metals following irradiation with self ions.

2   Faulted vacancy and interstitial loops are commonly observed in fcc metals, whereas in bcc metals both types of loops are predominantly perfect, although there is clear evidence that they nucleate as faulted loops on {110} planes.

3   In fcc metals, interstitial loops grow either by the biased flow of interstitial point defects to the loops or, at high temperatures, by vacancy emission from the loops. In contrast, because the loops are perfect and generally glissile in bcc metals, they may also grow by a process of glide and self-climb. This mechanism is dominant at low temperatures and can result in the development of a network at low homologous temperatures.

4   Results from irradiated Cu and Mo have shown that the influence of temperature on vacancy loop formation in displacement cascades is very different in fcc and bcc metals. This difference has the important practical consequence that recoil energy plays an important role in void formation growth in fcc metals and alloys, whereas the available evidence suggests that it does not in bcc metals.

With regard to hcp metals, considerably less work has been done and a complex picture emerges from the results. Nevertheless, it is clear that they exhibit significant differences in behaviour from that in fcc and bcc metals. For example, the steps involved in the formation of vacancy and interstitial loops and how the observed loop geometries develop appear to vary from one metal to another and are as yet not understood. Moreover, the observation of large vacancy loops in irradiated hcp metals is not consistent with the results from the cubic metals and appears at first sight to violate the principle of preferred interstitial flow to dislocation loops. Lastly, impurities again play an important role in the point defect clustering processes and it is possible that their effects may account for some of the apparent differences between hcp and cubic metals.

ACKNOWLEDGMENTS

The authors wish to thank a number of their colleagues for permission to include their unpublished results in this paper. One of us (B.L.E.) is also grateful to the University of Wisconsin for the provision of facilities during the preparation of this paper.

REFERENCES

1 M. WILKENS: 'Modern diffraction and imaging techniques in material science', 233, 1970, Amsterdam, North-Holland
2 M. WILKENS: 'Vacancies and interstitials in metals', 485, 1970, Amsterdam, North-Holland
3 M. RÜBLE: Proc. Symp. on 'Radiation damage in reactor materials', 1, 113, 1969, Vienna, IAEA
4 B. L. EYRE: J. Physics F., 1973, 3, 422
5 L. M. BROWN, et al.: Phil. Mag., 1969, 19, 721
6 N. YOSHIDA AND M. KIRITANI: J. Phys. Soc. Japan, 1973, 35, 1418
7 M. R. HAYNES: J. Nucl. Mat., 1975, 56, 267
8 R. BULLOUGH et al.: Nucl. Appl Technol., 1970, 9, 346
9 S. D. HARKNESS AND CHE-YU LI: Metall. Trans., 1971, 2, 1457
10 A. D. BRAILSFORD AND R. BULLOUGH: J. Nucl. Mat., 1972, 44, 121
11 M. KIRITANI: 'Fundamental aspects of radiation damage in metals, Vol. II, 695, Proc. Conf. held at Gatlinburg, Tennessee, USA, Oct. 1975, Conf. 75 1006-P2, 1976
12 M. K. HOSSAIN AND L. M. BROWN: in 'High voltage electron microscopy', 360, 1974, London, New York, Academic Press
13 P. J. BARTON et al.: 'Irradiation behaviour of fuel cladding and component materials', Conf. organised by KTG/BNES, Dec. 1974
14 T. M. WILLIAMS: unpublished results
15 H. R. BRAGER AND J. L. STRASSLAND: J. Nucl. Mat., 1973, 46, 134
16 E. E. BLOOM et al.: Rad. Effects, 1972, 14, 231
17 B. L. EYRE, 'Fundamental aspects of radiation damage in metals', Vol. II, 729, Proc. Conf. held at Gatlinburg, Tennessee, USA, Oct. 1975, Conf. 751006-P2, 1976
18 R. BULLOUGH et al.: Proc. Roy. Soc., 1975, A346, 81
19 D. J. MAZEY et al.: Phil. Mag., 1962, 7, 1861
20 D. I. R. NORRIS: Phil. Mag., 1970, 22, 1273
21 K. URBAN: Physica Status Solidi (a), 1971, 4, 761
22 K. URBAN AND M. WILKENS: Physica Status Solidi (a), 1971, 6, 173
23 B. L. EYRE AND R. BULLOUGH: Phil. Mag., 1965, 12, 31
24 B. C. MASTERS: Phil. Mag., 1965, 11, 881
25 E. A. LITTLE AND B. L. EYRE: Met. Sci. J., 1973, 7, 100
26 A. J. E. FOREMAN AND J. D. ESHELBY: Harwell Report, AERE, R-4170
27 B. L. EYRE AND D. M. MAHER: Phil Mag., 1971, 24, 767
28 J. BRIMHALL AND B. MASTEL: Rad. Effects, 1970, 3, 203
29 B. L. EYRE et al.: Phil. Mag., 1971, 23, 409
30 J. BENTLEY: Ph.D. Thesis, University of Birmingham, 1975
31 S. N. BUCKLEY AND S. A. MANTHORPE: 'Physical metallurgy of reactor fuel elements', (Eds. J. E. HARRIS and E. C. SYKES), 127, 1975, London, The Metals Society; J. Nucl. Mat., 1977, 65, 295.
32 D. S. GELLES AND J. E. HARBOTTLE: CEGB Report, 1974, to be published
33 T. D. GULDEN AND I. M. BERNSTEIN: Phil. Mag., 1966, 14, 1087
34 P. M. KELLY AND R. G. BLAKE: Phil. Mag., 1973, 28, 415
35 J. L. BRIMHALL et al.: Rad. Effects, 1971, 9, 273
36 S. KARIM: unpublished work, University of Birmingham
37 D. S. GELLES: 'High voltage electron microscopy', 348, 1974, London, New York, Academic Press
38 M. E. WHITEHEAD: to be published
39 E. G. TAPETADO et al.: Crystal Lattice Defects, 1974, 5, 199
40 P. B. HIRSCH et al.: Phil. Mag., 1958, 3, 897
41 J. SILCOX AND P. B. HIRSCH: Phil. Mag., 1959, 4, 72
42 M. H. LORETTO et al.: Phil. Mag., 1966, 13, 953
43 Y. SHIMOMURA: J. Phys. Soc. Japan, 1965, 20, 965

44 J. W. EDINGTON AND R. E. SMALLMAN: *Phil. Mag.*, 1965, **11**, 955

45 K. H. WESTMACOTT AND R. L. PECK: *Phil. Mag.*, 1971, **23**, 611

46 M. DE JONG AND J. S. KOEHLER: *Phys. Rev.*, 1963, **129**, 49

47 A. A. HUSSEIN AND R. A. DODD: *Phil. Mag.*, 1971, **24**, 1441

48 K. P. CHIK: *Physica Status Solidi*, 1965, **10**, 675

49 L. M. CLAREBROUGH *et al.*: *Phil. Mag.*, 1966, **13**, 1285

50 P. HUMBLE *et al.*: *Phil. Mag.*, 1967, **15**, 297

51 V. LEVY *et al.*: *Phil. Mag.*, 1975, **31**, 145

52 P. S. DOBSON AND R. E. SMALLMAN: *Proc. Roy. Soc.*, 1966, **A293**, 423

53 R. HALES *et al.*: *Met. Sci. J.*, 1968, **2**, 224

54 J. S. LALLY AND P. C. PARTRIDGE: *Phil. Mag.*, 1966, **13**, 9

55 J. HILLAIRET *et al.*: *Acta Met.*, 1970, **18**, 1285

56 M. KIRITANI: *J. Phys. Soc. Japan*, 1965, **20**, 1834

57 Y. SHIMOMURA AND SH. YOSHIDA: *J. Phys. Soc. Japan*, 1965, **20**, 1667

58 E. HAQUE *et al.*: 'Developments in electron microscopy and analysis', EMAG 1975, 429

59 A. J. MORTON AND L. M. CLAREBROUGH: *Australian J. Phys.*, 1969, **22**, 293

60 R. L. SEGALL AND L. M. CLAREBROUGH: *Phil. Mag.*, 1964, **9**, 865

61 K. H. WESTMACOTT: to be published

62 R. M. J. COTTERILL AND R. L. SEGALL: *Phil. Mag.*, 1963, **8**, 1105

63 I. A. JOHNSTON *et al.*: *Phil. Mag.*, 1968, **17**, 1289

64 Y. SHIMOMURA: *Phil. Mag.*, 1969, **19**, 773

65 K. H. WESTMACOTT: to be published

66 M. KIRITANI *et al.*: *J. Phys. Soc. Japan*, 1966, **21**, 22

67 L. M. CLAREBROUGH *et al.*: *Acta Met.*, 1967, **15**, 1007

68 J. SILCOX AND M. J. WHELAN: *Phil. Mag.*, 1960, **5**, 1

69 P. S. DOBSON *et al.*: *Phil. Mag.*, 1967, **16**, 9

70 L. M. CLAREBROUGH *et al.*: *Can. J. Phys.*, 1967, **45**, 1135

71 J. WASHBURN AND M. J. YOKOTA: *Crystal Lattice Defects*, 1969, **1**, 23

72 I. A. JOHNSTON *et al.*: *Crystal Lattice Defects*, 1969, **1**, 47

73 H. L. FRASER *et al.*: *Phil. Mag.*, 1973, **28**, 1043

74 K. H. WESTMACOTT *et al.*: *Met. Sci. J.*, 1968, **2**, 177

75 T. E. VOLIN AND R. W. BALLUFFI: *Physica Status Solidi*, 1968, **25**, 163

76 H. G. BOWDEN AND R. W. BALLUFFI: *Phil. Mag.*, 1969, **20**, 1001

77 R. E. SMALLMAN AND K. H. WESTMACOTT: *Mater. Sci. Eng.*, 1972, **9**, 249

78 R. L. PECK AND K. H. WESTMACOTT: *Met. Sci. J.*, 1971, **5**, 155

79 M. KIRITANI *et al.*: 'Fundamental aspects of radiation damage in metals', Vol. II, 889, Proc. Conf. held at Gatlinburg, Tennessee, USA, Oct. 1975, Conf. 75, 1006-P2, 1976

80 M. IPOHORSKI AND M. SPRING: *Phil. Mag.*, 1969, **20**, 937

81 J. M. LANORE *et al.*: 'Fundamental aspects of radiation damage in metals', Vol. II, 1169, Proc. Conf. held at Gatlinburg, Tennessee, USA, Oct. 1975, Conf. 75, 1006-P2, 1976

82 T. M. WILLIAMS AND B. L. EYRE: *J. Nucl. Mat.*, 1976, **59**, 18

83 R. BULLOUGH AND R. C. PERRIN: 'Radiation induced voids in metals', Proc. Conf. held at Albany, New York, Jun. 1971. USAEC, 1972

84 J. J. LAIDLER *et al.*: 7th Int. Symp. on 'Radiation effects on structural materials', Gatlinburg, Tennessee, USA, Jun. 1974, ASTM

85 M. RÜHLE *et al.*: *Physica Status Solidi (a)*, 1970, **39**, 609

86 J. MUNCIE: unpublished results

87 M. M. WILSON AND P. B. HIRSCH: *Phil. Mag.*, 1972, **25**, 983

88 K. MERKLE: *Physica Status Solidi*, 1966, **18**, 173

89 A. STATHOPOULOS: unpublished results

90 C. A. ENGLISH *et al.*: *Phil. Mag.*, in press

91 C. A. ENGLISH *et al.*: 'Fundamental aspects of radiation damage in metal', Vol. II, 910, Proc. Conf. held at Gatlinburg, Tennessee, USA, Oct. 1975, Conf. 75, 1006-P2, 1976

92 J. A. HUDSON: Harwell Report, AERE-R7922, 1975

93 R. S. NELSON AND J. A. HUDSON: this volume

94 M. J. MAKIN: unpublished results

95 E. A. LITTLE: unpublished results

96 D. M. MAHER *et al.*: *Phil. Mag.*, 1971 **23**, 409

97 D. M. MAHER *et al.*: *Phil. Mag.*, 1971, **24** 181

98 D. M. MAHER: Proc. 7th Int. Congress on Electron Microscopy, France, 1970

99 C. A. ENGLISH *et al.*: *Phil. Mag.*, in press

100 F. HAUSSERMANN: *Phil. Mag.*, 1972a, **25**, 1561; 1972, **25**, 583

101 C. A. ENGLISH *et al.*: *Nature*, in press

102 F. HAUSSERMANN *et al.*: *Physica Status Solidi (a)*, 1972, **50**, 445

103 W. JAGER AND M. WILKENS: *Physica Status Solidi (a)*, 1975, **32**, 89

104 P. SIGMUND: *Appl. Phys. Lett.*, 1974, **25**, 169

105 R. BULLOUGH *et al.*: unpublished results

106 B. L. EYRE AND A. F. BARTLETT: *J. Nucl. Mat.*, 1973, **47**, 143

107 J. H. EVANS: *Nature*, 1971, **229**, 403

108 D. J. MAZEY *et al.*: *J. Nucl. Mat.*, in press

109 V. K. TEWARY AND R. BULLOUGH: *J. Phys. F.*, 1972, **2**, L69

110 J. D. ELEN *et al.*: *J. Nucl. Mat.*, 1971, **39**, 194

111 B. A. COOPIES *et al.*: 'Fundamental aspects of irradiation damage', Vol. II, 1245, Proc. Conf. held at Gatlinburg, Tennessee, USA, Oct. 1975, Conf. 75, 1006-P2, 1976

112 V. R. SIKKA AND J. MOTEFF: *J. Nucl. Mat.*, 1974, **54**, 325

113 A. F. BARTLETT *et al.*: 'Radiation effects and tritium technology for fusion reactors', Vol. I, P.1-122, Proc. Int. Conf. held at Gatlinburg, Tennessee, Oct. 1975, Conf. 750989, 1976

114 S. N. BUCKLEY: unpublished work

Secondary defects in non-metallic solids

K. H. G. Ashbee and L. W. Hobbs

This paper points out features of secondary defect formation which are peculiar to non-metallic solids (excluding elemental semiconductors). Most of the materials of interest are compounds of two or more (usually more or less ionic) atomic species, an immediate consequence of which is a need to maintain both stoichiometry (or accommodate non-stoichiometry) and order. Primary defects in these solids, whether produced thermally, chemically or by irradiation, seldom are present or aggregate in exactly stoichiometric proportions, and the resulting extended defect structures can be quite distinct from those found in metallic solids. Where stoichiometry is maintained, it is often convenient to describe extended defects in terms of alterations in the arrangement of 'molecular' units. The adoption of this procedure enables several novel features of extended defect structures in non-metals to be explained. There are several ways in which a range of non-stoichiometry can be accommodated, which include structural elimination of point defects, nucleation of new coherent phases of altered stoichiometry, and decomposition.

K. H. G. Ashbee is at the University of Bristol and L. W. Hobbs was formerly at AERE, Harwell, and is now at the Case Western Reserve University, Cleveland, USA

There are altogether some 80 elements whose solid states exhibit properties which we call metallic. These combine easily with 13 or so non-metallic elements to generate over 1 000 binary combinations whose solid forms are to some extent ionically bonded. Add to this semiconducting compounds, ternary combinations, and the inestimable number of hydrocarbon solids, and one begins to appreciate that metals are a very special class of material whose comparatively simple behaviour is illustrative but by no means universal. In this paper, we shall consider some of the simpler non-metallic solids, excluding semiconductors and solidified gases, and investigate the ways in which we can describe and classify the structure and behaviour of extended defects in these solids. Such extended defects can often be thought of as arising from the aggregation of *point* defects (such as those produced by irradiation) in the form of defect complexes which may or may not preserve the stoichiometric proportions of the constituent elements of the solid.

FEATURES PECULIAR TO DEFECTS IN NON-METALLIC SOLIDS

It is instructive to consider first some fundamental differences between metallic and non-metallic solids, and to point out how these differences are reflected in the structure and behaviour of extended lattice defects.

Polyatomicity

With the exception of carbon, silicon, germanium, phosphorus, arsenic, sulphur, selenium, tellurium, iodine, and solidified gases, non-metallic solids are diatomic or multiatomic, and one must consider defects not only on a single sublattice but on two or several sublattices (lattices defined by a single element), as well as interactions between defects on different sublattices.

Charge trapping and valency alterations

The inability to delocalize electrons in the conduction band in non-metallic solids means that defects can acquire and retain charge and can often exist in several charge states. The perfect lattice can even trap excess charge, as in the case of the self-trapped holes in halides or changes in the valency of lattice atoms (usually cations). Simple point defects in ionic lattices are often by definition charged. An anion vacancy in rocksalt, for example, is a positively charged defect (charge of +1 with respect to the ionic lattice) which may trap one or more electrons to become neutral or negatively charged. Point defect charge states have a large influence on defect interactions and defect mobilities, and therefore on defect aggregation. For example, the electrically neutral F centre (anion vacancy and trapped electron) in KCl has a migration energy of 1·5 eV; the mutual interaction of two F centres to form a divacancy, however, proceeds with about 1 eV, and without its trapped electron, the empty anion vacancy moves with 0·7 eV. The influence may also be structural. Oxygen vacancies (E′ centres) in more covalently bonded SiO_2 may be considered to leave 'dangling bonds' which alter the stability of neighbouring $[SiO_4]$ tetrahedra, whereas alteration of the valence state of ions on occupied cation sites affects preferred coordination and site symmetry.

Preservation of electrical neutrality and order

The dominance of Coulombic binding or directed bonds introduces the necessity to preserve overall electrical neutrality in non-metallic solids in equilibrium. Excess charge is costly thermodynamically and will not persist uncompensated in equilibrium. For this reason equilibrium thermal disorder in ionic solids consists of neutral Schottky multiplets (ion vacancies on each respective sublattice in stoichiometric proportions) or Frenkel pairs on single sublattices. For the same reason, introduction of aliovalent ions results in generation of charge compensating defects, often on other sublattices. A corollary for Coulombic-bonded solids is that long-range atomic order must be maintained. There is always the tendency to maintain an ordered relationship between sublattices, and in particular antiphase disorder involves large electrostatic fault energies.

Stoichiometry

The presence of two or more atomic species in non-metallic solids means that, in order to reproduce defect structures wholly analogous to those in elemental solids (for example dislocations), it is necessary to accumulate equivalent defects from all sublattices in stoichiometric proportions. Where stoichiometric proportions are not maintained, extended defect structures may serve as vehicles for locally accommodating non-stoichiometry in the solid. As an example, clustering of anion vacancies in an ionic solid results in a local anion deficiency. Conversely, imposing an anion deficiency (say by chemical reduction) can lead to extended disorder involving anion vacancies; equally an imposed cation excess could effect the same result. We can often isolate a single simple defect complex which represents a collection of lattice defects in either stoichiometric or non-stoichiometric proportions. In the case of stoichiometric extended defects, stacking together of such units can introduce linear or planar faults or new phases of identical composition but altered crystal structure or geometry; in the case of non-stoichiometric extended disorder, stacking together of such units generates new phases of altered stoichiometry.

Radiolysis and radiation-induced bias

Lattice defects at high densities are often the unavoidable consequence of irradiation. In metallic solids, electronic excitations (arising from interaction of radiation with electrons) are quickly delocalized (in times of order 1 fs) throughout the conduction band. In insulating non-metallic solids, excitations can not only be localized (in the form of separated or associated electron-hole pairs) but can persist for times sufficiently long (seconds, even) that the lattice may become unstable at the site of the localization and respond by creating permanent atomic displacements, a process called radiolysis. In this way, it is possible to create sublattice disorder with the (usually dominant) ionizing component of the radiation. The

mechanism is efficient in a large number of systems (halides, hydrides, azides, quartz and silicates, and all organic solids), and it is possible to achieve quickly large defect densities (displacement rates can be as high as 1 dpa s^{-1}). Radiolysis is a selective process usually involving initially only a single sublattice in ionic crystals (usually the anions) or certain vulnerable bonds in organics. Thus, the *primary* defect spectrum becomes biased, and as discussed already the imbalance can lead to non-stoichiometry in the *secondary* defect structures and local loss of the chemical identity of the solid.

For non-metallic solids which do not undergo radiolysis (for example most oxides), displacement cross-sections for direct displacement of atoms (interaction of radiation with atom nuclei) are seldom identical for all sublattices. In extreme examples, displacement of certain atomic species may prove impractical (for example heavy metal ions such as U or Th in UO_2 or ThO_2 cannot be displaced with electrons <3 MeV). Displacement rates will therefore usually differ according to ion species, and the primary defect spectrum may again become biased. The resulting secondary defect structures consolidating the damage will usually reflect the non-stoichiometry of this initial disorder, particularly as defect mobilities for each sublattice will in general also differ.

It should be remembered, too, that point defects created by irradiation of non-metallic solids are frequently charged and often highly reactive in the chemical sense, whereas vacancies or interstitials in metals can do little more than eventually occupy lattice sites. Anion interstitials in ionic solids, for example, can react with lattice anions or other interstitials to form molecular defects. In radiolysis of organic solids, highly reactive bare protons or other radicals are invariably released, and the molecular fragments left behind can react with neighbouring molecules or other fragments to form new chemical entities.

STRUCTURES OF STOICHIOMETRIC EXTENDED DEFECTS

Most classical extended defect structures found in metals are reproduced in non-metallic crystalline solids but, if stoichiometry is to be maintained and in particular if order is to be preserved, certain conceptual modifications are necessary.

Dislocations in rocksalt and sapphire

A simple example is the structure of an edge dislocation in alkali halides with the rocksalt structure (Fig. 1). The interionic spacing vector $\frac{1}{2}\langle 100 \rangle$ cannot serve as a Burgers vector because it does not preserve ionic order, the resulting antiphase disorder being remarkably costly in energy. The shortest perfect Burgers vector is thus between diagonally opposite ions of the same kind, $\frac{1}{2}\langle 110 \rangle$. Stoichiometric planes of the type {220} are composed of equal numbers of anions and cations and, owing to the size of the Burgers vector necessary to preserve order, there are *two* 'extra' {220} planes above the slip plane. Equally, the perfect dislocation could be described in terms of lattice points in the unit cell, with an anion and a cation associated with each lattice point. There is theoretical evidence for appreciable relaxation[1] in the core due to polarization, but this does not appear to lead to the extended core structure once suggested[2] for this dislocation.

1   Edge dislocation in rocksalt structure with $b = \frac{1}{2}[110]$ and symmetrical core configuration viewed along the [001] dislocation line

Such a dislocation can climb by absorbing or emitting point defects, either vacancies or interstitials, but these must arrive in stoichiometric proportions. Dislocation loops in alkali halides and alkaline earth oxides have thus usually been supposed to consist of equal numbers of condensed anion and cation vacancies or interstitials. The presence of more than one extra half plane has some interesting consequences for climb. It is possible in some ionic structures (but not rocksalt) for each half plane to climb independently (but still stoichiometrically) and thus produce two partial dislocations. Such climb dissociation has been observed[3] for dislocation dipoles in sapphire. Each of the $\frac{1}{3}\langle 11\bar{2}0 \rangle$ component dislocations dissociates into $\frac{1}{3}\langle 10\bar{1}0 \rangle$ partials (Fig. 2), and annihilation of the two inner partials leads to a faulted dipole. The fault is equivalent to removal of one of the two original half planes.

It is also possible in principle to condense in planar form defects of only one sublattice on close-packed planes (in analogy to vacancy or interstitial precipitation in metals), for example the {222} planes in the rocksalt structure are alternately populated by ions of entirely

one species. However, the resulting fault energy is large because ions of like charge are brought into apposition across the faulted loop plane which is electrostatically undesirable. An alternative mode of condensation of anion vacancies is indicated in the section on F centre aggregation in halides, p. 85. Dislocations can however absorb defects of a single sublattice, for example cation vacancies present as charge compensators for aliovalent impurities, but they then acquire electrically charged jogs; the presence of dislocation charge[4] can affect subsequent interaction with other point defects. Stoichiometric dislocation cores in polyatomic solids can also serve as vehicles for transport of defects of a single sublattice. A classic example is the transport of F centres in alkali halides along dislocations.

Dislocation glide in ice

In some quasi-molecular crystals, such as ice, while the necessity for overall stoichiometry remains, the requirement of order can be at least partially relaxed. The structure of ice I-h, the stable form at low pressures, consists of a hexagonal network of oxygen atoms, illustrated by open circles in Fig. 3a, here viewed along $[10\bar{1}0]$, and hydrogen atoms so distributed that they obey the so-called Bernal–Fowler rules,[5] namely that each oxygen has two hydrogen neighbours and there is

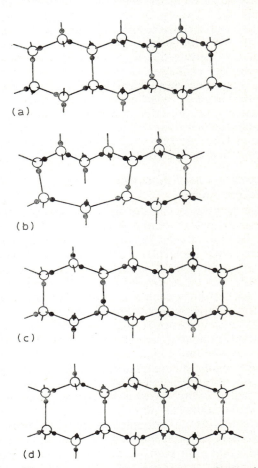

a perfect lattice; *b* core structure of an edge dislocation with Burgers vector $\frac{1}{3}[11\bar{2}0]$; *c* and *d* alternative hydrogen atom arrangements created by glide of dislocation in *b*

3   Projection of the structure of I-h ice along $[10\bar{1}0]$ (reproduced from ref. 7)

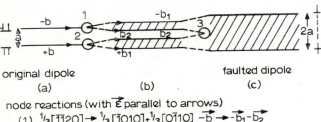

original dipole   faulted dipole
   (a)   (b)   (c)

node reactions (with $\vec{\xi}$ parallel to arrows)

(1) $\frac{1}{3}[\bar{1}\bar{1}20] \rightarrow \frac{1}{3}[\bar{1}010] + \frac{1}{3}[0\bar{1}10]$ $-\vec{b} \rightarrow -\vec{b_1} - \vec{b_2}$

(2) $\frac{1}{3}[11\bar{2}0] \rightarrow \frac{1}{3}[10\bar{1}0] + \frac{1}{3}[01\bar{1}0]$ $+\vec{b} \rightarrow \vec{b_1} + \vec{b_2}$

(3) $\frac{1}{3}[10\bar{1}0] + \frac{1}{3}[\bar{1}010] \rightarrow 0$ $+\vec{b_2}$ $-\vec{b_2} \rightarrow 0$

2   Climb dissociation of $b = \frac{1}{3}\langle 11\bar{2}0 \rangle$ dislocation dipoles in sapphire into b_1, $b_2 = \frac{1}{3}\langle 01\bar{1}0 \rangle$ partials, the inner two ($\pm b_2$) annihilating to leave a $b_1 = \frac{1}{3}\langle 01\bar{1}0 \rangle$ faulted dipole (reproduced from ref. 2)

only one hydrogen atom between any two adjacent oxygen atoms. Assuming that the oxygen–hydrogen bond distance is about that for the free water molecule, the hydrogen atom distributions must be similar to that denoted by the closed circles in Fig. 3a. Of particular importance to dislocation glide is the fact that the Bernal–Fowler rules do not require the arrangement of hydrogen atoms to be ordered. Primary slip of the form $(0001)\langle 11\bar{2}0\rangle$ can be illustrated by the passage of an edge dislocation (Fig. 3b) with Burgers vector $\frac{1}{3}[11\bar{2}0]$, defined by the oxygen–oxygen repeat distance, which leaves the molecular structure intact. However, glide of this dislocation creates hydrogen atom arrangements (Fig. 3c) which do not obey the Bernal–Fowler rules, viz. it creates oxygen–oxygen neighbours which are separated either by two hydrogen atoms or by no hydrogen atoms. These so-called Bjerrum[6] defects could be avoided by first allowing the hydrogen atoms to slide along their bonds, but such an event would create ionic defects either of the type $(H_3O)^+$ or $(OH)^-$ as in Fig. 3d. In either case, the structure of I-h ice is not preserved.

Glen[7] has proposed that glide in ice is accompanied by proton diffusion. Maintenance of the Bernal–Fowler rules during glide requires a three-stage motion of protons that is geometrically identical to the proton motion responsible for dielectric relaxation, and if this motion controls the velocity of dislocations, then glide is expected to be characterized by the same relaxation time. In fact, the dislocation velocities required to account for measured deformation rates in ice are at least an order of magnitude too fast for the proposed proton migration mechanism to be rate-controlling.[8] Enhanced availability of protons, such as might be afforded by the presence of Cottrell atmospheres of Bjerrum defects is required. Such a local excess concentration of Bjerrum defects could occur as charge compensation to a charged dislocation core.

Dislocations and disorder in other molecular solids

A convenient way of visualizing stoichiometric extended defects is to regard the solid as a regular arrangement of molecular units and the defect as altering the disposition of these units. Molecular units can often be defined which are geometrically so stable that the introduction of disorder can be expected to leave them more or less intact. An extreme example is found in the structure of most non-polymeric organic molecular solids in which the molecular units are strongly bonded internally but only weakly bonded to neighbouring units. It is unlikely in such cases to find completely absent or excess whole molecular units, analogous to point defects in atomic solids, but extended defects such as dislocations and planar faults certainly exist in which the displacements of molecular units can be regarded as for atoms in an atomic solid. As an example, an edge dislocation in anthracene ($C_{14}H_{10}$) is illustrated in Fig. 4a. The basic molecular unit is the planar group of three edge-sharing benzene rings, and the symmetry of the unit cell is monoclinic ($P2_1/a$) though nearly rhombohedral. The principal reason for the low symmetry in such solids is the *shape* of the molecular unit. The fact that the molecular units have shape means that different orientations are possible, with the result that dislocation Burgers vectors may be larger than if the molecular units

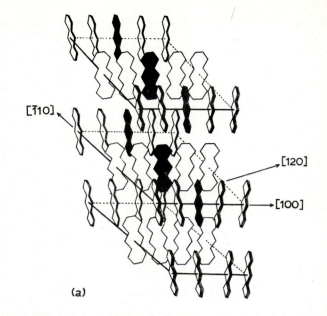

(a)

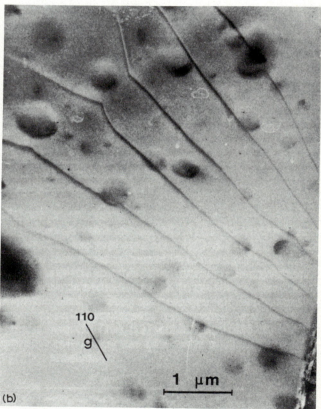

(b)

a extra half-plane of molecules (shaded) in the core of a $\frac{1}{2}$[120] dislocation on the (001) basal plane (reproduced from ref. 9); *b* TEM images of basal dislocations (L. W. Hobbs, unpublished)

4   Dislocations in anthracene (C_4H_{10})

were considered as single atoms without shape. The treatment of dislocation structures is then analogous to, say, that for ordered metal alloys, and the observed images (Fig. 4b) are straightforward.[9]

The directional bonding of many inorganic covalent crystals, such as quartz, permits a similar approach. The crystalline modifications of silica (quartz, tridymite, cristobalite) can each be described in terms of channels defined by helical chains of corner-sharing [SiO_4] tetrahedra (Fig. 5a). The observed slip directions are parallel to the axes of these channels, and the presence of dislocations can be conveniently described in terms of alterations to the pitch of the helices.[10] However, in contradistinction to organic molecular crystals in which the molecular units are only weakly bound to each other, dislocation motion in quartz requires the disruption of individual tetrahedra, in particular scission of strong silicon–oxygen bonds. It is known experimentally that the presence of water is essential to slip, and models in fact have been proposed[11] in which, instead of breaking and remaking covalent bonds, the very much weaker hydrogen-bonded linkages of the form

$$-Si-OH\cdot HO-Si-$$ temporarily replace the normal

linkages between tetrahedra at dislocation cores.

Unlike large molecules in organic molecular solids, the smaller [SiO_4] tetrahedral units apparently can be readily removed or added to form prismatic dislocations. Figure 5b illustrates the removal (or addition) of a single

rhombohedral layer of tetrahedra in α-quartz to form (half) a perfect prismatic dislocation loop on a rhombohedral plane. Such loops, it is believed, are observed to form during irradiation of quartz,[12,13] although it is by no means sure that the mechanism is as simple as that illustrated. Quartz, in common with organic molecular crystals, undergoes a crystalline-to-amorphous transition under the influence of ionizing radiation.[12,13] This can be understood in terms of extensive, random reorientation of [SiO_4] tetrahedra in the presence of many disrupted bonds.

ACCOMMODATION OF ANION DEFICIENCY

The inter-relationship between deviations from stoichiometry and secondary defect structures is best illustrated in systems exhibiting a degree of ionicity. According to one's bias, it is possible to speak of an excess (or deficiency) in one (or the other) ionic species. The choice of terminology often depends on the method by which the excess (or deficiency) was introduced, but a better basis for nomenclature is the nature of the secondary defect structure which accommodates the excursion from stoichiometry. We first consider two very different ways of accommodating a deficiency of anions.

Crystallographic shear in reduced transition metal oxides

Many simple and mixed transition metal oxides crystallize in open network structures which accommodate an astonishing range of stable metal/oxygen ratios. For example, stoichiometric WO_3 can be reduced to a number of distinct but structurally related phases listed in Table 1. The Nb_2O_5 system similarly exhibits a large number of stable compounds, as do mixed oxides such as Nb_2O_5–TiO_2.

The basic structural units for these systems are [MO_6] metal–oxygen octahedra which, in the simplest case (the ReO_3 structure of WO_3) are linked together by sharing apexes. In reduced compounds, the crystals must accommodate excess anion vacancies, and these may be accommodated structurally by altering the mode of octahedra linkage, for example from apex-sharing to edge-sharing, along certain planes known as 'crystallographic shear'* planes,[14] in this way eliminating sets of anion sites. The regular spacing of these defect planes (Fig. 6) thus governs the overall oxygen/metal ratio. It is

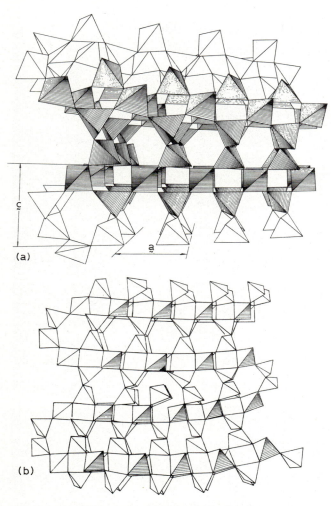

a perspective view of β-quartz showing channels defined by helical chains of tetrahedra; b formation of a perfect prismatic dislocation on a rhombohedral plane in α-quartz

5 Arrangement of [SiO_4] tetrahedra in quartz (reproduced from refs. 10, 12)

TABLE 1 Discrete compositions in two reduced transition metal oxide systems (oxygen-to-metal ratio in parentheses)

WO_3			
WO_3	(3·000)	$W_{20}O_{58}$	(2·9000)
$W_{40}O_{118}$	(2·950)	$W_{18}O_2$	(2·885)
$W_{24}O_{70}$	(2·917)		

Nb_2O_5–TiO_2			
Nb_2O_5	(2·500)	$22Nb_2O_5\cdot 5TiO_2$	(2·449)
$19Nb_2O_5\cdot TiO_2$	(2·487)	$5Nb_2O_5\cdot 2TiO_2$	(2·417)
$12Nb_2O_5\cdot TiO_2$	(2·480)	$2Nb_2O_5\cdot TiO_2$	(2·400)
$17Nb_2O_5\cdot 3TiO_2$	(2·459)	$Nb_2O_5\cdot TiO_2$	(2·333)
		TiO_2	(2·000)

* This is not shear in the normal sense in that the displacement must have a component normal to the shear plane in order to eliminate anion sites; an analogy with intrinsic faults is closer.

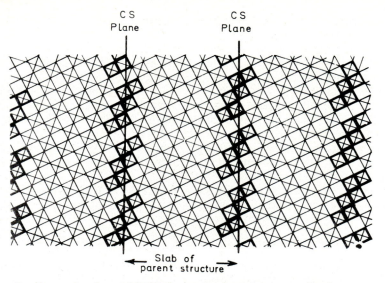

6   Face-sharing of [WO₆] octahedra along regularly spaced crystallographic shear planes in $W_{20}O_{58}$ (reproduced from ref. 18)

similarly possible for 'crystallographic shear' to manifest itself in two dimensions, generating 'block' structures[15] (Fig. 7a) which are in turn joined to other 'blocks' through shared sites. Variations in block size and block linkage provide still further possibilities for stable oxygen/metal ratios. In addition, ordering of aliovalent cations can occur at the cation sites (for example in the Nb_2O_5–TiO_2 system). A third topological variant is rotary shear manifested in the complex tunnel structures of the tungsten bronzes such as Nb_2O_5–WO_3 (Fig. 7c). Crystallographic shear is also a feature of silicate minerals.[16] Most of these shear structures are only two-dimensional and can be imaged more or less directly by high-resolution transmission electron microscopy; many-beam images from very thin (a few unit cells thick) crystal sections viewed along the invariant axis display

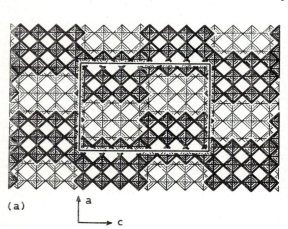

(a)

a idealized 3×4 block structures in $5Nb_2O_5 \cdot 2TiO_2$ (reproduced from ref. 15); *b* 3× 3 blocks in $VNb_{12}O_{25}$ with two intersecting planes of 4×3 blocks (Wadsley defects) reproduced by courtesy of Dr J. L. Hutchison; *c* rotary shear in the tetragonal tungsten bronze of composition $2Nb_2O_5 \cdot 7WO_3$ (reproduced from ref. 18)

7   Two-dimensional crystallographic shear

(b)

1 nm

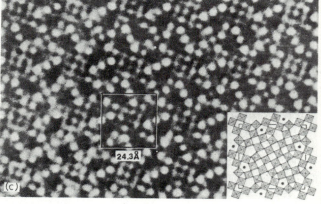

(c)

24.3Å

the projected lattice charge density and permit elucidation of both structure and structural defects. Figure 7b, for example, illustrates the 3×3 block structure of VNb_9O_{25} together with several interposed layers of 4×3 blocks (Wadsley[17] defects).

The crystal chemistry of these transition metal oxide structures has been expanded into a complex and sophisticated science, and several comprehensive reviews[14,16-18] are available; we shall therefore not pursue its complexities further, other than to remark that the general principles involved in the construction of these crystallographic shear structures have application to many other ionic defect structures where valency alterations are possible or mixed valency phases present. Some analogous processes in cation deficient and anion excess systems are discussed later (*see also* ref. 19).

F centre aggregation in halides

A second mode of anion vacancy accommodation results from the rigid valency constraints imposed by the more closely packed alkali and alkaline earth halides. Anion vacancies in such systems readily trap electrons to become neutral F centres. These may be produced in thermodynamic excess by the process of additive coloration at high temperature under partial pressure of metal vapour. During coloration, alkali metal atoms occupy new cation sites as ions, and the consequent anion vacancies (with electrons) diffuse into the halide crystal. The maximum F centre concentrations achieved by this method are typically dilute, of order 10^{-3}. The limit is set by the vapour pressure of the metal at the alkali (or alkaline earth) halide melting temperature. If the crystal is quenched to room temperature, these F centres remain isolated in dilute solution.

Such F centres are not in equilibrium with the crystal of their surroundings and, given sufficient mobility above room temperature, they either leave the crystal or interact with each other and aggregate to form regions of the crystal containing only metal cations and excess electrons. These regions transform (Fig. 8) into precipitates of alkali (or alkaline earth) metal whose sizes (1 nm–1 μm) lie in the 'colloidal' range. The volume

misfit with the matrix is, coincidentally, very small, and because the metal is readily compressible relative to the stiffer surrounding matrix, the accommodation strain in the matrix is even smaller. The evidence[20,21] is that the precipitates retain the face-centred cubic structure of the original cation sites rather than assuming, in the case of alkali halides, the normal body-centred cubic structure of the alkali metal. Since they are of colloidal size, these inclusions scatter visible light, but in addition excitation of surface plasmons in the free electron metal gives rise to absorption in the visible region; the two effects produce an optical absorption band whose peak position is characteristic of colloid size, width largely related to size distribution, and total integrated area proportional to the volume fraction of metal precipitated. These inclusions can be observed by transmission electron microscopy by virtue of their structure factor and absorption contrast, and are seen to nucleate preferentially along dislocation lines (Fig. 9). The evolution of colloid size distributions suggests that coalescence may be initially important during nucleation and initial precipitation, while Ostwald ripening dominates at longer times. A further discovery is that, when large, these metal particles acquire facetted shapes.

F centres can also be created by irradiation through a remarkably efficient radiolysis process[22] in which the dissociation of a self-trapped exciton (bound electron-hole pair) leads to an interstitial halogen atom and an

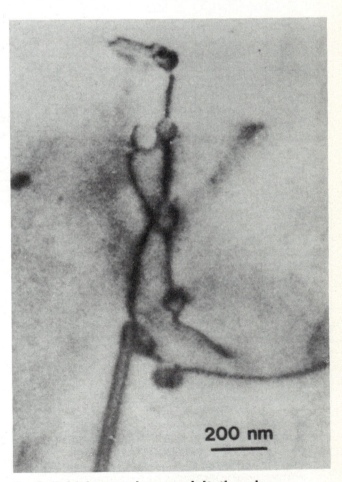

9 Colloidal potassium precipitation along dislocation lines in KCl additively coloured with potassium excess to an F centre density of 10^{25} Fm^{-3} and annealed 1 h at 673 K

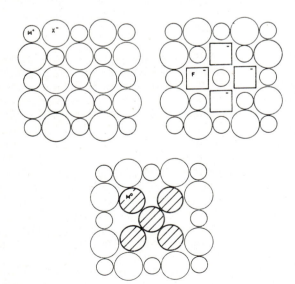

8 Transformation of F centre aggregate into coherent (fcc) precipitate of alkali metal in rocksalt-structure alkali halides

F-centre halogen vacancy. At room temperature and below, the rate of stable F centre production decreases with increasing irradiation dose; the kinetics are governed by the immobility (thus isolation) of F centres which serve as recombination sites for halogen interstitials and whose concentration eventually saturates[23] at $\sim 10^{-3}$. Irradiating at higher temperatures where F centres are mobile is equivalent to irradiating metals in the regions where voids are produced, in that F centres can diffuse directly to (neutral) alkali metal colloid sinks. Since the interstitial halogen excess at the same time diffuses to biased sinks (*see* section on 'Anion interstitial stabilization in alkali halides'), there is little recombination; the growth of the stabilized defect fraction is linear or even supralinear (Fig. 10*a*) with dose, and a large anion deficiency, of the order of several percent, can ultimately be accommodated in the form of precipitated alkali metal. The growth kinetics of these alkali metal colloids can be followed by optical absorption,

transmission electron microscopy, and small-angle scattering. The observed kinetics of colloid growth are adequately modelled[24] using the rate-theory approach[25] developed for void growth in irradiated metals; growth is found to be proportional to $(dose)^{1.5}$ for short times and eventually $(dose)^{1.0}$ for long times. For all systems studied so far (NaCl, LiF, NaF, KBr, KI, CaF_2, BaF_2, SrF_2) there exists a temperature of maximum production efficiency (Fig. 10*b*) which in many cases is only slightly removed from room temperature. The initial rise is due to the onset of F centre mobility, but the fall-off at higher temperature (which corresponds to thermal annealing of colloids produced at lower temperature of irradiation) involves recombination of F centres and interstitials. Electron microscopy (Fig. 11*a*) and small-angle neutron scattering reveal two additional features of colloid behaviour; when the particles become large they develop facets, just as do voids in metals, and in at least one system (the alkaline earth fluorides, CaF_2 and SrF_2) appear to establish an ordered array (Fig. 11*b*).

ACCOMMODATION OF CATION DEFICIENCY

We consider here two modes of accommodating cation vacancies induced in the first system by an intrinsic cation valency change, and in the second by introduction of aliovalent cation impurities.

Structure of wustite

The structure of cation-deficient wustite ($Fe_{1-x}O$) and its further oxidation to the higher oxides of iron (e.g. Fe_3O_4) is a classical problem of short-range order.[26] X-ray diffraction studies,[27] neutron diffraction studies[28,29] and Mossbauer studies[30] have indicated that hypostoichiometric FeO has a cation-deficient NaCl structure involving iron atoms in tetrahedral interstitial sites and cation vacancy clustering. The simplest stable defect cluster structure consistent with these features is the 4:1 cluster (Fig. 12*a*) comprising a trivalent Fe^{3+} interstitial (tetrahedral site) surrounded by four cation vacancies. Additional trivalent Fe^{3+} ions on normal substitutional (octahedral) sites are required to maintain charge balance. Model calculations[31] indicate that the 4:1 cluster forms exothermically and most likely constitutes the basic structural unit for still larger defect clusters.

Such defect units can associate in ways entirely analogous to those considered in the section on Crystallographic shear (p. 83) for $[MO_6]$ structural units in transition metal oxides, the allowed linkage modes being either corner sharing or edge sharing (face sharing is not compatible with the NaCl structure). These aggregation modes gradually decrease the vacancy/interstitial ratio as is found experimentally with increasing defect concentrations. Model calculations[31] favour the further formation of small (up to three) stacks of edge-sharing 4:1 defect clusters (i.e. an '8:3' cluster) for small defect concentrations. Superstructure peaks observed in X-ray and electron diffraction suggest that these defect clusters are arranged in an array with periodicity 2.5 times the host unit cell spacing.

Recent lattice-resolution transmission electron microscopy results[32] in $Fe_{0.92}O$ show that such clusters are arranged in a regularly twinned orthorhombic unit cell (Fig. 13) and that the number of vacancies per cluster is about 5, consistent with the '8:3' cluster. For

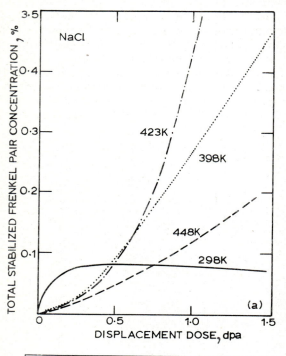

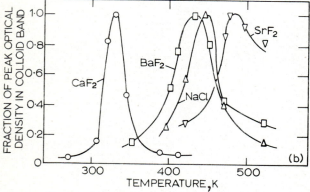

a linear growth of sodium metal fraction with dose in NaCl; *b* efficiency of radiation-induced metal precipitation in several halides showing temperatures of maximum production efficiency

10   Two aspects of alkali metal precipitation in irradiated halides

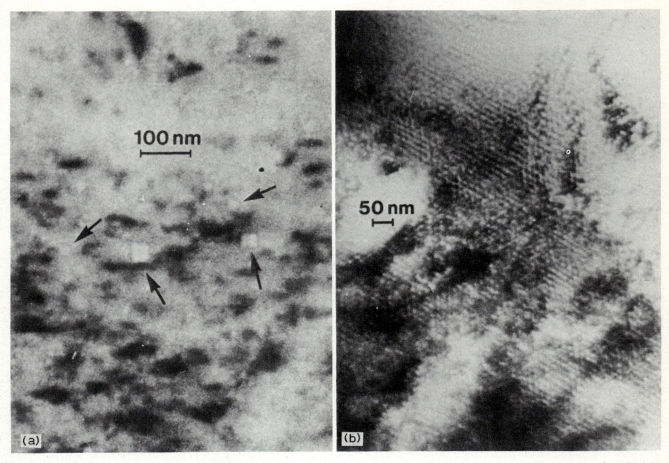

a {100} facets on sodium colloids in NaCl produced by irradiation at 423 K; *b* ordered array (probably calcium metal) in CaF_2 irradiated near room temperature

11 Two microscopical features of irradiation-induced metal precipitation in halides

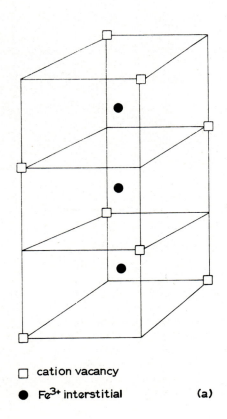

☐ cation vacancy

● Fe^{3+} interstitial   **(a)**

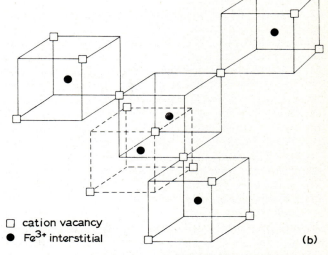

☐ cation vacancy

● Fe^{3+} interstitial   **(b)**

a '8:3' cluster constructed from two stacked '4:1' clusters sharing a common cation vacancy–cation vacancy edge; *b* '16:5' spinel-like aggregate constructed from '4:1' clusters sharing cation vacancy corners

12 Vacancy clustering in $Fe_{1-x}O$ (reproduced from ref. 31)

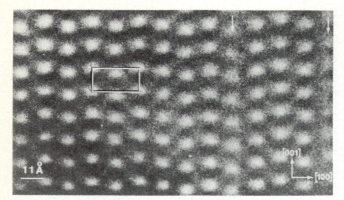

13 Regularly twinned orthorhombic lattice (unit cell indicated) of defect clusters in Fe$_{0.92}$O imaged by TEM at high resolution (reproduced from ref. 32)

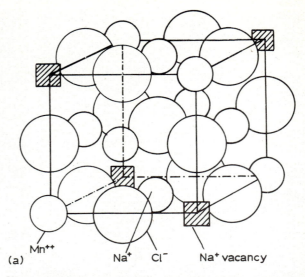

(a)   Mn^{++} Na$^+$ Cl$^-$ Na$^+$ vacancy

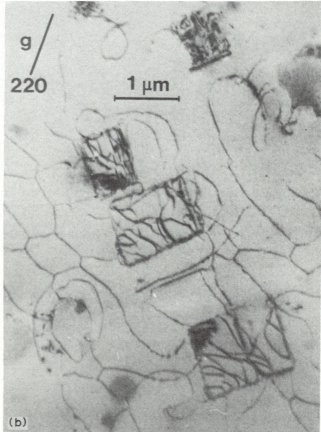

(b)

a one-eighth unit cell; *b* nearly cubic Suzuki-phase precipitates in NaCl:Mn showing interface dislocation structure (reproduced from ref. 37)

14 Suzuki-phase 6NaCl·MnCl$_2$

higher defect concentrations, growth of spinel-like corner-sharing aggregates (the '16:5' aggregate, Fig. 12*b*) appears to be favoured and leads directly to nucleation of the Fe$_3$O$_4$ inverse spinel structure.

Suzuki-phase precipitation in alkali halides

Cation vacancies are created as charge compensators for the introduction of substitutional divalent cation impurities into alkali halides; therefore doping with divalent cations can be used as a means of introducing a controlled concentration of (excess) extrinsic cation vacancies. In solid solutions MX:NX$_2$ the divalent impurity ions and cation vacancies can remain dissociated or exist as associated pairs. Calculations indicate about equal energies for dipoles in ⟨110⟩ orientation (nearest neighbour) and in ⟨100⟩ orientation (intervening anion). At temperatures and concentrations below the solubility limit (below ~573 K in NaCl:Mn for example) these associated pairs may cluster into dimers, trimers or higher aggregates. Further aggregation leads to precipitation, in some cases (e.g. NaCl:CaCl$_2$, NaCl:SrCl$_2$ NaCl:BaCl$_2$) as the divalent metal halide NX$_2$, in others (e.g. NaCl:CdCl$_2$, NaCl:MnCl$_2$, LiF:MgF$_2$) involving smaller divalent cations ($r^{2+}/r^+ < 1.2$) as the Suzuki[33] phase 6 MX·NX$_2$. The structure of the metastable Suzuki phase (one-eighth of the unit cell is illustrated in Fig. 14*a*) has been established by X-ray diffraction and confirmed by model calculations. It consists of an ordered array of divalent metal cations and alkali metal cation vacancies on alternate (200) NaCl planes, a structure which can arise from successive additions of ⟨100⟩ type impurity–vacancy dipoles anti-aligned along ⟨100⟩.[34] The Suzuki phase is stabilized by small anion displacements towards the divalent cation.[35] Having a low surface energy and a structure so similar to the matrix, this phase nucleates readily.

Suzuki-phase precipitates have been observed directly by optical microscopy, replica electron microscopy, and gold-decoration replication[36] and most recently by direct transmission electron microscopy.[37] The morphology (Fig. 14*b*) is in the form of facetted cubes or rectangular parallelopipeds with sides approximately parallel to {100} matrix planes. The isotropic matrix misfit strains, calculated using measured lattice parameters, and assuming similar elastic properties for matrix and precipitate, are small ($\epsilon \approx 0.1\%$ for 6NaCl·CdCl$_2$, $\epsilon \approx 0.2\%$ for 6NaCl·MnCl$_2$); this is con-

firmed by the TEM observations. Such misfits can be accommodated either elastically or by sets of misfit dislocations of screw character at the interface.[38] Figure 14*b* in fact shows what appear to be widely spaced interface dislocations, and cleavage ledges corresponding to intersection of a cleavage crack with misfit dislocations of the appropriate spacing (about 60 nm for 6NaCl·MnCl$_2$) have been observed by gold-decoration replication.[39] A combination of shear in two equivalent directions gives rise also to a shear distortion of the precipitate shape[40] which can be seen in Fig. 14b.

ACCOMMODATION OF ANION EXCESS

We consider finally two modes of accommodating an excess of anion species, the first in two systems capable of local cation valency changes, either intrinsically or by additions of substitutional aliovalent cations, and the second in a system with, again, a rigidly imposed valency constraint.

Anion excess in the fluorite structure

The fluorite structure, which forms the basis for both UO_2 and alkaline earth fluorides, provides a normally vacant octahedral site (Fig. 15) which may be easily occupied by an anion interstitial charge-compensating a nearest-neighbour oxidized lattice cation (e.g. $U^{4+} \rightarrow U^{5+}$ in UO_2) or substitutional aliovalent impurity (e.g. Y^{3+} in CaF_2). The resulting associated pairs and their aggregates form the basis for accommodating considerable anion excess in the systems UO_{2+x} for $0 < x < 0.25$ and $CaF_2 : M^{3+}$ to high dopant concentrations (many percent).

Experimental neutron diffraction data exist for both these systems,[41,42] while the local distortions accompanying the fluorine interstitial in the systems $CaF_2 : (YF_3, TmF_3, LaF_3)$ have been deduced by paramagnetic resonance[43] and X-ray lattice parameter measurements.[44] Model calculations[45] have investigated in detail several basic defect cluster configurations suggested by the neutron diffraction results and the interaction of neighbouring clusters.

The first cluster can form when two nearest-neighbour substitutional cation–interstitial anion pairs dimerize to form an aggregate ('2:2:2' cluster, Fig. 16a) which is stabilized by a coupled lattice–interstitial relaxation, the interstitials moving along $\langle 110 \rangle$ and lattice anions along $\langle 111 \rangle$. This structure requires a degree of covalent bonding between anion interstitials. The next larger aggregate arises from a substitutional–interstitial pair trimer which, when relaxed in analogous fashion, results in the '4:3:2' cluster (Fig. 16b). Still larger aggregates in the alkaline earth fluorides are unlikely due to the immobility of trivalent substitutional ions, but extended structures can form by interaction of the terminal $\langle 111 \rangle$ interstitials in neighbouring clusters. In UO_{2+x}, the 4:3:2 cluster forms a basis for nucleation of the ordered structure of the next higher oxide U_4O_9.

The high density of intermediate interstitial disorder may be the explanation for curious extended defects

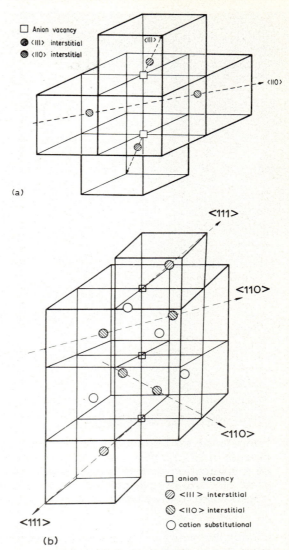

a dimer, showing coupled relaxations that give a '2:2:2' cluster; b trimer, giving a '4:3:2' cluster

16   Anion interstitial–aliovalent cation pair clustering in fluorite structure (reproduced from ref. 45)

(Fig. 17) observed some time ago by TEM in UO_{2+x}, the fault contrast from which wanes and vanishes into perfect lattice without abrupt termination.

Anion interstitial stabilization in alkali halides

We conclude by considering excess anion accommodation in an intrinsic system in which changes in valency or coordination are not possible. The fate of the anion interstitial in irradiated alkali halides has been a mystery ever since it was first established that radiolysis of these solids led to production of anion Frenkel pairs. Early classical work also concerned itself with incorporation of a chemically induced halogen excess. Parallels exist between the two cases since aggregation of interstitial halogen leads to local regions of excess halogen content.

The primary (anion) interstitial product of radiolysis, a halogen atom, exists in the form of an X_2^- molecular ion (X = halogen) situated on a single halogen lattice site.[47] This defect is called an H centre. The anisotropy of the

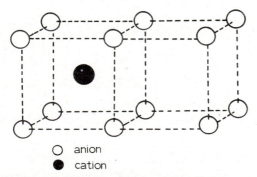

15   Portion of the fluorite structure unit cell showing vacant octahedral site occurring in every other cube of anions

17 Fringe contrast associated with extended defects in UO$_{2+x}$; the contrast is terminated by a dislocation on only one side and merges gradually into the perfect lattice on the other (reproduced from ref. 46)

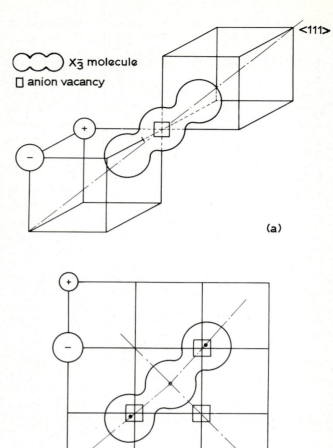

a [X$_3^-$]$_-^0$ interstitial molecule; *b* [X$_3^-$]$_{+--}^0$ substitutional molecule, right-angled configuration

18   Molecular halogen defects in NaCl

elastic strain field surrounding an H centre suggests that for certain mutual orientations there will be an attractive interaction between H centres. That such associated H centre pairs are formed at high H centre density appears well confirmed experimentally[48] in that the H optical absorption band saturates at the expense of other bands called V bands, which are similar to bands deriving from halogen molecules in solution. Possible configurations for the resulting [X$_2^0$] molecule are the ⟨100⟩ body centre–body centre position and the [X$_3^-$]$_-^0$ complex (Fig. 18a) in which the molecule is bound to a lattice ion along ⟨111⟩. (We adopt here the convention that a superscript outside the brackets enclosing the defect entity refers to the defect charge relative to the lattice, and a subscript to the lattice site, if any ((−) anion, (+) cation), occupied by the defect.) In some cases [X$_3^-$]$_-^0$ has the lower formation energy, but in both cases the large increase in elastic distortion energy in accommodating the molecule is more than offset by the large molecular binding energy (2·5 eV per chlorine molecule).

An important alternative site for the molecules is an associated anion–cation vacancy pair, occupation of which leads to an [X$_2^0$]$_{+-}$ defect. Formation of this defect in the perfect lattices (Fig. 19) involves ejection of an anion and a cation interstitial, the ejected interstitials forming the perfect prismatic interstitial dislocation loops or causing the stoichiometric climb of existing dislocations observed in irradiated alkali halides.[49] The contribution of the elastic energy (line energy) of the

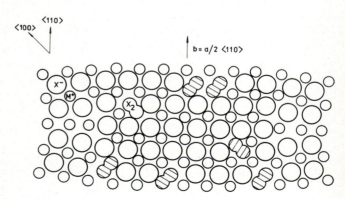

19   Interstitial dislocation loop formed in rocksalt-structure alkali halides as a result of [X$_2^0$] molecular defects occupying anion–cation sites substitutionally (reproduced from ref. 49)

dislocation per interstitial ion pair absorbed to the over-all formation energy of $[X_2^0]_{+-}$ defects is negligible for large loops, so formation of these defects constitutes a highly exothermic process.[50] The $[X_2^0]_{+-}$ molecular defect can bind with a lattice ion to form an X_3^- molecule ion in a trivacancy, i.e. $[X_3^-]_{+--}^0$, a defect proposed some time ago[51] for the V centres. Two sterically different alternatives are possible; the right-angled configuration (Fig. 18b) is the more stable and has lower formation energy than $[X_2^-]_{+-}^0$ defects, at least in chlorides. Similar structures must arise after chemical additions of excess halogen under high pressure (~5 MPa) of halogen gas.[52] In this case, the same V bands are produced as in irradiated crystals and dislocation loops are observed (Fig. 20a).[53] The proportionality of the V band absorption with pressure of the (diatomic) halogen gas[52,54] (Fig. 20b) confirms that the absorbed halogen is ultimately accommodated in molecular form.

Dislocation loops are thus extremely good sinks for halogen interstitials. These loops are responsible for all the usual observed changes in macroscopic properties after irradiation, for example irradiation hardening, decreased thermal conductivity, diffuse scattering, lattice parameter changes, and volume expansion. With increasing dose, the density of loops decreases continuously through coalescence of neighbouring loops.[55] Annealing above room temperature further coarsens the loop structure by a glide and self-climb mechanism. Consequently, there may be little correlation between the final distribution of interstitial loops and the distribution of molecular centres of the form $[X_2^0]_{+-}$ originally responsible for loop nucleation. The high defect densities, achieved by irradiating at elevated temperatures under conditions where F centres are also stabilized in large aggregate sinks (section on 'F-centre aggregation in halides'), lead to enormous loop growth and thus repeated intersections to form dense dislocation networks (Fig. 21) just as in metals.

Annealing crystals heavily irradiated at or below about 423 K results in two-stage recovery.[56] The first stage (Fig. 22) beginning at about 473 K results in a large reduction in anion vacancy centre concentration accompanied by corresponding changes in the V bands, thermoluminescence[57] and stored energy release.[58] There is little doubt that this stage corresponds to vacancy–interstitial recombination, probably from mobile F centres encountering dispersed molecular centres. The end products of this recombination stage should thus be anion–cation vacancy pairs which are complementary to the interstitial dislocation loops. There is no corresponding alteration in the distribution of interstitial dislocation loops during this first stage, however; these do not anneal out until a second stage well above 700 K,[55] presumably through absorption of vacancy pairs, with an additional release of stored energy.[58]

In heavily irradiated crystals, not all vacancy centres are removed in the first stage; about 30% remain (Fig. 22) in the form of alkali metal colloids which finally anneal out only above 600 K. This suggests that a substantial fraction of interstitial centres is unavailable for recombination, most likely because they are firmly clustered. In crystals containing a chemical excess of halogen (and thus no complementary anion vacancies), the first annealing stage does not exist; instead, the V bands evolve into a band at longer wavelength which corres-

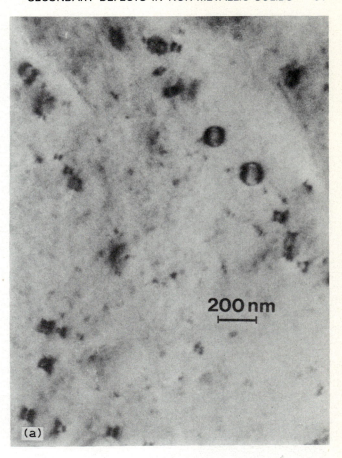

200 nm

(a)

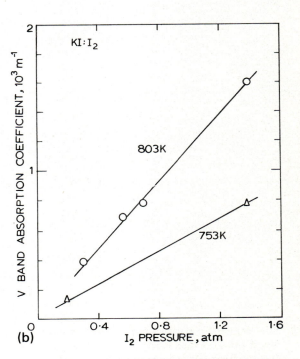

(b)

V BAND ABSORPTION COEFFICIENT, 10^3 m^{-1}

I_2 PRESSURE, atm

$KI:I_2$

803K

753K

a interstitial dislocation loops formed after iodine incorporation under 5 MN m^{-2} iodine vapour pressure at 773 K (reproduced from ref. 53); b proportionality of V band absorption with (diatomic) iodine pressure (reproduced from ref. 52)

20 Incorporation of a chemical excess of iodine in KI

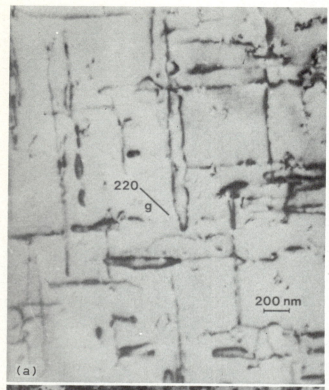

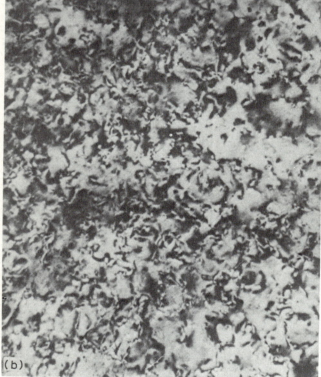

21   Dislocation network formation in NaCl
     irradiated at 423 K to; a ~0·1 dpa; b ~1 dpa
     (reproduced from ref. 55)

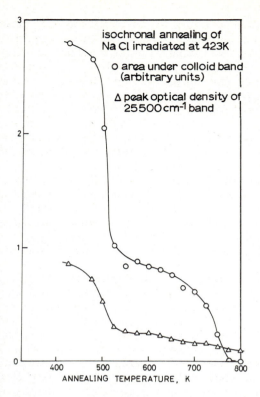

22   Recovery in NaCl heavily irradiated at 423 K
     as deduced from annealing of the alkali metal
     colloid optical absorption band; the first
     stage corresponds to recombination of F
     centres with relatively isolated halogen
     molecular interstitials; the second stage
     suggests some hydrogen clustering;
     interstitial loop shrinkage is not observed
     until ~600 K

The internal pressure and matrix accommodation strain for such precipitated inclusions, assuming no additional condensation of vacancy pairs, can be calculated.[55] Unstrained inclusions are expected for iodine in KI, and these are observed by transmission electron microscopy (Fig. 23a) in annealed crystals of KI additively coloured with excess iodine. Highly strained inclusions are expected for chlorine in KCl or NaCl and fluorine in LiF. Thin foils of KCl, NaCl, and LiF irradiated in the electron microscope[55] all develop such strained inclusions (Fig. 23b); the measured accommodation strains for KCl (~7%) agree well with the calculated strains. These results suggest that the ultimate mode of accommodating excess halogen in alkali halides is halogen precipitation, though the mechanism is by no means a straightforward one and involves several intermediate defect structures.

SUMMARY

Secondary defect structures in polyatomic non-metallic crystalline solids can significantly differ from those normally considered for metals. The differences are due to the presence of more than one atomic sublattice, the nature of the interatomic bonding, the tendency to maintain an ordered relationship between sublattices, and the necessity to preserve overall charge neutrality. Within these constraints, extended defects may preserve stoichiometry or serve to accommodate deviations from

ponds to absorption by free molecular halogen, suggesting that aggregation of molecular centres of the form $[X_2]^0_{+-}$ can precipitate halogen from solid solution. This is reasonable considering that the $[X_2]^0_{+-}$ defects represent fully substitutional halogen, i.e. substitutional on both sublattices.

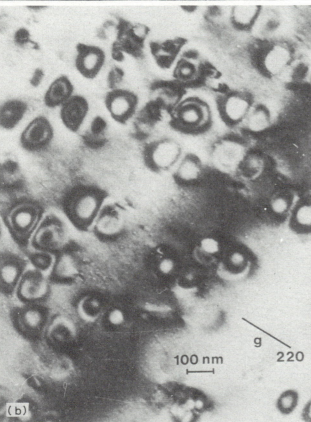

a unstrained inclusions (presumably iodine precipitation) in KI containing chemical excess of I$_2$; *b* strained inclusions in heavily irradiated KCl

23   Halogen inclusions in alkali halides

stoichiometry. In either case, the structural alterations may be described conveniently in terms of certain arrangements of 'molecular' units, in the first case units which preserve stoichiometric composition, and in the second defect units whose aggregation into defect complexes alters the chemical composition of the solid.

REFERENCES

1 M. PULS AND M. J. NORGETT: *J. Appl. Phys.*, 1976, **47**, 466

2 G. FONTAINE AND P. HAASEN: *Physica Status Solidi*, 1969, **31**, K67; L. A. DAVIS AND R. B. GORDON: *Physica Status Solidi*, 1969, **36**, K133

3 T. E. MITCHELL *et al.*: *Phil. Mag.*, 1976, **34**, 441

4 R. W. WHITWORTH: *Adv. Phys.*, 1975, **24**, 203

5 J. D. BERNAL AND R. H. FOWLER: *J. Chem. Phys.*, 1933, **1**, 515

6 N. BJERRUM: *Kgl. Danske Videnskab. Selskab. Mat. Fys. Medd.*, 1951, **27**, 56; *Science*, 1952, **115**, 385

7 J. W. GLEN: *Physik. Kondens Mat.*, 1968, **7**, 43

8 R. W. WHITWORTH *et al.*: *Phil. Mag.*, 1976, **33**, 409

9 W. JONES *et al.*: *J. Chem. Soc. Faraday Trans. II*, 1975, **71**, 138

10 K. H. G. ASHBEE: *Amer. Min.*, 1973, **58**, 947

11 D. T. GRIGGS AND J. D. BLACIC: *Science*, 1965, **147**, 292; D. T. GRIGGS: *Geophys. J. Roy. Astron. Soc.*, 1967, **14**, 19

12 R. D. BAËTA AND K. H. G. ASHBEE: *Physica Status Solidi (a)*, 1973, **18**, 155; in Proc. EMAG '75, (Ed. J. A. VENABLES), 307, 1976, Academic Press

13 G. DAS AND T. E. MITCHELL: *Radiation Effects*, 1974, **23**, 49

14 L. A. BURSILL AND B. G. HYDE: *Progress in Solid State Chemistry*, 1972, **7**, 178

15 J. G. ALLPRESS: *J. Solid State Chem.*, 1969, **1**, 66

16 J. E. CHISOLM: in 'Surface and defect properties of solids' (Eds. M. W. ROBERTS AND J. M. THOMAS), Vol. 4, 126, 1975, London, The Chemical Society

17 A. D. WADSLEY: in 'Non-stoichiometric compounds' (Ed. L. MANDELCORN), Chap. 3, 1964, New York, Academic Press

18 J. S. ANDERSON: in 'Surface and defect properties of solids' (Ed. M. W. ROBERTS AND J. M. THOMAS), Vol. 1, 1, 1972, London, The Chemical Society; J. S. ANDERSON AND R. J. D. TILLEY: *ibid.*, Vol. 3, 1, 1974

19 J. S. ANDERSON: in 'Defects and transport in oxides' (Ed. M. S. SELTZER AND R. I. JAFFEE), 25, 1974, New York, Plenum Press

20 L. W. HOBBS *et al.*: *Nature*, 1974, **252**, 383

21 G. CHASSAGNE *et al.*: *Physica Status Solidi (a)*, 1977, **40**(2), and **41**(1), in press

22 M. N. KABLER: in 'Radiation damage processes in materials' (Ed. C. H. S. DUPUY), 171, 1975, Leyden, Noordhof

23 A. E. HUGHES AND D. POOLEY: *J. Physics C*, 1971, **4**, 1963

24 U. JAIN AND A. B. LIDIARD: UKAEA Research Report, 1976, AERE-TP650; *Phil. Mag.*, 1977, **35**, 245

25 R. BULLOUGH AND R. S. NELSON: *Physics Technol.*, 1974, **5**, 29

26 R. W. WYCOFF AND E. D. CRITTENDON: *J. Amer. Chem. Soc.*, 1925, **47**, 2876

27 F. KOCH AND J. B. COHEN: *Acta Cryst.*, 1969, **B25**, 275

28 W. L. ROTH: *Acta Cryst.*, 1960, **13**, 140

29 A. K. CHEETHAM *et al.*: *J. Physics C*, 1971, **4**, 2160

30 N. N. GREENWOOD AND A. T. HOWE: *J. Chem. Soc. (Dalton)*, 1972, **1**, 110

31 C. R. A. CATLOW AND B. E. F. FENDER: *J. Physics C*, 1975, **8**, 3267

32 S. IIJIMA: in 'Diffraction studies of real atoms and real crystals', 217, 1974, Melbourne, Australian Academy of Sciences

33 K. SUZUKI: *J. Phys. Soc. Japan*, 1961, **16**, 67

34   J. E. STRUTT AND E. LILLEY: *Physica Status Solidi (a)*, 1976, **33**, 229

35   C. J. J. VAN LOOM AND D. J. W. IJDO: *Acta Cryst.*, 1975, **B31**, 770

36   A. I. SORS AND E. LILLEY: *Physica Status Solidi (a)*, 1975, **32**, 533; D. L. KIRK *et al.*: *J. Physics D*, 1975, **8**, 2013

37   M. J. YACAMAN, M. J. GORINGE AND L. W. HOBBS: *Physica Status Solidi (a)*, 1977, **39**(2), in press

38   C. M. SARGENT AND G. R. PURDY: *Phil. Mag.*, 1975, **32**, 27

39   M. J. YACAMAN AND J. P. HIRTH: 'Thin solid films', in press

40   J. W. MATTHEWS: *Phil. Mag.*, 1973, **29**, 797

41   B. T. M. WILLIS: *Proc. Brit. Ceram. Soc.*, 1964, **1**, 9

42   A. K. CHEETHAM *et al.*: *Solid-State Commun.*, 1970, **8**, 171; D. STEELE *et al.*: *J. Physics C*, 1972, **5**, 2677

43   J. M. BAKER: in 'Crystals with the fluorite structure' (Ed. W. HAYES), 341, 1974, Oxford, Oxford University Press; J. M. BAKER *et al.*: *Contemporary Physics*, 1972, **13**, 145

44   J. P. STOTT *et al.*: *J. Physics C*, 1977, **10**

45   C. R. A. CATLOW: *J. Physics C*, 1973, **6**, L64; UKAEA Research Reports, 1976, AERE-TP576, TP617, TP618

46   K. H. G. ASHBEE: *Disc. Faraday Soc.*, 1964, **38**, 309; in 'The chemistry of extended defects in non-metallic solids' (Ed. L. EYRING AND M. O'KEEFE), 373, 1970, Amsterdam, North-Holland

47   W. KÄNZIG AND T. O. WOODRUFF: *Phys. Rev.*, 1958, **109**, 220; *J. Phys. Chem. Solids*, 1958, **9**, 70

48   N. ITOH AND M. SAIDOH: *Physica Status Solidi*, 1969, **33**, 649; M. SAIDOH AND N. ITOH: *J. Phys. Soc. Japan*, 1970, **29**, 156; *J. Phys. Chem. Solids*, 1973, **34**, 1165

49   L. W. HOBBS *et al.*: *Proc. Roy. Soc.*, 1973, **A332**, 167

50   C. R. A. CATLOW *et al.*: *J. Physics C*, 1975, **8**, L34

51   T. P. ZALESKIEWICZ AND R. W. CHRISTY: *Phys. Rev.*, 1964, **A135**, 194

52   Y. UCHIDA AND Y. NAKAI: *J. Phys. Soc. Japan*, 1953, **8**, 795; 1954, **9**, 928; Y. UCHIDA *et al.*: *J. Opt. Soc. Amer.*, 1957, **47**, 246

53   D. LLEWELYN *et al.*: UKAEA Research Report, 1976, AERE-R8202, to be published

54   E. MOLLWO: *Nachr. Akad. Gottingen Math.-Phys. Kl.*, 1937, **1**, 97, 215; *Ann. Phys.*, 1937, **29**, 394

55   L. W. HOBBS: in 'Surface and defect properties of solids' (Ed. M. W. ROBERTS AND J. M. THOMAS) Vol. 4, 152, 1975, London, The Chemical Society

56   M. SAIDOH *et al.*: UKAEA Research Report, 1976, AERE-R8482, to be published

57   J. L. ALVAREZ-RIVAS AND V. AUSÍN: *J. Physics C*, 1972, **5**, 82; *Phys. Rev. B*, 1972, **1**, 4828

58   J. M. BUNCH AND E. PEARLSTEIN: *Phys. Rev.*, 1969, **181**, 1290

Point defect cluster hardening

P. B. Hirsch

Mechanisms of interaction of glide dislocations with loops (faulted and unfaulted) and tetrahedra are reviewed. It is shown that screws interact strongly with the defects, on making contact, to form helical configurations. The sweeping up of loops observed in quenched and irradiated metals is considered to be due to these mechanisms and to some involving other dislocations. The sweeping mechanisms operate only if the helical and jogged configurations are glissile; for large defects at low temperatures athermal processes operate when the helical or jogged segments cannot glide; little sweeping up occurs and the yield stress is temperature independent. Small loops glide more easily, channels occur readily, and the yield stress is determined by the need for some of the dislocations to overcome the elastic long-range stress repulsion from the defects before contact can be made. Some of the proposed interactions have been observed directly by electron microscopy, and the yield stress, its temperature dependence, and the slip structure of quenched and irradiated metals are discussed in some detail in terms of these mechanisms.

The author is at the University of Oxford

It is well known that clusters of point defects are generated in crystals by quenching or irradiation, the nature of the aggregates depending on the crystal structure, purity, annealing treatment, temperature of irradiation, etc. The defects identified by electron microscopy include dislocation loops, perfect and faulted, tetrahedra of stacking faults, and voids. The first step in any estimate of the hardening produced by these defects must be the determination of the mechanisms of interaction with the glide dislocations. Most attempts made so far to explain the strengthening have used a 'dispersed barrier' model in which the defects are treated as randomly distributed obstacles to dislocation motion, the defects being assumed to remain unchanged after the passage of the dislocation except for any jogs (or ledges in faults) produced. Relevant reviews have been published by Diehl,[1] Koppenaal and Arsenault,[2] Little,[3] and Kimura and Maddin.[4] It is known, however, that unlike the case for precipitation and dispersion hardened alloys, to which similar models are applied (for review *see* Brown and Ham[5]), in quench- and irradiation-hardened metals the glide dislocations combine with the defects and remove them. This phenomenon leads to local softening and is responsible for the coarse slip line structure typical of deformed quenched and irradiated metals. The formation of channels clear of defects is well established.[6-9] The question therefore arises as to whether the mechanisms responsible for sweeping up of the loops also affect the yield stress, or whether independent processes occur. The aim of the present paper is to review briefly the previously suggested interactions, to propose others, and to discuss and reconcile the observed strengthening and sweeping-up effects in terms of these processes. The interactions will be considered specifically for fcc metals, and the interaction of dislocations with voids will not be considered.

DISLOCATION-DEFECT INTERACTIONS

Intersection interactions with loops

We distinguish between 'intersection' and 'coalescence' interactions; 'intersection' in this context means that after the dislocation has passed the defect, it reverts back to its original configuration, apart from the jogs produced by intersection of the loop. A 'coalescence' reaction is one which leads to a permanent change in the configuration of the defect, or its combination with and removal by the glide dislocation. In this section we shall consider only intersection interactions.

Elastic interaction

Kroupa[10] and Kroupa and Hirsch[11] calculated the long-range elastic force of a prismatic loop on a glide dislocation. The total force on the dislocation decreases as $\sim r^{-2}$, where r is the distance from the loop, $r \gg R$, where R is the loop radius. The elastic interaction is therefore only important very close to the loop itself; the interaction is generally a maximum when the dislocation intersects the loop. An example for a screw dislocation intersecting a prismatic loop along its diameter is shown

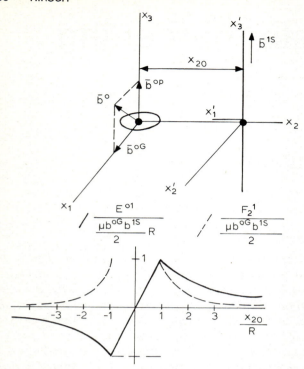

1   Interaction of screw dislocation, Burgers vector b^{lS} with a prismatic dislocation loop, Burgers vector $b°$; both the interaction energy E^{01} and total force F_2^1 are shown (from Kroupa[10])

in Fig. 1. In this case the maximum force is $F_{max} \sim Gb^2/2$ if the loop and the glide dislocations both have Burgers vectors of the same type. For the fcc case of prismatic loops on {111} with Burgers vectors $b = \frac{1}{2}[110]$ and glide dislocations on {111} with $b = \frac{1}{2}[110]$, taking account of all possible relative positions, the *average maximum* interaction energy is found to be $E_{max} \sim Gb^2R/4$, and the *average maximum* component of the force on the slip plane $F_{max} \sim Gb^2/4$, where G is the shear modulus. Since F falls off rapidly with distance (αr^{-2}), it is sufficient to consider only loops whose centres lie within a slab of thickness $\sim 2R$ about the slip plane. The force averaged over this thickness is estimated as $F \sim Gb^2/8$, and the interaction energy as $E \sim Gb^2R/8$. Thus, if l is the distance between obstacles along the dislocation, the yield stress τ_0 is

$$\tau_0 \sim \frac{Gb}{8l} \qquad (1)$$

Kroupa and Hirsch used the Mott statistics for l appropriate to a zig-zag dislocation configuration in the absence of stress. As Foreman[12] has pointed out it is more appropriate to use the Friedel criterion for a dislocation under stress (*see also* Brown and Ham,[5] and Nabarro[13]). Using the line tension value $T \sim Gb^2/2$, it can be shown readily that for n loops per unit volume:

$$\tau_0 \sim \frac{Gb\sqrt{Rn}}{16} \qquad (2)$$

The force distance curve for $r \gg R$ is of the form

$$F \sim \frac{Gb^2}{8}\left(\frac{R}{r}\right)^2 \qquad (3)$$

TABLE 1   Dislocation—loop interactions

Interaction	F_{max}	Yield stress	
Long-range elastic	$\dfrac{Gb^2}{8}$	$\dfrac{Gb\sqrt{Rn}}{16}$	Kroupa and Hirsch
Tetragonal distortion	$\dfrac{Gb^2}{3}$	$\dfrac{Gb\sqrt{Rn}}{5\cdot2}$	Fleischer
Junction reaction	$\dfrac{Gb^2}{4}$	$\dfrac{Gb\sqrt{Rn}}{5\cdot7}$	Friedel
		$\dfrac{Gb\sqrt{Rn}}{2\cdot8}$	Foreman

using the average value of the interaction. This law is in fact identical, apart from the numerical constant, to Fleischer's[14,15] approximation to the interaction between a dislocation and a tetragonal distortion. The activation energy is found to be

$$\Delta G \sim \frac{Gb^2}{8}R\left(1-\frac{\sqrt{\tau_T}}{\sqrt{\tau_0}}\right)^2 \qquad (4)$$

where τ_T is the yield stress at temperature T, and the temperature dependence is given by the Fleischer law

$$\tau_T^{\frac{1}{2}} = \tau_0^{\frac{1}{2}} - BT^{\frac{1}{2}} \qquad (5)$$

where

$$B = \left(\frac{kn^{\frac{1}{2}}\log(\dot{\varepsilon}/A)}{2bR^{\frac{1}{2}}}\right)^{\frac{1}{2}} \qquad (6)$$

$\dot{\varepsilon}$ is the strain rate, A is the pre-exponential parameter, and k is the Boltzmann constant. (Note that B depends on R.) (*See also* Ohr[16].) However, as Fig. 1 shows clearly the force–distance curve inside the loop is quite different; in this case the force is independent of distance, and in other cases a nearly linear variation is found. Depending on the relative signs of the dislocation and the loop, either the long-range stress field outside the loop or the short-range stress inside the loop can be controlling.

Fleischer's theory, which is an approximate treatment of the interaction between a dislocation and a tetragonal distortion gives:

$$\tau_0 \sim \frac{Gb}{3l} = \frac{Gb\sqrt{Rn}}{5\cdot2} \qquad (7)$$

the last term being derived using Friedel statistics, for a screw dislocation interacting with a Frank loop, and taking into account that only half the loops interact. The assumption is made that any loop lying with its centre within a slab of width $2R$ about the slip plane exerts the maximum force.

Fleischer[14,15] approximates the force–distance curve by the inverse square law, obtaining thereby the well known form.[5]

Since both the above treatments deal with the long-range stresses it is surprising that the estimates differ by a factor 3. The reason lies in the different assumptions made in determining the averages. Fleischer assumes that all loops lying within $\pm R$ of the slip plane interact with the maximum force $\sim Gb^2/3$; Kroupa and Hirsch assume that the average interaction within this slab is *half* the maximum force, the latter being $F_{max} \sim Gb^2/4$, which is close to the Fleischer estimate. The Kroupa and Hirsch F_{max} is already an average over various relative orientations of loops and dislocations, and includes cases where the interactions are small, whereas in Fleischer's case the loops with zero interaction are not included in the value F_{max}. Since with Friedel statistics $\tau_0 \propto F_2^3$, (*see* Brown and Ham[5]) a lower value of F has a more than proportional effect on τ_0. It appears then that the two basic estimates of F_{max} are quite close, and that a more thorough investigation is needed of the variation of F within the slab $\pm 2R$ (and possibly outside this region), which is responsible for the main difference between the two estimates of τ_0.

With regard to the temperature dependence of τ, this is a consequence of the inverse square law for the force–distance curve. The Fleischer approximation to the actual force–distance curve for tetragonal distortions has been criticized (e.g. Barnett and Nix[17]), but the inverse square law is a reasonable approximation for the long-range stress field from a dislocation loop. It does not apply however to the region inside the loop (*see* Fig. 1).

Various attempts have been made to interpret the magnitude and temperature dependence of quench and irradiation hardening using the above treatments. The Fleischer estimate for τ_0 gives reasonable agreement with the magnitude of hardening observed in quenched Al (*see* Maddin and Kimura[18], and Westmacott[19]), to within a factor of two or so. The experimental parameters are however rather uncertain.

It now seems generally accepted that irradiation hardening is due to interactions between glide dislocations and prismatic loops in displacement spikes (Silcox and Hirsch[20]). Frank *et al.*[21] interpreted the temperature dependence of the critical resolved shear stress of neutron-irradiated Cu in terms of the elastic interaction between a distribution of Frank vacancy loops on {111}, determined by electron microscopy,[22] and the glide dislocations intersecting them, using an analysis appropriate to a spectrum of obstacle strength.[23] The observed inhomogeneity of the deformation was taken into acount. Good agreement was obtained between theory and experiment without using any adjustable parameters. It should be noted, however, that the loops actually intersected by the dislocations could not be included in the calculations. Frank *et al.* justify this by showing that if these loops are assumed to act in the same way as those just not intersected by the dislocations, the agreement is equally good. Figure 1 shows, however, that the force–distance curve inside the loop differs from that outside, and since in an intersection the former may be rate controlling in some cases, the argument of Frank *et al.* does not seem tenable.

On the other hand, Koppenaal and Arsenault[2] found good agreement with Fleischer's theory without needing a spectrum of obstacle strengths, but the interpretation has been criticized on the grounds that adjustable parameters, particularly the athermal component of the stress, were used.[21]

Apart from the problem of adjustable parameters, Koppenaal and Arsenault have been criticized because the Fleischer treatment is not appropriate to loops, and because a distribution of obstacle strength is present and needed to explain all the observed effects.[21,24] On the first point, the Frank *et al.* treatment only considers part of the stress field of loops, i.e. the force field outside the loops, which is reasonably approximated by the Fleischer inverse square law; the part inside the loops is quite different and is not included. We therefore conclude that both treatments interpret the results in terms of basically similar models in which only the long-range stress field outside the loop is considered in an approximate manner, and the main difference is that one treatment takes a spectrum of obstacle strengths into account and the other does not. Neither theory deals with the interaction within the loop, and they both describe only the cases where the forces outside the loop oppose the dislocation motion.

Junction reaction

Friedel[25] was the first to suggest that a junction reaction might occur at a loop (*see* Fig. 2), and by analogy with forest hardening obtained an estimate:

$$\tau_0 \sim \frac{Gb}{4l} = \frac{Gb}{5 \cdot 7}\sqrt{Rn} \qquad (8)$$

(using Friedel statistics). Foreman[12] carried out a computer calculation, and showed that for {110} loops with $b = \frac{1}{2}[110]$, two of the six possible configurations lead to very strong junction reactions. The critical stress is found to be

$$\tau_0 \sim \frac{Gb}{2 \cdot 8}\sqrt{Rn} \qquad (9)$$

the numerical factor being independent of loop radius. If only the reactive loops are counted τ_0 is about twice as large as in equation (9). Foreman's estimate for perfect loops gives results of the correct order for quenched Al. Makin, *et al.*[26] correlated yield stress in irradiated and subsequently partially annealed Cu with the density of small defects <50 Å diameter, determined by electron microscopy. A good linear correlation was obtained with $n^{\frac{1}{2}}$ at 4 K showing that the strength is independent of the loop radius; the magnitude of the strengthening is about twice that predicted by equation (9), and close to that for the reactive loops in Foreman's calculations.

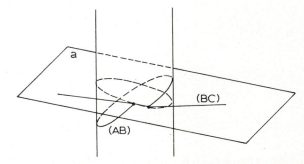

2 Junction reaction between a glide dislocation (*BC*) and a perfect prismatic loop gliding on its glide cylinder (from Saada and Washburn[29])

A criticism of Foreman's calculation is that it neglects the stress field of one part of the loop on another, and this may be particularly important for very small loops.

In common with the estimates from the long-range elastic stresses, F_{max} (in the line tension approximation) is independent of loop size; it is of the same order as that found for a random array of obstacles with a flat spectrum of breaking angle (Foreman and Makin[27]), and its value is about 30% below the Orowan stress. However, for very small loops the line tension approximation used cannot be correct and the strength of the interaction must decrease.

Coalescence reaction with loops

Perfect prismatic loops

In this discussion it is convenient to consider first the interaction between a triangular prismatic loop on (111) bounded by [110] directions with a glide dislocation. This establishes the type of reactions which can occur, using the Thompson tetrahedron.

In Fig. 3a a screw dislocation EF with Burgers vector **BC** on a interacts with perfect prismatic loops lying on any of the planes b, c, d, in Thompson's notation. The following cases may be distinguished:

Loop on b; Burgers vector BC

In this case EF cross slips on d, eliminating the part of the loop along AC. The result is shown in Fig. 3b. The screw has become helical; segments GA, CH are glissile on the cross-slip plane d; DC glides on a, the original slip plane of the dislocation. AD is a sessile jog in the sense that its glide plane is (100). The helix will of course try to straighten as much as possible by outward motion of G, C, H, to lower the energy.

Loop on b, Burgers vector BD (or BA)

In this case EF cross-slips as before, interacting with the loop along AC to produce the resultant Burgers vector **DC**. This sweeps across plane b reacting with the

remaining parts of the loop to form dislocations with Burgers vector **BC**; the net result is the same as in Fig. 3b. The result is the same for loops on b with Burgers vector **BA**.

Loop on c

The reactions are similar to those for loops on b.

Loop on d; Burgers vector BD (or CD)

EF can cross-slip again to react along AC, forming the Burgers vector **CD**; the loop then effectively collapses by glide of the three segments AC, BC, AB on planes b, a, c respectively towards D, leaving resultant segments along AD and DC with Burgers vector **BC**. Thus the final configuration is the same as in Fig. 3b. Similar reactions take place with loops with Burgers vector **CD**.

Loop on d; Burgers vector AD

In this case there is no first-order junction interaction since the Burgers vectors are at right angles. This case however needs more detailed consideration taking into account the dislocation dissociation.

Loop on a

The glide dislocation does not intersect these loops and no junctions are formed unless the loop rotates on its glide cylinder.

This discussion shows that in 8 out of 12 cases considered, the loop is removed by combination with the screw to form an angular helix. If the loops are originally on (110) planes normal to the Burgers vectors, then assuming that they can rotate into the appropriate (111) planes, this mechanism should occur in 5 out of 6 cases.

In practice the loops may be round or perhaps hexagonal. In that case the mechanism can still operate provided the loop can glide on its glide cylinder on to the appropriate (111) plane. The situation is sketched in Fig. 4. Suppose we consider EF with Burgers vector **BC** interacting with a loop with Burgers vector **BD** (case 2

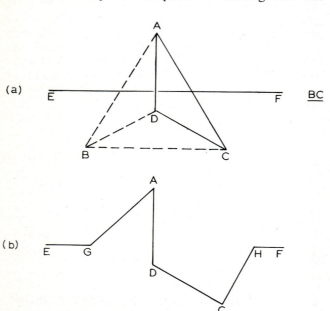

3   Screw dislocation with Burgers vector BC interacts with loop on _b_ with Burgers vector BC (_see_ text)

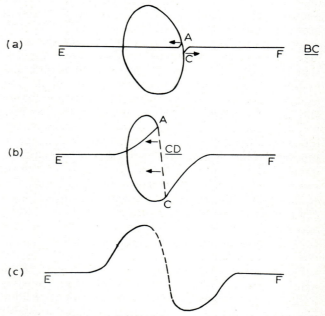

4   Screw dislocation with Burgers vector BC interacts with loop with Burgers vector BD (_see_ text)

above). The initial reaction takes place at the point of contact, and a short length *AC* is nucleated with Burgers vector **CD** (Fig. 4a). Nodes *A* and *C* are unstable and move out as shown in Fig. 4b, the dislocation *AC* sweeping across the plane of the loop, the assumption being that the loop will progressively turn into the glide plane *b* of dislocation *AC* (Burgers vector **CD**). Eventually *AC* reaches the opposite side of the loop and a helix will be generated. Clearly this mechanism depends on the possibility of nucleating the dislocation with **b** = **CD** at the junction, and on the ability of the dislocations to glide on their glide cylinder. The nucleation will be easier if the dislocation intersects the loop at a point where the loop is nearly tangent to the intersection of planes *b* and *d*, but the line tension forces at the junction are so large that the nucleation is likely to occur under more unfavourable conditions. In Al the loops are often hexagonal, and the chance of an intersection along the appropriate close-packed direction is good.

Observations on the formation of helices in quenched Al by dislocation-perfect loop interactions have been made by Strudel and Washburn.[28] They assume that the interaction occurs only if the Burgers vector of the perfect loop is the same as that of the glide dislocation; here we are suggesting that helices should also be formed if the Burgers vectors are different.

In the special case when the Burgers vector of the glide dislocation and the loop are the same, edge dislocations, or dislocations with intermediate character, can interact with the prismatic loop and carry away part of the loop (Saada and Washburn[29]).

Foreman and Sharp[30] have suggested an alternative mechanism for 'sweeping up' the loops. This is illustrated in Fig. 5, taken from their paper. The dislocation *G* interacts with the loop *L* to form a junction dislocation *J*. Then *J* and *L* rotate towards one another on their glide cylinders and combine to form a segment of the original dislocation *G*. Cross-slip of the junction dislocation is needed to effect complete combination. This mechanism is clearly related to that for the screw described above, and in both cases the loop has to glide on its glide cylinder and the dislocations must cross-

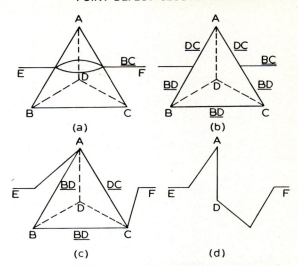

6 Screw dislocation with Burgers vector BC interacts with faulted loop on *d* with Burgers vector Dδ

slip. In our model the screw dislocation cross-slips; in the Foreman and Sharp model the junction dislocation cross-slips.

In conclusion, the mechanisms discussed in this section lead to the sweeping up of most of the dislocation loops by one set of glide dislocations, and this explains the clear channels observed on the primary glide plane, e.g. in quenched Al.[9,30] The combination of loops with screws to form helices has been observed experimentally.[29]

Faulted loops

Screw dislocation BC; loop on *d* with Dδ

The screw will cross-slip on *d* as shown in Fig. 6a. The two Shockley partials run across the face forming the resultant Burgers vectors indicated (Fig. 6b). Cross-slip of *EF* generates the configuration shown in Fig. 6c, which collapses towards *D* to form the helical dislocation in Fig. 6d, which is similar to that of Fig. 3b.

Screw dislocation BC; loop on *b* with Bβ

In this case the screw cross-slips on *d* and interacts along *AC*, forming the partial **βC** which sweeps across the loop plane *b* forming again the helical dislocation of Fig. 6d. Similar reactions occur with a loop on γ with **Cγ**.

60° dislocation BD; loop on *d* with Dδ

In this case the dislocation dissociates into the Frank partial **Dδ** plus the Shockley **Bδ** which sweeps over one part of the triangle. The resultant configuration, shown in Fig. 7b, consists of a jogged 60° dislocation, plus a smaller Frank loop left behind.

Mechanism *2* is described by Strudel and Washburn,[28] who have obtained evidence for it in Al. Mechanism *1* has been proposed by Silcox and Hirsch[31]; a similar mechanism occurs with tetrahedra of stacking faults (see below).

These mechanisms apply to quenched Al when the loops are faulted, and to the Frank loops in neutron-irradiated copper. These reactions probably occur spontaneously when the dislocations have made contact since they lead to a reduction in energy, the line tensions at the

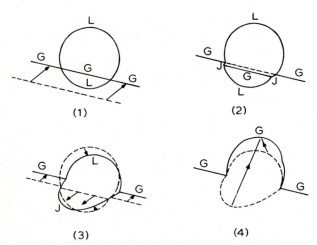

5 Coalescence of prismatic loop *L* with glide dislocation *G*; a junction dislocation *J* is formed; *J* and *L* then rotate towards one another on their glide cylinders and combine and form a segment of the original dislocation *G* (from Foreman and Sharp[30])

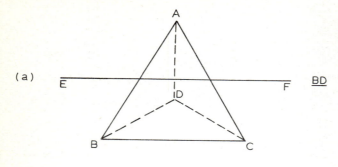

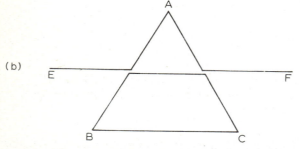

7   60° dislocation with Burgers vector BD interacts with faulted loop on *d* with Burgers vector Dδ

nodes providing the force for the various stages of the reactions. This would explain why in neutron-irradiated Cu, clear channels are produced on deformation even at 4 K.[32] Figure 8 shows a slip band in neutron-irradiated Cu deformed at room temperature in a high-voltage electron microscope.[33] The density of defects is substantially reduced in the slipped region (Fig. 8*a*), while the glide dislocations moving in the slip bands are found to be screws (Fig. 8*b*). (The geometry is such that edge dislocations would move out of the specimen close to the source.) The screws are heavily jogged and there is direct evidence for the generation of dipoles at superjogs, formed at pinning points which appear to be small loops. These experiments provide direct evidence for the sweeping up of loops by screws, but it should be emphasized that, unlike the bulk case, in the thin foils the glissile jogs can run out easily to the surface.

Tetrahedra of stacking faults

Kimura and Maddin[4] considered the following interaction of a screw with a tetrahedron. Figure 9 taken from their paper shows the sequence of events. The screw **AB** dissociates on *c*, and the Shockley partials formed remove the stacking fault on *c*, and then by combining with the stair-rods remove the faults on the other planes. The final helical configuration is shown in Fig. 9*c*; it should be noted that all segments except for *DC* are glissile either on *c* or *d*, and the helix will tend to straighten.

They also showed that a 60° dislocation will cause the defect to collapse to form a Frank loop on one of the (111) planes. Figure 10 shows the case where *EF* with **b** = **CA** meets the face *c*, dissociates on *c*, with **γA** sweeping over the plane to remove the fault. Interaction with stair-rods eventually results in the configuration in Fig. 10*b* where the tetrahedron has collapsed on to *a*. If the dislocation now turns round to be parallel **BC**, the last interaction discussed in the previous section occurs.

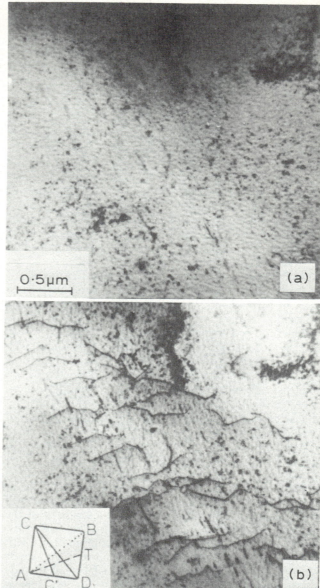

a dislocations out of contrast showing depletion of defects; *b* primary screw dislocations; note the pinning of the screws

8   Slip band in neutron-irradiated single crystal of Cu (10^{18} neutrons cm^{-2} at 80°C) formed by straining the crystal in a 1 MeV electron microscope

The small Frank loop remaining behind will redissociate to form a smaller tetrahedron or truncated tetrahedron. The first part of this reaction, i.e. the collapse, will presumably only occur if the tetrahedron is large and metastable.

This discussion shows that tetrahedra combine with screw dislocations or are partially removed leaving smaller defects. This explains the formation of channels in quenched gold deformed at elevated temperatures (Yoshida *et al.*,[34] Bapna *et al.*[7]). Figure 11 shows such a channel after 10% deformation at 260°C (Corbett and Hirsch[35]). Figure 12 shows a number of examples of the characteristic configuration expected from the combination of a screw with a tetrahedron. The three-dimensional configuration was checked by

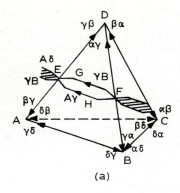

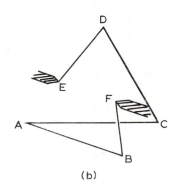

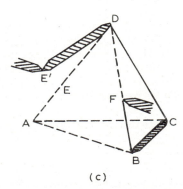

9 Interaction between screw dislocation *EF* on *c* with Burgers vector AB and a tetrahedron; the final configuration is an angular helix (*see* text; diagram from ref. 4)

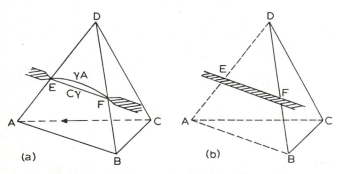

10 Interaction between 60° dislocation *EF* on *c* with Burgers vector CA and a tetrahedron; the final result is that the tetrahedron collapses into a Frank dislocation on *a* (*see* text; diagram from ref. 4)

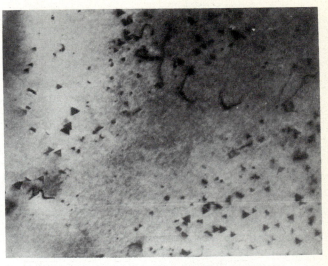

11 Slip channel in quench-hardened gold deformed 10% at 260°C[35] ×80 000

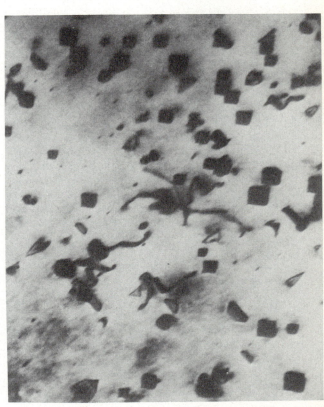

12 Jogged dislocations in quench-hardened gold deformed 2% in tension at 260°C; the angular helical shape of the dislocations is as expected from the interaction of screws with tetrahedra[35] ×103 000

stereomicroscopy and contrast experiments (Corbett and Hirsch[35]). Evidence for the 60° dislocation interaction was obtained by Silcox and Hirsch,[31] and Bapna *et al.*[7] reported a greater than average concentration of small tetrahedra in the channels.

MOBILITY OF JOGGED DISLOCATIONS

It is clear from the section on coalescence reaction with loops that once contact is made, screw dislocations coalesce with loops and tetrahedra forming helical configurations, and edge or 60° dislocations remove part of some of the defects. These interactions are not only confined to glide dislocations with these particular characters, since other dislocations will be able to change their directions at the junctions (by an amount depending on the applied stress and on the strength of the interactions) to make the reactions geometrically possible. It is likely therefore that these reactions, as well as those of the Foreman–Sharp[30] mechanism, apply to a wide range of configurations of the glide dislocations.

The reaction product is always a jogged dislocation, and jogs, particularly if they lie on planes other than the normal slip planes, may provide resistance to slip. In the case of dislocation loops in fcc crystals with sides parallel to [110] directions, or of tetrahedra, one (or more) segment of the resultant dislocation is a Lomer–Cottrell dislocation (LCD) and therefore difficult to move. It might be noted that the mechanism of coalescence of clusters of point defects with screws to form helices in quenched or irradiated metals, and the drag by sessile jogs, were discussed by Cottrell[37] at the 1957 meeting on 'Vacancies and other point defects in metals and alloys'.

We shall discuss the mobility of jogged dislocations specifically for the case in which the jogs are LC dislocations. Such dislocations may move in a number of different ways.

Athermal processes

Figure 13a shows the angular helical configuration characteristic of the coalescence reaction for a screw ($\mathbf{b} = \mathbf{BC}$), where AD is the sessile jog, EG, DC and HF are glissile on plane a, AG, CH on plane d. The screw can advance by reforming a prismatic loop, with nodes G and H coming together (Fig. 13b). The stress for this to occur is of the order of the Orowan stress. The reformed loop can fault or dissociate into a tetrahedron should this

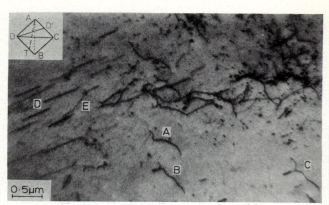

14   **Formation of dipoles by jogged screw dislocations (A, B, C) in a primary slip band in a neutron-irradiated crystal of Cu deformed in the HVEM at room temperature; both perfect (D) and faulted (E) dipoles are formed (Johnson and Hirsch[33])**

be energetically favourable. The stress is

$$\tau_0 \sim 0.8Gb/L \tag{10}$$

where L is the mean distance between obstacles in the slip plane.

If one or both jogs on the cross-slip plane had moved away (the local stresses near the defect, as for a misfitting inclusion, may favour this), for example the segment CH, a dipole can be formed. The stress for this to occur will be of the same order but somewhat larger than the Orowan stress. We believe that this is the mechanism responsible for the generation of dipoles by screws moving in slip bands in neutron-irradiated Cu, and subsequently deformed in an HVEM at room temperature; Fig. 14 shows examples of such dipoles.[33]

Thermally activated processes

Glide process

Consider a dislocation $GACH$ gliding on a, with Burgers vector $\mathbf{BD}$ at an angle ϕ to the line of the dislocation, with a LC segment AC (Fig. 15). The LCD can glide in the 100 plane, but only generating constrictions and kinks at the nodes A or C which then travel along AC from one end to the other. Figure 15a shows one possible mode of advance where a constricted kink is generated at C, the kink travelling to a position EE' during the activation, where any interaction between the double constriction and that at F can be overcome by the applied stress component on the 100 plane (τ_{100}), the

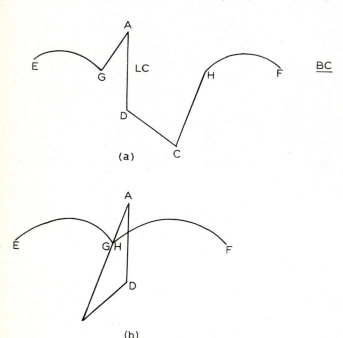

13   **a Angular helical turn on screw dislocation with Burgers vector BC with LC segment AD; b bypassing of loop by pressing nodes G and H together, reforming a prismatic loop**

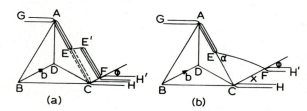

15   **a Dislocation GACH advances by glide by generating a kink EE'F in the LC dislocation which glides on the (100) plane (see text); b dislocation GACH advances by transforming the LC dislocation into a constricted glissile configuration on the (100) plane (see text)**

kink then travelling spontaneously towards A. The saddle point configuration is difficult to determine, but ignoring the work done by τ_{100} on the 100 plane, the activation energy for the process must be at least

$$\Delta H \sim U_c - \tfrac{1}{2}\tau b^2 l \sin \phi \qquad (11)$$

where U_c is the energy of the constricted kink EE', and l the distance between obstacles on the dislocation line. U_c includes the energy of the double constriction as well as the energy of the kink EE' and the reduction in energy due to shortening of the glide dislocation by $CF \cos \phi$. Values of U_c are not known, but using Stroh's[38] estimate of constriction energy of glide dislocations, $U_c \sim \tfrac{1}{4} \to \tfrac{1}{2}Gb^3$ for dislocation widths of ~ 5–$10b$, so that for Cu, Ag, and Au this mechanism is only going to be of any importance for dislocations near edge orientations.

Alternatively, the LCD may be changed into a glissile configuration as shown in Fig. 15b, where EF is the constricted part of the dislocation on (100). If we define the line tensions of the LCD, the constricted configuration EF, and the ordinary glide dislocation E_{LC}, E_C, and E_0 respectively, the energy difference between $AEFH'$ and $AECH$ is

$$\sim U_{CI} + \frac{E_C x}{\sin \alpha} - \frac{E_{LC} x}{\tan \alpha} - E_0 x \cos \phi - \tfrac{1}{2}\tau blx \sin \phi$$
$$- \tfrac{1}{2}\frac{\tau_{100}bx^2}{\tan \alpha} \qquad (12)$$

where U_{CI} is the interaction energy between the two half constrictions E and F and where α is the critical angle such that E will spontaneously move towards A. The condition for α is that

$$E_C \cos \alpha \lesssim E_{LC} \qquad (13)$$

where U_{CI} and the term $E_0 x \cos \phi$ have been neglected. The activation energy is obtained by maximizing (12) with respect to x, and neglecting the two terms just mentioned:

$$\Delta H \sim \left(\frac{E_C}{\sin \alpha} - \frac{E_{LC}}{\tan \alpha} - \tfrac{1}{2}\tau bl \sin \phi\right)^2 / 2\tau_{100}b \cot \alpha \qquad (14)$$

and the critical x is given by

$$x_c = \left(\frac{E_C}{\sin \alpha} - \frac{E_{LC}}{\tan \alpha} - \tfrac{1}{2}\tau bl \sin \phi\right) / \tau_{100}b \cot \alpha \qquad (15)$$

For reasonable values of $\Delta E = E_C - E_{LC} \sim (E_C/10)$, appropriate to Cu, Ag, Au, $\sin \alpha/2 \sim (1/20)^{\frac{1}{2}}$, and for $\phi \to 0$, x_c takes large values, say $l/5$, and ΔH is very large (several eV). The lengths of the jogs corresponding to such large values of x_c often exceed those found in quenched and irradiated metals. For jog lengths $AC = h < x_c \cot \alpha$, from (12) and (13)

$$\Delta H \sim hE_C \sin \alpha \tan \alpha - \tfrac{1}{2}\tau blh \tan \alpha \sin \phi \qquad (16)$$

the glissile configuration becoming stabilized when E reaches A. The sessile–glissile transformation can occur athermally at $T = 0$ K if

$$\tau_{t0} \sim \frac{2E_C \sin \alpha}{bl \sin \phi} = \frac{2\sqrt{2}\sqrt{E_C \Delta E}}{bl \sin \phi} \qquad (17)$$

For $\Delta E \sim \tfrac{1}{10}E_C \sim \tfrac{1}{20}Gb^2$,

$$\tau_{t0} \sim \frac{Gb}{2 \cdot 2 l \sin \phi} \qquad (18)$$

Thus, for dislocations with substantial edge character $\tau_{t0} < \tau_0$, but for near screw dislocations the Orowan stress is the smaller of the two stresses. When the jog length h is less than the normal dissociated width of the dislocation, the jog will not dissociate to its full width because of the constraining effects from the constriction at the two ends. Consequently ΔE will be less than the estimate made above, and τ_{t0} will be smaller. It should also be noted that there is a dependence of τ_{t0} on stacking fault energy through $\Delta E/E_C$, this parameter being smaller for, say, Al than for Cu, resulting in a smaller τ_{t0} for Al than that given in (18). At higher temperatures

$$\tau_t = \tau_{t0} - \frac{2KT}{blh \tan \alpha \sin \phi} \log(A/\dot{\varepsilon}) \qquad (19)$$

where A is the pre-exponential factor in the usual equation for the strain rate, $\dot{\varepsilon}$. Since $\sin \phi$ must be fairly large for this mechanism to be rate controlling, and with $\tan \alpha$ typically $\sim \tfrac{1}{2}$, the rate of change of τ_t with T is rather small since the activation volume $v = \tfrac{1}{2}blh \tan \alpha \sin \phi$ is large, for all but very small jog lengths.

Comparing (17) with the corresponding estimate from (11) for the process described in Fig. 15a, it seems that the stresses for the two processes are of the same order, and the sessile–glissile transformation would therefore be more likely at low temperatures.

This discussion assumes that the dislocation can glide easily when constricted on the (100) plane. This is, however, quite uncertain and the velocity of the dislocations on the (100) plane may be a function of temperature and stress.

Finally, it should be noted that the glissile configuration is inherently unstable and could revert back to the sessile configuration if the local dislocation configuration and stresses permit this. It would seem likely that this would happen as the advancing dislocations accumulate further segments of this type, leading to a reduction of the distance l between obstacles along the dislocations, i.e. to work hardening. The rate of work hardening depends on the mean free path λ before the reverse transformation occurs; λ is not known, and it is not clear whether this mechanism can contribute to the deformation other than in the microstrain region. The minimum $\lambda \sim L$, the mean distance between obstacles in the slip plane, and the minimum strain is $\rho bL \sim 10^{-4}$, where $\rho =$ dislocation density is taken as 10^8 cm^{-2}, and $L \sim 3 \times 10^{-5}$ cm. At the present time it is uncertain whether this mechanism can control the yield stress as normally measured.

Climb and glide process

Consider the dislocation $EFACG$ with Burgers vector **BD**, glissile segments EF and CG in plane a, and AF in the cross-slip plane c, and with the LCD along AC (Fig. 16a). This configuration is typical of the products of the coalescence reactions. The line tension forces at F, A, and C will tend to make the dislocations climb to straighten the helical configuration. The force exerted by the gliding parts of the dislocation also cause the segments to climb. This is clear from Figs. 16a and b, the latter being a projection on to the glide plane a. By climbing upwards from CG to $C'G'$ this dislocation can move forwards as shown in Fig. 16b. This process involves climb of both AC and CG, in directions

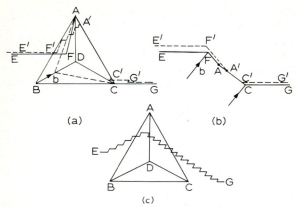

16   a Jogged dislocation *EFACG* (Burgers vector *BD*, glide plane *a*) advances by climb and glide; the LC and glide dislocations at *C* climb in directions shown; similarly the parts of the dislocation at *A* and *AF* on the cross-slip plane, **b** projection of configuration on to the glide plane *a* showing that climb at *C* and *A* in the directions shown in (*a*) permits the dislocation to advance; **c** final jogged but glissile configuration

shown. Similarly by *AF* climbing in the directions indicated by the dotted configuration, the dislocation *EF* can advance to *E'F'*. Since the climb process involves moving vacancies over only short distances from one part of the dislocation to another, pipe diffusion is likely to be the controlling process, particularly at low temperatures. The mechanism also requires the generation of jogs at the corners. The helical segment becomes glissile when the sessile dislocation is changed into a completely jogged configuration as shown in Fig. 16*c*; the LCD has effectively been changed by climb into a glissile configuration on planes *a* and *c*. Assuming that the time, t, taken to reach this configuration is controlled by the activation energy for the formation of jogs at *C* and *A*,

$$t^{-i} \sim \frac{\nu h}{2b} A \exp - \left[U_p + U_j - \frac{\tau b^2 l}{2\sqrt{3}} \cos \phi - \frac{2Eb^2}{h} \right] / kT \quad (20)$$

where ν is the Debye frequency, A an entropy factor, U_p the activation energy for pipe diffusion, U_j the jog energy, E the line tension; the line tension term varies of course as the dislocation climbs and the radius of curvature changes, and the last term in the bracket takes account of this force in an approximate manner only. Assuming that the dislocation element can advance over an area of slip plane $\sim L^2$ when the glissile configuration is obtained, we may write

$$\dot{\varepsilon} = \frac{1}{2}NL^2 h\nu \exp - \left[U_p + U_j \right.$$
$$\left. - \frac{\tau_c b^2 l \cos \phi}{2\sqrt{3}} - \frac{2Eb^2}{h} \right] / kT \quad (21)$$

whence

$$\tau_c = \frac{1}{v}\left(U_p + U_j - \frac{2Eb^2}{h} \right) - \frac{KT}{v} \log\left(\frac{NL^2 h\nu}{2\dot{\varepsilon}} \right) \quad (22)$$

where N = number of dislocation elements per unit volume, and v = activation volume = $b^2 l \cos \phi / 2\sqrt{3}$. The stress at $T = 0$ K is given by

$$\tau_{co} = \frac{2\sqrt{3}(U_p + U_j - 2Eb^2/h)}{b^2 l \cos \phi} \quad (23)$$

In contrast to the glide mechanism (equation 17), we note that (*a*) the obstacle is removed permanently and the mechanism can therefore control the yield stress; (*b*) the climb mechanism is particularly favourable for dislocations with large screw components, and these are likely to take part in coalescence reactions; (*c*) τ_{co} depends on the size of the jog, and for very small jogs only a few atoms long the line tension term is likely to be sufficient to compensate for the jog energy term, and even for the formation energy of the vacancy at the dislocation, which forms part of the U_p term; on the other hand for large h, τ_{co} can be large and exceed the Orowan stress for widely dissociated dislocations; (*d*) the activation volume v is considerably smaller for the climb mechanism for all but the smallest jogs.

Conclusions

Figure 17 summarizes the results of this section. Curve *a* gives the athermal Orowan stress; curve *b* gives the stress for the sessile–glissile transformation by glide for dislocations with edge components with long jogs. While this mechanism should be important in the microstrain region, it is not clear whether it can control the yield stress as normally measured. Curves *c* show the expected variation of τ with T for the climb and glide transformation for dislocations with screw components. For small jogs it is possible only to make guesses, and these are sketched schematically in the figure for various jog sizes. It is suggested in Fig. 17 that small jogs only a few Burgers vectors long should be able to move with the dislocations by transforming into glissile configurations by climb at stresses less than the Orowan stress. For long jogs, however, the athermal Orowan stress should apply at low temperatures, or possibly the sessile–glissile transformation by glide if the mean free path is sufficiently large, followed by a decrease in stress at higher temperatures when the climb mechanism for dislocations with screw components becomes favourable (Fig. 17).

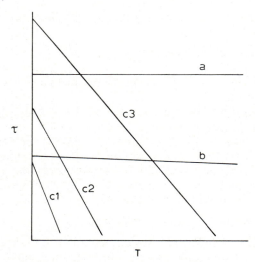

a Orowan or dipole formation stress; **b** sessile–glissile transformation for edge dislocations with long jogs; **c** climb-glide mechanisms for screws with (1) short jogs (*h* ~ 3*b*), (2) *h* ~ 5*b*, (3) long jogs (*h* ⩾ 50*b*)

17   Schematic diagram indicating the variation of stress with temperature for various mechanisms

YIELD STRESS AND CONDITIONS FOR SWEEPING UP OF DEFECTS

The conditions for sweeping up the defects are:

(i) the dislocation must make contact with the defects
(ii) cross-slip must occur at junctions or faces
(iii) loops not lying in glide planes must be able to rotate into the glide plane of the reaction product of the intermediate configuration
(iv) the reaction product of the intermediate configuration must be able to glide across the loop
(v) the helical turns or jogs must be able to move with the dislocation.

With regard to (i), some of the dislocations will be attracted to the defects, others repelled by the long-range stress. This repulsive interaction must be overcome first. Condition (ii) is probably automatically satisfied, since cross-slip is likely to occur spontaneously at dislocation junctions where line tension forces are very large, or on stacking fault faces, where the stacking fault provides the driving force. Conditions (iii) and (v) are similar, and we shall consider only (v), since for the two examples to be discussed (iii) does not apply. Condition (iv) is satisfied for fcc crystals, but for the bcc case, in which the intermediate reaction products are dislocations with [100] Burgers vectors, the implication is that these dislocations must be glissile.

We shall now apply the above principles to two cases: neutron-irradiated copper, and quenched gold.

Neutron-irradiated copper

The temperature dependence of the yield stress of neutron-irradiated and partially annealed copper single crystals is shown in Fig. 18 (Rühle[22]). In this case the defects in the irradiated metal are known to be Frank type loops with a distribution of sizes, with most of the measurable defects being in the size range up to ~50 Å diameter (Rühle[22]). For many of the smallest loops (<25 Å diameter) the Frank loops are unlikely to be significantly dissociated. Consequently, the stress

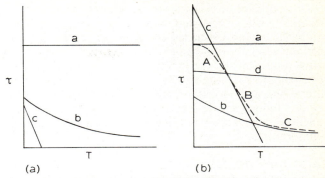

19 Schematic variation of stress with temperature for various processes occurring on deforming neutron-irradiated copper *a* as-irradiated, and *b* partially annealed after irradiation. In (*a*) the τ *v. T* curve follows curve *b*; in (*b*) it follows the dashed curve, assuming Orowan mechanism in region *A*. *a*, Orowan stress; *b*, elastic repulsion from loops; *c*, climb-glide process for screws with jogs; *d*, sessile–glissile transformation for dislocations with edge components with sessile jogs

required to move the jogged dislocations or to transform them into glissile configurations is expected to be low. In this case therefore it is quite likely that the rate-controlling stress is the long-range repulsive stress from the loops, and this may explain the reasonable agreement between experiments and the theories based on long-range elastic stresses (Frank *et al*,[21] Koppenaal and Arsenault[2]). The situation envisaged in this case is sketched in Fig. 19*a*. It should be noted however that there is considerable uncertainty about the magnitudes of the long-range stresses because of the averaging difficulty described in the second section. Figure 19*a* should be considered as schematic. Since the stress to move the jogged dislocation is less than the long-range elastic stress which controls the yield stress, sweeping up of loops should occur at 4 K, in agreement with experiment.[32,39]

After partial annealing the smallest loops will anneal out first,[24] so that the size distribution is shifted towards larger sizes, and there will be an increasing tendency for the loops to be hexagonal giving rise to sessile jogs in the glide dislocations following the coalescence reaction. (There is evidence that small Frank loops in irradiated fcc metals are dissociated–for review *see* Eyre[40].) We envisage the situation sketched in Fig. 19*b*, where at low temperatures (region *A*) the yield stress is likely to be controlled by the Orowan mechanism, (or possibly by the sessile–glissile transformation by glide, if the mean free path for the glissile configurations is sufficiently large); in region *B* the yield stress is determined by the climb and glide process, and in region *C* the long-range elastic stresses are rate controlling. The temperature dependence of the yield stress is qualitatively of this type (*see* Fig. 18). In the athermal region *A* little sweeping should occur since the loops are not removed by screws, although part removal of loops by edge (the 'chopping' process) should still occur. Experimentally fine slip lines and no 'channelling' are observed.[24,39] In regions *B* and *C* the sweeping process should occur, in agreement with experiment.

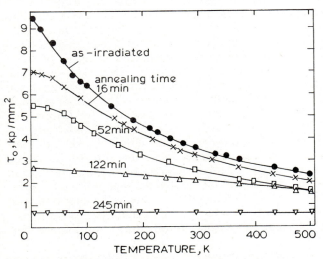

18 Temperature dependence of critical resolved shear stress of neutron-irradiated and subsequently partially annealed crystals of copper (nvt = 10^{19} neutrons cm^{-2}; annealing temperature 325°C; strain rate = 3×10^{-6} s^{-1}) (Rühle[22])

This explanation is similar to one already proposed by Makin,[24] who suggested that after annealing the strengths of the obstacles exceeds Gb^2 (but did not explain why). It should be noted that the long-range stress repulsion and the junction reaction give strengths independent of loop size (see section on intersection interaction with loops); in our model the large strengths are due to long sessile jogs arising from the coalescence reactions, which prevent the glide of dislocations at low temperatures.

Quenched gold

The temperature dependence of the flow stress of quenched polycrystalline gold is shown in Fig. 20, taken from Yoshida et al.[34] At low temperatures (region A) τ is nearly independent of T, when corrected for the temperature dependence of the shear modulus, but depends on the strain rate $\dot{\varepsilon}$; above 500 K (region B) there is a rapid decrease in τ with increasing T, while at even higher temperatures (region C) τ becomes nearly independent of T. The specimens were found to contain 9×10^{14} tetrahedra per cm^3, of average size 500 Å. Channels free of dislocations were observed after deformation in region B but not in region A. This has been confirmed by Corbett and Hirsch,[35] and Fig. 11 shows an example of a channel in a sample with tetrahedra in the same size range (~ 500 Å) after deformation at 260°C. Bapna and Meshii[36] and Imashimizu and Kimura[41] found similar behaviour for quenched single and polycrystals respectively, but in their case the onset of the rapid decrease of τ with T was found to be at ~ 300 K. Bapna and Meshii[36] determined the density and size of the tetrahedra to be $\sim 8.5 \times 10^{15}$ cm^{-3} and 135 Å respectively. These workers observed fine slip after deformation in region A, and coarse slip in region B, but channelling was observed to occur even in region A, as well as in region B.

The following interpretation is suggested for these results. In this case the sessile jogs are so long that the yield stress is determined by the glide of jogged dislocations in regions A and B. At low temperatures, region A, the yield stress should be determined by the Orowan stress. In the Bapna and Meshii[36] experiments the yield

stress at 0 K is ~ 2.9 kg/mm^2, compared with a calculated Orowan stress of ~ 6.2 kg/mm^2. Similarly, in the experiments of Yoshida et al.[34] the tensile stress at 0 K at the slowest strain rates used is ~ 4 kg/mm^2, compared with an estimated Orowan (tensile) stress (assuming tensile stress = twice shear stress) of ~ 7.8 kg/mm^2. At high strain rates, in the latter experiments the yield stress reaches a limit of about 7 kg/mm^2, in good agreement with the calculated Orowan stress. The reason for the smaller yield stress at low strain rates is not clear. It is tempting to identify this with the stress required for the sessile–glissile transformation for dislocations with edge components (equation 18), provided the mean free path is sufficiently large. The stress at 0 K was estimated to be about half the Orowan stress (see section on the glide process), in good agreement with the experimental values at low strain rates. In that case the strain rate dependence may be due to the resistance to slip of the constricted configuration on the (100) planes. Alternatively, however, the lower values of yield stress may be associated with an inhomogeneous distribution of tetrahedra, a more representative structure of the specimens being sampled at higher strain rates, the strain rate dependence being perhaps connected with the multiplication of dislocations. Further studies are clearly required on this point. In region A there should only be limited sweeping up of defects by the chopping mechanism for 60° dislocations. Bapna and Meshii[36] found channels after 5% deformation in region A, with an increase in the density of small tetrahedra remaining in the channel, in agreement with the 60° dislocation reaction. However, Yoshida et al.[34] did not find any channels in their specimens, but their observations appear to have been made after smaller extensions (1%); no well defined channels were observed by Corbett and Hirsch[35] in region A, but some limited sweeping up mechanism could not be ruled out.

Region B is thought to be controlled by the climb and glide mechanism. For large tetrahedra the activation energy (at $\tau = 0$) should be $U_0 \sim U_p + U_j$ (see equation 21); Yoshida et al.[34] report $U_0 \sim 1.7$ eV; this seems entirely reasonable, since U_p is expected to be ~ 1 eV, suggesting a reasonable value for U_j. The activation volume v calculated for the climb and glide process (see equation 22) for screws is $\sim 7 \times 10^{-21}$ cm^3, compared to an experimental value (interpreted in terms of shear rather than tensile stress) of $\sim 8 \times 10^{-21}$ cm^3. The reasonable agreement may be fortuitous in view of the inhomogeneity of the deformation. The shift of region B to lower temperatures, as observed by Bapna and Meshii[36] and by Imashimizu and Kimura[41] is as expected qualitatively from the smaller tetrahedra sizes in their specimens (see equation 22), but quantitatively the effect cannot be explained on the basis of the line tension term $2Tb^2/h$ and the quoted average tetrahedra sizes. This suggests that the effect is due to the distribution of tetrahedra sizes present, the beginning of stage B corresponding to the small tetrahedra within the distribution.

Region C is considered to be controlled by long-range internal stresses. The different regions and mechanisms are summarized in Fig. 21.

Since in regions B and C the jogs glide or climb with the dislocations, channelling should be efficient, as observed in practice.[34-36]

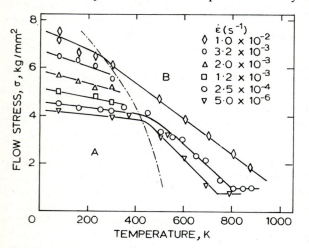

20   Flow stress of quenched gold polycrystals as a function of temperature and strain rate; the specimens contain 9×10^{14} cm^{-3} tetrahedra, size 500Å (from Yoshida et al.[34])

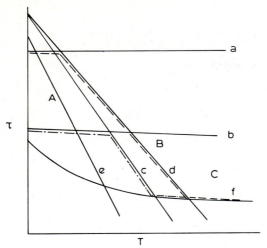

21 Schematic variation of stress with temperature for various processes occurring on deforming quenched gold: *a*, Orowan stress; *b*, sessile–glissile transformation by glide for dislocations with edge components with sessile jogs; *c*, *d*, climb-glide mechanism for screws for slow and fast strain rates; *e*, climb-glide mechanism for screws with smaller jogs; *f*, long-range elastic repulsion. The dashed-dotted and dashed curves show schematically the suggested τ/T curves at low and high strain rates, where at slow strain rates region *A* is tentatively identified with *b*

7   At high temperatures, long sessile jogs can change into glissile configurations by a climb and glide process, the climb being controlled by pipe diffusion. The sweeping up process is efficient under these conditions.

CONCLUSIONS AND SUMMARY

Although firm quantitative estimates of the parameters characterizing the various processes (i.e. stress, temperature dependence, activation volumes) have not been made, and must await further detailed calculations, the mechanisms which have been identified seem to account for the main features of the 'channelling' phenomena and for the temperature dependence of the yield stress in quenched and irradiated metals. Our conclusions are as follows.

1   Strong coalescence interactions occur between screws (and some other dislocations) and loops and tetrahedra, forming helical (and jogged) configurations when the dislocations and defects make contact.

2   The 'sweeping up' of loops and tetrahedra on deformation of quenched or irradiated metals is due to these mechanisms.

3   The 'sweeping up' process occurs only if the helical and jogged configurations can glide easily. Resistance to glide is due to jogs lying on abnormal slip planes and in fcc crystals to Lomer–Cottrell dislocations in particular.

4   Long Lomer–Cottrell jogs arising from reactions with large loops or tetrahedra will not be glissile at low temperatures, and athermal processes (i.e. Orowan stress or possibly the sessile–glissile transformation by glide mechanism) control the flow stress. Only limited sweeping up is possible under these conditions.

5   Short jogs tend to be glissile and the sweeping up will be efficient. The yield stress will be controlled by the long-range repulsive stresses from the defects which have to be overcome before contact is made.

6   The present theoretical treatments of the temperature dependence of the yield stress of neutron-irradiated metals take account of the long-range elastic stresses, and because of 5 have some justification.

REFERENCES

1   J. DIEHL: 'Vacancies and interstitials in metals', (Eds. A. SEEGER, *et al.*), 739, 1970, Amsterdam, North-Holland
2   T. J. KOPPENAAL AND R. J. ARSENAULT: *Metallurgical Rev.*, 1971, **157**, 175
3   E. A. LITTLE: *Metallurgical Rev.*, 1976, **21**, 25
4   H. KIMURA AND R. MADDIN; 'Lattice defects in quenched metals', 319, 1965, New York, Academic Press
5   L. M. BROWN AND R. K. HAM: 'Strengthening methods in crystals' (Eds. A. KELLY AND R. B. NICHOLSON), 12, 1971, London, Applied Science
6   J. V. SHARP: *Phil. Mag.*, 1967, **16**, 77
7   M. S. BAPNA *et al.*: *Phil. Mag.*, 1968, **17**, 177
8   U. ESSMANN AND A. SEEGER: *Physica Status Solidi*, 1964, **4**, 177
9   M. MESHII: 'Lattice defects in quenched metals', 430, 1965, New York, Academic Press
10   F. KROUPA: *Phil. Mag.*, 1962, **7**, 783; *Freiberger Forschungsh.*, 1965, **B101**, 181
11   F. KROUPA AND P. B. HIRSCH: *Disc. Faraday Soc.* 1964, **38**, 49
12   A. J. E. FOREMAN: *Phil. Mag.*, 1968, **17**, 353
13   F. R. N. NABARRO: 'The physics of metals 2', (Ed. P. B. HIRSCH), 152, 1975, Cambridge University Press
14   R. L. FLEISCHER: *Acta. Met.* 1962, **10**, 835
15   R. L. FLEISCHER: *J. Appl. Phys.*, 1962, **33**, 3504
16   S. M. OHR: *J. Appl. Phys.*, 1969, **39**, 5335
17   D. M. BARNETT AND W. D. NIX: *Acta Met.*, 1973, **21**, 1157
18   R. MADDIN AND H. KIMURA: Report of AERE Symposium on the 'Nature of small defect clusters', 503, 1966, London, HMSO
19   K. H. WESTMACOTT: Report of AERE Symposium on the 'Nature of small defect clusters', 521, 1966, London, HMSO
20   J. SILCOX AND P. B. HIRSCH: *Phil. Mag.*, 1959, **4**, 1356
21   W. FRANK *et al.*: *Physica Status Solidi*, 1968, **26**, 671
22   M. RÜHLE: *Physica Status Solidi*, 1968, **26**, 661
23   W. FRANK: *Physica Status Solidi*, 1968, **26**, 197
24   M. J. MAKIN: *Phil. Mag.*, 1968, **18**, 1245
25   J. FRIEDEL: 'Electron microscopy and strength of crystals' (Eds. G. THOMAS AND J. WASHBURN), 391, New York, Interscience
26   M. J. MAKIN *et al.*: *Phil. Mag.*, 1966, **13**, 729
27   A. J. E. FOREMAN AND M. J. MAKIN: *Can. J. Phys.*, 1967, **45**, 511
28   J. L. STRUDEL AND J. WASHBURN: *Phil. Mag.*, 1964, **9**, 491
29   G. SAADA AND J. WASHBURN: *J. Phys. Soc. Japan Suppl. 1*, 1963, **18**, 43
30   A. J. E. FOREMAN AND J. V. SHARP: *Phil. Mag.*, 1969, **19**, 931
31   J. SILCOX AND P. B. HIRSCH: *Phil. Mag.*, 1959, **4**, 72
32   J. V. SHARP: 4th European Reg. Conf. on Electron Microscopy, Rome, 1968, **1**, 385
33   E. JOHNSON AND P. B. HIRSCH: 1977, in preparation
34   S. YOSHIDA *et al.*: *Trans. Japan Inst. Metals*, 1968, **9**, Suppl., 83
35   J. M. CORBETT AND P. B. HIRSCH: 1977, in preparation
36   M. S. BAPNA AND M. MESHII: *Metall. trans.*, 1972, **3**, 3205
37   A. H. COTTRELL: 'Vacancies and other point defects in metals and alloys', Institute of Metals Monograph and Report Series No. 23, 1, 1958
38   A. N. STROH: *Proc. Roy. Soc.*, 1954, **A223**, 404
39   L. M. HOWE: *Radiat. Effects*, 1974, **23**, 181
40   B. EYRE: *J. Phys. F.* 1973, **3**, 422
41   Y. IMASHIMIZU AND H. KIMURA: *Trans. Japan Inst. Metals*, 1968, **9**, Suppl., 89

Vacancy aggregation and dislocation sink strengths in non-fissile materials

A. D. Brailsford and R. Bullough

The development of a reliable physical model of the evolution of damage structures in irradiated materials is essential for the extrapolation of existing data. The necessary ingredients of such a model are outlined and a particular discussion is given of the appropriate dislocation sink strengths. The relative simplicity of vacancy aggregation under electron irradiation is emphasized and critical calibration experiments are suggested.

A. D. Brailsford is with the Ford Motor Co., Michigan, USA, and R. Bullough is at AERE, Harwell

The development of a physical model of the formation and evolution of damage structures in irradiated materials is essential for the identification of the dominant physical and irradiation variables and for the subsequent extrapolation outside the existing data domain. This requirement is important, not only within one type of irradiation,* but is absolutely essential if simulation data obtained from electron or heavy-ion bombardment are to be used to predict successfully fast-neutron effects.[3] The rigorous and systematic exploitation of such simulation data for the design or discovery of low-swelling alloys suitable for either the structural components of the fast-reactor core or for the blanket material of a fusion reactor is of obvious technological importance. Moreover, the situation in the fusion environment is particularly crucial since the whole development programme for the foreseeable future must depend completely on simulation studies because of the complete lack of a fusion materials testing reactor facility.

Such a physical model of the evolution of damage structures under irradiation, based on chemical rate theory, has been developed and refined by various workers over the last few years.[3,5–9] A brief summary of the current situation will be given in the next section where, in particular, the significance of the dislocation sink strengths will be highlighted. The need for further

experimental information is here emphasized. Recent criticisms by Heald and Speight[10] of our previous work, concerning the dependence of the dislocation bias on the dislocation density, will also be introduced in this section. In the third and fourth sections we discuss various analyses of the effective sink strength for straight dislocations and dislocation loops respectively which should both help to clarify and amplify our previous position. In the fifth section we extend the present rate theory to include the presence of 'grown in', 'impotent' or solute-trapped dislocations; thus we obtain the sink strength for a dislocation that is unable to climb (trapped by solute or precipitate, etc.) but can nevertheless act as a saturable trap for interstitials and hence constitutes a region of enhanced vacancy–interstitial recombination. The effect of such dislocations on the swelling kinetics will also be discussed and evidence for such dislocation behaviour will be cited. Finally, in the sixth section, we reiterate some suggestions for critical experiments and indicate some of the avenues down which we believe useful theoretical progress could be made.

RATE THEORY OF VOID SWELLING

Point defects produced by displacement events under irradiation can migrate to existing sinks such as dislocations or grain boundaries or can form defect aggregates such as interstitial dislocation loops and voids, or form small vacancy loops by direct cascade collapse. At low temperatures when the vacancies are relatively immobile the point defect concentrations are reduced by mutual recombination between the vacancies and interstitials, while at higher temperatures the vacancy concentration can be supplemented by thermal vacancy emission from most of the point defect sinks.

---

* For example the large change in swelling rate (cavity growth rate), associated with the high-temperature peak in M316 steel under fast-neutron irradiation,[1,2] occurs because of the gradual accumulation of transmutation gas within the cavities with an eventual transition from slow gas-driven swelling to relatively rapid bias-driven swelling.[3] The high dose behaviour of this peak could not therefore be predicted by simple mathematical extrapolation from the low-dose data; an awareness that such a change of mechanism might occur can only be achieved when the extrapolation of the low-dose data is carried out with the aid of a physical model.[3,4]

If K is the irradiation-produced point-defect generation rate (in displacements/atom/second, dpa/s), K^e is the total vacancy emission rate from all sinks, and α is the recombination parameter, then the steady-state fractional vacancy concentration c_v and interstitial concentration c_i are given by the conservation equations[7]

$$K - k_i^2 D_i c_i - \alpha c_i c_v = 0$$
$$K + K^e - k_v^2 D_v c_v - \alpha c_i c_v = 0 \tag{1}$$

In this pair of simultaneous quadratic equations for c_v, c_i the quantities D_v and D_i are the vacancy and interstitial diffusion constants respectively and k_v^2 and k_i^2 are the total sink strengths for vacancies and interstitials, respectively, for all the sinks present in the effective medium. If, for simplicity, we consider the only sinks to be voids, interstitial loops, and network dislocations then the total sink strengths may be written (see Brailsford and Bullough[9] for a discussion of the grain-boundary sink and Bullough et al.[3] for the vacancy loop and recoil spectra effects):

$$k_i^2 = k_{ic}^2 + k_{iL}^2 + k_{iN}^2$$
$$k_v^2 = k_{vc}^2 + k_{vL}^2 + k_{vN}^2 \tag{2}$$

where the first subscript indicates the point defect type and the second the sink type (thus, for example, k_{ic}^2 is the sink strength of cavities* for interstitials, k_{iL}^2 of interstitial loops for interstitials, and k_{vN}^2 of network dislocations for vacancies). Similarly the thermal vacancy emission rate may be written

$$K^e = K_c + K_L + K_N \tag{3}$$

where

$$K_c = k_{vc}^2 D_v \bar{c}_c \tag{4}$$

with

$$\bar{c}_c = c_v^e \exp\left[\left(\frac{2\gamma}{r_c} - p_g\right)\Omega/kT\right], \tag{5}$$

$$K_L = k_{vL}^2 D_v \bar{c}_L \tag{6}$$

with

$$\bar{c}_L = c_v^e \exp[-(\gamma_{SF} + F_{el}(r_L))b^2/kT] \tag{7}$$

and

$$K_N = k_{vN}^2 D_v c_v^e \tag{8}$$

In the above expressions c_v^e is the equilibrium vacancy concentration away from the sinks, γ is the surface energy of the spherical cavity of radius r_c, p_g is the gas pressure within the cavity, Ω is the atomic volume, b is the magnitude of the Burgers vector of the dislocation loops, γ_{SF} is the stacking fault energy, and $F_{el}(r_L)$ is the line energy of the loop; both the stacking fault energy and the line energy reduce the probability of vacancy emission from an interstitial loop.

If C_c is the volume concentration of voids and r_c is the mean void radius, then the fractional swelling

$$S = \tfrac{4}{3}\pi r_c^3 C_c = q_{vc} \tag{9}$$

where q_{vc} is the fractional number of vacancies in voids, and by definition

$$\frac{dS}{dt} = \frac{dq_{vc}}{dt} = 4\pi r_c^2 C_c \frac{dr_c}{dt} = k_{vc}^2 D_v c_v - k_{ic}^2 D_i c_i - K_c \tag{10}$$

when the voids have all nucleated and C_c is constant.

* Here we follow a suggestion of Russell and adopt the subscript c for cavity instead of V for void to eliminate confusion with v for vacancies.

Similarly if q_{iL} is the fractional number of interstitials in interstitial loops of average radius r_L and volume density N_L, then

$$q_{iL} = \pi r_L^2 b N_L \tag{11}$$

and, again by definition,

$$\frac{dq_{iL}}{dt} = k_{iL}^2 D_i c_i - k_{vL}^2 D_v c_v + K_L \tag{12}$$

thus, if the loop nucleation has ceased

$$\frac{dq_{iL}}{dt} = 2\pi b r_L N_L \frac{dr_L}{dt} = b\rho_L \frac{dr_L}{dt} \tag{13}$$

where ρ_L is the transient loop dislocation density. Finally, if q_{iN} is the fractional number of interstitials lost to the network dislocations,

$$\frac{dq_{iN}}{dt} = k_{iN}^2 D_i c_i - k_{vN}^2 D_v c_v + K_N \tag{14}$$

We note that for consistency and atomic conservation we must have

$$\frac{dS}{dt} = \frac{dq_{iL}}{dt} + \frac{dq_{iN}}{dt} \tag{15}$$

in the absence of other sinks, as follows, directly from (1), (10), (13), and (14).

The sink strengths required for the voids and dislocations have been discussed by various authors. In the original paper by Brailsford and Bullough[7] we deduced sink strengths for neutral sinks such as voids, for fixed bias sinks such as dislocations which are deemed to have a preferential interaction with interstitials compared with vacancies, and for so-called variable bias sinks. The last sink type will be referred to again in the fifth section when we discuss sessile dislocation sinks. The void sink strength was deduced by a careful embedding procedure in which a composite body is envisaged where a spherical void is defined, surrounded by a defect-free annular zone which is in turn surrounded by the infinite effective medium containing all the other sinks in addition to the voids. The spatial diffusion equations in the sink-free and surrounding infinite region are then solved subject to an appropriate boundary condition at the void surface and continuity on the spherical surface between the sink-free zone and the effective medium. The flux of point defects into the central void is then calculated and, for self consistency, equated to the assumed flux of point defects to an equivalent sink in the effective medium. If the other sink densities are not large compared with the void densities this procedure yields the well-known,[7,11,12] first-order result

$$k_{ic}^2 = k_{vc}^2 = 4\pi r_c C_c \tag{16}$$

The bias sink strength for the dislocation loop was obtained[7] by surrounding the toroidal loop by an effective sphere and replacing the drift field arising from the interstitial–dislocation interaction energy by an enhanced transfer velocity for interstitials across the toroidal core/matrix interface. By this procedure with suitable jog limitation considerations* we obtained the result for the dislocation loop sink strengths

$$k_{iL}^2 = Z_i \rho_L \tag{17}$$

$$k_{vL}^2 = Z_v \rho_L \tag{18}$$

* These precise boundary conditions will be discussed in some detail in the two following sections.

where the bias parameters are independent of the dislocation density ρ_L and the relative bias $[Z_i - Z_v]$ should be of the order of a few percent. The same sink strengths were also adopted for the network dislocations:

$$k_{iN}^2 = Z_i \rho_N \tag{19}$$

$$k_{vN}^2 = Z_v \rho_N \tag{20}$$

and the only distinction between the loops and network arose from the different vacancy emission rates (6) and (8).

In a more recent study by Brailsford et al.[13] (henceforth referred to as BBH) the second-order corrections for the void sink strength and for the network dislocations using the rate limitation model have been discussed fully, and in particular the necessary justification for embedding the sink directly into the effective medium with no sink-free zone was established. The important result is that the simple void sink expressions (16) should be replaced by:

$$k_{ic}^2 = 4\pi r_c C_c [1 + k_i r_c] \tag{21}$$

$$k_{vc}^2 = 4\pi r_c C_c [1 + k_v r_c] \tag{22}$$

where

$$\begin{aligned} k_i^2 &= k_{ic}^2 + k_{iD}^2 \\ k_v^2 &= k_{vc}^2 + k_{vD}^2 \end{aligned} \quad (k_{iD}^2 = k_{iL}^2 + k_{iN}^2, \text{etc.}) \tag{23}$$

These correction terms are only important for the high dislocation density limit when

$$r_c^2 \rho_D \gtrsim 1 \tag{24}$$

However, their inclusion is essential if one wishes to include all the cellular model results[14,15] in the rate theory. Thus, using the dislocation sink strengths (17)–(20), then (21) and (22) may be written

$$k_{ic}^2 = 4\pi r_c C_c [1 + (Z_i \rho_D)^{\frac{1}{2}} r_c] \tag{25}$$

$$k_{vc}^2 = 4\pi r_c C_c [1 + (Z_v \rho_D)^{\frac{1}{2}} r_c] \tag{26}$$

The influence of the other sinks on the dislocation sink strengths was obtained by BBH with the rate limitation model, which for the long straight cylindrical geometry and no other sinks present reduced to the well known result for the dislocation sink strengths[13]

$$k_{iN}^2 \simeq 2\pi \rho_N / \ln[1/b(Z_i \rho_N)^{\frac{1}{2}}] \tag{27}$$

$$k_{vN}^2 \simeq 2\pi \rho_N / \ln[1/b(Z_v \rho_N)^{\frac{1}{2}}] \tag{28}$$

Heald and Speight[10] have recently argued that such explicit logarithmic dislocation dependencies must be included in Z_i and Z_v and that by taking Z_i, Z_v constant we have committed 'an error of physical interpretation' since they argue that Z_i, Z_v should be strongly temperature dependent through the temperature variation of the dislocation density in irradiated materials. We disagree with this criticism and the analyses in the next two sections constitute a quantitative response. However before proceeding to calculation we would make several more qualitative reactions to the criticism.

1   In irradiated materials the irradiation-produced dislocations almost always evolve as growing interstitial loops or as a population of very small vacancy loops. The logarithmic form (27) and (28) arises from the cylindrical geometry—it does not appear in the loop–loop situation (see the fourth section for an accurate loop solution).

2   There is no observed discontinuity in swelling behaviour when the transition from loop structure to network occurs. In any case the tangled network configuration is probably better modelled as a tangle of loops than as an array of parallel like dislocations.

3   We were fully aware of the logarithmic variation for such straight dislocations but considered its omission was physically superior to its inclusion as the basis of a model of the real dislocation situation.

4   The dual averaging procedures implicit in incorporating both long-range stress effects and diffusion to a random sink array in an effective medium formalism has yet to be justified rigorously, although we have now made some progress in this regard (see next section). Cellular approaches to the calculation of sink strengths, for subsequent use in a spatially homogeneous chemical rate theory, are entirely ad hoc.

5   The implications of adopting such a logarithmic variation in the bias include a huge drop in the bias between $0 \cdot 2$ and $0 \cdot 3 T/T_m$ as the dislocation density is supposed to rise to incredibly high values at low temperatures.* In fact such behaviour is certainly not universal since molybdenum swells at temperatures down to $0 \cdot 2 T/T_m$ under neutron irradiation.[17] It leads to a temperature variation of irradiation creep that is almost exactly contrary to observations.

6   Finally, in a more constructive vein, if we substitute the void sink strengths (25), (26) into the general swelling rate equation (10) we obtain a simple result appropriate for temperatures below the vacancy emission temperature, namely

$$\frac{dS}{d(Kt)} = \frac{D_i D_v}{2\alpha K} [-1 + (1+\eta)^{\frac{1}{2}}] 4\pi r_c C_c \rho_D$$

$$\times \{[Z_i - Z_v] + \rho_D^{\frac{1}{2}} r_c (Z_i Z_v)^{\frac{1}{2}} [Z_i^{\frac{1}{2}} - Z_v^{\frac{1}{2}}]\} \tag{29}$$

where

$$\eta = 4\alpha K / D_i D_v k_i^2 k_v^2 \tag{30}$$

If we are near the peak swelling temperature or ρ_D is large, so that recombination can be neglected, then (29) simplifies to

$$\frac{dS}{d(Kt)} = 4\pi r_c C_c \rho_D \{[Z_i - Z_v]$$

$$+ \rho_D^{\frac{1}{2}} r_c (Z_i Z_v)^{\frac{1}{2}} [Z_i^{\frac{1}{2}} - Z_v^{\frac{1}{2}}]\} / k_i^2 k_v^2 \tag{31}$$

The ideal experiment is thus to measure the swelling rate, r_c, C_c, and ρ_D at various doses and appropriate temperatures in the HVM and see if Z_i, Z_v has to be a strong function of ρ_D for consistency. Of course such an experiment must be carried out in very pure material to avoid complex trapping effects, etc. The theory does give a reliable description of the instantaneous swelling rate; the problems arise when one wishes to integrate an equation such as (31) since then it is essential to include the dose dependent changes in ρ_D, via ρ_L, and r_c. The linear dose dependencies of swelling[18–20] frequently observed in the HVM provide good evidence for such ρ_L variations with dose.

---

* The densities of interstitial loops obtained by Kiritani[16] for a wide range of metals at temperatures considerably below $0 \cdot 2 T/T_m$ never exceed $5 \cdot 10^{16}$/cc and thus loop–loop contact should preclude dislocation densities much greater than $5 \cdot 10^{11}$ cm/cm^3.

SINK STRENGTH FOR A STRAIGHT DISLOCATION

In this section we shall adopt the rather extreme model for the dislocation network, namely that it can be represented by an array of parallel like edge dislocations. This is the model used by Heald and Speight[10] as the basis for their view that the logarithmic dislocation density dependence of the dislocation–interstitial (Z_i) and dislocation–vacancy (Z_v) bias parameters as used in the rate theory is essential. We begin by outlining the drift model used by them to obtain explicit results for Z_i and Z_v and point out some of its shortcomings; in particular the absence of rate limitation at the core and the importance of the presence of other sinks—both dislocations and voids. New analyses of both these features will then be given and the possibility of representing such spatial field effects by effective rate limitation without a field will be discussed. This then enables our previous rigorous sink strength calculation[13] for a purely rate-limited straight dislocation sink to be exploited. The conditions under which, even with this straight dislocation model, the sink dependencies of Z_i and Z_v can be dropped may then be seen.

Heald–Speight model

An edge dislocation is taken along the axis of a cylindrical cell of dimension $R = (\pi \rho_D)^{-\frac{1}{2}}$ where ρ_D is the average dislocation density. A point defect a distance r from such a dislocation is assumed to have a size effect interaction energy/unit length given by[21–23]

$$E = \frac{\mu b}{\pi} V e \frac{\sin \theta}{r} \qquad (32)$$

where θ is the angle between the radius vector $\mathbf{r}$ defining the position of the defect and the direction of the Burgers vector $\mathbf{b}$, V is the volume of the effective elastic (spherical) inclusion that represents the particular point defect, and e is the relaxation volume strain associated with the isolated point defect. Appropriate subscripts i and v can be added to V, e and other relevant quantities later to identify an interstitial or vacancy as necessary. The continuous production of the point defects in the cellular volume is simulated by a constant concentration boundary on $r = R$.

$$c = c^\infty \qquad (33)$$

which is taken arbitrarily to represent the mean concentration in the effective medium. At the dislocation core $r = r_0$ the concentration is set equal to the thermal equilibrium value

$$c = c^e \exp(-E(r_0, \theta)/kT) \qquad (34)$$

where c^e is the thermal equilibrium concentration of the point defects away from the sinks. The steady-state concentration in the cellular region $r_0 \leqslant r \leqslant R$ is given by the differential equation[22]

$$\nabla^2 c + \frac{1}{kT} \operatorname{grad} c \cdot \operatorname{grad} E = 0 \qquad (35)$$

since $E(r, \theta)$ given by (32) is a harmonic function. This equation, subject to the boundary conditions (33) and (34), has been solved by Margvelashvili and Saralidze.[24] The essential change of variable is

$$c = e^{-E/2kT} \psi \qquad (36)$$

which transforms the differential equation (35) for c into

$$\frac{1}{r} \frac{\partial}{\partial r} \left(r \frac{\partial \psi}{\partial r} \right) + \frac{1}{r^2} \frac{\partial^2 \psi}{\partial \theta^2} - \frac{1}{4} \frac{L^2}{r^4} \psi = 0 \qquad (37)$$

for $\psi(r, \theta)$, where

$$L = \frac{\mu b V e}{\pi kT} \qquad (38)$$

A further change of variable:

$$z = L/2r \qquad (39)$$

then transforms (37) into Bessel's equation with a general solution in terms of Bessels functions $K_n(z)$, $I_n(z)$ of imaginary arguments. The loss rate $\mathscr{L}$ of defects into unit length of the dislocation is

$$\mathscr{L} = \frac{1}{\Omega} \int_0^{2\pi} d\theta \, \mathbf{r} \cdot \mathbf{J} \qquad (40)$$

where $\mathbf{r}$ is *any* radius vector within $r_0 \leqslant r \leqslant R$ since there are no other sinks in the cell, Ω is the atomic volume, and

$$\mathbf{J} = D \operatorname{grad} c + \frac{Dc}{kT} \operatorname{grad} E \qquad (41)$$

It is convenient to choose $r = R$, since c there is a constant, from (33), and $(\operatorname{grad} E)_r \propto \sin \theta$ implies that the second term in (41) does not contribute in (40). If

$$z_R = \frac{L}{2R} \ll 1 \qquad (42)$$

which it will be for low dislocation densities, we can neglect all but the first term in the series solution for $\mathscr{L}$ and obtain the good approximation for $\mathscr{L}$

$$\mathscr{L}^0 = \frac{2\pi D}{\Omega K_0(z_R)} [c^\infty - c^e] \qquad (43)$$

when

$$z_0 = \frac{L}{2r_0} \gg 1 \qquad (44)$$

Thus, from the small argument expansion of K_0 in (43),

$$\mathscr{L}^0 \simeq \frac{2\pi D}{\Omega} [c^\infty - c^e] \Big/ \ln \left(\frac{2R}{L} \right) \qquad (45)$$

which is equated to the equivalent loss of defects to unit length of each dislocation

$$\frac{D}{\Omega} [c^\infty - c^e] Z \qquad (46)$$

in the equivalent rate theory medium, to yield the result quoted by Heald and Speight:[10]

$$Z = 2\pi \Big/ \ln \left(\frac{2R}{L} \right) \qquad (47)$$

$$= 2\pi / \ln[2/L(\pi \rho_D)^{\frac{1}{2}}] \qquad (48)$$

The validity of this result for the effective biases for defining the sink strengths of dislocations for interstitials and vacancies may be immediately questioned since:

(i) there is no rate limitation at the dislocation core; this aspect will be discussed in the next subsection

(ii) no other sinks are present in the cellular region $r < R$ and the long-range interaction field (32) from any one dislocation is allowed to bathe the whole cell without any field interaction effects—clearly the r^{-1} (long-range) field (32) must be modulated by the other dislocations present in the adjacent cells.

(iii) it is not obvious that the effect of a spatially homogeneous source of point defects K can be simulated by the constant outer boundary condition $c = c^\infty$. However, if one takes the single dislocation in a cell with zero flow of point defects across $r = R$ and a source term K then, unless other sinks are also present in the cell, the rate of loss of point defects at the dislocation is simply given by the rate of creation and the field is irrelevant; hence the importance of discussing (ii)

(iv) last, but by no means least, the dislocations produced by irradiation are not long, straight and parallel; we shall discuss the dislocation loop sink strength in the next main section.

Effect of rate limitation

A rate limitation boundary condition on the dislocation core $r = r_0$ may be written

$$D \frac{\partial c}{\partial r} = n\eta D[c - c^e e^{-E(r_0,\theta)/kT}] \tag{49}$$

where again subscripts can be introduced later to distinguish interstitials and vacancies; n is the number of jogs per unit length on the dislocation and η has the general form

$$\eta = [\alpha\nu/\alpha_p\nu_p]f \exp\left[\frac{E_m - E'_m}{kT}\right] \tag{50}$$

where the probability of a point defect jumping into the core is

$$\nu e^{-E'_m/kT} \alpha nb$$

Here ν is the attempt frequency (possibly lower than the perfect lattice attempt frequency ν_p), α is the geometrical factor defining the number of ways a point defect could jump in (which might be higher than the corresponding perfect lattice number α_p because of lattice distortion near the core), nb is the number of jogs within a jump distance of the point defect (note that the latter is the real basis of the interface control at $r = r_0$ since it implies that a point defect can only be absorbed on to the dislocation if a suitable job is present), $e^{-E'_m/kT}$ is the probability that the actual jump will occur, and E'_m is the modified migration energy at the dislocation core arising from the coupling between the dislocation hydrostatic field and the migration volume of the point defect.[25] Finally in (50) f is the fraction of dislocation core circumference over which a point defect can possibly enter the core. This factor, having a minimum value of

$$f = b/2\pi r_0 \simeq \tfrac{1}{2}\pi \tag{51}$$

can also be used to describe partially the neglected angular variation in $[E_m - E'_m]$.

With the minimum value of f (implying entry of point defects from one side of the dislocation core only) the parameter η introduced by Brailsford et al.[13] (and henceforth denoted by η_{BBH}) in their discussion of dislocation sink strengths with only rate limitation at the core

should be replaced by the significantly smaller parameter η where

$$\eta = \eta_{BBH} nb/2\pi \tag{52}$$

This leads to the replacement of equation (32) of BBH by

$$Z_\eta = \frac{2\pi\eta bkbK_1(kb)}{2\pi kbK_1(kb) + \eta nbK_0(kb)} \tag{53}$$

which, even for such a field free pure rate limitation analysis, removes almost all the dislocation dependence in Z_η suggested by (27) (this is obvious from Fig. 2 of BBH for η_{BBH} values < 1 and arises from (53) by the dominance of the K_1 term in the denominator of (53) for small kb).

Returning now to the field-dependent drift problem, if we assume $z_R \ll 1$ and $z_0 \gg 1$ the equation (35) can be solved subject to the boundary conditions (33) and (49) to yield for Z_η:

$$Z_\eta \simeq \frac{\pi\epsilon\sqrt{\beta/2}}{1 + \epsilon\sqrt{\beta/2}\,K_0(z_R)} \tag{54}$$

where

$$\epsilon = \frac{I_0(z_0)}{K_0(z_0)} \simeq \frac{e^{2z_0}}{\pi} \gg 1, \tag{55}$$

$$\beta = \frac{2r_0^2}{L} n\eta, \tag{56}$$

and an outline of the analysis leading to (54) is given in the Appendix. Thus we see that with the field term present the dislocation dependence in Z_η can only be lost if there is low jog density. However, as we have previously commented there is no justification for embedding the sink in the effective medium when a field is also present—such direct embedding has been justified by BBH when rate limitation alone prevails at the sink. Thus we describe now a pseudo-effective medium approach which attempts to deal with the presence of such fields in a more rigorous fashion.

Pseudo-effective medium approach

If other sinks are present in the body then the controlling equation for $\psi(r, \theta)$ has the form

$$\frac{1}{r}\frac{\partial}{\partial r}\left(r\frac{\partial\psi}{\partial r}\right) + \frac{1}{r^2}\frac{\partial^2\psi}{\partial\theta^2} - \left[\frac{1}{4}\frac{L^2}{r^4} + k^2\right]\psi = 0 \tag{57}$$

where $-Dk^2c$ is the additional loss rate to the other sinks and

$$c = e^{-E(r,\theta)/2kT}\psi \tag{58}$$

We can thus follow the spirit of the effective medium analysis[7,13] for calculating k_D^2 by surrounding the dislocation along $r = 0$ by a cylindrical sink-free region of radius r_c where

$$r_c = [L/2k]^{\frac{1}{2}} \tag{59}$$

is suggested from the last term in (57) and within which the field is present. In the outer region $r > r_c$ the field is dropped. Thus consider the system defined by $r \leqslant r_c$

$$\nabla^2 c + \frac{1}{kT}\,\text{grad}\,c\,.\,\text{grad}\,E = 0 \tag{60}$$

$r \geqslant r_c$

$$\frac{D}{r}\frac{\partial}{\partial r}r\frac{\partial c}{\partial r} - Dk^2 c = 0 \tag{61}$$

with

$$c = c^\infty \qquad \text{as } r \to \infty \tag{62}$$

and the rate-limited core boundary condition

$$D\frac{\partial c}{\partial r} = n\eta D[c - c^e\, e^{-E(r_0,\theta)/kT}] \tag{63}$$

At the interface $r = r_c$ we ensure continuity of concentration. Again for $z_R \ll 1$ and $z_0 \gg 1$ we find for Z_η in this instance

$$Z_\eta \simeq \frac{2\pi\epsilon\sqrt{\beta/2}}{1 + 2\epsilon\sqrt{\beta/2}\,K_0(z_c)} \tag{64}$$

where $\epsilon \simeq e^{2z_0}/\pi$ and

$$z_c = \frac{L}{2r_c} = \left(\frac{kL}{2}\right)^{\frac{1}{2}} \tag{65}$$

This is identical to (54) with $K_0(z_R) \to 2K_0(z_c)$. The result (64) therefore provides a very convenient expression for including the effects on Z_η of the other sinks in addition to dislocations, since in (64):

$$z_c \sim \left(\frac{L}{2}\right)^{\frac{1}{2}}[Z\rho_D + k_c^2]^{\frac{1}{4}} \tag{66}$$

If the dislocations (long and straight) are the dominant sinks then (64) reverts to (54).

We have thus discussed the straight dislocation with three models:

(i) *the cellular model* with or without rate limitation at the dislocation core, the latter being the Heald–Speight approach
(ii) *the pseudo-effective medium approach* with rate limitation at the core
(iii) *the effective medium approach* up to the core with the field effects subsumed into a rate limitation at the dislocation core (BBH).

Method (i) with no rate limitation at the core yields

$$Z^{(a)} \simeq \frac{2\pi}{K_0(L/2R)} \simeq \frac{2\pi}{\ln(1/b\sqrt{\rho_D}) + \ln(4b/e^\gamma\sqrt{\pi}L)} \tag{67}$$

where $\gamma = 0.559$ is Euler's constant. Then, if the difference in relaxation volumes for interstitials and vacancies is *small*, $L_i = L_v(1+\epsilon)$, $Z_i = Z_v(1+\delta)$ and

$$\left(\frac{Z_i - Z_v}{Z_v}\right)^{(a)} \simeq \frac{Z_v^{(a)}\epsilon}{2\pi} \tag{68}$$

Method (ii) with no rate limitation:

$$\left(\frac{Z_i - Z_v}{Z_v}\right)^{(b)} \simeq \frac{2\epsilon Z_v^{(b)}}{4\pi - Z_v^{(b)}} \tag{69}$$

where

$$Z_v^{(b)} \simeq \pi/\ln[2\sqrt{2}/e^\gamma(Z_v^{(b)}\rho_d)^{\frac{1}{4}}L_v^{\frac{1}{2}}] \tag{70}$$

Method (iii) with no rate limitation:

$$\frac{Z_i^{(c)} - Z_v^{(c)}}{Z_v^{(c)}} \simeq \frac{2\epsilon Z_v^{(c)}}{4\pi - Z_v^{(c)}} \tag{71}$$

Thus we see that the 'last jump' model of BBH duplicates the *form* of the bias obtained from the pseudo-effective medium method when $(L_i - L_v)/L_v \ll 1$. Since we consider that the pseudo-effective medium approach is unquestionably the best procedure it is very reassuring to see this correlation with the previous effective medium approach. It thus lends further support to the remarks made after equation (53), namely that a combination of jog limitation and entry restriction to the core can remove the explicit dependence of Z_η on the dislocation density. However, as we have previously emphasized the purpose of the present discussion on the sink strengths of straight dislocations is primarily to put such calculations on a more detailed physical footing and not to suggest they have strong relevance to the irradiation situation. The next section, which concerns itself with the sink strength for a dislocation loop is, we believe, much more relevant.

SINK STRENGTH FOR A DISLOCATION LOOP

The sink strength for the dislocation loop was obtained by Brailsford and Bullough[7] by surrounding the loop (torus) with an equivalent sphere of radius r_L and adjusting the rate limitation on the sphere to ensure that the flux of point defects at the sphere was equal to the flux into the torus. The effective medium method was then used to derive the sink strength for the effective sphere and hence for the torus.

An edge dislocation loop has a spatial drift field interaction with point defects that is almost negligible at distances greater than a loop radius from the loop perimeter; it is thus reasonable to subsume these field effects into a rate limitation at the loop.

The boundary condition on the sphere of radius r_L is therefore

$$D\frac{\partial c}{\partial r} = \frac{\pi r_0}{r_L}\cdot n\eta D[c - c^e] \tag{72}$$

where field effects have been dropped and η is given by (50). If such a sphere is put into the effective medium we obtain

$$Z_\eta(k) = \frac{2nb\eta(1 + r_L k)}{nb\eta + 2(1 + r_L k)} \tag{73}$$

where if dislocation loops are the dominant sinks k is given by

$$k^2 = \frac{2nb\eta(1 + r_L k)}{nb\eta + 2(1 + r_L k)}\rho_L \tag{74}$$

where

$$\rho_L = 2\pi r_L N_L \tag{75}$$

and N_L is the volume density of such loops. Solving (74) and substituting k into (73) yields

$$Z_\eta \simeq \left[\frac{2nb\eta}{2 + nb\eta}\right]\left\{\frac{\sqrt{2}(nb\eta + 2)^{\frac{3}{2}} + r_L(4 + nb\eta)(nb\eta\rho_L)^{\frac{1}{2}}}{\sqrt{2}(nb\eta + 2)^{\frac{3}{2}} + r_L(4 - nb\eta)(nb\eta\rho_L)^{\frac{1}{2}}}\right\} \tag{76}$$

Thus for Z_η to depend significantly on ρ_L the average loop radius must be very large; for vacancy loops formed by cascade collapse ($r_L < 20$ Å), ρ_L would have to exceed

$\sim 10^{14}\,\text{cm/cm}^3$! If we normalize the vacancy value so that

$$nb\eta_v = 2 \tag{77}$$

then

$$Z_{\eta_v} \simeq 1 + \frac{r_L}{2}\rho_L^{\frac{1}{2}} \tag{78}$$

and we can replace (76) by

$$Z_\eta \simeq \frac{2nb\eta}{2 + nb\eta} \tag{79}$$

This result can be improved by exploiting the known electrical capacitance C_t of a charged metal torus of the same dimension:

$$C_t = \pi r_L / \ln(8r_L/r_0) \tag{80}$$

to scale the far diffusion field of the torus.[26] This procedure replaces (79) by:

$$Z_\eta \simeq \frac{2nb\eta}{\{\pi/\ln(8r_L/r_0) + nb\eta\}}\,\frac{\pi}{\ln(8r_L/r_0)} \tag{81}$$

The dependence of Z_η on ρ_L is thus, in general, very weak and it is a sensible working hypothesis to take Z_η independent of the dislocation density.

TRAPPED DISLOCATION

It is frequently observed that some dislocations do not appear to move during irradiation. Thus, for example, in the Harwell LINK experiments on N4 aluminium[27] it is observed that under 200 keV Al$^+$ ion irradiation the dislocation density rapidly builds up and then 'freezes' into a sensile dense tangle. Such immobility of the dislocations in the N4 alloy is probably caused by the trapping of the dislocations by solute (presumably Mg in this case), thereby inhibiting absorption of interstitials or vacancies by the dislocations. It is thus interesting to ask what effect on the void swelling such solute trapping of dislocations should have, and thereby contribute to our general understanding of possible solute trapping effects.

Sink strength

We envisage a dislocation which has sufficient solute atoms (or a continuous precipitate) in its core so that the jogs are completely immobile and climb is not possible. Such a dislocation will still retain some residual hydrostatic field which will enable it to act as a saturable trap for interstitial atoms, having a characteristic binding energy E_B, say. We expect therefore a flow of interstitials into the trap, a flow of interstitials out by thermal emission, and a flow of vacancies in *when* the trap is occupied. It is thus a centre of enhanced recombination with a steady-state occupation probability.

If we denote the sink strength of such a trap by k_{ig}^2 and k_{vg}^2 for interstitials and vacancies respectively (the subscript g indicates 'grown in' dislocations) then the effective medium method[7,13] with the saturable condition at the trap (that is no accumulation of either vacancies or interstitials around the dislocation) yields the results

$$k_{vg}^2 = \rho_g \tag{82}$$

$$k_{ig}^2 = \frac{(K + K^e)k_{i0}^2}{Kk_{v0}^2 - K^e\rho_g}\cdot\rho_g \tag{83}$$

where ρ_g is the density of such dislocations, k_{i0}^2, k_{v0}^2 are given by equation (2) and represent the total sink strengths of all the *other* sinks. K^e is the total *vacancy* emission rate, given by (3), from all the *other* sinks and the interstitial thermal emission rate from the ρ_g traps has been subsumed into k_{ig}^2. The occupation probability p is given by

$$p = \frac{D_ic_i - D_vc_v}{D_ic_i - D_vc_v + D_ic_B} \tag{84}$$

where

$$c_B = \exp[-E_B/kT] \tag{85}$$

The steady-state fluxes D_ic_i and D_vc_v in (84) are given by the simultaneous quadratic equations (1) in terms of the total sink strengths

$$\begin{aligned}k_i^2 &= k_{i0}^2 + k_{ig}^2 \\ k_v^2 &= k_{v0}^2 + k_{vg}^2\end{aligned} \tag{86}$$

and the total vacancy emission rate given by (3).

We note that the saturable bias sink strength k_{ig}^2 is not an intrinsic quantity but depends on the magnitudes of the other total sink strengths k_{i0}^2, k_{v0}^2, together with the total thermal vacancy emission rate K^e from the other sinks. The imposed constant occupation probability ($dp/dt = 0$) at such dislocations has forced this sink strength to depend on the other sink strengths and although the binding energy E_B does not appear explicitly in its sink strength it does define the occupation probability of the available sites around the dislocation given by (84). Finally, we have only formally excluded an explicit thermal interstitial emission from such dislocations; such emission is implicitly included in the *net* absorption sink strength k_{ig}^2.

Effect on swelling of ρ_g

In general, the sink strengths (82) and (83) can be simply included in the rate equations (10) and (12) which can then be integrated numerically to yield $S(t)$, etc. However, it is useful to proceed analytically if recombination (αc_ic_v) can be neglected in (1). In the temperature range near the peak swelling temperature and above such an assumption is sometimes reasonable[7] and then we obtain, from (1), (10), (12) and (82) and (83), the explicit swelling rate (using the first-order void sink strengths (16) for simplicity):

$$\begin{aligned}\frac{dS}{dt} &= \frac{K4\pi r_cC_c[Z_i - Z_v]\rho_0}{(4\pi r_cC_c + Z_i\rho_0)(4\pi r_cC_c + Z_v\rho_0 + \rho_g)} \\ &+ \frac{4\pi r_cC_cZ_vD_v}{(4\pi r_cC_c + Z_v\rho_0 + \rho_g)}[\rho_N(c_v^e - \bar{c}_c) + \rho_L(\bar{c}_L - \bar{c}_c)] \\ &+ \frac{4\pi r_cC_c\rho_gD_v}{(4\pi r_cC_c + Z_i\rho_0)(4\pi r_cC_c + Z_v\rho_0 + \rho_g)} \\ &\times [\rho_N(Z_vc_v^e - Z_i\bar{c}_c) + \rho_L(Z_vc_L - Z_i\bar{c}_c)]\end{aligned} \tag{87}$$

where

$$\rho_0 = \rho_N + \rho_L \tag{88}$$

and the other quantities are defined in the second section. In this result the first term is the direct bias swelling term, the second term is the usual thermal vacancy emission term, and the third term is a thermal emission term arising directly from the presence of ρ_g. We note that the saturability of ρ_g has produced an interesting coupling of the bias parameters Z_i, Z_v and the thermal vacancy emission quantities c_v^e, $\bar{c}_c$, and $\bar{c}_L$. Furthermore ρ_g will decrease the swelling rate even if the temperature is too low for any vacancy emission. In that case

$$\frac{dS}{dt} = \frac{K4\pi r_c C_c[Z_i - Z_v]\rho_0}{(4\pi r_c C_c + Z_i\rho_0)(4\pi r_c C_c + Z_v\rho_0 + \rho_g)} \quad (89)$$

Finally at such temperatures we can explicitly integrate (89) since

$$\rho_0 = \rho_L + \rho_N = \left[\rho_N^2 + \frac{4\pi N_L}{b}S\right]^{\frac{1}{2}} \quad (90)$$

If $\rho_N = 0$ (growing dislocation loops and ρ_g only) the integral of (89) is elementary and yields:

$$\frac{9}{10N}S^{5/6} + (Z_i + Z_v)S + \tfrac{8}{7}NS^{7/6}$$

$$+ \frac{1}{2}\left(\frac{3}{a}\right)^{\frac{2}{3}}\frac{\rho_g S^{\frac{1}{3}}}{N} + \frac{3}{8}\left(\frac{3}{a}\right)^{\frac{2}{3}}\frac{\rho_g Z_i S^{2/3}}{N}$$

$$= K(Z_i - Z_v)t \quad (91)$$

where

$$N = 3[4\pi N_L/b]^{\frac{1}{2}}/4[4\pi C_c\sqrt{3}]^{\frac{2}{3}} \quad (92)$$

and

$$a = 4\pi C_c \quad (93)$$

The bias parameters Z_i, Z_v have been assumed independent of ρ_L. The first three terms in (91) are the usual ones[28] and provide a good explanation of linear swelling frequently observed[18-20] in the HVM. The extra terms do not have a swelling power near unity and if dominant would lead to a dose dependence power between 1·5 and 2 and a swelling magnitude independent of void density that decreases with increasing ρ_g.

DISCUSSION

The results of the analyses in the third section for the straight dislocation sink strength indicate that if sufficient rate limitation on the absorption of point defects at the dislocation cores prevails then, even for such pure cylindrical cellular geometry, the dependence of the dislocation bias parameters Z_i, Z_v on the dislocation density can be neglected. However, such jog limitations and restricted jump entry to the core, though attractive, are assumptions and the only reasonable procedure is to perform careful experiments on the HVM, as indicated, to discover whether such variations of Z are important. It is particularly advantageous to use the HVM for such calibration experiments since problems of recoil spectra sensitivity are thereby avoided.

In our opinion the irradiation-produced network is best considered as a tangle of dislocation loops and therefore the relatively weak dependence of Z_i, Z_v for such loops on ρ_L indicated in the fourth section is support for neglecting any such explicit dependence for ρ_N. If there was a strong difference in bias between dislocation loops and dislocation network one would expect a transition in swelling behaviour when the growing loops impinge on each other and the network begins to evolve. As far as we can discern no such discontinuous behaviour with dose has been reported. However, again careful experiments on pure metals in the HVM would be most welcome.

The enormous importance of solute trapping on the void swelling problem is well known. Many kinds of traps can be envisaged ranging from single atom traps for vacancies or interstitials to the saturable dislocation trap discussed in the previous section. The 'dual' trap to this last trap is a vacancy trap that allows vacancy emission and interstitial capture only when a suitable vacancy is in the trap with which it can recombine. Such a *vacancy* emitting trap has been discussed previously by us and is envisaged for example as a coherent precipitate. The analysis of the simultaneous presence of both types of trap constitutes an important theoretical problem.

Finally, we reiterate our plea for more data on the quantitative microstructure in pure and well defined doped metals. We need to know, in addition to the variation of void size and density with temperature and dose, how the dislocation density, in the form of loops and network, varies since any interpretation and calibration must make careful provision for overall *conservation* of interstitials and vacancies.

REFERENCES

1 H. J. BUSBOOM et al.: General Electric Rep. GEAP-14062, 1975
2 E. E. BLOOM et al.: Nucl. Technol., to be published
3 R. BULLOUGH et al.: Proc. Roy. Soc., 1975, **A346**, 81
4 A. D. BRAILSFORD AND R. BULLOUGH: J. Nucl. Mat., 1973, **48**, 87
5 S. D. HARKNESS AND CHE-YU LI: Metall. Trans., 1971, **2**, 1457
6 H. WIEDERSICH: Conf. Proc., Vol. 2, 2nd Int. Conf. on 'Strength of metals and alloys', 784, 1970, ASM
7 A. D. BRAILSFORD AND R. BULLOUGH: J. Nucl. Mat., 1972, **44**, 121
8 A. D. BRAILSFORD AND R. BULLOUGH: Nucl. Met., 1973, **18**, 493
9 A. D. BRAILSFORD AND R. BULLOUGH: 'Physical metallurgy of reactor fuel elements', 148, 1975, London, The Metals Society
10 P. T. HEALD AND M. V. SPEIGHT: Acta Met., 1975, **23**, 1389
11 A. C. DAMASK AND G. J. DIENES: 'Point defects in metals', 1971, New York, Gordon and Breach
12 G. W. GREENWOOD et al.: J. Nucl. Mat., 1959, **4**, 305
13 A. D. BRAILSFORD et al.: J. Nucl. Mat., 1976, **60**, 246
14 R. BULLOUGH AND R. C. PERRIN: Proc. Conf. on 'Voids formed by irradiation of reactor materials' (Eds. S. F. PUGH et al.), 1971, Brit. Nucl. Energy Soc.
15 R. BULLOUGH AND R. C. PERRIN: Proc. Conf. on 'Radiation induced voids in metals' (Ed. J. W. CORBETT AND L. C. IANNIELLO), USAEC, 1972
16 M. KIRITANI: Proc. Conf. on 'Fundamental aspects of radiation damage in metals', II, p. 695 (Eds. M. T. ROBINSON AND F. W. YOUNG, JR.), Oct. 1975, Gatlinburg, Tenn.
17 J. BENTLEY et al.: ibid., p. 925
18 T. M. WILLIAMS AND B. L. EYRE: J. Nucl. Mat., 1976, **59**, 18
19 D. I. R. NORRIS: J. Nucl. Mat., 1971, **40**, 66
20 J. J. LAIDLER et al.: Proc. Conf. on 'Properties in reactor structural alloys after neutron or particle irradiation', ASTM. STP-570, 1975
21 A. H. COTTRELL AND B. A. BILBY: Proc. Phys. Soc., 1949, **62**, 49
22 R. BULLOUGH AND R. C. NEWMAN: Rep. Prog. Phys., 1970, **33**, 101
23 R. BULLOUGH AND J. R. WILLIS: Phil Mag., 1975, **31**, 855

24 I. G. MARGVELASHVILI AND Z. K. SARALIDZE: *Sov. Phys. Solid St.*, 1974, **15**, 1774
25 H. K. BIRNBAUM *et al.*: *Phil. Mag.*, 1971, **23**, 495
26 D. N. SIEDMAN AND R. W. BALLUFFI: *Phil. Mag.*, 1966, **13**, 649
27 D. J. MAZEY *et al.*: *J. Nucl. Mat.*, 1976, **62**, 73
28 A. D. BRAILSFORD AND R. BULLOUGH: Proc. Conf. on 'Defects and defect clusters in BCC metals and their alloys', *Nuclear Metallurgy*, 1973, **18**, Gaithersburg, Nat. Bureau of Standards

APPENDIX

Mathematical analysis for the bias of an edge dislocation with core rate limitation with a drift field present

We consider an edge dislocation, as described in the third section, to lie along the axis of a cylindrical coordinate system with a core radius r_0 at the centre of a cell of radius R. The object is to obtain the loss rate of point defects $\mathscr{L}$ into unit length of the dislocation where

$$\mathscr{L} = \frac{1}{\Omega} \int_0^{2\pi} d\theta \, \mathbf{r} \cdot \mathbf{J} \tag{A1}$$

$$\mathbf{J} = D \, \text{grad} \, c + \frac{Dc}{kT} \, \text{grad} \, E \tag{A2}$$

and

$$E = A \frac{\sin \theta}{r} \tag{A3}$$

The point defect concentration $c(r, \theta)$ satisfies equation (35) and the boundary conditions are given by equation (33) on $r = R$ and equation (49) at the dislocation core $r = r_0$.

The change of variables

$$z = A/2kTr = L/2r \tag{A4}$$

and

$$c = e^{-E/2kT} \psi \tag{A5}$$

as described in the section on the Heald–Speight model leads to the core boundary condition on the function $\chi(z, \phi)$ (at $z = z_0 = L/2r_0$)

$$\frac{\partial \chi}{\partial z} + (\beta + \cos \phi)\chi - \beta c^e e^{-z_0 \cos \phi} = 0 \tag{A6}$$

where

$$\chi(z, \phi) = \psi(z, \theta) \tag{A7}$$

with the angular variable

$$\phi = \frac{\pi}{2} - \theta \tag{A8}$$

and

$$\beta = \frac{2r_0^2 n\eta}{L} \tag{A9}$$

The outer boundary condition (at $z = z_R = L/2R$) is

$$\chi = c^\infty e^{z_R \cos \phi} \tag{A10}$$

and $\chi(z, \phi)$ satisfies

$$z \frac{\partial}{\partial z}\left(z \frac{\partial \psi}{\partial z}\right) + \frac{\partial^2 \chi}{\partial \phi^2} - z^2 \chi = 0 \tag{A11}$$

Since the function

$$\chi = c^e e^{-z \cos \phi} \tag{A12}$$

satisfies (A11) identically, the boundary conditions (A6) may be simplified by introducing a final new variable $\xi(z, \phi)$, where

$$\chi = c^e e^{-z \cos \phi} + \xi \tag{A13}$$

$\xi(z, \phi)$ now also satisfies (A11) and is subject to the boundary conditions

$$\xi(z_R, \phi) = c^\infty e^{z_R \cos \phi} - c^e e^{-z_R \cos \phi} \tag{A14}$$

and

$$\frac{\partial \xi}{\partial z}(z_0, \phi) + (\beta + \cos \phi)\xi(z_0, \phi) = 0 \tag{A15}$$

The general solution of (A11) is

$$\xi(z, \phi) = \sum_{n=0}^{\infty} \{A_n I_n(z) + B_n K_n(z)\} \cos n\phi \tag{A16}$$

where I_n, K_n are Bessel functions of imaginary argument. The boundary condition at $r = R$ yields, assuming $z_R \ll 1$:

$$A_0 I_0(z_R) + B_0 K_0(z_R) = c^\infty - c^e \tag{A17}$$

$$A_n I_n(z_R) + B_n K_n(z_R) = 0 \qquad n \neq 0 \tag{A18}$$

The core boundary condition (A15) yields

$$A_0\{I_0'(z_0) + \beta I_0(z_0)\} + \tfrac{1}{2}A_1 I_1(z_0) \\ + B_0\{K_0'(z_0) + \beta K_0(z_0)\} + \tfrac{1}{2}B_1 K_1(z_0) = 0 \tag{A19}$$

$$A_1\{I_1'(z_0) + \beta I_1(z_0)\} + A_0 I_0(z_0) + \tfrac{1}{2}A_2 I_2(z_0) \\ + B_1\{K_1'(z_0) + \beta(z_0)\} + B_0 K_0(z_0) + \tfrac{1}{2}B_2 K_2(z_0) = 0 \tag{A20}$$

$$A_n\{I_n'(z_0) + \beta I_n(z_0)\} + \tfrac{1}{2}\{A_{n+1}I_{n+1}(z_0) + A_{n-1}I_{n-1}(z_0)\} \\ + B_n\{K_n'(z_0) + \beta K_n(z_0)\} + \tfrac{1}{2}\{B_{n+1}K_{n+1}(z_0) \\ + B_{n-1}K_{n-1}(z_0)\} = 0 \qquad n \geq 2 \tag{A21}$$

In the limit $z_0 \gg 1$,

$$I_n'(z_0) \simeq I_n(z_0) \quad \text{and} \quad I_n(z_0) \simeq I(z_0) \sim \frac{e^{z_0}}{\sqrt{2\pi z_0}}$$

$$K_n'(z_0) \simeq -K_n(z_0) \quad \text{and} \quad K_n(z_0) \simeq K(z_0) \sim e^{-z_0}\sqrt{\frac{\pi}{2z_0}}$$

If we use these large argument results in (A19) to (A21) and eliminate B_n ($n \neq 0$) from (A18) the set of equations may be reduced to

$$A_0(1+\beta) + \tfrac{1}{2}A_1 + \frac{B_0(\beta-1)}{\epsilon} = 0$$

$$A_0 + A_1(1+\beta) + \tfrac{1}{2}A_2 + \frac{B_0}{\epsilon} = 0 \tag{A22}$$

and

$$A_n(1+\beta) + \tfrac{1}{2}(A_{n+1} + A_{n-1}) = 0 \qquad n \geq 2 \tag{A23}$$

where $\epsilon = I(z_0)/K(z_0) \gg 1$. The solution of (A23) is

$$A_n = A_1 q^{n-1} \tag{A24}$$

where

$$q = -(\beta + 1) + [(\beta + 1)^2 - 1]^{\frac{1}{2}} \tag{A25}$$

The required solution for ξ is thus given by (A16) with

$$A_0 = \tfrac{1}{2}A_1\{(\beta - 1)(1 + \beta + \tfrac{1}{2}q) - \tfrac{1}{2}\}$$
$$B_0 = -\tfrac{1}{2}\epsilon A_1\{(\beta + 1)(1 + \beta + \tfrac{1}{2}q) - \tfrac{1}{2}\} \tag{A26}$$

where A_1 follows from the outer boundary condition (A17):

$$A_1 = 2(c^\infty - c^e)/\{[(\beta - 1)(1 + \beta + \tfrac{1}{2}q) - \tfrac{1}{2}]I_0(z_R)$$
$$- \epsilon[(\beta + 1)(1 + \beta + \tfrac{1}{2}q) - \tfrac{1}{2}]K_0(z_R)\} \tag{A27}$$

The coefficients for $n \geqslant 2$ are given by (A24). When $\beta \to 0$, $q \simeq -1 + \sqrt{2\beta}$ and

$$A_1 \simeq -2(c^\infty - c^e)\Big/\left\{I_0(z_R) + \epsilon\sqrt{\tfrac{\beta}{2}}K(z_R)\right\}$$

$$A_n \simeq A_1(-1)^{n-1}\left\{1 - (n-1)\sqrt{\tfrac{\beta}{2}}\right\}$$

$$B_n = -A_n I_n(z_R)/K_n(z_R)$$

$$A_0 \simeq \frac{(c^\infty - c^e)(1 - \sqrt{\beta/2})}{I_0(z_R) + \epsilon\sqrt{\beta/2}\,K_0(z_R)}$$

and

$$B_0 \simeq \frac{(c^\infty - c^e)\epsilon\sqrt{\beta/2}}{I_0(z_R) + \epsilon\sqrt{\beta/2}\,K_0(z_R)}$$

and thus

$$\xi(z, \phi) \simeq \frac{(c^\infty - c^e)}{I_0(z_R) + \epsilon\sqrt{\beta/2}K_0(z_R)}\left\{e^{-z\cos\phi} - \sqrt{\tfrac{\beta}{2}}I_0(z)\right.$$
$$\left. - \sqrt{2\beta}\,I_2(z)\cos 2\phi + \ldots + \sum_{n=0}^{\infty} B_n K_n \cos n\phi\right. \tag{A29}$$

The loss rate of point defects $\mathscr{L}$, given by (A1), has the general form (after integrating over ϕ):

$$\mathscr{L} = -\frac{DL\pi}{R\Omega}\sum_{n=0}^{\infty}(-1)^n I_n(z_R)\{A_n I_n'(z_R) + B_n K_n'(z_R)\} \tag{A30}$$

and from (A29) and (A28) gives the final result

$$\mathscr{L} \simeq \frac{2\pi D}{\Omega}\frac{\epsilon\sqrt{\beta/2}}{1 + \epsilon\sqrt{\beta/2}\,K_0(z_R)} \tag{A31}$$

The result (54) then follows.

Theories of nucleation and growth of bubbles and voids

M. V. Speight

The application of classical nucleation theory to the formation of voids from a supersaturated concentration of vacancies is reviewed. The effect of a dissolved concentration of barely soluble gas on the nucleation rate of voids is emphasized. Exposure to a damaging flux of irradiation is the most effective way of introducing a vacancy supersaturation, but interstitials are produced at an equal rate. The concentration of interstitials inhibits the nucleation of voids which can occur only in the presence of dislocations since they preferentially absorb interstitials. It is well known that a definite value of internal gas pressure is necessary to stabilize a bubble so that it shows no tendencies to either shrink or grow. The arguments are reviewed which conclude that this pressure is determined by the specific surface free energy of the solid rather than the surface tension. While the former property refers to the energy necessary to create new surface, the latter is a measure of the work done in elastically stretching a given surface. The presence of an equilibrium gas bubble leaves the stresses in the surrounding solid unperturbed only when surface energy and surface tension are numerically equal. A bubble with internal pressure greater than the restraint offered by surface energy tends to grow to relieve the excess pressure. The mechanism of growth can involve the migration of vacancies from remote sources to the bubble surface or the plastic straining of the solid surrounding the bubble. The kinetics of both mechanisms are developed and compared. The theory of growth of grain-boundary voids by vacancy condensation under an applied stress is also considered. In the absence of stress the voids shrink but it is shown that the effective sintering pressure is a minimum for voids with a particular ratio of radius to spacing. Both relatively smaller and larger voids experience a greater collapse force.

The author is with the CEGB, Berkeley Nuclear Laboratories

Cavities may be introduced into a solid material in a variety of ways. Principal causes of cavitation are stress and strain, whether externally imposed or internally developed. For instance, during creep formation of cavities occurs mainly at grain boundaries and their total number has been found to increase linearly with the overall strain.[1] This implies that, irrespective of the exact mechanism, nucleation originates with the relative movement of grains along their intervening boundaries.[2] Such grain-boundary sliding is necessary to preserve strain compatibility through the deforming body when the applied stress is in the range where diffusion creep predominates.[3] Although it is not then an essential part of the deformation process sliding does occur[4] at higher applied stresses when the individual grains deform by a dislocation mechanism. Following nucleation there is experimental evidence[5] that subsequent growth of cavities relies on the formation and diffusion of vacancies in the boundaries.

Real materials seldom behave in an isotropic and homogeneous manner and respond to an imposed physical change by developing internal stresses which can result in cavity formation. For example, the rapid heating of a metal containing ceramic precipitates with a much lower thermal expansion coefficient can induce unrelaxed tensile stresses of sufficient magnitude to cause particle decohesion (for example the work of Olsen et al.[6] on thoria in a nickel matrix). Additionally, internal stresses can be readily generated in an agglomerate of randomly oriented crystals of an anisotropic substance. Thus the orientation dependence of the lattice thermal expansion coefficient leads to intergranular stresses which are thought responsible for the growth of voids during the thermal cycling of polycrystalline uranium.[7] Similarly internal stresses resulting from the irradiation-induced dimensional changes of individual grains along prescribed crystallographic directions (i.e. irradiation growth) appear to play a vital role in the swelling of irradiated uranium.[8,9]

Even nominally pure and crystallographically isotropic materials can generate a system of internal stresses with ensuing void formation. Under applied uniaxial compression the existence of grain boundaries as planes of easy shear enables intergranular tensile stresses and associated voids to develop.[10,11]

Precipitation from a supersaturation of vacancies is an obvious source of void formation. The process is aided by the presence of insoluble gases which co-precipitate with the vacancies. The contained gas exerts a pressure, in addition to the effective chemical stress provided by the vacancy supersaturation, stabilizing the void against collapse from the need to reduce surface energy. Without the vacancy supersaturation or other factors influencing void size stable gas bubbles are formed when the gas and surface energy pressures exactly balance. Bubbles are formed extensively in nuclear fuels where at sufficiently high irradiation temperatures the uniformly created fission gases can diffuse and agglomerate (see Speight[12] for a review of this subject). Conversely gas-deficient voids are formed in non-fissile materials, where compared to fuels gas generation is small, irradiated at temperatures sufficiently low that the irradiation damage process can sustain vacancy concentrations above their thermal equilibrium levels (see Norris[13] for a review).

A further source of hole growth in solids is associated directly with the fundamental process of atomic diffusion. In a solid solution binary alloy of non-uniform concentration any disparity between atomic mobilities of the components, both diffusing by a vacancy mechanism, implies a net current of vacancies along the concentration gradient. Under suitable conditions this vacancy flow leads to the precipitation and growth of cavities, known as Kirkendall Voids. Similar vacancy flows arise under different physical circumstances. In particular, during oxidation the removal of metal atoms into the surface scale without movement of the oxide/metal interface is equivalent to the continuous injection of vacancies into the metal. Evidence for the process has been obtained by Dobson and Smallman[14] who followed the growth of vacancy loops during the oxidation of zinc. The aggregation of vacancies to form voids within the metal has been observed by Evans et al.[15] during the oxidation of stainless steels.

In this paper the thermodynamics and kinetics of the nucleation, stability and growth of voids by some of the above mechanisms are examined in detail.

HOMOGENEOUS VOID NUCLEATION

Precipitation from vacancy supersaturation

The approach generally adopted to this problem is consistent with classical nucleation theory first developed to describe the formation of liquid droplets from their vapours. Thus there is a critical size at which void nuclei can be classed as thermodynamically stable when further growth results in a net decrease in free energy of the system. Smaller nuclei are unstable since on average the probability of growth through capture of a diffusing vacancy is less than the probability of shrinkage by vacancy emission. However, statistical fluctuations ensure, that in the time taken for vacancy emission, there is a finite probability that a number of vacancies strike and coalesce with a subcritical nucleus sufficient to bring it to the critical size. In this random manner a stable nucleus is formed.

The arguments here imply that only single vacancies are mobile and that growth of nuclei does not occur by the diffusion and coalescence of vacancy clusters. The influence of this latter process on nucleation has been considered by Whapham[16] and Singh and Foreman[17] but there is serious doubt concerning a quantitative description of cluster mobility. Initially[18] small voids were postulated to move by a process controlled by the diffusion of atoms over their internal surfaces. Subsequent experimental results[19-21] and analysis[22] have shown that mobility may be severely limited by the slowness of surface nucleation events which are essential for void movement. Since the magnitude and radius dependence of cluster mobility is uncertain it is assumed here, for the purposes of illustration, that only single vacancies can diffuse significantly. The kinetics of void nucleation can be then treated in a completely analogous manner to the approach of Damask et al.[23] who applied their calculations to the precipitation of carbon in iron.

The total number, n, of vacancies comprising a sub-critical void nucleus defines its volume as $n\Omega$ where Ω is the atomic volume of the material. The probability per unit time, p_n, of the cluster absorbing an extra vacancy to reach a total of $n+1$ is

$$p_n = \beta_n D_v C_v$$

where β_n is an effective capture cross-section of the void and D_v the diffusion coefficient of dissolved vacancies of atomic concentration C_v. Conversely the probability per unit time, q_n, of forming a cluster of size n through emission of a single vacancy from the next higher size is

$$q_n = \alpha_n D_v$$

where

$$\alpha_n \approx \beta_n \exp(\delta \Delta F_n / kT).$$

Here $\delta \Delta F_n$ is the net increase in free energy on forming a cluster of size $n+1$ from one of size n, and kT is the thermal energy. In dynamic equilibrium the ratio (C_{n+1}/C_n) of the concentrations of clusters of size $n+1$ and size n becomes

$$(C_{n+1}/C_n) = (p_n/q_n) = (\beta_n C_v/\alpha_n)$$
$$= C_v \exp - (\delta \Delta F_n / kT) \quad (1)$$

Similarly

$$(C_n/C_{n-1}) = (\beta_{n-1} C_v/\alpha_{n-1}) \quad (2)$$

and equivalent relationships apply between successive cluster sizes down to the expression which relates the concentration, C_2, of di-vacancies to that of single vacancies, $C_1 = C_v$. Thus

$$C_2 = \beta_1 C_v^2/\alpha_1$$

and so equation (2) can be written[23]

$$C_n = \prod_{i=1}^{n-1} (\beta_i/\alpha_i) C_v^n = C_v^n \exp - \left(\sum_{i=1}^{n-1} \delta \Delta F_i / kT \right) \quad (3)$$

This concentration of subcritical nuclei is sustained through the finite random chance that a net number, n, of individual vacancies have of combining together.

The quantity $\sum_{i=1}^{n-1} \delta \Delta F_i$ appearing in equation (3) is the total increase in free enrgy when $(n-1)$ vacancies are successively combined with a single vacancy to form a cluster of size n. While recognizing the limitations[24] the excess free energy of a cluster, assumed spherical, is usually described by its effective surface free energy.

This is the so-called capillarity model. Thus

$$\Delta F_n = \sum_{i=1}^{n-1} \delta\Delta F_i = (4\pi)^{\frac{1}{3}}(3\Omega n)^{\frac{2}{3}}\sigma - n\,\Delta H_f \quad (4)$$

where σ is the specific surface free energy and ΔH_f is the vacancy formation energy. By differentiation with respect to n

$$\delta\Delta F_n = (2/3)(4\pi/n)^{\frac{1}{3}}(3\Omega)^{\frac{2}{3}}\sigma - \Delta H_f \quad (5)$$

A cluster is deemed to have attained a critical size, beyond which its existence is permanent, when the probabilities of capturing and emitting vacancies are equal. Above the critical size vacancy capture is favoured relative to emission leading to enhanced stability as growth proceeds. The critical nucleation size n^* is obtained by equating the ratio (p_n/q_n) to unity in equation (1). Thus, using equation (5)

$$n^* = (32\pi\Omega^2\sigma^3)/3(kT)^3[\ln(C_v/C_e)]^3 \quad (6)$$

where C_e, the thermal equilibrium vacancy concentration, is given by the usual expression $C_e = \exp(-\Delta H_f/kT)$.

The concentration of nuclei of critical size is obtained by first substituting equation (6) into equation (4) and using the result in equation (3) to give

$$C_{n^*} = \exp-(\Delta F_{eff}/kT) \quad (7)$$

where ΔF_{eff} is the effective free energy of nucleation given by[25]

$$\Delta F_{eff} = 16\pi\Omega^2\sigma^3/3(kT)^2[\ln(C_v/C_e)]^2$$

This quantity and hence the rate of production of stable voids, which is proportional to $D_v C_{n^*}$, depends critically on the degree of vacancy supersaturation (C_v/C_e).

Effect of gas atoms

The presence of gas within subcritical vacancy clusters inhibits their tendency to eject vacancies and so aids the void nucleation process. A quantitative assessment of the effect requires consideration of the increase in Helmholtz free energy ΔF_g when m gas atoms are removed from solution in the matrix and placed, at pressure p, within a void whose volume $n\Omega$ remains constant. The appropriate expression is readily derived from the work of Greenwood:[26]

$$\Delta F_g = mkT\ln(p/p_e) - pn\Omega$$

which becomes

$$= mkT\ln(p/p_e) - mkT \quad (8)$$

if the simplifying, though not altogether realistic, assumption of perfect gas behaviour is assumed. The quantity p_e is the pressure of gas in the void in equilibrium with the dissolved concentration of gas in the lattice. In other words p_e is the enclosed gas pressure at which the differential $(\partial\Delta F_g/\partial m)$, evaluated at constant void volume, is zero.

The total increase in free energy when n individual vacancies and m separate gas atoms co-precipitate to form a void is then

$$\Delta F_n^1 = \Delta F_n + \Delta F_g \quad (9)$$

The differential increase in free energy when a further

vacancy is added at constant number of gas atoms is:

$$\delta\,\Delta F_n^1 = \delta\,\Delta F_n + (\partial\,\Delta F_g/\partial n)_m$$
$$= \delta\,\Delta F_n - p\Omega$$

and the condition defining the critical nucleus size is now

$$C_v \exp-(\delta\,\Delta F_n^1/kT) = 1.$$

Following the same calculational procedure as before the size of the critical void nucleus, containing gas at pressure p, becomes

$$n^* = (32\pi\Omega^2\sigma^3)/3[kT\ln(C_v/C_e) + p\Omega]^3 \quad (10)$$

which reduces to equation (6) when p is zero. Conversely, when the vacancy supersaturation is zero equation (10) predicts the radius of the critical nucleus to be $(2\sigma/p)$ in agreement with the usual condition for unlimited bubble growth at constant internal gas pressure.

Equation (10) emphasizes the profound influence of precipitated gas on the formation of stable nuclei. To quantify the effect, the problem is to estimate the pressure of gas developed in vacancy clusters. This depends primarily on the relative rates at which gas atoms and vacancies can diffuse and coalesce with clusters. A comprehensive kinetic calculation must be done numerically[27] but, at the opposite ends of extrapolation corresponding to extremely immobile and mobile gas, the effects of gas can easily be demonstrated. In the former case gas atoms may provide fixed sites for heterogeneous void nucleation but, if finer scale homogeneous nucleation occurs, it should be at the same rate as in the absence of gas. There is a limit to the influence of very mobile gas atoms on the homogeneous nucleation of voids. This applies where the maximum gas pressure p_e, in thermodynamic equilibrium with the dissolved concentration of gas, is sustained in vacancy clusters throughout their growth. The critical void nucleation size is then a minimum and is defined by substituting $p = p_e$ in equation (10) to obtain the expression quoted by Russell.[28] Under these conditions the contribution, ΔF_g, that gas precipitation makes to the total free energy change has a minimum (corresponding to $(\partial\Delta F_g/\partial m)_n = 0$) value of $(-p_e n^*\Omega)$ and the effective free energy of nucleation ΔF_{eff} is a minimum. From equations (3), (4), (7), (8), (9), and (10) this minimum value is given by[28]

$$\Delta F_{eff} = (16\pi\Omega^2\sigma^3/3)/[kT\ln(C_v/C_e) + p_e\Omega]^2$$

This defines the free energy barrier to nucleation when the gas is sufficiently mobile to maintain the equilibrium pressure in developing clusters. If, in addition, the gas is relatively insoluble then p_e is large and void nucleation is comparatively easy. In the limit of negligible solubility the agglomeration of two gas atoms and associated vacancies which enable them to diffuse can be sufficient to form a stable nucleus. The rate of nucleation is then controlled by the chance meeting of diffusing gas atoms and this defines the conditions envisaged by Greenwood et al.[29] in their treatment of the problem. The conclusions of these latter authors become progressively less applicable with decreasing gas mobility and increasing gas solubility (i.e. decreasing binding energy of complexes of two gas atoms).

Effect of interstitial concentration

Exposure to a damaging flux of irradiation is the most effective way of maintaining a vacancy supersaturation in a material. Since interstitials are also produced in equal numbers the criterion for void nucleation cannot be simply deduced by considering the vacancies in isolation. The capture of mobile interstitials by subcritical nuclei diminishes their chance of reaching a stable size. This inhibiting effect of interstitials on nucleation can be readily quantified.[25]

If C_I is the concentration of dissolved interstitials with diffusion coefficient D_I, the probability per unit time of a cluster of size $(n+1)$ capturing an interstitial is

$$\beta_{n+1}D_I C_I$$

and, including thermal vacancy emission, the total probability of such a cluster shrinking to size n in unit time is

$$q_n + \beta_{n+1}D_I C_I$$

By analogy with previous arguments the criterion for stable void nucleation of equal probabilities for further growth and shrinkage is

$$p_n/(q_n + \beta_{n+1}D_I C_I) = 1$$

or

$$\beta_n D_v C_v/(\alpha_n D_v + \beta_{n+1}D_I C_I) = 1$$

where

$$\alpha_n \approx \beta_n \exp(\delta \Delta F_n^1/kT)$$

Re-arranging and substituting for $\delta \Delta F_n^1$ gives the critical nucleus size as [25]

$$n^* = (32\pi\Omega^2\sigma^3)/3\{kT \ln[(\beta_n D_v C_v$$
$$- \beta_{n+1}D_I C_I)/\beta_n D_v C_e] + p\Omega\}^3$$

Instead of simply the vacancy supersaturation it is the parameter $(D_v C_v - D_I C_I)$ which determines the ease of nucleation. Its value depends on the equilibrium conditions maintained by the creation of defects by irradiation and their disappearance at suitable sinks. If all point defect sinks accept vacancies and interstitials equally readily then[29]

$$D_v C_v - D_I C_I = D_v C_e$$

and the logarithmic term in the above equation is identically zero. It only adopts a positive value because dislocations, through a stronger elastic interaction,[30] preferentially absorb interstitials and suppress their concentration compared to vacancies. The critical size for void nucleation depends sensitively on the degree of bias dislocations show for attracting interstitials and this varies with the dislocation density.[30] The dislocation density also determines the absolute equilibrium concentrations of defects[31] and thus is a crucial factor in deciding the level of void nucleation in irradiated non-fissile materials. The rate of void nucleation is proportional to $D_v C_{n^*}$ where from equivalent equations to (1), (2), and (3) C_{n^*} is given by

$$C_{n^*} = \prod_{i=1}^{n^*-1} (\beta_i D_v C_v^n)/(\alpha_i D_v + \beta_{i+1}D_I C_I)$$

which reduces to equation (3) when C_I is zero.

GAS BUBBLES IN THERMODYNAMIC EQUILIBRIUM

It was deduced earlier, following equation (10), that, in the absence of other factors such as irradiation and applied stress, growth of a bubble of radius a and internal gas pressure p is energetically favoured when $a > 2\sigma/p$. Here σ is the specific surface free energy which, in general, differs in magnitude from the surface tension or surface stress γ. This latter parameter defines the mechanical resistance of a solid surface to elastic stretching[32] while σ is the energy required to create plastically unit area of new surface at constant elastic surface strain. If U is the excess free energy associated with a surface of area A then:

$$U = \sigma A$$

An incremental change in energy accompanying an elastic expansion of the area at constant number of surface atoms N is described by:

$$\left(\frac{\partial U}{\partial A}\right)_N = \gamma = \sigma + A\left(\frac{\partial \sigma}{\partial A}\right)_N$$

which defines the relationship between surface tension and specific surface energy.

In an elastic solid the effective mechanical pressure acting inwards over a bubble surface is $2\gamma/a$. It is not then immediately obvious that the state $p = 2\sigma/a$ defines a condition of equilibrium for a bubble containing a given number of gas atoms since (because $\gamma \neq \sigma$) there is an associated strain field in the surrounding crystal lattice. Only when $p = 2\gamma/a$ does this strain field disappear. The true equilibrium state must be defined by minimizing the total energy of the system including contributions from elastic stored energy. The mechanism which establishes bubble equilibrium in an elastic solid is the flow of lattice vacancies to or from available remote sources (for example grain boundaries).

The state of gas bubble equilibrium has been examined in detail by Lidiard and Nelson[33] for a spherical bubble in an isotropic elastic solid. In polar co-ordinates (r, θ) general elastic solutions for the radial and circumferential stresses (s_r and s_θ respectively) in the material are:

$$s_r = X - Y/r^3 \qquad s_\theta = X + Y/2r^3 \qquad (11)$$

where X and Y are constants determined by the states of stress at the bubble surface, $r = a$, and the external surface, $r = b$, of the supposedly spherical body. The net radial tensile stress exerted on the bulk solid by the bubble surface is $(2\gamma/a) - p$ while a zero externally applied stress is assumed. These conditions define a stress system in the bulk solid with an associated total elastic strain energy E_1 given by [33]

$$E_1 = \{\pi a^3[p - (2\gamma/a)]^2/2\mu(b^3 - a^3)\}$$
$$\times \{b^3 + 2a^3(1 - 2\nu)/(1 + \nu)\}$$

where μ is the shear modulus. A further component of stored energy E_2 originates with the elastic strain in the bubble surface. In its equilibrium configuration the bubble surface is displaced elastically a distance v in the radial direction and the surface strain energy is simply the work done against the surface stress during this displacement.

Hence

$$E_2 = (2\gamma/a)4\pi a^2 v$$

where, if the superscript a denotes stresses evaluated at

the bubble surface, v is given by the usual elastic formula

$$v/a = (1-\nu)s_\theta^a/M - \nu s_r^a/M$$

with M as Young's modulus. Evaluating the stresses for the given boundary conditions to determine v enables the energy E_2 to be derived as

$$E_2 = \{4\pi\gamma a^2[p-(2\gamma/a)]/\mu(b^3-a^3)\}$$
$$\times \{(b^3/2)+a^3(1-2\nu)/(1+\nu)\}$$

Other components of the total energy are the excess energies of the free surfaces E_3 and the free energy of the gas enclosed in the bubble E_4. These are respectively

$$E_3 = 4\pi(a^2+b^2)\sigma$$

and from equation (8) $E_4 = mkT \ln(p/p_e) - mkT$.

The equilibrium radius of the bubble corresponds to a minimum in total energy achieved through the transport of vacancies between the bubble and the external surface of the solid while the number of contained gas atoms remains constant. Within the expression for total energy the stored energy terms E_1 and E_2 vary inversely as the elastic modulus and for realistic stresses are comparatively small as shown by the calculations of Lidiard and Nelson.[33] It follows that the state of minimum energy is dictated by the appropriate balance between surface energy and enclosed gas pressure. For a small bubble in an extensive solid $b \gg a$ and the condition of equilibrium is simply $p = 2\sigma/a$.

The discrepancy between σ and γ means that, in an otherwise stress-free solid, internal bubbles, even when they are in thermodynamic equilibrium, have an associated stress field which is sufficiently large may be detected by electron microscopy.[34] Speight[35] has shown that the presence of this stress field is the sole reason why equilibrium bubbles experience a force when the solid is subject to an applied stress gradient. Such a gradient should not cause equilibrium bubbles to migrate if the surface properties σ and γ are identically equal.

KINETICS OF BUBBLE GROWTH

Vacancy diffusion

In an elastic solid without irradiation the absorption of thermal vacancies diffusing from remote sources is the mechanism permitting the growth of bubbles which contain gas at a pressure in excess of that offered by surface energy restraint. Here the kinetics of this growth process are examined. The procedure entails solving the diffusion equation describing the motion of vacancies subject to the boundary conditions of fixed vacancy chemical potentials at the bubble surface and at the sources. These sources can be taken as grain boundaries since there is experimental evidence (for example Brett and Seigle[36]) that dislocations have a limited and readily exhaustible capacity to absorb or emit vacancies.

The bubble of radius a is considered to lie at the centre of a spherical grain of radius b. In the steady state the total flow of vacancies through any spherical surface lying at radius r between the boundary and bubble is invariant. Thus if J is the vacancy flux towards the bubble, and $\dot{N}_v$ the rate of vacancy collection by the bubble

$$4\pi r^2 J = \dot{N}_v = \text{constant} \tag{12}$$

and

$$J = (D_v C_v/\Omega kT)\nabla f$$

where f is the vacancy chemical potential given by:

$$f = kT \ln(C_v/C_e) + E \tag{13}$$

Here, E is the elastic interaction energy between the vacancy and the stress field around the bubble; it is the work done on the system when the vacancy is taken from a standard state outside the influence of the bubble stress field (since usually $b \gg a$ the grain boundary can be taken to define the ground state) and introduced to its particular location. The energy, E, comprises two parts; the first E_f is the net work done against hydrostatic tension because the vacancy occupies a smaller volume than the atom it displaces. For the case of a spherical bubble equation (11) shows the surrounding hydrostatic stress to be everywhere constant with a value X. Thus E_f is identically zero since there is no net work done, due to the size difference, on interchanging an atom with a vacancy whatever their relative positions with respect to the bubble. This situation contrasts sharply with the strong first-order size interaction experienced by vacancies in the non-uniform hydrostatic stress field of a dislocation.[37] With zero applied stress the chemical potential f_b of a vacancy at the boundary is zero and since $b \gg a$ the hydrostatic stress X is also zero.

The second component of the interaction, E_s, arises from the fact that a region containing a vacancy is more easily deformable than an equal portion of perfect crystal and, for a given triaxial stress system, experiences larger strains in the three orthogonal directions (see for instance Heald et al.).[38] These strain differences, each multiplied by the corresponding stress, sum together to give a measure of the work done when a vacancy in the stress-free boundary is exchanged with an atom near the bubble. Since each differential strain is proportional to the corresponding stress the quantity E_s varies as the sum of the squares of the components of local stress and hence, from equation (11), inversely as the sixth power of distance from the bubble centre.[39] The intensity of the stress field and so the absolute magnitude of the interaction, E_s, depends on the bubble surface net stress $(p - 2\gamma/a)$.

Substituting for J and subsequently for f the appropriate solution to the diffusion equation (12) becomes

$$(\dot{N}_v\Omega/4\pi r^2 D_v)\exp(E_s/kT) = \frac{d}{dr}\{C_v\exp(E_s/kT)\}$$

which from equation (13)

$$= C_e\frac{d}{dr}\{\exp(f/kT)\}$$

and on integration yields

$$\int_a^b (\dot{N}_v\Omega/4\pi r^2 D_v)\exp(E_s/kT)\,dr$$
$$= C_e\{\exp(f_b/kT)-\exp(f_a/kT)\}$$

where f_a is the chemical potential of a vacancy in the lattice immediately adjacent to the bubble. In dynamic equilibrium it is equal to the chemical potential of a vacancy in the bubble and this latter quantity is simply the increase in free energy when a vacancy is added at constant number of gas atoms. From the previous arguments only the contributions from surface energy and gas pressure are significant so that

$$f_a = (2\sigma/a - p)\Omega$$

and under conditions of zero applied stress $f_b = 0$.

Writing $E_s = -\alpha(a/r)^6$ (ref. 39) where α like f_a is $\ll kT$, series expansions of the exponentials yield

$$(4\pi D_s a/kT)(p - 2\sigma/a) = \dot{N}_v[1 - (\alpha/7kT)] \qquad (14)$$

where D_s ($= D_v C_e$) is the self-diffusion coefficient. Putting $\alpha = 0$ equation (14) reduces to the expression of Greenwood et al.[29] who considered the growth of an overpressurized bubble solely by the random diffusion of vacancies from grain-boundary sources. The bubble growth rate is increased by the existence of the second-order elastic interaction between its stress field and the vacancy but the effect is very small. The rate of condensation of vacancies at growing bubbles is then virtually independent of the elastic properties of the material.

Plastic deformation

The stress field surrounding a bubble has been assumed here to be completely determined by the elastic properties of the material with the absence of any plastic stress relaxation process. This probably well describes the conditions around a very small bubble where the stress field is of too limited an extent to activate any plastic deformation mechanism. With larger bubbles approaching $\sim 1\mu$m the range of the stress field is of the same order as the dislocation spacing and plastic flow of the matrix becomes possible. A bubble with excess gas pressure is then able to expand simply by plastically deforming the surrounding material. The question is whether this growth mechanism can compete with the capture of lattice vacancies flowing to the bubble from available sources.

To evaluate the growth rate of the bubble by plastic flow it is usual to consider that the surrounding material deforms according to its macroscopic steady-state creep properties.[40] The problem to be treated is that of a spherical bubble with given excess pressure lying at the centre of a sphere of material whose outer surface of radius b is stress free or, more generally, subject to some fixed externally applied pressure. The dimension $2b$ corresponds to the distance between adjacent bubbles. Since the whole material, including the bubble surface, behaves plastically the effective pressure that the bubble exerts on the surrounding solid is $(p - 2\sigma/a) \times$ $[c/f(p - 2\gamma/a)$ in an elastic solid] and so long as this exceeds the applied restraint bubble growth occurs.

Wilkinson and Ashby[41] have recently considered the shrinkage rates of voids through the creep deformation of the surrounding solid. The present problem is identical except with the sign of the stresses reversed to allow bubble growth. From their analysis, assuming $b \gg a$ and zero applied pressure, the rate of increase in bubble volume becomes

$$4\pi a^2 (da/dt) = 2\pi Z a^3 (3/2\omega)^\omega (p - 2\sigma/a)^\omega \qquad (15)$$

where Z and ω are the parameters occurring in the equation describing secondary creep of the material under a uniaxial stress, s viz, creep rate $= Zs^\omega$.

Equation (15) can be compared directly with equation (14) to assess the relative features of the two growth processes. The creep mechanism is progressively more favoured with increasing bubble size, but whether it can become the dominant mode of growth at realistic bubble sizes depends on the exact material parameters, notably D_s and Z. Unless the matrix deforms at a rate linear with imposed stress, increasing excess gas pressure preferentially aids the creep process.

VOID GROWTH UNDER AN APPLIED STRESS

Growth of voids is a phenomenon which occurs generally during the creep testing of most materials. The cavaties develop preferentially at grain boundaries and, after nucleation, there is evidence[5] that they subsequently grow by the absorption of vacancies which are continuously created in the boundary by the action of the normal component of applied tensile stress. The kinetics of this process have been considered theoretically by a variety of authors.[42–46] Most of these treatments, while broadly correct, neglect some physical aspect of the diffusion problem. The exception is the work of Speight and Beere whose analysis of void growth is now reviewed.

The concept of a grain boundary is that it is able to behave as a perfect and inexhaustible source of vacancies. If the stress at every point on the boundary is to remain constant with time then vacancies must be generated at a uniform rate (ϵ per unit area) over the entire surface. Any spatial variation in vacancy production rate would lead to a continuous redistribution of stress over the boundary plane. After creation vacancies flow in the plane of the boundary to the nearby voids and in the steady state the concentration of vacancies is everywhere constant independent of time. The appropriate diffusion equation describing the flow is then[45]

$$\nabla^2 f + (\epsilon kT\Omega/D_g d) = 0$$

where f is the vacancy chemical potential, D_g the grain-boundary self-diffusion coefficient and d the boundary thickness. This equation must be solved, assuming cylindrical symmetry around each void, subject to the appropriate boundary conditions. These are:

(i) fixed vacancy chemical potential at each void surface; this value is $\Omega(2\sigma/a)$ if voids are assumed virtually spherical, implying a low grain-boundary energy, with radius a

(ii) zero gradient of vacancy chemical potential midway between adjacent voids spaced $2b$ apart.

The solution of the above equation giving f as a function of the radial co-ordinate, r, measured from the void centre is then

$$f = (2\sigma\Omega/a) - (\epsilon kT\Omega/4D_g d)[r^2 - a^2 - 2b^2 \ln(r/a)]$$

The quantity f represents the work which must be done on the system to precipitate a vacancy at any point in the boundary and allow adjacent crystals to collapse together. If a complete atomic layer, of thickness λ, of vacancies is plated out over the entire boundary area then the total work done per void, W, is

$$W = (\lambda/\Omega)\int_a^b 2\pi r f \, dr$$

This work is necessary to overcome the opposition to collapse from the net tensile force on the boundary. An obvious contribution to this force is provided by the applied stress but a further component acting in the opposite direction originates with surface energy. Thus if the boundary completely collapses by one atomic layer then the total surface energy of each void is reduced by $2\pi a\lambda\sigma$ and this is a source of effective compression on the boundary. If s is the applied tensile stress it follows that

$$\int_a^b 2\pi r f \, dr = (s\pi b^2 - 2\pi a\sigma)\Omega$$

Substituting for f enables ϵ, the parameter describing the generation rate of vacancies and hence the growth of voids, to be determined. Algebraic manipulation yields

$$4\pi a^2 (da/dt) = \pi b^2 \epsilon \Omega$$
$$= 2K[\ln(b/a) - (1 - a^2/b^2)$$
$$\times (3 - a^2/b^2)/4]^{-1}$$

where $K = \pi\Omega D_g d(s - 2\sigma/a)/kT$. The formula predicts that, above the critical size at which $s = 2\sigma/a$, voids grow at an increasing rate as they become bigger with respect to their spacing. The rate of increase in growth rate accelerates as voids become relatively larger.

In the absence of an applied stress voids shrink under the sole influence of surface energy. The effective sintering pressure consists of two parts; one stems directly from the surface curvature of the void while the second arises from the compressive stress that each void exerts on the boundary. This compression enhances the ability of the boundary to act as a sink for the vacancies which are ejected from each void by the action of surface curvature. The total sintering pressure is then

$$2\sigma/a + 2\sigma a/(b^2 - a^2)$$

The first term describes the surface curvature effect which decreases with increasing void size. The second represents the compressive stress with which the voids clamp the boundary and this term increases with larger voids. These opposing tendencies combine to produce a minimum in the shrinkage pressure for voids with a particular relative size of $(b/a) = \sqrt{3}$. This conclusion applies to a simple array of voids on a grain boundary, but Beere[47] has shown that for the more complex arrangement of porosity in powder compacts a broad minimum exists in the sintering force throughout a particular range of densities. Higher and lower density materials experience greater sintering forces.

CONCLUDING REMARKS

The present analysis deals with various kinetic and thermodynamic aspects of the nucleation, stability and growth of bubbles or voids. The existence of a vacancy supersaturation, the presence of insoluble gases, or the action of stresses, whether internally developed or externally applied, are all factors which assist the formation and growth of cavities. The influence of a vacancy supersaturation is most pronounced in the development of voids during the irradiation of non-fissile materials, while the creation of large quantities of fission gas is mainly responsible for the appearance of bubbles in nuclear fuels. In the absence of irradiation, tensile stress is the most obvious and important source of cavity growth. One of its principal actions is to create vacancies at grain boundaries. These subsequently flow to voids which are above the critical size at which their vacancy chemical potential is lower than the grain boundaries.

Cavity growth is opposed by the requirement to reduce total surface free energy. In general the characteristic material parameters of specific surface free energy and surface tension or stress are not equal. The former controls the tendency of a cavity to shrink by vacancy ejection while the latter determines the magnitude of the surrounding elastic stress field. It follows that an equilibrium gas bubble whose internal pressure is just sufficient to suppress net vacancy emission has nevertheless an associated elastic stress field. With a non-equilibrium pressure this stress field is more intense and interacts with any lattice vacancy in the vicinity of the bubble or void. However, the magnitude of the interaction is slight and has a negligible influence on the void rate of volume change, which is principally determined by the flux of vacancies occurring by random diffusion.

ACKNOWLEDGMENT

This paper is published by permission of the Central Electricity Generating Board.

REFERENCES

1 G. W. GREENWOOD: *Phil. Mag.*, 1969, **19**, 423
2 J. E. HARRIS: *Trans. Met. Soc. AIME*, 1965, **233**, 1509
3 I. M. LIFSHITZ: *Soviet Physics JETP*, 1963, **17**, 909
4 R. L. BELL AND T. G. LANGDON: Proceedings Conference on Interfaces, Melbourne, 115, 1969
5 R. T. RATCLIFFE AND G. W. GREENWOOD: *Phil. Mag.*, 1965 **12**, 59
6 R. J. OLSEN, G. JUDD AND G. S. ANSELL: *Metall. Trans.* 1971, **2**, 1353
7 R. C. LOBB *et al.*: *J. Inst. Metals*, 1968, **96**, 262
8 J. W. HARRISON: *J. Nucl. Mat.*, 1967, **23**, 139
9 M. V. SPEIGHT AND J. E. HARRIS: Proc. 10th Metallurgical Colloquium, Saclay, 171, 1966
10 R. RAJ AND M. F. ASHBY: *Metall. Trans.*, 1971, **2**, 1113
11 G. L. REYNOLDS *et al.*: *Acta Met.*, 1975, **23**, 573
12 M. V. SPEIGHT: 'Physical metallurgy of reactor fuel elements', 222, 1975, London, The Metals Society
13 D. I. R. NORRIS: *Radiat. Effects*, 1972, **14**, 1
14 P. S. DOBSON AND R. E. SMALLMAN: *Proc. Roy. Soc.*, 1966, **A293**, 423
15 H. E. EVANS, *et al.*: CEGB Internal Report RD/B/N3416; 1976
16 A. D. WHAPHAM: *Phil. Mag.*, 1971, **23**, 987
17 B. N. SINGH AND A. J. E. FOREMAN: *Scripta Met.*, 1975, **9**, 1135
18 G. W. GREENWOOD AND M. V. SPEIGHT: *J. Nucl. Mat.*, 1963, **10**, 140
19 M. E. GULDEN: *J. Nucl. Mat.*, 1967, **23**, 30
20 L. E. WILLERTZ AND P. G. SHEWMON: *Metall. Trans.* 1970, **1**, 2217
21 K. Y. CHEN AND J. R. COST: *J. Nucl. Mat.*, 1974, **52**, 59
22 W. B. BEERÉ: *J. Nucl. Mat.*, 1972/73, **45**, 91
23 A. C. DAMASK *et al.*: *Acta. Met.*, 1965, **13**, 973
24 J. J. BURTON: *Acta Met.*, 1973, **21**, 1225
25 K. C. RUSSELL: *Acta Met.*, 1971, **19**, 753
26 G. W. GREENWOOD, *J. Mater. Sci.*, 1969, **4**, 320
27 H. WIEDERSICH *et al.*: *J. Nucl. Mat.*, 1974, **51**, 287
28 K. C. RUSSELL: *Acta Met.*, 1972, **20**, 899
29 G. W. GREENWOOD *et al.*: *J. Nucl. Mat.*, 1959, **4**, 305
30 P. T. HEALD: *Phil. Mag.*, 1975, **31**, 551
31 A. D. BRAILSFORD AND R. BULLOUGH: *Phil. Mag.*, 1973, **27**, 49
32 P. R. COUCHMAN AND W. A. JESSER: *Surface Sci.*, 1973, **34**, 212
33 A. B. LIDIARD AND R. S. NELSON: *Phil. Mag.*, 1968, **17**, 425

34 L. M. BROWN AND D. J. MAZEY: *Phil. Mag.*, 1964, **10**, 1081

35 M. V. SPEIGHT: *J. Nucl. Mat.*, 1975, **58**, 55

36 J. BRETT AND L. SEIGLE: *Acta Met.*, 1963, **11**, 467

37 F. S. HAM: *J. Appl. Phys.*, 1959, **30**, 915

38 P. T. HEALD *et al.*: *Scripta Met.*, 1971, **5**, 543

39 W. G. WOLFER AND M. ASHKIN: *Scripta Met.*, 1973, **7**, 1175

40 J. F. NYE: *Proc. Roy. Soc.*, 1953, **A219**, 477

41 D. S. WILKINSON AND M. F. ASHBY: *Acta Met.*, 1975, **23**, 1277

42 D. HULL AND D. E. RIMMER: *Phil. Mag.*, 1959, **4**, 673

43 M. V. SPEIGHT AND J. E. HARRIS: *Metal Sci. J.*, 1967, **1**, 83

44 R. RAJ AND M. F. ASHBY: *Acta Met.*, 1975, **23**, 653

45 M. V. SPEIGHT AND W. BEERÉ: *Met. Sci.*, 1975, **9**, 190

46 J. WEERTMAN, *Scripta Met.*: 1973, **7**, 1129

47 W. BEERÉ, *Acta Met.*: 1975, **23**, 139

Irradiation-induced diffusion and re-solution phenomena in metals

J. A. Hudson and R. S. Nelson

The irradiation of a metal results in different identifiable physical processes. Such processes have been generally known for some time; however, a complete understanding of the underlying principles is now emerging with the current interest in irradiation environments where materials receive large damage doses (up to 100 displacements per atom) at temperatures where both vacancies and interstitials are mobile. Displacement of atoms leads to point defect supersaturations which can result in enhanced diffusion and precipitation effects. In addition, atomic displacement may result in dissolution and disordering so that in a given system a dynamic equilibrium is established which can be different from the normal thermal equilibrium state. In some cases this includes the appearance of phases 'out of place' on the equilibrium phase diagram and other phases, as yet unidentified. In addition, segregation effects are observed where solute atoms preferentially migrate towards or away from point defect sinks such as voids and dislocations. No unified theory has yet been developed which can successfully predict all the microstructural changes that will occur in a given alloy during irradiation.

The authors are at AERE, Harwell

The principal characteristic of irradiation damage in metals is the permanent displacement of atoms from their equilibrium sites to produce equal numbers of vacancies and self-interstitials. Such displacement can result in concentrations of point defects and their aggregates far in excess of the equilibrium values at a particular temperature and the physical and, in some cases, the chemical properties of an alloy may be significantly altered during irradiation.

Current technological interest in irradiation damage is centred around the environment of fast and fusion reactor structural materials where the total damage intensity is of the order of 100 displacements per atom (dpa) at temperatures where both vacancies and interstitials are mobile. Two phenomena of paramount importance under these conditions are void formation and irradiation creep and these subjects are reviewed separately by other authors. In this paper we will concentrate on other phenomena resulting from the displacement of atoms and the build-up of non-equilibrium defect concentrations although in some cases the processes occurring have the same origin. We shall also compare and contrast the effectiveness of different irradiation damage regimes in producing these effects. This is particularly important since both ion and electron bombardment techniques are widely used to simulate neutron-induced irradiation damage phenomena. Enhanced vacancy supersaturations can lead to both enhanced nucleation and growth of precipitates at temperatures where neither process would occur in the absence of irradiation. Often the effects of enhanced diffusion and enhanced nucleation are difficult to separate. In addition the atomic displacement events can result in precipitate shrinkage to oppose the enhanced nucleation and growth. Thus a balance may be achieved between dissolution and growth leading to a steady-state precipitate distribution quite different from that developed at the same temperature in the absence of radiation damage. In many cases the phases present in an alloy during irradiation appear at temperatures different from those expected from thermal equilibrium. In addition, there are several examples of phases not yet identified appearing in irradiated commercial alloys such as austenitic stainless steels and Ni-based alloys. Several mechanisms have been proposed to explain specific observations and as yet the relative importance of these mechanisms are not fully understood. In this paper we will discuss our present knowledge of the physical processes giving rise to enhanced diffusion, re-solution, and phase change phenomena.

DEFECT PRODUCTION AND SUPERSATURATION

Let us assume, in the first instance, that a particular radiation environment gives rise to a uniform production rate of vacancies and interstitials, K dpa s^{-1}. The point defect concentrations generated at steady state are determined by balancing the defect production rate with the loss due to intrinsic recombination and the loss at fixed sinks such as a dislocation network and a population of voids:

$$K - D_i c_i k_i^2 - \alpha c_v c_i = 0 \quad \text{(interstitials)} \quad (1)$$

$$K + K_e - D_v c_v k_v^2 - \alpha c_v c_i = 0 \quad \text{(vacancies)} \quad (2)$$

Here suffixes i and v refer to interstitials and vacancies respectively, c is the defect concentration, D the diffusivity, and k^2 the sink strength for the defect (k^{-1} is the mean free path of the point defect in the presence of the sinks). In the case of dislocation sinks $k_i^2 = Z_i \rho_d$ and $k_v^2 = Z_v \rho_d$ where $(Z_i - Z_v)$ is the dislocation bias for interstitials (1–10%) and ρ_d is the dislocation density. In the case of voids which are assumed to be neutral sinks $k_i^2 = k_v^2 = 4\pi r_v C_v$ where r_v and C_v are the void radius and concentration respectively. More details of accurate values for these and other sink strengths will be given in the accompanying papers on void formation. The value of the recombination coefficient α is in the range 10^{16}–$10^{17}D_i$ in close-packed metals. The additional production term K_e is the thermal vacancy production rate, the equivalent term for interstitials being ignored due to the high formation energy of the latter. Simple expressions can be obtained for c_i and c_v from equations (1) and (2) in conditions where one or other of the defect loss rates dominates. For example, in many cases of practical interest the irradiation induced sinks dominate in which case the enhanced vacancy concentration is given by

$$c_v \simeq \frac{K}{D_v k_v^2} \quad (3)$$

In cases where recombination dominates it can be shown that[1]

$$c \simeq \left(\frac{K\nu_i}{\alpha\nu_v}\right)^{\frac{1}{2}} \quad (4)$$

where ν_i and ν_v are the interstitial and vacancy jump frequencies. The general solution of (1) and (2) has been derived by Brailsford and Bullough[2] in their theory of void growth. Figure 1 shows an example of how the vacancy concentration during irradiation follows the opposite temperature dependence of the normal thermal equilibrium value. Values of c_v during irradiation were calculated from experimental data obtained from the dislocation and void microstructures of samples of solution-treated AISI 316 stainless steel bombarded to 40 dpa with 46·5 MeV Ni^{6+} ions at a damage rate of about 10^{-3} dpa s^{-1}.[3] A value of $5 \times 10^{16}D_i$ was taken for the recombination parameter α in the calculations and voids and dislocations were the only defect sinks considered.

Although there is some uncertainty in our knowledge of α, the uncertainty in the value of K is now being recognized as a more important problem in the calculation of defect supersaturations. Great strides have been

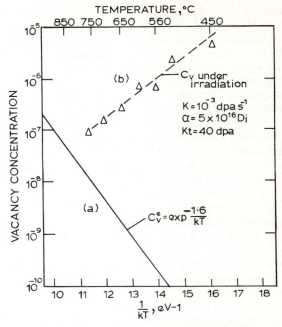

a thermal concentration; b concentration under irradiation after 40 dpa

1   Vacancy concentration as a function of temperature in solution-treated 316 stainless steel

made in calculations of K values for different radiation damage regimes[4] but these values represent the initial atomic displacement rates which may differ significantly from the supply rate of freely migrating defects, particularly in complex alloys containing a variety of defect traps. Irradiation regimes may be broadly classified into two categories: those which produce isolated Frenkel pairs in collisions where the elastic energy transferred is of the order of the displacement energy E_d (20–50 eV in metals) and those which produce the bulk of the displaced atoms in localized collision cascades 1–10 nm in size. High-energy electron bombardment, as in a high-voltage electron microscope is an example of the first type whereas heavy-ion and fast-neutron bombardment are examples of the second. In the first type of environment the effect of defect trapping can be treated straightforwardly, provided we know the defect-trap binding energy, as discussed for example by Schilling and Schroeder.[5] In the cascade case there are the additional problems of cascade collapse to vacancy loop together with defect trapping and recombination within the cascade where the local defect concentrations may initially be as high as 10^{-2}–10^{-1}, i.e. in some cases of the same order as the trap concentration. Vacancy loop production from collapsed collision cascades and its effect on void growth has been discussed by Bullough et al. who invoked the phenomenon to explain the different void growth behaviour of stainless steels under 1 MeV electron bombardment on the one hand and ion or neutron bombardment on the other.[6] The experimental evidence on vacancy loop production in all the metals investigated is limited to relatively small doses ($\ll 1$ dpa) and low temperatures. There is evidence that under these conditions heat treatment and impurity content do influence the survival rate of vacancies in collapsed loops from collision cascades.[7] However, there is no experimental evidence that vacancy loops occupy a substantial volume of material at higher doses after a well defined irradiation-induced dislocation network has developed. There is some evidence that cascade collapse into loops cannot occur in the vicinity of a dislocation strain field. Furthermore, loops which do form have been observed to rapidly slip into the network, especially at elevated temperatures. On the other hand, the possibility of substantially reducing the calculated values of K by trapping and enhanced recombination at traps inside cascades still exists.[8] Significant effects may be achieved at such locally high point defect concentrations with lower defect-trap binding energies than are required to produce comparable effects with the freely migrating defects in the bulk.

A further complication in the determination of effective K values can arise from the contribution of subthreshold events. Under normal circumstances such events constitute only a small fraction of the total number of defects; however, in special cases they must be taken into account. Subthreshold events which can add to the total defect concentration occur predominantly at surfaces or dislocations which are themselves defect sinks. The mechanism is thought to arise in the following way. Near threshold displacements occur within the bulk as a consequence of replacement collision sequences which separate the interstitial and vacancy sufficiently for them not to spontaneously recombine; this separation is of the order of 3 or 4 atomic spacings. However, if such a replacement sequence occurs near to a dislocation line or a surface (as in Fig. 2) recombination is prohibited and a vacancy is injected into the system. The process described can result in vacancy production at energies below the bulk displacement threshold but will of necessity require the collision sequence to be of the replacement variety, i.e. with initial energy $>\frac{1}{2}E_d$. The significance of subthreshold vacancy injection at a free surface is of course masked by the role of the surface as a defect sink, but in special circumstances such injection is manifest in the production of vacancy tetrahedra.[9,10] On the other hand, vacancy injection at dislocations is a reasonably efficient process and can play an important role in those circumstances where bulk displacement is either negligible (e.g. subthreshold) or subjected to heavy mutual recombination.

Vacancy injection can simply be included as an additional production rate G into the basic equations (2) and the extra vacancy supersaturation, which is not affected

by recombination is given by

$$c_v = \frac{G}{D_v k_v^2} \tag{5}$$

It can be shown that $G = n\eta K\Omega^{2/3}\rho_d$ where n is the effective number of sites around a dislocation which can result in vacancy injection and η relates the number of recoils relevant to subthreshold events at $\frac{1}{2}E_d$ to those at E_d.[11] In general G is two or three orders of magnitude lower than K and vacancy injection can therefore only have a significant effect on phenomena which rely on differential diffusion fluxes between interstitials and vacancies, such as void swelling. Examples of this effect have been given by Nelson and Minter.[11]

RADIATION-ENHANCED DIFFUSION

The development of point defect concentrations in excess of those dictated by thermal equilibrium was discussed in the previous section. In all cases when the point defects are mobile, irradiation-enhanced diffusion is possible. The basic equations for determining the enhanced diffusion coefficient D' have been described by several authors[1,12-14] and are equivalent to (1) and (2) given earlier. Values of D' vary linearly or as the square root of the damage rate K depending on whether loss to fixed sinks or recombination dominates. For vacancy-controlled diffusion

$$D' \simeq \frac{K}{k_v^2} \tag{6}$$

for loss to fixed sinks and

$$D' \simeq \lambda^2 \left(\frac{K\nu_v}{Z}\right)^{\frac{1}{2}} \tag{7}$$

when recombination dominates where λ is the vacancy jump distance, ν_v is the vacancy jump frequency, and Z is the recombination probability ($= \alpha/\nu_i$). Although D' will be approximately temperature independent in certain cases, in many cases of practical interest this will not be so since the major contributions to the sink strength, k_v^2, will be a result of the irradiation and will, therefore, be both temperature and dose dependent. For example, in the Ni^{6+} bombarded 316 steel referred to in the previous section and Fig. 1, k_v^2 varies from 2×10^{11} cm cm^{-3} to 2×10^{10} cm cm^{-3} over the void formation temperature range 450°–750°C for a damage rate of about 10^{-3} dpa s^{-1}.

Accurate measurement of enhanced diffusion coefficients would provide an ideal method of investigating the effects of different defect traps on the effective value of K discussed in the previous section and be of considerable practical importance, for example, in designing void-resistant alloys. There have been many experiments aimed at determining values of D' and the theoretical predictions of damage rate and temperature dependencies have been generally verified although as yet there has been no accurate determination of D' and k^2 in the same irradiated specimen. A review of most of these determinations has recently been presented by Adda et al.[15] A list of some determinations and the methods used in the experiments, covering a wide range of bombarding species and damage rates, is presented in Table 1. Within these experiments there are still disagreements and uncertainties, for example as to the relative roles of vacancy and interstitial enhanced diffusion

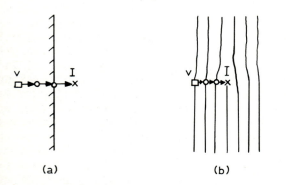

(a)   (b)

2 Schematic illustration of sub-threshold vacancy injection *a* at a free surface and *b* at a dislocation

TABLE 1 Some measured values of enhanced diffusion coefficients in metals

System	Method	Temperature range, °C	Range of thermal diffusion coefficient, cm^2 s^{-1}	Bombarding particles	Calculated damage rate, dps s^{-1}	Enhanced diffusion coefficient, cm^2 s^{-1}	Ref.
Cu 30% Zn	Short-range order (resistivity)	$0 \rightarrow 150$	$10^{-32} \rightarrow 10^{-19}$	neutrons	$10^{-10} - 10^{-9}$	$\sim 10^{-20}$	13
Ni into Cu	Loss of surface activity	$210 \rightarrow 285$	$10^{22} \rightarrow 10^{-19}$	3 MeV α	6×10^{-9}	$\sim 10^{-18}$	19
Au and Cu in Al	Secondary ion analysis	$36 - 200$	$10^{-22} \rightarrow 5 \times 10^{-15}$	fast neutrons	$1 - 2 \times 10^{-8}$	$10^{-15} \rightarrow 10^{-14}$	23
Ag in Ag	Penetration of Ag	$-60 \rightarrow 240$	$< 10^{-18}$	270 keV H$^+$	$0.8 \rightarrow 6.5 \times 10^{-6}$	$10^{-17} - 10^{-16}$	18
Al 4% Cu	Electron microscopy of precipitate growth	$20 \rightarrow 120$	$< 10^{-18}$	650 keV electrons	5×10^{-3}	2×10^{-16}	20
Cu 10% Zn	Loss of Zn by optical reflectivity	$207 \rightarrow 316$	$< 10^{-16} \rightarrow 10^{-13}$	30 keV Ne$^+$	$\sim 10^{-2}$	$10^{-14} \rightarrow 10^{-12}$	21
Zn in Al	Rutherford backscattering to give Zn profile change (2 MeV He)	$50 \rightarrow 130°C$	$< 10^{-12}$	80 keV Ne$^+$	~ 1	$10^{-15} \rightarrow 10^{-14}$	22

in particular processes such as the short-range ordering of α brass revealed by resistivity changes.[16,17] One of the most recent investigations by Lam *et al.*[18] on radiation-enhanced self-diffusion in Ag indicates good agreement with simple theory with contributions from both vacancy and interstitial diffusion. Some of these results are shown in Fig. 3 where the two curves labelled 'simple theory' are calculations of enhanced vacancy + interstitial diffusion coefficients determined using calculated values of K and estimated values of k^2. The latter will hopefully be subsequently replaced by measured values after electron microscopy of the irradiated specimens.

Recent approaches to the determination of point defect concentrations and enhanced diffusion coefficients are centred around the effects of point defect gradients around sinks on the movement of solute atoms. Marwick and Piller have studied the effect of ion bombardment on the concentration profile of Mn below the surface of single crystals of Ni over a range of temperatures and related the results to a model of the solute, vacancy, and interstitial diffusion.[24] At temperatures where the vacancy is mobile a solute peak develops below the surface due to solute migration up the vacancy gradient. At temperatures where the vacancy is immobile, solute is driven to the surface by the dominant interstitial gradient leaving a depleted region below the surface. Kinetic models of solute segregation were pre-

sented by Anthony[25,26] in which solute may either flow in the opposite direction to the vacancy flow or be 'dragged' in the direction of the vacancy flow in the case of strong vacancy–solute binding as discussed by Howard and Lidiard.[27] Following their observations of strain contrast around voids during high dose rate ($\sim 10^{-3}$ dpa s^{-1}) radiation damage experiments, Okamoto and Wiedersich identified the segregation of undersized atoms such as Si to the free surfaces.[28] They formulated a simple model of so-called 'solute drag' in which steady state is achieved when the solute flux to a void surface is balanced by the reverse flow induced by the enhanced concentration. A more comprehensive theoretical treatment of solute segregation in dilute fcc alloys has recently been developed by Johnson and Lam.[29] The model includes a $\langle 100 \rangle$ split interstitial binding to impurities to second neighbour distances, vacancy binding to impurities to first neighbour distances and the migration of bound impurity-defect complexes. The behaviour of thin foils bombarded at different damage rates over a range of temperatures was investigated. With a solute–interstitial binding energy of 0.2 eV the interstitial mechanism was the predominant process of solute migration to the foil surface and such segregation decreased as the vacancy–solute binding energy increased. The temperature dependence predicted for the solute segregation is similar to the measured temperature dependence of void swelling in alloys irradiated in the temperature range $0.2-0.5 T_m$ (K). An example of the segregation of γ' (Ni$_3$Ti, Al) precipitates at void surfaces and along edge dislocations is shown in Fig. 4 which comprises bright and dark field electron micrographs of the same area of a specimen of Nimonic PE16 bombarded with 46.5 MeV Ni^{6+} ions at 625°C to a dose of 60 dpa.[30] Several experimental determinations of radiation-enhanced diffusion coefficients have involved measurements of solute concentrations at bombarded surfaces. Great care is obviously needed in interpretation of such experiments in view of the possibility of the segregation effects currently predicted.

PRECIPITATE STABILITY

The radiation-enhanced diffusion discussed in the previous section can, as expected, lead to both enhanced nucleation and growth of precipitates in many alloys. However, additional irradiation-enhanced phenomena

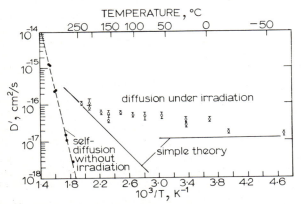

3   Comparison of thermal and radiation-enhanced diffusion coefficients in Ag during proton bombardment (after Lam *et al.*[18])

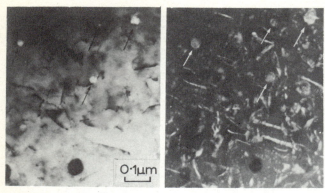

4 **Bright and dark field electron micrographs of 46·5 MeV Ni^{6+} bombarded Nimonic PE16 showing segregation of γ' to void surfaces and dislocations.**

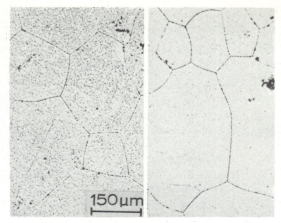

left unirradiated; *right* irradiated

5 **Optical micrograph illustrating dissolution on Mg$_2$Si precipitates by 250 keV Al$^+$ bombardment at room temperature (after Vaidya and Böhm[32])**

have been observed which perturb the effect of simple enhanced diffusion. Mechanisms have been proposed to explain specific radiation-enhanced processes but as yet no over-riding theory capable of predicting the behaviour of a given alloy under irradiation has been developed.

Recoil dissolution and disordering

The collisions which occur as a result of energetic displacement cascades can cause both the dissolution of precipitates and the disordering of ordered structures. The recoil dissolution of a large precipitate can be considered as the internal analogue of surface sputtering in which atoms are ejected from the surface of bombarded solids via interaction with collision cascades. In the sputtering case the number of atoms crossing the interface in a given time is readily calculated from a knowledge of the surface binding energy of the atoms and the energy spectrum of the collision cascades in which the number of atoms with energy E in the range dE, $N(E)\, dE \propto 1/E^2$.[31] The energy required to permanently displace an atom from a precipitate to the surrounding matrix is clearly of the same order as the displacement energy and will be in the range 20–100 eV/atom. Then for a spherical precipitate of volume V and radius r

$$\frac{dV}{dt}(\text{recoil}) = -\frac{4\pi r^2}{N}RK \qquad (8)$$

where R, which relates the flux of atoms crossing unit area within a solid to the damage rate, is about 10^{14}–10^{15} cm^{-2} and N is the atomic density. There are several examples of complete dissolution consistent with a recoil dissolution mechanism. For example, Vaidya and Böhm observed the dissolution of incoherent Mg$_2$Si precipitates in an Al–Mg–Si alloy during 250 keV Al$^+$ bombardment at room temperature as illustrated in the optical micrograph of Fig. 5.[32]

In experiments on the effects of ion bombardment on coherent γ' precipitates (Ni$_3$Al) in a binary Ni–Al alloy, Nelson *et al.* found evidence of significantly higher dissolution rates than expected on the recoil dissolution model.[33] These authors proposed a mechanism of more efficient dissolution for the case of ordered coherent precipitates. In this disorder dissolution mechanism the destruction of the ordered lattice by displacement cas-

cades creates local regions of high solute concentration. In the absence of diffusion such disorder will persist but when diffusion can occur small disordered regions within a precipitate will re-order while those near the surface will lose solute to the surrounding matrix by diffusion. If we suppose that only those displacements which occur in a shell of thickness l at the precipitate surface can result in this loss of solute atoms and that a fraction f of the solute atoms in the shell are lost, then

$$\frac{dV}{dt}(\text{disorder}) = -4\pi r^2 \psi K \qquad (9)$$

where $\psi = lf$. We expect l to be approximately the cascade dimension, i.e. 1–10 nm. This gives a dissolution rate of 10–100 times that of the direct recoil mechanism for modest values of f. There have been many direct and indirect experiments illustrating the disordering effect of different irradiation regimes.[34–37] Cascade regimes are more efficient, disordering 10–100 times more atoms than they permanently displace although ordered structures are certainly disordered during electron irradiation at sufficiently low temperatures. In general we would expect precipitate dissolution during electron bombardment to be far less efficient than during ion or neutron bombardment. Recoil dissolution will be somewhat restricted due to the low transfer energies and disorder dissolution should occur at relatively low rates, ψ being of the order of one atomic spacing only.

For irradiation when the solute is mobile a balance may be achieved between dissolution and enhanced diffusion. Even at temperatures where the thermal diffusion coefficient exceeds the radiation-enhanced value dissolution mechanisms can still operate. An example of the development of a new precipitate distribution during irradiation at an elevated temperature is shown in Fig. 6, which comprises dark field transmission electron micrographs of γ' precipitates in a Ni–Al alloy. The specimen had been aged to produce distribution of relatively large precipitates and was then bombarded at 550°C to about 5 dpa with low-energy ions at a damage rate of about 10^{-2} dpa s^{-1}. After irradiation a fine distribution of small precipitates had replaced the original distribution (the fragments of large precipitates remaining were

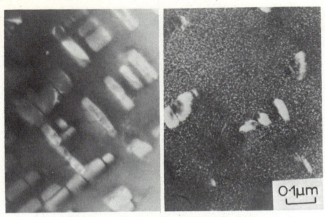

6   Dark field electron micrographs showing the change in γ' distribution during ion bombardment of a Ni–Al alloy at 550ρC

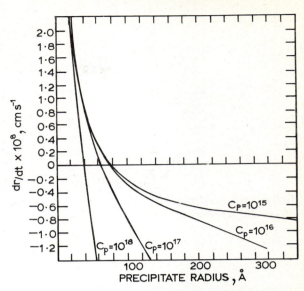

7   Calculated growth rate of γ' precipitates as a function of radius for various concentrations c_p for a damage rate of 10^{-2} dpa s^{-1} at 550°C

beyond the range of the bombarding ions). Similar distributions were observed in other specimens given different pre-irradiation aging treatments, after the same irradiation. The authors interpreted these and other results in terms of a balance between disorder dissolution and enhanced growth via enhanced diffusion given by [38]

$$\frac{dV}{dt} = \frac{3D'c_x r}{p} \qquad (10)$$

where c_x is the solute concentration in solution and p the atomic fraction of solute atoms constituting the precipitate phase. From a range of experiments at the same damage rate but different temperatures it was found that above 300°C redistribution of γ' occurred whereas below this temperature disordering and dissolution dominated, total dissolution occurring after 1–10 dpa. Equating the growth and dissolution terms gives the time dependence of the precipitate radius:

$$\frac{dr}{dt} = \frac{D'}{r}\left\{\frac{3C}{4\pi p} - c_p r^3\right\} - \psi K \qquad (11)$$

where C is the total solute concentration and c_p the final precipitate concentration. A value of ψ was estimated by equating dr/dt to zero at 300°C for precipitates of a few atomic dimensions and using calculated values of K and the enhanced diffusion coefficient D'. Figure 7 shows the predicted behaviour of the γ' precipitates where it is seen that under irradiation all precipitates below a certain size grow and all those above this size will shrink. The energy associated with the increase in surface area of the precipitates during shrinkage is of course provided by the irradiation. An analysis of the effect of small dissolution rates on the coarsening kinetics of a dispersion of spherical particles has been presented by Bilsby.[39] He also predicts that under certain circumstances dissolution can limit the maximum precipitate size developed during irradiation.

Point defects as chemical components

The stability of non-coherent precipitates, whose atomic volume differs substantially from that of the matrix, has been considered by Maydet and Russell[40] by treating the irradiation-induced point defects as chemical components in the system. A similar treatment had been applied to the case of a quenched-in vacancy concentra-

tion on the precipitation of incoherent particles and the particular case of matrix nucleation of NbC in an austenitic stainless steel which has $\sim 24\%$ misfit in the austenite matrix has been investigated by Shepherd and the role of quenched-in vacancies on the precipitate nucleation identified.[41]

Incoherent spherical particles of solute dilute in solvent are considered in a solvent dilute in solute. A precipitate is described in terms of the number of solute atoms (x) and the number of excess vacancies (n) and thus

$$n = a - x \qquad (12)$$

where a is the number of matrix atoms displaced by the precipitate.

Precipitate stability is investigated in terms of behaviour in (n, x) phase space, where

$$\frac{dx}{dt} = A_x - L_x \qquad (13)$$

and

$$\frac{dn}{dt} = A_v - L_v - A_i \qquad (14)$$

where A and L are arrival and loss rates and the subscript i refers to interstitials, x to solute atoms, and v to vacancies. Loss due to recoil dissolution is ignored in the current theory. Values of A are determined by the concentrations and mobilities of the respective point defects as discussed in the second section

$$\left(\frac{A_i}{A_v} = \frac{Z_v}{Z_i} \text{ at an unbiased sink}\right)$$

Loss rates L are determined by balancing the reactions

$$(n, x - 1) + x \rightleftharpoons (n, x) \qquad (15)$$

and

$$(n - 1, x) + n \rightleftharpoons (n, x) \qquad (16)$$

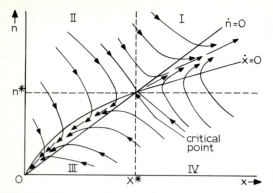

8 Schematic particle trajectories in (n, x) phase space (after Maydet and Russell[40])

at equilibrium to give

$$L_v \simeq A_v \exp\left(\frac{1}{kT}\frac{\partial \Delta G^0}{\partial n}\right) \tag{17}$$

$$L_x \simeq A_x \exp\left(\frac{1}{kT}\frac{\partial \Delta G^0}{\partial x}\right) \tag{18}$$

where ΔG^0 is the free energy of formation of the precipitate.

These relations have been analysed by Maydet and Russell in terms of nodal lines (dx/dt and $dn/dt = 0$) and their points of intersection. Detailed description is out of place here but an example of the analysis is illustrated in Fig. 8 for a case of a precipitate with positive misfit ($\delta > 0$). The arrows show the direction of motion of particles in the various regions of (n, x) space. By definition the solute and vacancy nodal lines must be crossed vertically and horizontally respectively. Particles between the nodal lines cannot escape. The nodal lines intersect at $(n^* x^*)$ the critical point, where, in the absence of statistical fluctuations, the precipitate is immobile, resembling a critical nucleus in nucleation theory. The particle radius at this point is given by

$$r^* = \frac{2\gamma\Omega}{kTS_i} \tag{19}$$

where S_i is an irradiation-modified supersaturation:

$$S_i = S_x\left\{S_v\left(1 - \frac{A_i}{A_v}\right)\right\}^\delta \tag{20}$$

where δ is the volume misfit and S_x and S_v are solute and vacancy supersaturations. Maydet and Russell use S_i to define a potential function

$$\Delta\phi = -kT \ln S_i \tag{21}$$

such that if $\Delta\phi < 0$ precipitates larger than r^* will be stable and if $\Delta\phi > 0$ the nodal lines do not intersect and all particles will decay.

The predictions of this 'chemical effect' theory are amenable to experiment and the important variables, vacancy supersaturation, and precipitate/matrix misfit can be chosen to cover relatively large ranges of values. An example is shown in Fig. 9 which is an electron micrograph of an Al–2 wt-% Ge foil ($\delta + ve$), bombarded to about 70 dpa at 125°C with 100 keV Al$^+$ ions at a rate of about 3×10^{-2} dpa s^{-1}. Voids and a population of small elongated precipitates and a low concentration of larger precipitates are evident in one half of the micrograph. The other half corresponds to an area of the

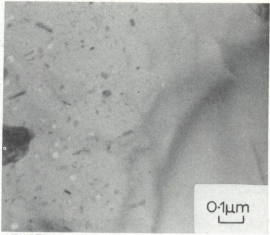

9 Electron micrograph of Ge precipitates and voids in Al–2 wt.% Ge bombarded to 70 dpa at 125°C with 100 keV Al$^+$ ions; one half of the specimen was shielded during irradiation[42]

specimen shielded during the irradiation by a Cu grid. Diffraction analysis has confirmed the nature of the precipitates as elemental Ge.[42]

There are many examples of enhanced precipitation of known and unknown phases during irradiation. In particular enhanced matrix precipitation of carbides in austenitic steels has often been reported.[43-45] Defect clusters introduced by irradiation have often been suggested as nucleation sites for this enhanced precipitation. Heterogeneously nucleated vacancy loops are certainly not a necessary prerequisite for enhanced carbide precipitation as shown by Fisher and Williams who found enhanced precipitation of NbC in an 18–10–Nb stainless steel during electron bombardment in the temperature range 350°–500°C.[46] Nevertheless, recent investigations by Williams[47] have shown that the amount of enhanced NbC precipitation is far greater in Ni ion bombarded FV548 steel than in material bombarded by 1 MeV electrons under the same conditions of damage rate and temperature. The appearance of 'non-equilibrium' phases should not be a surprise bearing in mind that the energy supplied by irradiation in the form of point defects is far in excess of the free energy differences between potential phases. The contribution of $\Delta\phi$ in the case of non-coherent precipitates is illustrated in the schematic phase diagram of Fig. 10 from the

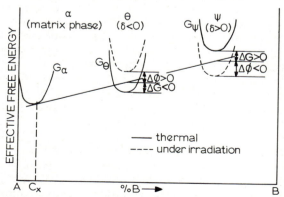

10 Schematic effect of irradiation on equilibrium phase diagram; phase ψ becomes stable at the expense of θ (after Maydet and Russell[40])

work of Maydet and Russell.[40] Here the phase $\psi(\delta + ve)$ becomes stable at the expense of $\theta(\delta - ve)$ during irradiation, the effective free energy curves being shifted by $-\delta kT \ln S_v(1 - A_i/A_v)$.

REFERENCES

1 J. V. SHARP: AERE Report R6267, 1969
2 A. D. BRAILSFORD AND R. BULLOUGH: *J. Nucl. Mater.*, 1972, **44**, 121
3 J. A. HUDSON: *J. Nucl. Mater.*, 1976, **60**, 89
4 A. D. MARWICK: *J. Nucl. Mater.*, 1976, **56**, 355
5 W. SCHILLING AND K. SCHROEDER: Proc. UKAEA Consultants Symposium on 'The physics of irradiation produced voids' (Ed. R. S. NELSON), AERE Report R7934, 1975
6 R. BULLOUGH et al.: *Proc. Roy. Soc. Lond.* 1975, **A346**, 81
7 C. A. ENGLISH et al.: *J. Nucl. Mater.*, 1975, **58**, 220
8 K. C. RUSSELL AND R. S. NELSON: to be published
9 D. CHERNS et al.: Proc. of 6th Int. Conf. on Atomic Collisions in Solids
10 D. CHERNS et al.: *Nucl. Inst. and Methods*, 1976, **132**, 369
11 R. S. NELSON AND F. J. MINTER: to be published
12 W. M. LOMER: AERE Report T/R 1540, 1954
13 G. J. DIENES AND A. C. DAMASK: *J. Appl. Phys.*, 1958, **29**, 1713
14 A. C. DAMASK: 'Studies of radiation effects in solids 2, 1967, Gordon and Breach
15 Y. ADDA et al.: Proc. Conf. on 'Low temperature diffusion and applications to thin films' IBM, 1974
16 G. J. DIENES: 'Interaction of radiation with solids, 445, 1967, Plenum Press
17 W. SCHULE et al.: *Rad. Effects*, 1970, **2**, 151
18 N. Q. LAM et al.: *Thin Solid Films*, 1975, **25**, 157
19 H. P. BONZEL: *Acta Met.*, 1965, **13**, 1084
20 P. S. SKLAD AND T. E. MITCHELL: *Scripta Met.*, 1974, **8**, 1113
21 R. A. ARNDT AND R. L. HINES: *J. Appl. Phys.*, 1961, **32**, 1913
22 S. M. MYERS AND S. T. PICRAUX: *J. Appl. Phys.*, 1975, **46**, 4774
23 D. ACKER et al.: *J. Nucl. Mater.*, 1974, **5²**, 281
24 A. D. MARWICK AND R. C. PILLER, unpublished
25 T. R. ANTHONY: in 'Atomic transport in solids and liquids', 1971, Verlag. Z. Naturforch Tübingen
26 T. R. ANTHONY: in 'Radiation-induced voids in metals and alloys' AEC Sym. Series, Conf. 701601, p. 630, 1972
27 R. E. HOWARD AND A. B. LIDIARD: *Rep. Prog. Phys.*, 1964, **27**, 161
28 P. R. OKAMOTO AND H. WIEDERSICH: in Proc. UKAEA Consultant Symp. on 'The physics of irradiation produced voids' (Ed. R. S. NELSON), AERE Report R7934, 1975
29 R. A. JOHNSON AND N. Q. LAM: *Phys. Rev.*, 1976, **B13**, 4364
30 J. A. HUDSON: *J. Brit. Nucl. Energy Soc.*, 1975, **14**, 127
31 M. W. THOMPSON: 'Defects and radiation damage in metals', 1969, CUP
32 W. V. VAIDYA AND H. BÖHM: Proc. European Conf. on 'Irradiation behaviour of fuel cladding and core component materials', KTG/BNES Karlsruhe, 1974
33 R. S. NELSON et al.: *J. Nucl. Mater.*, 1972, **44**, 318
34 S. SIEGEL: *Phys. Rev.*, 1949, **75**, 1923
35 L. R. ARONIN: *J. Appl. Phys.* 1955, **25**, 344.
36 P. R. EGGELSTON AND F. E. BOWMAN: *J. Appl. Phys.*, 1953, **24**, 229
37 A. CORDIER et al.: *Rad. Effects*, 1973, **17**, 127
38 F. S. HAM: *J. Phys. Chem. Sol.*, 1958, **6**, 335
39 C. F. BILSBY: *J. Nucl. Mater.*, 1975, **55**, 125
40 S. I. MAYDET AND K. C. RUSSELL: *J. Nucl. Mater.*, in press
41 J. P. SHEPHERD: *Met. Sci.*, 1976, May, 174
42 K. BERTRAM et al.: to be published
43 D. R. ARKELL AND P. C. L. PFEIL: *J. Nucl. Mater.*, 1964, **12**, 145
44 J. S. WATKIN et al.: ASTM STP 529, p. 509, 1973
45 P. R. B. HIGGINS AND A. C. ROBERTS: *J. Iron Steel Inst.*, 1966, **204**, 489
46 S. B. FISHER AND K. R. WILLIAMS: *Phil. Mag.*, 1972, **25**, 355
47 T. M. WILLIAMS: private communication

Irradiation creep due to point defect absorption at dislocations

P. T. Heald and R. Bullough

The stress-induced preferential absorption (SIPA) mechanism of irradiation creep is briefly reviewed. Its origin is explained in terms of the first-order size effect interaction and the induced interaction between the intrinsic point defects and the dislocations. These two interactions combine in a body under uniaxial stress to ensure that preferential absorption of interstitials will occur at those edge dislocations oriented with their extra planes normal to the stress axis and vice-versa for the vacancies. A simplified rate theory argument is used to deduce the magnitude and microstructural dependence of the resulting creep strain. It is shown that the calculated creep rate is linear in applied stress and irradiating flux and is fairly independent of temperature.

P. T. Heald is with the CEGB, Berkeley Nuclear Laboratories, and R. Bullough is at AERE, Harwell

When non-fissile cubic materials are subject to a small applied stress while being bombarded with energetic particles it is found experimentally that they creep at a rate directly proportional to the applied stress; furthermore, it is found that, after some incubation dose during which transient rates occur, the creep rate is linearly proportional to the irradiating flux. Accordingly the radiation-induced creep strain may be written as the sum of the transient and steady-state components and the steady-state creep rate may be written in the form

$$\dot{\epsilon} = B\epsilon\phi \qquad (1)$$

where ϵ is the applied strain, ϕ the irradiating flux*, and B is a numerical constant of order unity. Steady-state irradiation creep is observed to be fairly independent of material and microstructure; for example, the values of B for the nimonic alloy PE16, for stainless steels, and for molybdenum are all within a factor ~ 2 of each other. Similarly, steady-state irradiation creep appears to be remarkably insensitive to temperature, perhaps varying by a factor ~ 2 over the temperature range 300°C–500°C. Certainly it does not have the dramatic temperature variation associated with irradiation swelling or thermal creep and this is illustrated schematically in Fig. 1.

In this article we shall describe the physical basis of the stress-induced preferential absorption (SIPA) mechanism of steady-state irradiation creep which was originally proposed by Heald and Speight[1] and Wolfer and Ashkin[2] and developed by these authors[3] and Bullough and Willis[4] and Bullough and Hayns.[5] We shall demonstrate that the creep rate resulting from this mechanism is linear in flux and stress, is fairly independent of microstructure and temperature, and furthermore is in reasonable quantitative agreement with the experimental values.

The SIPA mechanism arises from certain well established point defect–dislocation interactions and we shall begin by outlining the physical origins of these interactions.

RELEVANT DISLOCATION–POINT DEFECT INTERACTIONS

The various ways in which a point defect can interact with a dislocation have been reviewed by Bullough and Newman.[6] Of these interactions we shall be particularly concerned with the induced interaction, originally analysed by Eshelby,[9] which arises because the point defect volume has a different harmonic response (local elastic moduli) from the surrounding matrix. However, the dominant interaction is usually the so-called first-order size effect interaction.[7,8] In the continuum model the point defect is represented by a misfitting inhomogeneous inclusion of volume V in an isotropic

---

* We shall use displacements per atom per second as the standard materials damage measure.

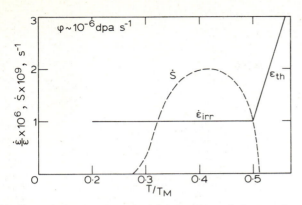

1   Schematic representation of the deformation rates

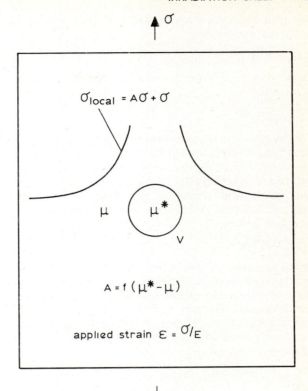

Interaction between inclusion and applied stress $E = \frac{1}{2} A \sigma \epsilon V$

3   Induced inhomogeneity interaction between a point defect and a stress system

elastic solid of bulk modulus K and shear modulus μ. If the inclusion misfits by a volume ΔV (which represents the relaxation volume around the defect) then the first-order size interaction is just the work done when the lattice suffers a uniform dilation in the hydrostatic stress field of the dislocation (Fig. 2), namely[7,8]

$$E^0 = p \, \Delta V \tag{2}$$

where

$$p = -\tfrac{1}{3}\sigma_{kk}^{d} = \frac{(1+\nu)\mu b}{3\pi(1-\nu)} \cdot \frac{\sin\theta}{r} \tag{3}$$

is the hydrostatic stress field around a dislocation. Combining equations (2) and (3) we may write

$$E_{\alpha}^{0} = \frac{(1+\nu)\mu b \cdot \Delta V_{\alpha}}{3\pi(1-\nu)} \cdot \frac{\sin\theta}{r} \tag{4}$$

where the subscript denotes either interstitials (I) or vacancies (V).

The induced interaction may be illustrated by the following example. An elastic body with shear modulus μ and bulk modulus K contains a perfectly fitting inclusion of volume V with elastic constants μ^* and K^*. If the body is subject to a uniform applied stress, σ^a, the field

in the vicinity of the inclusion is perturbed from the uniform value it would have if the inclusion and matrix had the same elastic properties. The local stress immediately adjacent to the inclusion may be written as (see Fig. 3)

$$\sigma_{local} = A\sigma^a + \sigma^a \tag{5}$$

where A is a function of the difference in elastic constants between the inclusion and the matrix $(\mu^* - \mu)$ and $(K^* - K)$. The stress field given in equation (5) may also be produced by an equivalent homogeneous $(\mu^* = \mu, K^* = K)$ misfitting inclusion provided that the inclusion 'misfits' by just the right amount and produces the same stress field everywhere as the original inhomogeneous inclusion. That is, the fictitious misfitting inclusion produces a local stress σ^T such that

$$\sigma^T = A\sigma^a \tag{6}$$

The interaction between the applied stress and the original inhomogeneous inclusion, U, is equal to the interaction between the applied stress and the equivalent homogeneous misfitting inclusion, namely, the product of the inclusion stress field σ^T and the applied strain ϵ^a integrated over the volume of the inclusion,[9]

$$U = \tfrac{1}{2}\sigma^T\epsilon^a V = \tfrac{1}{2}A\sigma^a\epsilon^a V = \frac{AV}{2E}(\sigma^a)^2 \tag{7}$$

where E is the Young's modulus of the matrix. This result holds for either a source of internal stress or externally applied stress and we are concerned with a situation when both are present: the internal stress σ^d

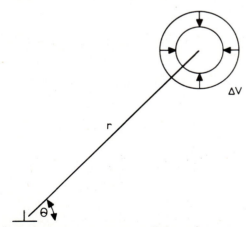

Hydrostatic stress due to dislocation

$$p = \frac{(1+\nu)\mu b}{3\pi(1-\nu)} \frac{\sin\theta}{r}$$

Interaction energy $E^0 = p \, \Delta V$

2   First-order size interaction between a dislocation and a point defect

due to a dislocation and a stress σ^e due to externally applied loads, so that

$$\sigma^a = \sigma^e + \sigma^d \qquad (8)$$

Combining equations (7) and (8) we have

$$U = \frac{AV}{2E}\{(\sigma^e)^2 + 2\sigma^e\sigma^d + (\sigma^d)^2\} \qquad (9)$$

Clearly only the second term on the right hand side of equation (9) mixes the direction of the applied stress and the orientation of the dislocation. It is an external stress-induced interaction energy that is linear in the dislocation stress and therefore significant. The relevant induced interaction between the dislocation and the inclusion is thus

$$\delta E = AV\sigma^d\epsilon \qquad (10)$$

in the presence of an applied strain $\epsilon(=\sigma^e/E)$. Heald and Speight[1] first indicated that the magnitude of this induced interaction would depend on the orientation of the dislocations Burgers' vector with respect to the stress axis (*see* Fig. 4) and Bullough and Willis[4] subsequently gave a rigorous analysis of the interaction and their result may be written in the form

$$\delta E_\alpha^{(\beta)} = \frac{\mu b\epsilon V_\alpha A_\alpha^{(\beta)}}{\pi} \cdot \frac{\sin\theta}{r} \qquad (11)$$

where the superscript (β) denotes the type of dislocation; those dislocations with Burgers' vectors parallel to the stress axis are denoted by $\beta = 1$ while those dislocations with Burgers' vector perpendicular to the stress axis are denoted by $\beta = 2$ or 3. The numerical constants $A_\alpha^{(\beta)}$ depend on both the type of defect and the type of dislocation involved. If the effective elastic constants of

the defect region are given by

$$\left.\begin{array}{l} K_\alpha^* = K + \Delta K_\alpha \\ \text{and} \\ \mu_\alpha^* = \mu + \Delta\mu_\alpha \end{array}\right\} \qquad (12)$$

then[4]

$$A_\alpha^{(1)} = -\left\{\frac{(1-2\nu)(1+\nu)\,\Delta K_\alpha}{3(1-\nu)K + (1+\nu)\,\Delta K_\alpha}\right.$$
$$\left. + \frac{5(1+\nu)(2-\nu)\,\Delta\mu_\alpha}{15(1-\nu)\mu + 2(4-5\nu)\,\Delta\mu_\alpha}\right\} \qquad (13a)$$

$$A_\alpha^{(2)} = A_\alpha^{(3)} = -\left\{\frac{(1-2\nu)(1+\nu)\,\Delta K_\alpha}{3(1-\nu)K + (1+\nu)\,\Delta K_\alpha}\right.$$
$$\left. - \frac{5(1+\nu)^2\,\Delta\mu_\alpha}{15(1-\nu)\mu + 2(4-5\nu)\,\Delta\mu_\alpha}\right\} \qquad (13b)$$

We would expect the vacancy region to be softer in compression and shear than the perfect lattice; consequently we have

$$\left.\begin{array}{l} \Delta K_v < 0 \\ \text{and} \\ \Delta\mu_v < 0 \end{array}\right\} \qquad (14)$$

From atomistic simulation studies of the vibrational modes arising from the split dumbell configuration of the self-interstitial, Dederichs *et al.*[10] have shown that the interstitial is soft in shear. Holder *et al.*[11] have experimentally confirmed this effect for interstitials in copper. Accordingly we have

$$\Delta\mu_I < 0 \qquad (15)$$

and normally one would expect the interstitial to be hard in compression; thus

$$\Delta K_I > 0 \qquad (16)$$

Consequently we have from equations (13) the following inequalities

$$\left.\begin{array}{l} A_V^{(1)} > 0 \\ A_I^{(2)} < 0 \\ A_I^{(1)} - A_I^{(2)} > 0 \\ A_V^{(1)} - A_V^{(2)} > 0 \end{array}\right\} \qquad (17)$$

The total relevant interaction energy is the sum $E_\alpha^0 + \delta E_\alpha^{(\beta)}$ of equations (4) and (11), namely

$$E_\alpha^{(\beta)} = \left\{\frac{(1+\nu)\mu b\,\Delta V_\alpha}{3\pi(1-\nu)} + \frac{\mu b\epsilon V_\alpha A_\alpha^{(\beta)}}{\pi}\right\} \cdot \frac{\sin\theta}{r} \qquad (18)$$

The relative climb directions due to the induced term are illustrated in Fig. 4. However, in order to calculate the climb velocities of the various dislocations we must relate the flow of point defects to a dislocation to the diffusivity and the interaction between the point defects and the dislocation.

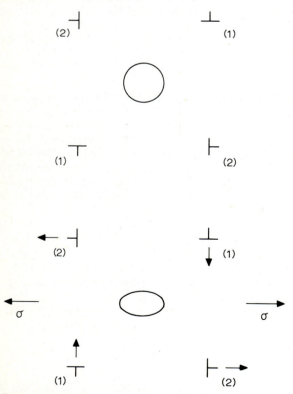

4   Schematic representation of the stress-induced asymmetry in the point defect–dislocation interaction

DIFFUSION OF POINT DEFECTS TO DISLOCATIONS

The flux, $\mathbf{J}$, of either species of point defect to a dislocation is given by

$$\Omega \mathbf{J}_\alpha^{(\beta)} = D_\alpha \nabla C_\alpha + \frac{D_\alpha C_\alpha}{kT} \nabla E_\alpha^{(\beta)} \tag{19}$$

where Ω is the atomic volume, C_α is the atomic concentration of points defects, D_α their diffusion coefficient, and kT is the thermal energy. The first term on the right hand side of equation (19) represents the flow of point defects to a dislocation by random diffusion while the second term represents the drift flow superimposed on the random motion by the interaction between the point defect and the dislocation. Only approximate or numerical solutions of the full diffusion equation are available; however, these details need not concern us here. In order to illustrate the physical mechanism it is sufficient to use the drift flow approximation, that is, neglect random diffusion and use only the second term in equation (19). The force on a point defect attracting it *towards* a dislocation is

$$\nabla E_\alpha^{(\beta)}$$

Thus, from Einstein's relation the flux of defects to a dislocation is just the defect mobility times the force:

$$\Omega \mathbf{J}_\alpha^{(\beta)} = \frac{D_\alpha C_\alpha^0}{kT} \nabla E_\alpha^{(\beta)} \tag{20}$$

where C_α^0 is some average concentration of defects in the lattice. This concentration, though not rigorously defined, will be used as a measure of the average concentration in the effective rate theory continuum of which a more extensive discussion has been given.[13] For simplicity we may drop the angular variation in equation (18) (or replace it by the average value $2/\pi$) and use the purely radial form of the potential

$$E_\alpha^{(\beta)} \approx - \left| \frac{(1+\nu)\mu b \, \Delta V_\alpha}{3\pi(1-\nu)} + \frac{\mu b V_\alpha \epsilon \, \Delta_\alpha^{(\beta)}}{\pi} \right| \cdot \frac{1}{r} \tag{21}$$

In this approximation the point defect flux is

$$\Omega \mathbf{J}_\alpha^{(\beta)} = \frac{D_\alpha C_\alpha^0}{kT} \frac{\partial E_\alpha^{(\beta)}}{\partial r}$$

$$= D_\alpha C_\alpha^0 \left\{ \frac{(1+\nu)\mu \, \Delta V_\alpha}{3\pi(1-\nu)kT} + \frac{\mu V_\alpha \epsilon \, \Delta_\alpha^{(\beta)}}{\pi kT} \right\} \frac{b}{r^2} \tag{22}$$

and the number of point defects reaching unit length of dislocation is

$$F_\alpha^{(\beta)} = 2\pi b \Omega \mathbf{J}_\alpha^{(\beta)}(r=b) = Z_\alpha^{(\beta)} D_\alpha C_\alpha^0 \tag{23}$$

where

$$Z_\alpha^{(\beta)} = \left\{ \frac{2(1+\nu)\mu \, \Delta V_\alpha}{3(1-\nu)kT} + \frac{2\mu V_\alpha \epsilon A_\alpha^{(\beta)}}{\pi kT} \right\} \tag{24}$$

in the drift flow approximation. Equation (24) may be written in the form[5]

$$Z_\alpha^{(\beta)} = Z_\alpha^0 \left\{ 1 + \frac{3(1-\nu)\epsilon V_\alpha A_\alpha^{(\beta)}}{\pi(1+\nu) \, \Delta V_\alpha} \right\} \tag{25}$$

where Z_α^0 is the value of $Z_\alpha^{(\beta)}$ in the absence of an applied stress and, apart from a small numerical difference in the

second term,[3] equation (25) is still appropriate when the full diffusion equation is used.

The total flow of point defects to all dislocations is, from equation (23),

$$(Z_\alpha^{(1)}\rho_d^{(1)} + Z_\alpha^{(2)}\rho_d^{(2)} + Z_\alpha^{(3)}\rho_d^{(3)}) D_\alpha C_\alpha^0$$

$$= \tfrac{1}{3}(Z_\alpha^{(1)} + Z_\alpha^{(2)} + Z_\alpha^{(3)})\rho_d D_\alpha C_\alpha^0 = \bar{Z}_\alpha \rho_d D_\alpha C_\alpha^0 \tag{26}$$

where $\rho_d^{(\beta)}$ are the densities of dislocations having their Burgers' vectors parallel to the three Cartesian axes ($\beta = 1, 2, 3$). For a random array of dislocations

$$\rho_d^{(1)} = \rho_d^{(2)} = \rho_d^{(3)} = \tfrac{1}{3}\rho_d \tag{27}$$

If, in addition to the dislocations, there are ρ_h voids of average radius r_h present, the total loss of point defects to the voids is[12]

$$4\pi r_h \rho_h D_\alpha C_\alpha^0 \tag{28}$$

Thus the total loss of defects to all sinks is

$$(\bar{Z}_\alpha \rho_d + 4\pi r_h \rho_h) D_\alpha C_\alpha^0 = k_\alpha^2 D_\alpha C_\alpha^0 \tag{29}$$

and, if we neglect mutual recombination, this is equal to the number of point defects produced by the irradiating flux so that

$$D_\alpha C_\alpha^0 = \phi/k_\alpha^2 \tag{30}$$

DEFORMATION RATES

It is convenient to make use of the formulation of chemical rate theory and write the flow of point defects to a dislocation in the form $Z_\alpha^{(\beta)} D_\alpha C_\alpha^0$ so that the net number of interstitials flowing to a dislocation is

$$Z_I^{(\beta)} D_I C_I^0 - Z_V^{(\beta)} D_V C_V^0 \tag{31}$$

This flow comprises two parts; one responsible for swelling and the other giving rise to a volume conserving creep process. The swelling component is simply the total excess number of interstitials absorbed at all dislocations:

$$\dot{S} = \sum_{\beta=1}^{3} (Z_I^{(\beta)} D_I C_I^0 - Z_V^{(\beta)} D_V C_V^0)\rho_d^{(\beta)} \tag{32}$$

The creep rate is given by the volume-conserving part of (31), namely,

$$\dot{\epsilon}^{(1)} = (Z_I^{(1)} D_I C_I^0 - Z_V^{(1)} D_V C_V^0)\rho_d^{(1)}$$

$$- \tfrac{1}{3}\sum_{\beta=1}^{3} (Z_I^{(\beta)} D_I C_I^0 - Z_V^{(\beta)} D_V C_V^0)\rho_d^{(\beta)}$$

$$= \tfrac{2}{3}(Z_I^{(1)} D_I C_I^0 - Z_V^{(1)} D_V C_V^0)\rho_d^{(1)}$$

$$- \tfrac{1}{3}\{(Z_I^{(2)} D_I C_I^0 - Z_V^{(2)} D_V C_V^0)\rho_d^{(2)}$$

$$+ (Z_I^{(3)} D_I C_I^0 - Z_V^{(3)} D_V C_V^0)\rho_d^{(3)}\} \tag{33}$$

since $Z_\alpha^{(2)} = Z_\alpha^{(3)}$ we may combine the last two terms to give

$$\dot{\epsilon}^{(1)} = \tfrac{2}{9}\rho_d\{\Delta Z_I D_I C_I^0 - \Delta Z_V D_V C_V^0\} \tag{34}$$

where

$$\Delta Z_\alpha = Z_\alpha^{(1)} - Z_\alpha^{(2)} = \frac{3\epsilon V_\alpha Z_\alpha^0 a_\alpha}{\Delta V_\alpha} \tag{35}$$

and

$$a_\alpha = \frac{(1-\nu)}{(1+\nu)}(A_\alpha^{(1)} - A_\alpha^{(2)})$$
$$= -\frac{15(1-\nu)\Delta\mu_\alpha}{15(1-\nu)\mu + 2(4-5\nu)\Delta\mu_\alpha} \quad (36)$$

Substituting for $D_\alpha C_\alpha^0$, from equation (30), into equation (35) we have finally

$$\dot\epsilon^{(1)} = \frac{2}{3}\epsilon\phi\left\{\left(\frac{Z_I^0\rho_d}{k_I^2}\right)\cdot\frac{V_I a_I}{\Delta V_I} - \left(\frac{Z_V^0\rho_d}{k_V^2}\right)\cdot\frac{V_V a_V}{\Delta V_V}\right\} \quad (37)$$

The creep rate is linear in both the applied stress and the irradiating flux and the creep strain sensitivity, B, is given by

$$B = \frac{2}{3}\left\{\left(\frac{Z_I^0\rho_d}{k_I^2}\right)\frac{V_I a_I}{\Delta V_I} - \left(\frac{Z_V^0\rho_d}{k_V^2}\right)\frac{V_V a_V}{\Delta V_V}\right\} \quad (38)$$

In most circumstances it is found experimentally that $Z_\alpha^0\rho_d/k_\alpha^2 \sim 1$ and discrete lattice calculations of point defect properties give $V_\alpha/|\Delta V_\alpha| \sim 1$. If, following Bullough and Hayns,[5] we take

$$\Delta\mu_I = -\mu \quad \text{and} \quad \Delta\mu_V = 0 \quad (39)$$

then, for $\nu = \frac{1}{3}$

$$a_I = 2 \quad \text{and} \quad a_V = 0 \quad (40)$$

consequently $B \sim 1$ and we have finally

$$\dot\epsilon \approx \epsilon\phi \quad (41)$$

Since B depends on the sink strengths through the slowly varying ratio $(Z_\alpha^0\rho_d/k_\alpha^2)$ it is insensitive to the precise values of the sink strengths. All other quantities appearing in equation (38) are only weakly temperature dependent. Consequently the SIPA mechanism predicts a creep rate which is insensitive to material and microstructural properties and to the irradiation temperature and agrees in absolute magnitude with the experimental value.

DISCUSSION

We have shown, in simple terms, how the stress-induced inhomogeneity interaction between a point defect and a dislocation gives rise to an irradiation creep mechanism. We have concentrated on the physical basis of the SIPA mechanism rather than details of the calculation, but even so we have been able to illustrate the salient features of the creep rate resulting from this mechanism:

namely, the linear dependence of the creep rate on applied strain and irradiating flux and its insensitivity to material properties and temperature. It is perhaps worth remarking on the magnitude of the SIPA creep rate when it occurs concurrently with void swelling. The SIPA mechanism depends on how a given number of point defects partition themselves amongst the dislocations. Reducing the number of defects available to the dislocations by removing some to neutral sinks, such as voids, necessarily reduces the creep rate from this mechanism. Consequently the SIPA mechanism cannot give rise to a creep rate which increases in proportion to the irradiation swelling. Indeed it must suffer a small decrease.[5] It follows that any strong temperature variation of the steady-state irradiation creep rate (whether directly proportional to the irradiation swelling rate or not) must arise from a different mechanism from that described here.

ACKNOWLEDGMENTS

It is a pleasure to thank our colleagues at BNL and Harwell for useful discussions, in particular M. V. Speight, J. R. Willis, and M. R. Hayns who were our co-authors on the papers upon which this review is based.

This paper is published by permission of the Central Electricity Generating Board and the United Kingdom Atomic Energy Authority.

REFERENCES

1 P. T. HEALD AND M. V. SPEIGHT: *Phil. Mag.*, 1974, **29**, 1075
2 W. G. WOLFER AND M. ASHKIN: *J. Appl. Phys.*, 1976, **47**, 791
3 P. T. HEALD AND M. V. SPEIGHT: *Acta Met.*, 1975, **23**, 1389
4 R. BULLOUGH AND J. R. WILLIS: *Phil. Mag.*, 1975, **31**, 855
5 R. BULLOUGH AND M. R. HAYNS: *J. Nucl. Mat.*, 1975, **57**, 348
6 R. BULLOUGH AND R. C. NEWMAN: *Rep. Prog. Phys.*, 1970, **33**, 101
7 A. H. COTTRELL: Report on Strength of Solids, 30, 1948, London, Physical Society
8 B. A. BILBY: *Proc. Phys. Soc.*, 1950, **A63**, 3
9 J. D. ESHELBY: *Proc. Roy. Soc.*, 1957, **A241**, 376
10 P. H. DEDERICHS *et al.*: *Phys. Rev. Lett.*, 1973, **31**, 1130
11 J. HOLDER *et al.*: *Phys. Rev.*, 1974, **B10**, 363
12 A. D. BRAILSFORD AND R. BULLOUGH: *J. Nucl. Mat.*, 1972, **44**, 121
13 A. D. BRAILSFORD AND R. BULLOUGH: this volume

Session III
Microstructure and creep

Chairman: S. F. Pugh (AERE)
Technical Secretary: B. Burton (CEGB)

Grain boundaries as vacancy sources and sinks

G. W. Greenwood

The conditions under which vacancies may be created or may disappear at grain boundaries have been clearly established in many cases but a variety of recent experimental results have led to increasing awareness of the significance of microstructure, segregation and the structure of the grain boundary itself. This review first defines the perfect behaviour of grain boundaries as vacancy sources and sinks. The extent of evidence for such behaviour is next explored and examples are given to illustrate that, when perfect source and sink behaviour is coupled with firmly based treatments of diffusion processes, the range of problems that can be solved is, at least in principle, very large. In several situations, however, there is evidence that grain-boundary action is less than perfect. This has far-reaching implications because new theoretical analyses are then required and there is an urgent need for a quantitative description of this behaviour. Microstructural observations often give important clues on the extent to which grain boundaries act as vacancy sources and sinks and an assessment is briefly given of the present level of understanding.

The author is at the University of Sheffield

It is well known that grain boundaries can supply, annihilate, and transport vacancies in a manner that may depend on the stress system, chemical potential gradient, and microstructure. In the present paper emphasis is placed on phenomena that are influenced by the emergence of vacancies from grain boundaries or from their disappearance at these boundaries and the terms source and sink behaviour are generally applied to these processes. The number of phenomena that are embraced is extremely wide and includes creep, sintering, reaction to quenching, and liability to fracture at elevated temperatures. Additionally there is relevance to the effects of vaporization from the solid solutions, to interdiffusion and to irradiation damage.

Whether grain boundaries act as sources or sinks depends on the conditions to which the material is subjected, but it is also important to appreciate that both vacancy source and sink behaviour are essentially competitive processes with dislocations, voids, gas bubbles, and sometimes interstitials, as well as grain boundaries, all acting in their own characteristic ways. Some of these ways were admirably summarized[1] in the report of the conference on 'Vacancies and other point defects in metals and alloys' which was the forerunner of the present one some 20 years ago. There has been no slackening of effort in recent years in this area of research and the value of these studies is becoming increasingly apparent.

In this review, first a definition is given of perfect source and sink behaviour of grain boundaries, with examples of theoretical analyses, and with experimental evidence that, in some situations at least, such behaviour occurs in practice. Secondly, some examples of imperfect source and sink operation are given and the problems of analysing this situation are outlined. Next, source and sink behaviour are related to diffusion and to vaporization processes and to the changes in microstructure brought about by stress. The agglomeration of gases at grain boundaries is shown to be particularly relevant to the source and sink behaviour of the boundaries and some examples of technological importance are given.

PERFECT SOURCE AND SINK BEHAVIOUR
The simplest description of perfect source and sink behaviour is that grain boundaries have a capacity for producing or absorbing any number of vacancies to provide a flux that is proportional to the chemical potential gradient. In practice, the rate of emergence from, or disappearance at, grain boundaries is often a critical feature, and a further description of 'perfect' behaviour is that the production or disappearance rate of vacancies is adequate to enable the vacancy flux to be determined entirely by the boundary conditions that the chemical potentials superimpose.

Where the grain boundaries emit or absorb vacancies that diffuse through the crystal lattice of the grains, and no vacancies are created or lost within the grains, then standard applications of Poisson's equation can be made, resulting in solutions exactly analogous with those well known in classical problems of heat conduction.[2] The situation, however, is somewhat different where both vacancy production and diffusion both occur within the narrow confines of the boundary plane.[3] These circumstances occur frequently in practice and are often particularly prominent at temperatures of less than two-thirds of the absolute melting temperature and for fine-grained materials. In the analyses of these situations a variety of approaches have been adopted. From these it is not always easy to gain a physical insight of their basis. More simplified forms of these approaches on a common basic theme will be presented here to clarify the situation.

The physical picture of vacancy production and diffusion along grain boundaries contributing to diffusion creep and to the growth of grain-boundary cavities can be appreciated by the following simplified approach.

First, considering the effect on creep, we assume a cylindrical grain with ends perpendicular to the axis and to an applied stress σ as in Fig. 1. Continuity at the grain boundary and steady-state creep implies a uniform rate of production of vacancies β per unit area per unit time at all points on the boundary. When the vacancy flux j is confined within the boundaries of width w in a metal of atomic volume Ω, the equation of continuity can be written

$$\operatorname{div} j + (\beta/w) = 0 \tag{1}$$

where $j = -D_v \nabla c/\Omega$, where D_v is the diffusion coefficient for vacancies in the grain boundary and $(\partial c/\partial r)$ is the vacancy concentration gradient. It is convenient to use cylindrical polar coordinates so that equation (1) can be rearranged and written as

$$D_v \nabla^2 c = D_v \left(\frac{\partial^2 c}{\partial r^2} + \frac{1}{r}\frac{\partial c}{\partial r} \right)$$
$$= \left(\frac{D_v}{r} \right) \frac{\partial}{\partial r} \left(r \frac{\partial c}{\partial r} \right) = -\frac{\beta\Omega}{w} \tag{2}$$

Integrating, we obtain

$$D_v r \left(\frac{\partial c}{\partial r} \right) = -\frac{\beta r^2 \Omega}{2w} + A_1 \tag{3}$$

where A_1 is a constant but, from Fig. 1, $(\partial c/\partial r) = 0$ for $r = 0$ and hence $A_1 = 0$. Further integration gives

$$D_v c = -(\beta r^2 \Omega/4w) + A_2 \tag{4}$$

Now the vacancy concentration $c = c_0$, the equilibrium concentration when $r = R$. It follows that

$$D_v(c - c_0) = \left(\frac{\beta\Omega}{4w} \right)(R^2 - r^2) \tag{5}$$

At a position on the grain boundary where the normal stress is σ_r, the excess vacancy concentration $\Delta c = (c - c_0) = c_0 \exp(\sigma_r\Omega/kT) - c_0$, where T is absolute temperature and k is Boltzmann's constant. Since $\sigma_r\Omega$ is usually very much less than kT, then

$$\Delta c = (c - c_0) = c_0\sigma_r\Omega/kT \tag{6}$$

However, from equation (5) it is clear that Δc is not constant at all points on the boundary. It is a maximum in the centre where $r = 0$ and tends to zero at the

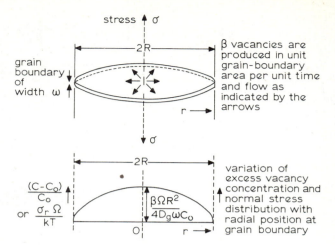

1 A schematic illustration of the radial flow outwards through a grain boundary of radius R and width w of vacancies produced uniformly in the boundary at a rate of β per unit area; the excess vacancy concentration $(C - C_0)/C_0$ and the normal stress σ_r at a distance r from the origin both follow the curve shown

periphery for $r = R$. It follows from equation (6) that the diffusion flux of vacancies is also associated with a stress redistribution at the grain boundaries since $\Delta c \propto \sigma_r$ at constant temperature. Δc can be evaluated by considering the total force to which the boundary is subjected, with the conditions imposed by equation (6). Thus

$$\int_0^R \sigma_r 2\pi r \, dr = \sigma\pi R^2$$

where σ is the applied stress, so that

$$\int_0^R \left(\frac{c - c_0}{c_0} \right) \frac{kT}{\Omega} 2\pi r \, dr = \sigma\pi R^2$$

and so

$$\int_0^R \left(\frac{\beta\Omega}{4D_v wc_0} \right)(R^2 - r^2) \frac{kT}{\Omega} 2\pi r \, dr = \sigma\pi R^2 \tag{7}$$

Integrating, we obtain

$$\frac{\beta kTR^2}{8D_v wc_0} = \sigma \tag{8}$$

Now the product $D_v c_0$ is equal to D_g the grain-boundary self-diffusion coefficient and the creep rate $\dot\epsilon$ is related to β since, in an opposite direction to the vacancy flow, a total volume of $\beta\pi R^2\Omega$ of atoms arrives at the grain boundary per unit time, producing a strain $\beta\pi R^2\Omega/2\pi R^3$ in this period. Hence $\dot\epsilon = \beta\Omega/2R$ and substituting for β in equation (8) we obtain

$$\dot\epsilon = 4D_g w\sigma\Omega/kTR^3 \tag{9}$$

If d is the grain size, then $d = R/2$ and so

$$\dot\epsilon = 32D_g w\sigma\Omega/kTd^3 \tag{10}$$

The coefficient 32 is about two-thirds of the value $150/\pi$ determined in the corrected[4] Coble analysis[5] but this coefficient depends on the grain geometry considered and the above derivation can be regarded as a simplified analysis of grain-boundary controlled diffusion creep which nevertheless contains a full description of the physical basis of the process.

A similar approach has also been employed to evaluate the contribution of grain-boundary sources and diffusion to the growth of cavities on grain boundaries normal to the applied stress.[6] Using equation (3) in the situation illustrated in Fig. 2, the constant A_1 is determined from the condition that mid-way between the cavities, where $r = x$, $(\partial c/\partial r) = 0$. The constant on second integration is found from the condition that

$$\left(\frac{c - c_0}{c_0}\right) = \frac{2\gamma\Omega}{kTr_0}$$

when $r = r_0$, the effective cavity radius, and γ is the surface free energy. The full solution then becomes

$$\Delta c = (2\gamma\Omega/kTr_0) - (\beta\Omega/4D_g w)$$
$$\times [r^2 - r_0^2 - 2x^2 \ln(r/r_0)] \qquad (11)$$

To make use of equation (11), the value of Δc must also be related to the stress distribution between the cavities, when the applied normal tensile stress on the boundary is σ and the stress acting at a position on the boundary at a distance r from the centre of a cavity is given by σ_r. Now a cavity of effective radius r_0 situated on a transverse grain boundary exerts a compressive stress $2\pi r_0\gamma$ and so the value of σ_r is obtained from the relationship

$$\int_{r_0}^{x} \sigma_r 2\pi r \, dr = \sigma\pi x^2 - 2\pi r_0\gamma \qquad (12)$$

Since, at any point on the boundary $\Delta c = c_0\sigma_r\Omega/kT$, the value of β in equation (11) can be determined in terms of the applied stress σ. The rate of volume increase of each

cavity $dv/dt = \beta\pi x^2\Omega$ and it follows that

$$\frac{dv}{dt} = \frac{\left[\dfrac{2\pi D_g w\{\sigma - (2\gamma/r_0)\}\Omega}{kT}\right]}{\left[\ln\left(\dfrac{x}{r_0}\right) - \dfrac{(1 - (r_0/x)^2)(3 - (r_0/x)^2)}{4}\right]} \qquad (13)$$

For $x/r_0 \geqslant 10$, dv/dt is not strongly dependent on this ratio, but it increases rapidly for $x/r_0 < 10$.

Equation (13) can also be applied to the case of sintering, e.g. when $\sigma = 0$ or by putting $\sigma = -P$ when a hydrostatic pressure P is superimposed.

Equations (10) and (13) and their consequences follow directly from the action of grain boundaries as perfect sources of vacancies and it is important to assess their range of applicability.

EXTENT OF EXPERIMENTAL EVIDENCE OF PERFECT SOURCE AND SINK BEHAVIOUR

The most direct ways in which the theories described above can be put to experimental test are in the fields of low stress creep, both through the lattice[7] and through grain boundaries,[5] by sintering experiments involving void shrinkage and by the converse process of the growth of voids during creep under tensile stress.

By far the most extensive work has been concentrated on copper, usually of the OFHC variety. The analysis first developed by Nabarro[7] and Herring[8] for creep involving vacancy fluxes without dislocation motion, but relying entirely on grain boundaries acting as perfect sources and sinks, has been amply supported[9] and all the parameters in their equation have been found experimentally to have the influence predicted. The only small discrepancy is in a general finding that the dimensionless constant B in the equation

$$\dot\epsilon = \frac{BD\sigma\Omega}{kTd^2} \qquad (14)$$

is somewhat greater than that anticipated.[10] There still remains some doubt why this should be so but it serves to emphasize the significance of the changes that can be brought about purely by vacancy fluxes.

Equally convincing is the observed transition for copper between the case where lattice diffusion is rate controlling to that where the diffusion along grain boundaries becomes predominant.[11] This transition is independent of the creep stress level but depends on temperature in the way predicted by the theory and at the low temperatures, where grain-boundary processes are most important, the creep rate is inversely proportional to the cube of the grain size. It follows further from this that there is a means of assessing the activation energy for grain-boundary diffusion. It is to be noted that this is the true grain-boundary self-diffusion and not the diffusion of vacancies which is being assessed in the experiments. Equation (10) is thus shown[11] to be supported in this instance on a firm experimental basis.

The experiments described are also well supported by studies of the shrinkage of voids on and away from grain boundaries.[12] Voids can readily be produced in OFHC copper during creep and high precision measurements of density changes can lead to accurate assessments of void shrinkage rates. While there has been a certain amount of confusion in this area, there now seems no doubt that

2 A schematic illustration of the flow of vacancies to a cavity of radius r_0 from an area πx^2 of grain boundary; the excess vacancy concentration $(C - C_0)/C_0$ and the normal stress σ_r vary as shown with radial distance r from the centre of the cavity

this is caused by residual gases and, when facilities are available for doing experiments under superimposed hydrostatic pressure, results become more consistent. These results are then in general agreement with a shrinkage rate given by equation (13) when σ is replaced by $(P_g - P)$ where the hydrostatic pressure P is partially opposed by the pressure P_g of the residual gas in the voids and from this there is an independent means of assessing the value of Q_g, the activation energy for grain-boundary self-diffusion. The values derived show that the void shrinkage process is governed by an activation energy similar to that involved in diffusion creep via grain boundaries according to equation (10).

The applicability of the approaches described has been challenged to a greater extent when they have been applied to void growth processes during creep[13] and to the prediction of creep life.[14] It now seems, however, that some of the difficulties have arisen because of uncertainties about void nucleation rather than about void growth. Since there is much evidence to suggest that cavities are continually nucleated during creep,[14] then the resultant volume change is governed by nucleation as well as by growth, and it is difficult to separate these processes because the size of individual voids cannot be followed. More recently, experiments on copper and on iron[15] have shown that nucleation is approximately proportional to the square of the stress level and is relatively independent of temperature. This can be used to cause a high nucleation of voids at high stresses and low temperature and subsequently these voids can be grown at very low stresses at elevated temperatures in a situation where a negligible number of new voids are nucleated. In this way, the factors depending on the nucleation of new voids are eliminated and a direct test can be made of equations that predict void growth where grain boundaries show perfect source and sink behaviour. Under these special circumstances direct tests become possible and preliminary information[16] again indicates that the approach based on the perfect source and sink behaviour of grain boundaries is entirely adequate to explain the situation.

The experiments mentioned, although indicating the role of grain boundaries as perfect sources and sinks, also imply that other potential sources and sinks are of negligible importance. This is not always the case in cavity formation during creep because there is a threshold stress for nucleation depending on the ability of the cavity to accept vacancies.[17] There is now ample evidence that a cavity will only act as a vacancy sink under the condition that

$$r_0 > 2\gamma/\sigma$$

but above this size continuous void growth can take place.

Rather special circumstances arise when gases are present in the voids because it may then be expected that a void would effectively act as a perfect sink for vacancies. This is not so, however. If the void grows by vacancy absorption then the gas pressure immediately is decreased and this decrease, depending inversely on the cube of the effective cavity radius, is more rapid than the change in surface tension forces. Thus the cavity can only continue to act as a vacancy sink providing the stress exceeds a given value which is not very much less than that which exists for cavities without gas.[18]

There is the further possibility that dislocations will also act as vacancy sources and sinks and indeed there is much evidence of their doing so in many metallurigical phenomena. In the experiments mentioned, however, it must be recalled that the chemical potential for vacancies is low and, although dislocations may accept vacancies, they can only do so to a very limited extent to cause them to bow out but not to act continuously as sources and sinks in the manner described by Bardeen and Herring.[19] We conclude from this that predictions can be made of the circumstances when dislocations become serious competitors to grain boundaries in their ability to provide or absorb a virtually infinite number of vacancies.

Although not providing quantitative evidence for vacancy source and sink behaviour of grain boundaries, the early evidence[20] of the growth of helium bubbles caused by alpha particle bombardment in copper illustrates clearly that the growth of bubbles in regions close to grain boundaries depends on the supply of vacancies from these boundaries, but not from nearby twins or from dislocations. This gives further support to the now widespread evidence that grain boundaries are highly effective vacancy sources, at least in some materials.

EVIDENCE FOR IMPERFECT BEHAVIOUR OF GRAIN BOUNDARIES AS VACANCY SOURCES AND SINKS

Perfect source and sink behaviour discussed above immediately implies that, at all finite stresses perpendicular to grain boundaries, no matter how small, source or sink behaviour will occur depending on whether the stress is tensile or compressive. More detailed explorations of diffusion creep phenomena which extended the range of experiments to lower temperatures and to smaller stable grain sizes have shown that there is a threshold stress below which creep does not occur.[21] The value of this threshold stress is not constant and it is generally found to be greatest at low temperatures and at the finer grain sizes. There is some evidence that this threshold stress, for a number of metals at a constant fraction of their melting temperature and for a given grain size, is related to the Young's modulus of the material.

Although this behaviour is clearly 'imperfect', it raises questions concerning the mechanism of vacancy generation at grain boundaries and this problem is now seen to be of central importance. Whereas for the situation of 'perfect' behaviour there was no need to search for an answer to this problem (except from the point of view of pure scientific curiosity), since this information is not required in theoretical treatments, the position is completely changed when imperfect behaviour is analysed. It is to be hoped that the recent progress[22] in identification and description of grain-boundary structures will ultimately converge with analyses of source and sink operation so that satisfactory models can be developed and predictions made. Before the emergence of such analyses, however, it is reasonable to propose, at the present time, that vacancies are emitted or absorbed[23,24] by grain boundaries at sites that correspond to dislocations with a component of their Burgers vectors perpendicular to the plane of the boundary so that an effective climb motion of the dislocation can take place. Just as in the corresponding situation of Bardeen–Herring sources[19] within grains, one can envisage the situation where dislocations in grain boundaries may produce or absorb an infinite number of vacancies but they will only do so if a sufficiently high stress or source of chemical potential gradient exists. A model such as this could go some way towards explaining why the threshold stress tends to be greater for small grain sizes although the influence of temperature is more difficult to interpret.

The assessment of diffusion creep in alloys presents a whole range of new problems[25] and a really systematic approach towards their analysis has hardly begun. First, since solutes may affect dislocation creep by reducing the stacking fault energy, there are many cases where it would be anticipated that the range of importance of diffusion creep would be extended and consequently of more importance. It is then necessary to determine the effect of solutes on vacancy fluxes.

Present evidence suggests that solid solutions may have either faster or slower diffusion creep than that of either of their components. There is the further complication that the range of predominance of lattice and grain-boundary diffusion is less easy to evaluate. In copper, for example, zinc and tin additions increase the rate of diffusion creep whereas nickel retards it.[26] From the experimental results available, the important conclusion has been drawn that mass transport diffusion coefficients evaluated from creep data are in reasonable agreement with interdiffusion coefficients although it is too early to regard this as a fully established relationship. For solid solution alloys it is now generally agreed that there may be both a change in the rate of diffusion creep by alloying and also of the magnitude of the threshold stress. Taking a dislocation model of grain boundaries it is not difficult to imagine a solute atmosphere around each grain boundary dislocation and the dragging of these solute atoms may give rise to the inability of the grain boundary to act as a perfect source and sink for vacancies.

Looking at this problem from a macroscopic point of view, if it is considered that there is interface control of vacancy emission or absorption at grain boundaries then the emergence or absorption of vacancies will simply be proportional to grain-boundary area.[24,27] Using this approach it is expected that the creep rate will be inversely proportional to the grain size rather than to the grain size squared or cubed as is the case where there is perfect source and sink behaviour and the vacancy fluxes are through the lattice or along the grain boundaries respectively.

For two-phase systems the situation is still more complex.[25] There is now much evidence to show that particles can inhibit the action of grain-boundary vacancy sources and sinks although not all particles act in this way and the particle size, shape, and distribution are of importance.[28] In magnesium alloys, for example, there is a marked effect of zirconium hydride particles in reducing the rate of diffusion creep[29] but particles of the alpha manganese phase have negligible effects.[30] Such observations have led to the view that vacancy source and sink behaviour at the interfaces between particles and matrix only occurs in a near perfect manner when the interface has an energy similar to that of the grain boundary on which the particle is situated. At particle/matrix interfaces where there is a large difference in energy, for example in metal–oxide systems, the influence of particles is very strong and it appears then that the particle/matrix interface cannot easily admit or absorb vacancies. For alloys with particles, the threshold stress for diffusion creep is usually very much greater than that under corresponding conditions for pure metals.[25] Very plausible ideas have been developed along the lines of the pinning of grain-boundary dislocations by particles and the difficulty of these dislocation lines bowing between such particles is then identified as the source of the threshold stress. This is clearly one area where much further work is required but the difficulties should not be underestimated.

One of the experimental problems is that the microstructure of the material is not constant during creep tests. Even when the particles are sufficiently stable not to coarsen, there is still the effect[31] that diffusion produces denuded zones near the grain boundaries approximately perpendicular to the tensile stress while particles progressively accumulate at the grain boundaries that are nearly parallel to this stress. The effect of these progressive changes is to reduce the creep rate continually and so an effective secondary creep rate is never achieved. There is some indication that the creep strain tends to attain a constant value at a rate that corresponds to the well established law of charge accumulation of a capacitor under constant voltage.

Many aspects of imperfect source and sink behaviour are also immediately relevant to sintering and to the build-up of creep damage. It is now well established that the final stages of sintering depend strongly on the ability of grain boundaries to absorb vacancies and where this ability is impeded then sintering rate is much reduced. This aspect gives rise to a whole series of problems. On the one hand, where no particles are present grain boundaries may move more freely and they may become detached from the voids and make sintering processes more difficult. In the converse situation where many particles are present on the grain boundaries the danger of the detachment of grain boundaries from areas of porosity is reduced but the particles may impede the capacity of the grain boundaries for vacancy absorption. Thus, there is every need, technologically, for a detailed understanding of the pinning of grain boundaries by those particles that will not impede the vacancy absorption process.

Correspondingly important areas of research remain in the field of creep damage assessment and creep failure prediction.[32] Arguments have long been pursued about the relative contributions of grain deformation by dislocation motion and of vacancy fluxes in contributing to void growth.[14] These arguments have recently been put in better perspective[33,34] but situations have also been identified where, although voids appear to have existed for a long period of time, they have nevertheless grown at a much lower rate than would have been predicted by vacancy flux models. Many high-strength nickel alloys and important commercial steels seem to fall into this category and the interpretation of this appears to have its roots in the inability of grain boundaries to supply vacancies at a rate anticipated from the theory of perfect behaviour. Unlike the situation that is needed to obtain satisfactory sintering, for materials to withstand stresses at elevated temperatures for considerable periods of time, the need is to block as effectively as possible a tendency for the grain boundary to produce vacancies.

MICROSTRUCTURAL EFFECTS PRODUCED BY THE ACTION OF GRAIN BOUNDARIES AS VACANCY SOURCES AND SINKS

The longest known effect of microstructural changes asociated with the absorption of vacancies by grain boundaries came from observations on precipitation hardened alloys. It is now clear that in many of these alloys there is a region free of precipitates adjacent to grain boundaries, and powerful evidence has been put forward to show that this lack of precipitation in these regions is due to a depletion of vacancies because they have been annihilated at the grain boundaries.[35] In practice this phenomenon is of great importance. The occurrence of these precipitate-free zones has been related to the poor mechanical properties of materials in this state and also of their greater likelihood of chemical attack and of stress corrosion. Where the grain boundary is incapable of absorbing vacancies then this problem does not occur. However, there have also been shown to be other solutions to this problem, such as the reduction of vacancy mobility by their attachment to solute atoms.[36]

More recently it has been shown that precipitate-free zones near some grain boundaries occur as a consequence of diffusion creep.[31] Indeed, this observation is one of the most convincing ways of detecting the extent to which diffusion creep has taken place. It is an essential feature of the process that, when grain boundaries are the only sources and sinks for vacancies (excluding external surfaces) then there can be no change in the spacing between precipitates within the grains. This situation is entirely different from that where dislocations move between the precipitates to shear the material such that the precipitates within the grains can be translated with respect to each other. The most important feature of the denuded zones produced by diffusion creep is that they are widest for those grain boundaries perpendicular to the applied stress and the zone width decreases with increase of the angle between the applied stress and the normal to the grain boundary. When this angle exceeds 45° then, on average, instead of denuded zones being formed, the boundaries effectively move inwards because of the loss of atoms through vacancy production and, as this process continues, precipitates progressively accumulate on the grain boundaries. This accumulation is greatest for those grain boundaries that are parallel to the applied stress. It is one of the consequences of these effects that diffusion creep is much more difficult to investigate in two-phase or multiphase structures. Because of the precipitate accumulation the ability of the grain boundary to emit vacancies may be substantially impaired and so a progressive decrease in the creep rate may be observed. It then appears that diffusion creep is essentially a primary creep process that may reach a saturation.

As well as grain boundaries acting as sources and sinks for vacancies, which is their major role in diffusion creep, it is now known that grain-boundary sliding is an essential concurrent process. Grain-boundary sliding can be looked upon either as a natural consequence of diffusion creep or, conversely, it can be considered that grain-boundary sliding occurs at a rate that is permitted by the diffusion creep process.[37] The important conclusions are that the occurrence of grain-boundary sliding of this form, which is essentially an accommodation process, does not modify the equations for diffusion creep in considering their applicability to equiaxed polycrystalline material.

There are many ways in which grain-boundary source and sink action can modify structure that relate to observations on materials in which pores initially exist or in which there are driving forces for their creation. One of the best known experiments[38] on the fundamentals of sintering has shown that pores disappear most rapidly on grain boundaries and when many pores have disappeared then the grain boundary can move and become attached to a separate set of pores that, in turn, can also begin to shrink, while the pores within the grain remain relatively unaltered. Observations such as these show that the grain boundary is acting effectively as a sink for vacancies and, in sintered materials, these microstructural features give important clues.

More recently further microstructural effects have been reported and many of these concern processes that take place in chemical potential gradients, e.g. the well-known Kirkendall effect of unequal diffusion of two components flowing in opposite directions across their common interface leads to void production and dimensional changes. In most situations grain boundaries are relatively unimportant since voids are formed within grains in regions that are parallel to the original interface. Quite recently, however, it has been shown that if the grains in the low melting component in which the voids are formed are sufficiently fine, then the grain boundaries can accept vacancies more efficiently than potential void nuclei[39] and so porosity is not developed and dimensional changes are then much reduced.

Grain-boundary source and sink behaviour has also recently been shown to be relevant to phenomena that occur during the vaporization of one component from an alloy. The well known case of the vacuum dezincification of brass leaving behind pores due to the excess removal of zinc atoms has been shown to be affected by the presence of grain boundaries. Since this process may take place at temperatures of about half the melting temperature then grain-boundary diffusion may also predominate and the effect is to produce voids that follow the solute concentration profiles and lead to the occurrence of voids in regions near to the grain boundaries which penetrate well below the surface of the alloy.[40]

When gas is present in materials grain-boundary source and sink behaviour can produce some of its greatest effects. The effects can be of two kinds. Where the gases are completely insoluble then total enlargement may only take place by coalescence and this depends on the possible sweeping action of the grain boundaries.[41] Bubble coalescence has an important effect because, unlike the situation for solid precipitates, gas bubbles in equilibrium are usually subject to the condition that the gas pressure is exactly balanced by the surface tension forces. When two bubbles merge this condition breaks down unless the resultant bubble becomes enlarged and it may only do so through the ability to acquire vacancies.[42] When vacancies are available then it is the total surface area of the bubbles that is conserved with an overall increase in the volume of the new bubble compared with the sum of the volumes of the two bubbles from which it was formed. The driving force for this expansion is entirely derived from the increase in entropy as the gas pressure is decreased. Gas bubbles have the effect of greatly reducing the mobility of the grain boundaries on which they are situated. This mobility however is related both to bubble size and to bubble spacing.[41] If a hydrostatic pressure is applied then bubble size is reduced in consequence because the gas pressure is then restrained both by surface tension forces and by the external pressure. The shrinkage of bubbles is then dependent on their ability to emit vacancies and for these to be absorbed at grain boundaries. Since sintering processes often involve high pressures, then this reduction in bubble size and alteration in grain-boundary mobility may be important. It has been shown that the binding energy of a gas bubble on a grain boundary is always rather close to the energy of the area of the grain boundary removed by the bubble and so calculations can be done on this basis.[43]

Bubbles can coarsen without coalescence if there is some solubility. The coarsening process is somewhat analogous to that of the overaging of precipitates, but it has the difference that, when one bubble disappears to contribute to growth of a neighbouring bubble, then, just as in the case of coalescence, the resultant volume of the new bubble is greater than the sum of its two constituents providing a supply of vacancies is available to permit this growth. Since it is the area of the bubble surfaces that remains constant it follows that, if only bubbles on grain boundaries are involved, then any grain-boundary coarsening process still leaves constant the fraction of grain-boundary area removed by bubbles. This is only true if bubbles within the grains have no effect. It may be, for example, that the growth of bubbles at grain boundaries causes their impingement on bubbles within the grains and a further growth process becomes possible.[44] Instances such as these may possibly account for the role of residual potassium being one of the causes of life limitation in tungsten lamp filaments operating close to their melting point.

Further instances of the effects of gases can occur when there is some reaction with an external gas. The best known case is the situation where copper is surrounded by an atmosphere with a significant partial pressure of hydrogen.[45] Even in the case of OFHC copper then there is often enough cuprous oxide in small particles to be attacked by the hydrogen that readily diffuses through the lattice to reduce the oxide to form water vapour. When this occurs then gas bubbles can be formed but only when there is a supply of vacancies. It is clear from observations of this kind that vacancies are readily produced by the grain boundaries of copper and the effect is illustrated in Fig. 3. If grain boundaries were not effective sources of vacancies then phenomena such as this would not arise.

CONCLUSIONS

It is clear that in some instances grain boundaries can act as vacancy sources and sinks in a manner that can be described as 'perfect'. Such action can provide or absorb any number of vacancies to maintain a vacancy flux that is proportional to the chemical potential gradient. Under these circumstances, well established theoretical analyses can be applied and two examples have been detailed in this review. At least in principle similar approaches can be applied in solving a wide range of problems. There is, however, a danger in expecting such analyses always to have useful application. This is not because there are flaws in the analytical methods. The difficulties arise because it now seems evident that grain

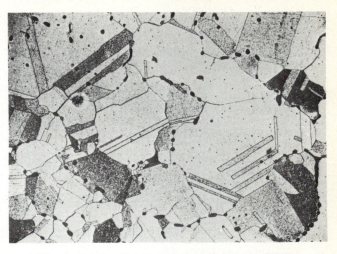

3 **Effect of a hydrogen atmosphere on OFHC copper held at 800°C for 240 h; bubbles of H$_2$O are formed by reduction of Cu$_2$O and these grow by the acquisition of vacancies from the grain boundaries ×100**

boundaries in many materials are incapable of displaying perfect source and sink behaviour. There are then limitations to the rate of production or absorption of vacancies that can severely restrict the effect of the chemical potential gradient and new approaches are required in dealing with these situations. These approaches are likely to be complex. There is much scope for further exploration and analysis in this area and increasing knowledge of grain-boundary structure may be valuable in this respect.

REFERENCES

1 Symposium on 'Vacancies and other point defects in metals and alloys', 1957, London, Institute of Metals
2 H. S. CARSLAW AND J. C. JAEGER: 'Conduction of heat in solids,' 2nd Edn, 191, 1959, Oxford University Press
3 P. G. SHEWMON: 'Diffusion in solids', 1963, McGraw-Hill
4 J. E. BIRD et al.: 'Quantitative relations between properties and microstructure', 1969, 255, Israel University Press
5 R. L. COBLE: J. Appl. Phys., 1963, **34**, 1679
6 M. V. SPEIGHT AND W. B. BEERÉ: Metal Sci., 1975, **9**, 190
7 F. R. N. NABARRO: Rep. on Conf. on the 'Strength of solids', 75, 1948, London, The Physical Society
8 C. HERRING: J. Appl. Phys., 1950, **21**, 437
9 R. B. JONES: J. Sheffield Univ. Met. Soc., 1973, **12**, 34
10 H. JONES, Mater. Sci. Eng., 1969, **4**, 106
11 B. BURTON AND G. W. GREENWOOD: Metal. Sci. J., 1970, **4**, 215
12 W. B. BEERÉ AND G. W. GREENWOOD: Metal Sci. J., 1971, **5**, 107
13 D. M. R. TAPLIN: 'The hot fracture story', Proc. Golden Jubilee Symposium, 1973, Benares Hindu University
14 A. J. PERRY: J. Mat. Sci., 1974, **9**, 1016
15 B. J. CANE AND G. W. GREENWOOD: Metal Sci., 1975, **9**, 55
16 P. F. MORRIS: to be published
17 A. H. COTTRELL: 'Structural processes in creep', 1, 1961, London, The Iron and Steel Inst.
18 E. D. HYAM AND G. SUMNER: 'Radiation damage in solids', 323, 1962, Vienna, IAEA
19 J. BARDEEN AND C. HERRING: 'Imperfections in nearly perfect crystals' (Ed. W. SHOCKLEY), 279, 1952, Wiley
20 R. S. BARNES: Phil. Mag., 1960, **5**, 635
21 D. J. TOWLE AND H. JONES: Acta Met., 1976, **24**, 399

22 B. RALPH: 'Grain boundaries', 11, 1976, Inst. of Metallurgists Spring Residential Conf.
23 B. BURTON: *Mater. Sci. Eng.*, 1972, **10**, 9
24 M. F. ASHBY: *Scripta Met.*, 1969, **3**, 837
25 G. W. GREENWOOD: 'Physical metallurgy of reactor fuel elements', 53, 1975, London, The Metals Society
26 B. BURTON AND B. D. BASTOW: *Acta Met.*, 1973, **21**, 13
27 G. W. GREENWOOD: *Scripta Met.*, 1970, **4**, 171
28 B. BURTON: *Metal Sci. J.*, 1971, **5**, 11
29 J. E. HARRIS *et al.*: *J. Australian Inst. Metals*, 1969, **14**, 154
30 R. B. JONES: 'Quantitative relations between properties and microstructure', 343, 1969, Israel University Press
31 R. L. SQUIRES *et al.*: *J. Nucl. Mat.*, 1963, **8**, 77
32 R. K. PENNY AND D. L. MARRIOTT: 'Design for creep', 1971, McGraw-Hill
33 B. F. DYSON AND D. M. R. TAPLIN: 'Grain boundaries', E23, 1976, Inst. of Metallurgists Spring Residential Conf.
34 R. RAJ AND M. F. ASHBY: *Acta Met.*, 1975, **23**, 653
35 R. B. NICHOLSON *et al.*: *J. Inst. Metals*, 1958–59, **87**, 429
36 J. T. VIETZ *et al.*: *J. Inst. Metals*, 1963–4, **92**, 327
37 M. F. ASHBY AND R. A. VERRALL: *Acta Met.*, 1973, **21**, 149
38 J. BRETT AND L. SEIGLE: *Acta Met.*, 1963, **11**, 467
39 J. D. WHITTENBERGER: *Metall. Trans.*, 1972, **3**, 3038
40 G. FRADE AND P. LACOMBE: *Rev. Mét.*, 1966, **63**, 649
41 M. V. SPEIGHT AND G. W. GREENWOOD: *Phil. Mag.*, 1964, **9**, 683
42 G. W. GREENWOOD: *J. Mat. Sci.*, 1969, **4**, 320
43 G. W. GREENWOOD *et al.*: *Phil. Mag.*, 1975, **31**, 39
44 G. W. GREENWOOD: *Phil. Mag.*, 1975, **31**, 673
45 S. HARPER *et al.*: *J. Inst. Metals*, 1962, **90**, 44

The intermediate stage of sintering

W. Beeré

The review covers the intermediate stage of sintering in single-phase materials. Initially, theoretical aspects are covered beginning with a comparison of several types of idealized model porosity. These consist of regular pore shapes placed on the edges of tetrakaidecahedra to form a continuous network. The driving forces for sintering are discussed and the densification rates calculated. Comparison between models shows substantial agreement on the parametric form of the densification rate although the absolute value does vary. Certain geometrical properties such as pore surface curvature can be calculated from models and this is presented along with criteria for the collapse of interlinked porosity into discrete porosity.

The author is with the CEGB, Berkeley Nuclear Laboratories

Powder-sintering is one of the earliest of technologies and is still being actively developed. Research in this subject covers a wide range of topics especially in the effects of alloying and trace elements as well as fabrication procedures for producing high-density compacts. The present short review covers sintering in single-phase materials during the intermediate stage. A comparison is made between theoretical models and experimental data. Special reference is made to the development of the microstructure during densification.

MODELS OF INTERCONNECTED POROSITY

The second stage of sintering is characterized by an interconnected system of pores which have complex geometry. The pore size is not uniform but varies with position and duration of sintering. During sintering the surface area of the porosity always decreases and in the majority of cases the pore volume is also reduced. Despite the complex nature of the porosity, experiments have shown that in many cases the density increases linearly with the logarithm of sintering time at constant temperature. This has prompted simplified models of pore morphology to enable a quantitative calculation of the densification rate. A feature of all the models described here is that the grain size and pore size are uniform throughout and it is assumed they can be related to the average values found during sintering.

One of the best known models is due to Coble[1] who fitted cylinders round the edges of a tetrakaidecahedron-shaped grain. The tetrakaidecahedron has the property of fitting together without leaving gaps between the faces, the porosity arising solely from the continuous network of cylinders (Fig. 1*a*). Coble calculated the rate of change of volume V of one grain from the vacancy flux between the pore surface and the grain boundaries. The flux arises from the difference in vacancy concentration which was taken as $C_0(1 + \gamma\Omega/kTr)$ at the pore and C_0 at the boundary, where C_0 is the equilibrium concentration at a plane surface, γ is the surface tension, Ω the atomic volume, r the cylinder radius (Fig. 1), and kT has its usual meaning. The rate of change of volume of one grain was given by Coble as

$$\frac{dV}{dt} = -\frac{112\pi D\gamma\Omega}{kT} \tag{1}$$

It is convenient to express the sintering rate in terms of the density change and so

$$\frac{1}{V}\frac{dV}{dt} = -457\frac{D\gamma\Omega}{kTd^3} \tag{2}$$

where d is the grain diameter. If ρ is the density and ρ_d the value when fully dense, then the left hand side of equation (2) is related to the relative density ρ/ρ_d by

$$\frac{1}{V}\frac{dV}{dt} = -\left(\frac{\rho_d}{\rho}\right)\frac{d}{dt}\left(\frac{\rho}{\rho_d}\right) \tag{3}$$

During sintering the grain diameter and pore radii are not constant, and yet equation (2) has been derived from the steady-state fluxes for a constant grain diameter and pore radius. The solution is correct however because the vacancy concentration gradient re-adjusts itself much

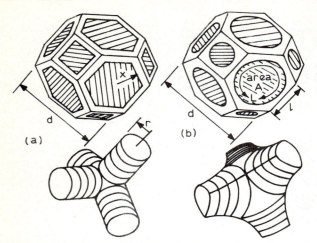

(a) cylindrical porosity; (b) porosity of constant surface curvature

1   The arrangement of porosity round a tetrakaidecahedron grain

more rapidly than the changes in microstructure. Equation (2) gives the sintering rate at an instantaneous value of the grain diameter.

Coble also derived an equation for the sintering rate due to grain-boundary diffusion. Expressed in the same form as above the rate is

$$\frac{1}{V}\frac{dV}{dt} = -40\frac{D_g\delta\gamma\Omega}{kTd^4}\left(\frac{l}{r}\right) \tag{4}$$

where D_g is the grain-boundary diffusion coefficient, δ the grain-boundary width, and l is the length of side of the tetrakaidecahedron (Fig. 1a).

The sintering rate of this type of cylindrical porosity has been calculated more thoroughly by Eadie and Weatherly.[2] The improvements arise from a better estimation of the driving force for sintering, applying the correct relationship between the diffusion fluxes and the sintering rate, and a more thorough analysis of the flux geometry.

Taking these points in order, the increase in vacancy concentration on the void surface was taken as $C_0(1 + \gamma\Omega(1/r - 1/x)/kT)$, where x is the radius of the grain boundary on a grain face, Fig. 1a. The curvature of the interconnected pore surface is anticlastic and a better approximation to the total curvature is given by $1/r - 1/x$. Also, the vacancy concentration at the boundary is not the equilibrium value since plating out a layer of vacancies will do work against the surface tension forces which act at the junction of the boundary and pore surface. The average vacancy concentration at the boundary is given by $C_0(1 - 2\gamma\Omega/kTx)$ and the difference in concentration between pore and boundary is given by $C_0\gamma\Omega(1/r + 1/x)/kT$.

Eadie and Weatherly calculate the linear shrinkage rate of a grain, dL/dt, where L is the grain diameter. The volume shrinkage rate is then given by $(1/V) \times (dV/dt) = (3/L)(dL/dt)$ and the expression for the shrinkage rate is

$$\left(\frac{1}{V}\frac{dV}{dt}\right) = -108\frac{D\gamma\Omega}{kTd^3}\left(\frac{l}{r}+\frac{l}{x}\right)\left(\frac{l}{x}\right) \tag{5}$$

where L has been put equal to d. Calculating the rate of transfer of volume per grain by diffusion and dividing by

the grain volume gives a slower sintering rate than equation (5). This is because the latter method underestimates the total volume change. The vacancy flux removes a certain volume from the porosity but the pore volume is reduced further when the flux annihilates on the boundary and moves the grain centres closer together.

Eadie and Weatherly[2] also calculate the sintering rate due to grain-boundary diffusion:

$$\frac{1}{V}\frac{dV}{dt} = -353\frac{D\delta\gamma\Omega}{kTd^4}\left(\frac{l}{x}+\frac{l}{r}\right)\frac{l^2}{x^2} \tag{6}$$

The flux geometry is solved exactly and the sintering rate is exact for the type of model porosity being considered.

A further development of grain-edge pore models is to replace the cylinders by complex shapes which have constant surface curvature and satisfy the balance of tensions at the pore/boundary intersection, Fig. 1b.[4,5] The driving force for sintering is again proportional to the difference in vacancy concentration at pore and grain boundary given by $C_0\gamma\Omega(H + L\sin\theta/A)/kT$, where H is the surface curvature, L is the circumference, and A is the area of a grain boundary, Fig. 1b. The dihedral angle θ is given by $\cos^{-1}(\gamma_g/2\gamma)$, where γ_g is the grain-boundary energy, and the $\sin\theta$ term arises because the boundary energy is now considered as non-zero. The sintering rate, corrected for a factor of 2 error, is given by[3]

$$\frac{1}{V}\frac{dV}{dt} = -36B\frac{D\gamma\Omega}{kTd^3}(H + L\sin\theta/A)l \tag{7}$$

for volume diffusion control. The factor B in equation (6) arises from putting the effective diffusion distance between pore and boundary equal to l/B. The shortest conceivable distance is half the grain-boundary radius and so the diffusion distance will lie between the limits $l > l/B > A/L$.

The sintering rate has also been calculated exactly for the case of grain-boundary rate control[3] and is

$$\frac{1}{V}\frac{dV}{dt} = -1\,187\frac{D_g\delta\gamma\Omega}{kTd^4}\left(H + \frac{L\sin\theta}{A}\right)\frac{l^3}{A} \tag{8}$$

This equation reduces to the basic form of equation (6) with essentially the same constant if the boundary area A is put equal to πx^2.

The three expressions for volume diffusion control sintering, equations (2), (5), and (7) are compared in Fig. 2. The values of cylinder radius r, Fig. 1a, were calculated from the volume fraction of porosity

$$1 - \rho/\rho_d = 12\pi r^2/8\sqrt{2}l^2 \tag{9}$$

and the grain-boundary radius x was found from the expression $x = \sqrt{3}l/2 - r$. The grain-boundary area A and boundary circumference L were taken from shape calculations of uniform curvature porosity. The dihedral angle θ was taken as 75 degrees.

The comparison between the three expressions has been made by compensating the sintering rate for temperature and material parameters such that it is dimensionless and depends only on the pore geometry. Figure 2 shows how the compensated sintering rate varies with porosity. Although the theories only show agreement to within about an order of magnitude the predicted sintering rate is almost constant within the range $0\cdot9 > \rho/\rho_d > 0\cdot6$.

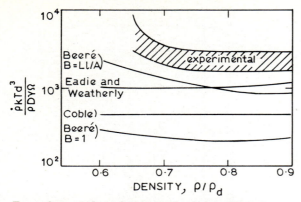

2 Experimental and theoretical densification rates adjusted for grain size and temperature v. density

It is also instructive to consider the driving forces assumed for the sintering process. Taking only the geometrical parts, the driving forces are $1/r$, Coble,[1] $(1/r + 1/x)$, Eadie and Weatherly,[2] and $(H + L \sin \theta/A)$, Beeré.[3] The variation with density is shown in Fig. 3 where the geometrical part has been multiplied by the grain edge length l to make it dimensionless and independent of grain size. Considered over the practical range of second-stage sintering $0.97 \rho/\rho_d > 0.65$ the driving force does not show great variation at constant grain size. At low densities the driving force depends almost solely on the contribution from the grain-boundary potential. This is illustrated by plotting the surface curvature considered by Eadie and Weatherly, $(1/r - 1/x)$, and that computed by Beeré, H. The contribution to the driving force from the pore surface curvature is seen to be negligible or even negative at a relative density of 0.6, Fig. 4.

The sintering rates due to a grain-boundary diffusion mechanism are illustrated in Fig. 5 according to the three expressions, equations (4), (6), and (8). The rates calculated by Eadie and Weatherly and Beeré are both significantly faster than that due to Coble. The difference in rates between the former is almost entirely due to the approximate method employed in the present paper to calculate the grain-boundary radius when the porosity is cylindrical. For instance when calculated by

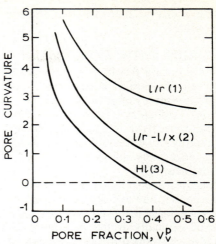

4 Surface curvature of porosity v. pore volume; key as for Fig. 3

the method outlined previously the grain boundary occupies an area of about one-third the area of a hexagonal face on the tetrakaidecahedron at a relative density of 0.6. Since as the density decreases the sintering rate depends on the third power of a decreasing and small boundary radius, any errors are magnified. An exact calculation of the pore volume at low densities involves a significant increase in complexity.

A fourth type of sintering rate analysis due to Johnson[5] relates the densification rate to the observable geometrical properties of the powder compact. If $\bar{H}$ and $\bar{x}$ are average values of the pore curvature and grain-boundary radius then

$$\frac{1}{V}\frac{dV}{dt} = -8\frac{\bar{H}\gamma\Omega}{\bar{x}kT}(DS_v + D_g\delta L_v) \qquad (10)$$

where S_v is the pore surface area per unit volume and L_v is the length of grain boundary/pore intersection per unit volume. During sintering the geometrical properties vary continuously. Equation (10) calculates the instantaneous sintering rate for the instantaneous values of $\bar{H}$, $\bar{x}$, S_v, and L_v.

It is also important to consider isolated as well as interlinked porosity. Over 85% dense an increasing

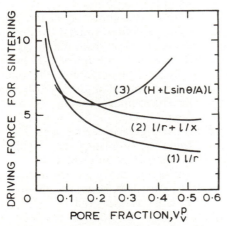

(1) Coble[1]; (2) Eadie and Weatherly[2]; (3) Beeré[3]

3 Geometrical part of the driving force for sintering v. pore volume fraction

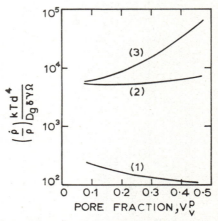

5 Theoretical densification rate for a grain-boundary diffusion mechanism v. pore volume fraction; the rate is compensated for grain size, temperature, etc., and the key is as for Fig. 3

volume of porosity becomes isolated and this has to be treated separately.[6]

COMPARISON WITH EXPERIMENT

Analysis of sintering results is complicated by several factors. Often the initial particle size is below $1\,\mu m$ making measurements of grain size and pore curvature difficult until sufficient grain growth has occurred towards the end of the intermediate stage. Secondly, surface energies and diffusion coefficients are often not sufficiently accurately known to make comparison between theories.

One of the best sets of data for quantitative analysis is still that on copper due to Coble and Gupta.[7] The initial particle size was large resulting in grain sizes which could be measured optically at the beginning of the intermediate stage. Also the diffusion coefficient in copper is known to within a factor of about $1\cdot5$,[8] and the surface energy is known accurately from zero creep measurements.[9] The data are shown in Fig. 6 with the sintering rate compensated for temperature plotted against grain diameter. The surface energy was taken as $1\cdot7\,J/m^2$, $k = 1\cdot38\times10^{-23}\,J/K$, $\Omega = 1\cdot18\times10^{-29}\,m^3$, $D = 7\cdot8\times10^{-5}\exp(-Q/kT)\,m^2/s$, and $Q = 2\cdot10\times10^5\,J/mole$. Except for a few points at 750°C, the data fall on a line with a slope of -3 as expected for a volume diffusion mechanism. Thus, the data can be represented over most of the intermediate stage by the equation

$$\frac{1}{V}\frac{dV}{dt} = -2\cdot4\times10^3\frac{D\gamma\Omega}{kTd^3}\mathrm{s}^{-1} \qquad (11)$$

The experimental data are compared with the theoretical rates in Fig. 2. Observed rates are at least a factor of two faster than the theories predict. This situation may be compared with diffusion creep rates. The same set of assumptions apply when calculating the creep rate and again experimental rates are consistently faster than theory.[10]

The sintering rate due to Johnson,[5] equation (10), has been used to analyse the densification of cobalt oxide.[11] The geometric properties of the compact were specially

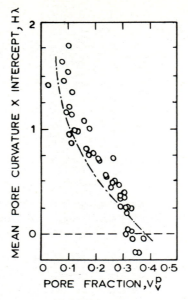

7 Surface curvature of porosity in copper compacts[12]

determined throughout sintering. The behaviour fitted the volume diffusion mechanism. The general applicability of equation (10) to the beginning of the intermediate stage may be in some doubt. It was previously mentioned that at low densities the driving force for sintering derives mainly from the suppression of the vacancy concentration on the grain boundary and not from the excess at the pore surface. This is borne out by measurements on copper[12] in which the average surface curvature $\bar{H}$ became zero between 60 and 70% dense, Fig. 7. Compacts of this density usually have a rapid sintering rate although equation (10) predicts a zero rate. The applicability of equation (10) may be extended by replacing $\bar{H}$ with $(\bar{H}+L_v\sin\theta/A_v)$, where A_v is the neck or grain-boundary area per unit volume.

The curvature measured for copper, Fig. 7, may be compared with the curvature calculated for model porosity, Fig. 4. The experimental values have been multiplied by the mean pore intercept $\bar{\lambda}$ given by $\bar{\lambda} = 2V_v/N_L$ where V_v is the volume fraction of porosity and N_L is the number of intercepts per unit length of test line on a polished section. This allows small and large grain-size compacts to be compared since it eliminates curvature changes resulting solely from differences of scale. The theoretical curvatures are multiplied by the grain edge length, l, which is proportional to the grain diameter. Since $\bar{\lambda}$ and l are not equal, theory and experiment cannot be accurately compared except when the curvature becomes zero.

It is usual to investigate sintering rates, but it is also possible to measure the sintering force.[13] This has been accomplished by imposing a uniaxial tensile stress which just stops the shrinkage in the direction of the stress. Ideally, negative hydrostatic stress should be applied in which case shrinkage is expected to cease when the tensile stress on the boundaries is equal to $(\bar{H}+L_v\sin\theta/A_v)$. The boundary stress must be related to the external stress through the pore geometry. It is assumed that when a uniaxial stress is applied the conditions are the same except only shrinkage in one direction is considered.

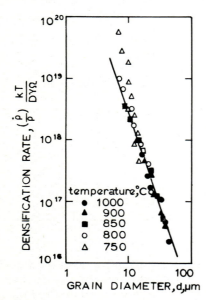

6 Experimental densification rates for copper sintered at five temperatures[7] v. grain size

Development of the microstructure

Grain size

During the intermediate stage the large decrease in the densification rate characterizes the kinetics. This results not from any change in the geometry of pore shapes but from the coarsening of the structure. The increase in grain size with time frequently follows the relationship

$$d^3 = d_0^3 + Kt \tag{12}$$

where K is temperature dependent.[1,7,14]

The sintering rate is relatively insensitive to change in density at constant grain size (Fig. 2) for a volume diffusion mechanism and so

$$\frac{1}{V}\frac{dV}{dt} = -\text{const}\frac{D\gamma\Omega}{kTd^3} \tag{13}$$

Substituting for the grain size gives

$$\int\frac{1}{V}dV = -\text{const}\frac{D\gamma\Omega}{kT}\int\frac{dt}{d_0^3 + Kt} \tag{14}$$

Integrating between $t = 0$ and t and rearranging gives

$$\ln(\rho/\rho_d) = \ln(\rho_0/\rho_d) + \text{const}\frac{D\gamma\Omega}{kTK}\ln(1 + Kt/d_0^3) \tag{15}$$

If only the latter part of the intermediate stage is considered then $\ln(\rho/\rho_d)$ behaves like ρ/ρ_d and $Kt/d_0^3 \gg 1$. Equation (15) then reduces to

$$\rho/\rho_d = \rho_0/\rho_d + \text{const}\frac{D\gamma\Omega}{kTK}\ln(Kt/d_0^3) \tag{16}$$

Equation (16) has been derived by Coble and is widely used to represent sintering data.

If equation (15) is rewritten substituting the grain size instead of time (equation (12)) we then have

$$\ln(\rho/\rho_d) = \ln(\rho_0/\rho_d) + \text{const}\frac{3D\gamma\Omega}{kTK}\{\ln(d) - \ln(d_0)\} \tag{17}$$

This type of behaviour has been observed for copper, zinc oxide, beryllium oxide, and alumina[15] and also for silver.[16] The density/grain size relationship was found to be independent of sintering temperature. This can arise when the temperature dependence is the same for the diffusion and grain growth processes, D and K respectively, equation (17). Sometimes the observations are plotted on linear axes but they probably form a better straight line when plotted logarithmically. The data of Coble and Gupta[7] on copper are illustrated in Fig. 8.

Surface area

The intermediate stage of sintering is characterized by a large decrease in the surface area of the compact. Figure 9 shows the decrease in surface area of three nickel compacts between 50 and 90% dense.[17] As expected the smallest 30 μm particles have the largest area per unit volume. Densification removes porosity thus reducing surface area but the decrease observed in compacts often results mainly from grain growth. This can be seen by considering the slope of the surface area, and volume fraction dS_v/dV_v where V_v is the volume fraction of porosity. If surface area is removed solely by the diffusion of vacancies to grain boundaries then $dS_v/dV_v = -2/r$ for a system of spherical cavities radius r, $-1/r$ for cylinders of radius r. The reciprocal of the gradient

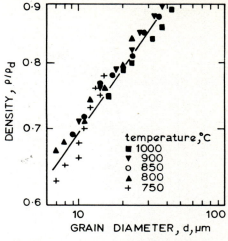

8   Variation of density with grain size in copper at five temperatures[7]

observed in Fig. 9 at 80% dense is 3, 6, and 9 μm for initial particle sizes of 30, 57, and 115 μm respectively. Thus, the observed distance is much smaller than can possibly be expected for this mechanism of surface reduction. The second stage is also characterized by an increase in the grain size or a general coarsening of the porosity. If for instance the density remained constant and the grain size coarsened then the reciprocal gradient would have zero length. This is in fact observed in some covalent compounds in which the surface area can reduce by a factor of 3 or 4 without any increase in density.[18] In copper, Coble and Gupta[7] observed the grain size to increase by a factor of ten between 60 and 90% dense, Fig. 8. Thus, the surface area per unit volume is also expected to reduce by a factor of ten solely as a result of grain growth.

Precompaction of the green compact also affects the surface area/density relationship. Figure 10 shows the results of Rhines, De Hoff, and Whitney[17] on uranium dioxide powder. The surface area per unit mass is constant immediately after precompaction although the surface area per unit volume obviously depends on the compact density. Figure 10 shows that specimens precompacted to different starting densities follow different microstructural paths during sintering. Unfortunately it is not possible to ascertain how far the pore geometry

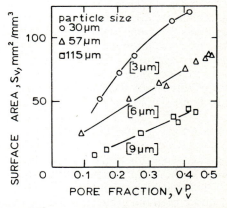

9   Variation of surface area per unit volume v. pore volume fraction for nickel particles of initial sizes of 30, 57, and 115 μm[17]

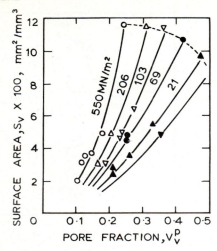

10 Variation of surface area per unit volume v. pore volume fraction for compacts of UO₂ precompacted at pressures up to 550 MN/m² [17]

deviates from an 'ideal' structure for a given pore fraction since the surface area per unit volume is not compensated for the scale of the pore structure. High green density UO₂ sinters much more rapidly than the low green density material[19] and the grain size at a given pore volume fraction may be dependent on precompaction pressure.

Ostwald ripening and grain growth

During the intermediate stage the reduction in total pore volume is accompanied by an increase in the average pore size. Previously the sintering kinetics have been described by a model compact having a uniform and ideally symmetrical pore structure. No mechanism was advanced for the way in which the average pore size increased, but this was incorporated through the empirical relationship $d^3 = d_0^3 + Kt$ for the increase in grain diameter. It was assumed the calculations could be compared with experiment by equating the mean of the distribution of grain sizes in the compact with the uniform grain size of the model. This procedure has not been justified theoretically but in practice it does offer a parametric description of a wide range of data.

Several authors have proposed mechanisms for grain growth in compacts. Kingery and Francois consider the effect of a pore on a moving boundary.[20] The physics of the process is not entirely explained but the authors arrive at a cubic relationship between grain size and time. Lay[21] has also reported on possible mechanisms although these were intended mainly for the initial and final stages. It is generally agreed that growth is by a different mechanism than in fully dense material.

So far the discussion has been concerned with the increase in grain size but provided these are sufficient boundaries it is the size of the pore structure which governs the sintering rate. First, the driving force for sintering was previously shown to be dependent on the pore size and secondly the diffusion distances were proportional to the size of the pore structure. If the pore structure size is represented by the tetrakaidecahedrons of Figs. 1 and 2 then the grain size may be smaller thus forming internal subgrains. The internal boundaries cannot accept vacancies from the porosity because the subgrains have no way of moving closer together. Thus,

the diffusion fluxes and hence the sintering rate are the same as the situation of equal grain and pore structure size.

A detailed atomistic model of Ostwald ripening has not yet been presented for the second stage. Since the evolution of the microstructure, the large decrease in surface area and the decreasing sintering rate are all intimately linked to the coarsening process it is worth considering ripening in precipitates and drawing analogies.

The vacancies concentration at the surface of a small pore is higher than that at a larger pore. If the two are neighbours there will be a diffusion flux which reduces the volume of the smaller pore. Likewise two precipitates will show a similar flux if they are of different sizes. If the precipitates exist in a distribution of sizes and the swelling rate is controlled by volume diffusion, then the average precipitate size $\bar{r}$ increases as

$$\bar{r}^3 - \bar{r}_0^3 = \frac{8\gamma D\Omega}{9kT}t \qquad (18)$$

where $\bar{r}_0$ is the average size at time zero.[22,23] If the diffusion path is along grain boundaries or interface surfaces then the growth law takes the form

$$\bar{r}^4 - r_0^4 = Kt \qquad (19)$$

where K is temperature dependent.[24,25]

Small precipitates are allowed to diminish by the same shrinkage process until they finally disappear. Interconnected porosity collapses before zero volume is reached[3] and the swelling laws may have to be modified to account for this. The swelling laws, equations (18) and (19), may be compared with the empirical grain growth law equation (12). Swelling by volume diffusion would account for the temperature independent variation of density with grain size, equation (17).

The problem of Ostwald ripening in compacts has also been considered on a statistical basis by Kuczynski.[26] By assuming only a quasi distribution function he showed that the grain size and porosity are related by the equation

$$\log(d) = \mathrm{const} + \mathrm{const}\,\log(\rho) \qquad (20)$$

where ρ is the porosity. Equations (20) and (17) are essentially identical.

A thorough understanding of the ripening process in powder compacts is essential for a complete description of densification.

ACKNOWLEDGMENT

This paper is published by permission of the Central Electricity Generating Board.

REFERENCES

1 R. L. COBLE: *J. Appl. Phys.*, 1961, **32**, 787
2 R. L. EADIE AND G. C. WEATHERLY: *Scripta Met.*, 1975, **9**, 285
3 W. BEERÉ: *Acta Met.*, 1975, **23**, 131
4 J. A. TURNBULL AND M. O. TUCKER: *Proc. Roy. Soc. A.*, 1975, **343**, 299
5 D. L. JOHNSON: *J. Amer. Ceram. Soc.*, 1970, **53**, 574
6 S. C. COLEMAN AND W. BEERÉ: *Phil. Mag.*, 1975, **31**, 1403
7 T. K. GUPTA AND R. L. COBLE: 'Sintering and related phenomena', 423, 1967, Plenum

8 D. B. BUTRYMOWICZ *et al.*: *Physical Chem. Ref. Data*, 1974, **2**, 643

9 H. JONES: *Metal Sci.*, 1971, **5**, 15

10 B. BURTON AND G. W. GREENWOOD: *Acta Met.*, 1970, **18**, 1237

11 P. KUMAR AND D. L. JOHNSON: *J. Amer. Ceram. Soc.*, 1974, **57**, 65

12 F. N. RHINES *et al.*: 'A topological study of the sintering process', Final Report AT (40-1) 2581

13 R. A. GREGG: PhD Thesis, University of Florida 69-10912, 1968

14 T. E. CLARE: *J. Amer. Ceram. Soc.*, 1966, **49**, 159

15 T. K. GUPTA: *J. Amer. Ceram. Soc.*, 1972, **55**, 276

16 S. C. SAMANTA AND R. L. COBLE: *J. Amer. Ceram. Soc.*, 1972, **55**, 583

17 F. N. RHINES *et al.*: 'Quantitative determination of the structure–property relationships in nuclear fuel element materials', University of Florida, ORO-4212-15, 1975

18 R. M. GERMAN *et al.*: *Ceramic Bull.*, 1975, **54**, 178

19 B. FRANCOIS AND W. D. KINGERY: 'Sintering and related phenomena', 499, 1967, Gordon and Breach

20 W. D. KINGERY AND B. FRANCIS: *J. Amer. Ceram. Soc.*, 1965, **48**, 546

21 K. W. LAY: 'Sintering and related phenomena', 65, 1973, Plenum

22 I. M. LIFSHITZ AND V. V. SLYOZOV: *J. Phys. Chem. Solids*, 1961, **19**, 35

23 C. WAGNER: *Z. Elektrochemie*, 1961, **65**, 243

24 M. V. SPEIGHT: *Acta Met.*, 1968, **16**, 133

25 YA. E. GEGUZIN *et al.*: *J. Phys. Chem. Solids*, 1969, **30**, 1173

26 G. C. KUCZYNSKI: 'Sintering and catalysis', 325, 1975, Plenum

Vacancy flow as a deformation mechanism in polycrystals

B. Burton

The early theory of diffusional creep is considered and the experiments which originally confirmed the predictions of the theory are briefly reviewed. These experiments were performed on thin wire or foil specimens having one grain per cross-section at temperatures near to the melting point and consequently represent rather special conditions. Diffusional creep of materials of technological importance is more complex. These materials generally have a fine polycrystalline structure with boundaries inclined at various angles to the stress axis. Consequently boundaries must carry shear as well as normal components. Grain-boundary sliding must occur to accommodate grain shape changes and preserve material coherency. These materials are also used at relatively low fractions of the melting point so that short circuiting diffusion along grain boundaries is possible and in addition they may contain impurities, solutes, precipitates or may be non-stoichiometric compounds. In this paper, these additional features necessary to extend the original theory to more practical situations are discussed. Conditions under which diffusional creep is likely to be the predominant deformation mode are indicated and also conditions under which diffusional creep can be suppressed are discussed. Two recent extensions to diffusional creep theory to include the additional contribution due to dislocations and cavities acting as vacancy sources and sinks are briefly considered.

The author is with the CEGB, Berkeley Nuclear Laboratories

ORIGINAL THEORY AND EARLY EXPERIMENTS

The vacancy chemical potential at a grain boundary across which a normal stress $\pm\sigma$ exists is $\pm\sigma\Omega$, where Ω is the atomic volume. (Tensile stresses are designated positive.) Thus, if tensile and compressive stresses are applied to perpendicular faces of a cubic grain of edge length d, then a potential gradient is set up given by:

$$\nabla\mu \simeq 2\sigma\Omega/d \tag{1}$$

The vacancy diffusional flux driven by this gradient is:

$$\phi_v = \nabla\mu (D_v C_v/kT) \tag{2}$$

where D_v is the vacancy diffusion coefficient, C_v the vacancy concentration, k is Boltzmann's constant, and T the absolute temperature. The flux of matter (equal and opposite to the vacancy flux) leads to depletion of material at compressive boundaries and deposition at tensile boundaries giving rise to a creep rate:

$$\dot{\varepsilon} = \phi_v \Omega/d \tag{3}$$

Noting that $D_v C_v = D C_{atom}$ where D is the atomic diffusion coefficient and $C_{atom}(=1/\Omega)$ is the concentration of atoms, gives the creep rate:

$$\dot{\varepsilon} = B\sigma\Omega D/d^2 kT \tag{4}$$

This equation was originally calculated independently by Nabarro[1] and Herring,[2] and in the approximate derivation given above the numerical constant $B = 2$. More exact solutions give higher values with the precise value depending upon grain geometry. For example, for a cubic grain, B takes the value 12,[3] and for a grain in a foil of thickness a and side length l, $B = 7\cdot5$.[3] In the latter case the appropriate grain size term in equation (4) is (al).

The earliest experimental observations of diffusional creep were on thin wire or foil specimens having one grain per cross-section, commonly called 'bamboo structures'. These experiments were designed to measure the surface energy of metals by the 'zero-creep' method. This involves the technique of balancing the tendency of thin specimens to shrink under the action of surface tension by applying an opposite tensile stress. At very low stresses, specimens shrink under the action of surface tension and creep positively at higher stresses. The zero creep stress (equivalent to the surface tension) is found by interpolation. The available data on bamboo structures have been reviewed by Jones.[4] In almost all respects, behaviour is typical of Nabarro–Herring creep. Creep rates are linearly proportional to the effective stress (the difference between the applied stress and that due to surface tension), they vary as the inverse product of the grain dimensions, and the activation energies for creep are close to those for lattice diffusion. The only discrepancy is the value of the proportionality constant between creep rate and stress when it is evaluated from tests performed for relatively short times. This is shown in Fig. 1 taken from Jones. The vertical axis shows the ratio of diffusion coefficients estimated from creep rates using the Nabarro–Herring equation and those measured by radio tracer methods. The horizontal axis represents a dimensionless test duration parameter $P = (Dt/al)^{\frac{1}{2}}$, where t is the test time and al the product of the grain dimensions. For values of $P > 3$ experimentally measured rates agree closely with those predicted, while those for values of $P < 3$ show enhanced rates. Jones suggests that these enhanced rates may be explained by a transient creep component due to the presence of dislocations which may act as additional vacancy sources and sinks in the early stages of testing.

These experiments on bamboo structures clearly demonstrated the success of the Nabarro–Herring equation in predicting creep rates at low stresses and also provided one of the few examples in materials science where the theoretical prediction of a phenomenon has preceded the experimental observation of it.

These first experiments were all primarily aimed at measuring surface energies of metals and, because the surface tension is so low, were performed at very high temperatures (0.95–$0.99\,T_m$) to ensure measurable rates of creep. As a consequence, it became the general impression that diffusional creep was only of importance under these extreme conditions of high temperature and very low stress. It was over a decade before its impor-

tance was realized for polycrystalline structured materials under more usual conditions.

APPLICATION TO POLYCRYSTALLINE MATERIALS

Zero creep experiments represent very idealized conditions. The experiments are mainly on single-phase pure metals and because all the boundaries are normal to the stress axis the stress distribution is relatively simple. Also temperatures are so high that contributions by short circuiting diffusion are unlikely. Diffusional creep of materials of technological interest is not so simple. These materials generally have a fine grain structure and boundaries are inclined at an angle to the stress axis. They are used at much lower temperatures and may contain impurities, solutes, precipitates or may be non-stoichiometric compounds. Nevertheless the main features of diffusional creep are exhibited by many polycrystalline materials; Figure 2 shows the linear dependence of creep rate upon stress for various materials and Fig. 3 indicates the strong dependence upon grain size. Thus it is clear that the diffusion creep equation can predict the correct parametric dependence of creep rate for polycrystals. However, the deformation characteristics differ from those of the simple bamboo structures in several important ways. These differences are now discussed.

Grain-boundary diffusion

Short circuiting diffusion along grain boundaries has a lower activation energy than lattice diffusion and thus tends to become important at lower temperatures. Most technologically important materials are used at relatively low fractions of their melting point and thus boundary diffusion can become important. Coble[15] was the first to recognize this and calculated the creep rate due to grain-boundary diffusion to be:

$$\dot{\varepsilon}_g = B'\sigma\Omega w D_g / d^3 kT \qquad (5)$$

where w is the grain-boundary width, D_g the grain-boundary diffusion coefficient, and $B' = 150/\pi$. A major

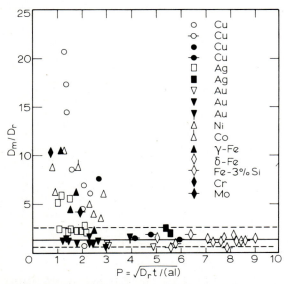

1 Ratio of the diffusion coefficient estimated from creep D_m to the radiotracer value D_r plotted as a function of a dimensionless test duration parameter $P = (D_r t/al)^{\frac{1}{2}}$, t being the test time and (al) the product of the grain dimensions; the figure indicates enhanced rates at low values of P (taken from Jones[4])

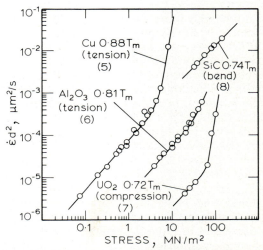

2 Variation of grain size compensated creep rate for various materials; the linear relationship between creep rate and stress at low stress levels indicates diffusional creep (the vertical scale is displaced vertically by a factor of 10 for SiC to avoid overlap)

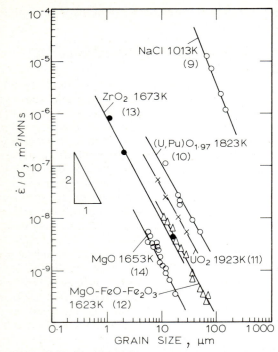

3   Variation of creep rate per unit stress with grain size for several ceramic materials

difference between this and the Nabarro–Herring equation is the grain size dependence. This difference arises because diffusion is now restricted to grain boundaries. The available area for diffusion across any plane is then reduced by the factor w/d, since the length of boundary of width w, intersected by a unit plane, is $\sim 1/d$.

Since both lattice and grain-boundary diffusion contribute independently to creep it is convenient to represent the overall creep rate by the sum of the two expressions. Thus:

$$\dot{\varepsilon} = \frac{B\sigma\Omega D}{d^2 kT}\left(1 + \frac{B'}{B}\frac{wD_g}{dD}\right) \qquad (6)$$

The stronger grain size dependence of Coble creep also makes it relatively more important in polycrystals since these tend to be finer grained than bamboo structures.

In 1965, two years after the prediction of creep by grain-boundary diffusion, Jones[16] reported the experimental verification of this creep mechanism in pure magnesium. When this material was tested at relatively high temperatures, behaviour was characteristic of lattice diffusional creep. Creep rate was proportional to stress, inversely proportional to the square of the grain size, and the activation energy was identical to that for lattice self diffusion ($Q_L = 1\cdot37 \times 10^5$ J/mol, from Shewmon and Rhines[17]). However, at temperatures $<0\cdot59T_m$, creep rates were faster than those predicted by Nabarro and Herring, the grain size dependence changed to reciprocal cube, and the activation energy for creep decreased to $\sim 0\cdot6Q_L$, indicating control by the grain-boundary diffusion mechanism of Coble. Subsequently, creep controlled by grain-boundary diffusion has been detected in other metallic systems; namely, zirconium,[18] copper,[19] silver and nickel[20] and cadmium.[21] Also a reinterpretation by Jones[22] of existing creep data on lead has shown that the stress, temperature and grain size dependence can all be described by

Coble creep. The characteristics of this type of creep are demonstrated in Figs. 4 and 5.[19,23]

Figure 4 clearly shows the transition from a region with high activation energy at high temperatures (lattice diffusion) to a regime having a lower activation energy at lower temperatures (grain-boundary diffusion). Note that the transition temperature from lattice to grain-boundary control is higher for the finer grained material owing to the different grain size dependence in the two regimes. The reciprocal cube dependence of Coble creep is indicated in Fig. 5 for copper[19] and a magnesium alloy.[23] (For the magnesium alloy the slope decreases to less than 3 at large grain size, indicating that experiments were performed in a range which extended into the region where lattice diffusion predominated.) These results showed that diffusional creep could be important at much lower temperatures than were previously supposed. Indeed, for a time it was the general opinion that diffusional creep represented the limiting creep strength of materials since attempts to prevent creep by dislocation movement such as solute additions or particle dispersions may be ineffective in retarding diffusional creep.

Geometrical aspects

The grain structure of polycrystals is more complex than bamboos and additional features have to be considered because of this difference in geometry. Boundaries in polycrystals are inclined at angles to the stress axis so that stresses at boundaries have both normal and shear components. This leads to the redistribution of stress during creep. In polycrystals it is also necessary to accommodate the shape changes of grains by relative movement of other grains in the vicinity and at large strains, grain rotation and grain switching events can take place under certain conditions.

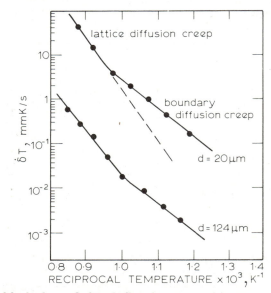

4   Variation of the deflection rate of copper helical springs with reciprocal temperature;[19] the springs had 5 active coils of 2·5 mm wire coiled to a radius of 5 mm. The surface shear stress was 0·3 MN/m². The diagram indicates a transition from Nabarro–Herring to Coble creep at lower temperatures; the transition is at a higher temperature for the finer grain size

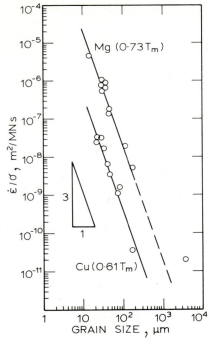

5   Variation of the creep rate per unit applied stress with grain size for copper[19] and magnesium;[23] in the grain size range 20–200 μm a reciprocal cube grain size dependence is indicated in agreement with the Coble equation

Grain-boundary sliding

Grain-boundary sliding is an inevitable consequence of the diffusional deformation of grains within a polycrystal. This is shown schematically in Fig. 6. Diagram (a) represents an idealized array of hexagonal grains and (b) shows the individual grains elongated by the diffusion process. If grains did not move relative to one another by sliding then cavities must open up. In order to maintain material coherency, sliding must occur such that a marker line across the boundary becomes offset as in diagram (c).

The difference in the role of sliding in diffusional and dislocation creep should be noted. In the former it is an inevitable consequence of the diffusional process; in the latter it need not occur (although often does), provided the material has a sufficient number of slip systems to satisfy the Von Mises criterion. If this criterion is satisfied, each grain can deform freely without gross separations appearing at grain boundaries.

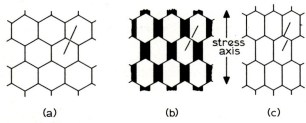

6   Schematic representation of grain-boundary sliding during diffusional creep; if sliding did not occur, cavities would open up as shown in (b); note the marker line offset

There has been much debate over the possibility of sliding being able to contribute independently to diffusional creep strain. Stevens,[24,25] Cannon,[26] Gates[27] and Aigeltinger and Gifkins[28] have argued that it can. However, as indicated by Raj and Asby,[29] Speight[30] and Beeré,[31] that since specimen strain can only develop when both sliding and diffusion operate simultaneously and in the absence of either is zero (apart from elastic accommodation) then the only logical description of the process is to attribute 100% of the creep *either* to sliding *or* to diffusion. So, in polycrystals creep may be considered to be diffusion-accommodated sliding or sliding-accommodated diffusion, and in no proper sense can they contribute independently.

Since the diffusional deformation of a polycrystal requires both sliding and grain shape changes, it is possible in principle for the sliding to control the overall deformation behaviour. The grain boundary's resistance to sliding may be defined by a viscosity η. This is the proportionality constant between the applied shear stress and the resulting shear rate $\dot\gamma$, assuming that such a linear relationship does hold. The overall creep behaviour can then be represented by the relationship:

$$\dot\varepsilon = B\sigma\Omega D/d^2 kT(1+A\eta) \qquad (7)$$

where $A = B\Omega D/d^2 kT$. The limiting cases are $A\eta \ll 1$ and $A\eta \gg 1$ giving the Nabarro–Herring equation in the former case and $\dot\varepsilon \propto \sigma/\eta$ in the latter.

In this section the magnitude of the grain-boundary viscosity is considered and it is shown that for most cases creep is controlled by diffusion across grains rather than by grain-boundary sliding.

For a macroscopically plane boundary, the obstacles to sliding are steps having a height approximating to atomic dimensions. The applied shear stress raises the chemical potential of certain atoms on one side of the boundary and lowers that of others on the other side. Atoms then diffuse across the boundary along the potential difference. This process is rapid, owing to the short diffusion distance and Ashby[32] calculates the boundary viscosity to be:

$$\eta = kT/8bD_g \qquad (8)$$

where D_g is the grain-boundary diffusivity.

In real materials grain boundaries are not plane but have for instance, macroscopic steps, wavy structures, or contain precipitates. For these boundaries the 'intrinsic' viscosity calculated above has little relevance. The simplest case of a boundary containing a double step of length L, height h, and separation λ has been treated by Ashby *et al.*[33] The boundary is shown schematically in Fig. 7a. The principle of the calculation is similar to that outlined above. The shear stress σ_s sets up normal stresses σ at the steps given by:

$$2\sigma h = \sigma_s\lambda \qquad (9)$$

This stress σ locally displaces the chemical potential of vacancies in the region of the steps by the amount $\sigma\Omega$. A chemical potential gradient is set up between ledges given by:

$$\nabla\mu = -2\sigma\Omega/L \qquad (10)$$

Diffusion, driven by this gradient, gives rise to a rate of sliding:

$$\dot\gamma = \frac{\sigma_s\Omega}{kT}\frac{w}{h^2}\frac{\lambda}{L}D_g \qquad (11)$$

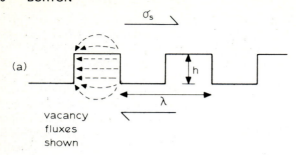

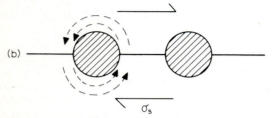

7 Grain-boundary sliding accommodated by vacancy diffusion around (a) grain-boundary steps, and (b) particles

Equation (11) predicts a sliding rate much slower than the 'intrinsic' sliding rate of equation (8). For example, taking $L \sim \lambda$ the rate is 10^6 times slower even for a relatively small step of height $h = 10^3 b$.

Diffusion from step to step can also occur by bulk diffusion. Ashby et al.[33] also include this contribution to the overall sliding rate, giving:

$$\dot{\gamma} = \frac{\lambda}{h^2} \frac{\sigma_s \Omega D}{kT} \left(1 + \frac{w}{L} \frac{D_g}{D}\right) \tag{12}$$

If the grain boundary contains an array of particles or precipitates as depicted in Fig. 7b, diffusion must occur around the particles in order for sliding to occur. Once again the calculation follows similar lines to the previous one;[29] the local stress changes the vacancy chemical potential at the particles and diffusion around the particles (by lattice and grain-boundary diffusion) enables sliding to occur. The sliding rate is given by:

$$\dot{\gamma} = 1 \cdot 6 \frac{\sigma_s \Omega}{kT} \frac{\lambda^2}{(2r)^3} D \left(1 + 2 \cdot 5 \frac{w}{r} \frac{D_g}{D}\right) \tag{13}$$

where λ is the particle spacing and r the particle radius.

Raj and Ashby[29] have also treated the general case of a non-planar two-dimensional boundary described by the cosine Fourier series:

$$x = \sum_{n=1}^{\infty} h_n \cos \frac{2\pi}{\lambda} n_y \tag{14}$$

where h_n is the Fourier coefficients describing the boundary shape. The steady-state diffusion-controlled sliding rate is calculated to be:

$$\dot{\gamma} = \frac{2}{\pi} \frac{\sigma_s \Omega}{kT} \frac{\lambda}{h^2} D \frac{1}{\sum_{n=1}^{\infty} \dfrac{h_n^2/h^2}{(1/n + \pi w/\lambda)(D_g/D)}} \tag{15}$$

where h is the total height of the boundary shape and λ the basic periodicity as defined in Fig. 7. This expression enables the sliding rate to be calculated for a variety of boundary shapes. For example, a sinusoidal boundary of wavelength λ and amplitude h is defined by the first term in the Fourier series of equation (14) i.e.:

$$x = \frac{h}{2} \cos \frac{2\pi}{\lambda} y \tag{16}$$

This gives a sliding rate:

$$\dot{\gamma} = \frac{8}{\pi} \frac{\sigma_s \Omega}{kT} \frac{\lambda}{h^2} D \left(1 + \frac{\pi w}{\lambda} \frac{D_g}{D}\right) \tag{17}$$

Other more complex shapes can also be treated but similar forms to the above equation are generally obtained.

All these calculations for the rate of grain-boundary sliding limited by diffusional flow around protruberences or particles give much higher viscosities than the 'intrinsic' grain-boundary velocity; nevertheless they give values of $A\eta \ll 1$ in equation (9) even for boundaries containing large steps or particles. Consequently the overall rate of deformation should be controlled by the diffusional deformation of the grains and not by sliding. This is intuitively obvious, since both processes occur by diffusion and the diffusion path length around particles or ledges is always much less than the grain size.

In cases of boundaries containing particles for which the particle/matrix interfacial diffusion coefficient is less than the grain-boundary diffusion coefficient, the previous statements may not be valid. It is possible to envisage a situation where boundary sliding controlled the overall rate. This may also be the case if the particle/matrix interface is not a good sink and source for vacancies. Then the sliding rate may be limited by the rate of plastic rather than diffusional accommodation at the particle. This plastic creep accommodation process is likely to have a stress dependence higher than unity and once again it would be possible for sliding to become rate limiting.

Stress redistribution during creep of polycrystals

During the diffusional creep process the applied stress is redistributed leading to the generation of internal stresses. These stresses enable the two processes, diffusion and sliding, to proceed at rates which are directly related and thus for the specimen to maintain coherency. As an illustration of the process a regular hexagonal array of grains is shown in two dimensions in Fig. 8, taken from Beeré.[31] The average stress acting on particular boundaries are shown and the subscripts s and n refer respectively to shear and normal stresses. In order to relate the stresses on the boundaries to the externally applied stress, sections can be removed from the hexagons as shown and forces can be balanced across their surfaces. The internal stress system varies with position in the grain but the average value over the dotted face of the element must equal the external stress. This is because the dotted lines cut the hexagons symmetrically and the internal stress system therefore repeats itself periodically on each element. Balancing forces in the 1 and 2 directions gives the relationships:

$$\sigma_{11} = -2\sigma_{nab}/3 - (\sigma_{nac} + \sigma_{nbc})/6 - (\sigma_{sbc} - \sigma_{sac})/2\sqrt{3} \tag{18}$$

$$\sigma_{21} = 2\sigma_{sab}/3 + (\sigma_{sbc} + \sigma_{sac})/6 + (\sigma_{nac} - \sigma_{nbc})/2\sqrt{3} \tag{19}$$

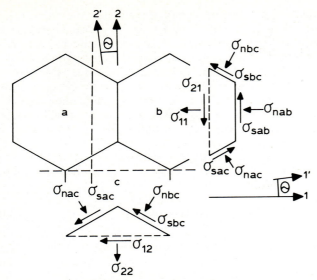

8   Stresses acting on the boundaries of a hexagonal array of grains; the subscripts s, n refer to shear and normal stresses

$$\sigma_{22} = -(\sigma_{nac} + \sigma_{nbc})/2 - (\sigma_{sac} - \sigma_{sbc})/2\sqrt{3} \qquad (20)$$

$$\sigma_{12} = (\sigma_{nac} - \sigma_{nbc})/2\sqrt{3} - (\sigma_{sac} + \sigma_{sbc})/2 \qquad (21)$$

The torque terms σ_{21} and σ_{12} must be equal and from equations (19) and (21) the sum of the boundary shear stresses is zero:

$$\sigma_{sab} + \sigma_{sbc} + \sigma_{sac} = 0 \qquad (22)$$

We require the boundary forces when a uniaxial compressive stress σ acts at an angle θ with respect to the l axis. Putting $\sigma'_{11} = -\sigma$ where the dashed axis is again rotated by an angle θ from the undashed axis we have

$$\sigma'_{12} = \sigma'_{21} = \sigma'_{22} = 0$$

and

$$\sigma_{11} = -\sigma \cos^2 \theta \qquad (23)$$

$$\sigma_{22} = -\sigma \sin^2 \theta \qquad (24)$$

$$\sigma_{12} = \sigma_{21} = -\sigma \sin \theta \cos \theta. \qquad (25)$$

Eliminating σ_{11} from equations (18) and (23) and also σ_{22} from (20) and (24) and adding shows that the sum of the normal boundary stresses is equal to three halves the applied stress.

$$\sigma_{nab} + \sigma_{nbc} + \sigma_{nac} = 3\sigma/2 \qquad (26)$$

The resulting stress redistribution on boundaries during creep depends on the relative ease of the diffusional process and of grain-boundary sliding.

In the limit of easy sliding, the boundary is unable to support shear stresses which thus tend to zero. The externally applied stress is then solely supported by normal boundary forces which can be calculated by putting $\sigma_{sab} = \sigma_{sbc} = \sigma_{sac} = 0$ in equations (18) to (21) and substituting into equations (23) to (25) giving:

$$\sigma_{nab} = \sigma(3/2 - 2 \sin^2 \theta) \qquad (27)$$

$$\sigma_{nbc} = \sigma(\sin^2 \theta + \sqrt{3} \sin \theta \cos \theta) \qquad (28)$$

$$\sigma_{nac} = \sigma(\sin^2 \theta - \sqrt{3} \sin \theta \cos \theta) \qquad (29)$$

The normal stress acting across the boundary between grains a and b is shown in Fig. 9a. The dotted line shows the stress for a completely elastic body with no plastic deformation. It should be noted that relaxing the shear stresses has doubled the amplitude of the normal stresses about their mean value of half the applied stress. Of particular importance is the fact that the normal stress can be of *opposite* sign to the applied stress. For example in Fig. 8, if the stress along the 2 axes is uniaxial compression, the stress across the ab boundary is tensile and equal to half the applied stress. This can be seen to arise from the wedging action of grain c. The development of internal tensile stresses in a component under compression can be particularly important since it can lead to formation of creep fracture cavities on these tensile boundaries.[34]

In this limit of easy sliding the creep rate depends upon the rate at which material plates out at boundaries. This plating rate depends only upon the normal stresses, equations (27–29), and simple manipulation reveals a creep rate identical to the Nabarro–Herring equation (*see* Beeré[31]). The trigonometrical terms in θ cancel and the creep rate is thus independent of orientation.

The opposite limiting case is when grain-boundary sliding becomes much slower than the diffusion process. In this case the compressive stress on all the boundaries becomes the same, i.e. $\sigma/2$ (from equation (26)). Substituting this value into equations (18)–(25) gives the shear stresses in terms of the applied stress and orientation:

$$\sigma_{sab} = -\sigma(2 \sin \theta \cos \theta) \qquad (30)$$

$$\sigma_{sbc} = \sigma(\sqrt{3} \cos^2 \theta + \sin \theta \cos \theta - \sqrt{3}/2) \qquad (31)$$

$$\sigma_{sac} = \sigma(-\sqrt{3} \cos^2 \theta + \sin \theta \cos \theta + \sqrt{3}/2) \qquad (32)$$

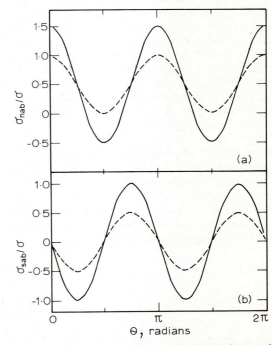

9   (a) Normal stress acting on a boundary when shear forces have been relaxed (easy sliding); (b) shear stress acting on a boundary if Nabarro–Herring creep is rapid compared to sliding; the dashed lines represent the stresses acting for a purely elastic body (after ref. 31)

The variation of the average shear stress between grains a and b is shown in Fig. 9b. The dotted line shows the boundary shear stress for a rigid elastic body. Relaxing the normal stresses has doubled the boundary shear stresses. However, as pointed out previously, sliding is usually rapid and this latter case is not usually relevant.

Grain rotation and switching events

When polycrystals are subjected to large diffusional strains, the grains remain equiaxed. This rather surprising observation was originally made by Hensler and Cullen[35] who observed that grains in MgO remained equiaxed during creep up to ~30% strain. Also it has been shown in a particularly striking way during superplastic creep by Holtz et al.[36] In situ observations of deformation of Pb/Sn were made in the scanning electron microscope. Grain-switching and grain-rolling events were observed and grains which were originally neighbours often became widely separated. Ashby and Verrall[37] have also shown that this type of switching event is a main feature in the flow of a simulated polycrystal in the form of an oil emulsion. This emulsion consisted of 'grains' of oil separated by thin 'grain boundaries' of detergent solution and its deformation was studied by forcing it to flow through a channel. The unit step of the deformation process is shown in Fig. 10. This represents an idealized hexagonal array of identical grains. In reality grains are not all the same size and are not hexagons. The rotation of the grains which in real materials occurs as well as the translation event depicted is related to the probability of one grain being surrounded by others of particular sizes and in particular angular positions.[37]

Ashby and Verrall note two interesting features of this grain-switching type of deformation. The first is that the diffusional accommodation requires a much shorter path length than envisaged in the Nabarro–Herring model since matter has to be transported only in the region of triple points rather than across entire grains. Ashby and Verrall calculate geometrically that the amount of matter transported per unit strain is $\sim 0.1d^3$ in grain-switching creep, compared to d^3 in Nabarro–Herring creep. Thus the rate is increased by ~10 times. The parametric dependence of creep rate is unchanged.

Effective diffusion coefficient for creep

Most materials of technological importance contain more than one atomic species and the choice of the diffusion coefficient to use in the creep equation becomes more complex. In general the individual species have different mobilities so that immediately after the stress is applied, the diffusion fluxes differ. This leads to a preferential flux of the fastest moving species to boundaries under tension. If this process continued, appreciable segregation would occur and in materials where the species have an associated charge (ionic solids) then an electric field can be set up. However, a steady state is quickly reached since the concentration gradient and electric field both oppose further segregation. The appropriate diffusion coefficients will now be discussed for the two examples, binary alloys and a two-component ionic solid.

In his original paper, Herring indicated that the appropriate diffusion coefficient for multi-component alloys is:

$$D^H = \left[\sum_{ij} (D^{-1})^H_{ji} X_i X_j \right]^{-1} \tag{33}$$

where X_i is the atom fraction of component i and $(D^{-1})^H_{ji}$ is the reciprocal of the diffusion coefficient matrix in the multi-component flux equations:

$$J_i = -\frac{N}{kT} \sum_i D^H_{ij} \nabla(\mu_i - \mu_v) \tag{34}$$

where N is the number of unit cells per unit volume, μ_i is the chemical potential of component i and μ_v is the vacancy chemical potential. For a binary system in the case where the D^H_{ij} matrix is symmetrical and when cross coefficients are not negligible the equation reduces to:

$$D^H_c = \frac{D^H_{11}D^H_{22} - D^H_{12}D^H_{21}}{D^H_{22}X_1^2 - 2D^H_{12}X_1X_2 + D^H_{11}X_2^2} \tag{35}$$

and if the D^H_{ij} matrix is diagonal the expression further reduces to:

$$D^H_c = \frac{D^H_{11}D^H_{22}}{D^H_{22}X_1^2 + D^H_{11}X_2^2} \tag{36}$$

If cross coefficients are zero and when the solution is ideal, the direct coefficient D^H_{ii} is related to the intrinsic coefficient D_{ii} by the expression $D^H_{ii} = X_i D_{ii}$ and equation (36) can be written:

$$D^W_c = \frac{D_1 D_2}{X_1 D_2 + X_2 D_1} \tag{37}$$

(see Weertman[38] for example).

There are only a few reported experiments on binary alloys. The earliest studies were on the dilute alloys Fe–Si,[39] Fe–Ni and Fe–P[40] in the form of bamboo structures. In these alloys $X_2 \ll X_1$, $D^H_c \simeq D^H_{11}$ and D^H_{11} was identified with the relevant tracer diffusivity D^*_{Fe}. Good agreement was found between theory and experiment. A further study on Zr–1% Sn[18] demonstrated that the addition of tin caused a significant decrease in creep rate and an increase in creep activation energy, although comparisons with tracer diffusion data were not possible in this material, which under the relevant test conditions was shown to deform by Coble creep.

Results on concentrated alloys having polycrystalline structure are scarce and even when they do exist[41–43] are not often in a form which enables them to be compared with theoretical predictions. In other cases[44] the diffusion data are not available. However, the creep of two binary copper alloys Cu–Zn and Cu–Ni have been studied[44,45] and appropriate diffusion data are available.

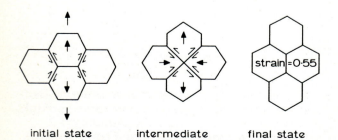

10   Unit step of the deformation process showing grain switching by grain-boundary sliding and diffusional accommodation [37]

initial state   intermediate   final state

strain = 0.55

The comparison between the diffusion coefficient estimated from creep rates using the Nabarro–Herring equation is compared with those calculated from equation (36) using published diffusion data (*see* Fig. 11). The influence of solute additions on diffusional creep is clearly shown. Zn, which increases atomic mobility, also increases creep rate; Ni, which decreases mobilities, has the opposite effect. However, the precise shapes of the curves are dissimilar. It is to be particularly noted that agreement between experiment and theory is not precise even for the pure metals Cu and Ni. In both cases experimental rates exceed those predicted.

In the case of ceramic materials it is necessary to account for the fact that a molecular group must be transferred by the diffusion process. Ruoff[46] has considered the case of an ionic compound $A_\alpha B_\beta$ where the diffusion occurs via A-vacancies and B-vacancies through the lattice. The defects move individually, but because they are charged, their motion must be restricted by the requirements of charge neutrality in a representative volume, of zero net current flow and at steady state, α A-vacancies must be transferred for all β B-vacancies transferred.

In a stress-free crystal the flux of the ith type point defect is given by:

$$J_i = B_i C_i (\nabla \mu_i + q_i \nabla \phi) \qquad (38)$$

where B_i $(= D_i/kT)$ is the mobility, C_i the concentration, $\nabla \mu_i$ the chemical potential gradient, q_i the charge of the ith species and $\nabla \phi$ the electric potential gradient.

Charge neutrality requires that:

$$\beta C_a = \alpha C_b \qquad (39)$$

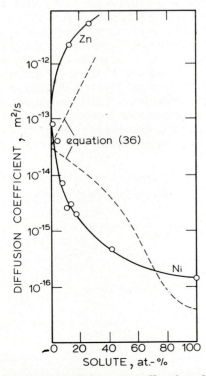

11   Comparison between the effective diffusion coefficient for diffusional creep of two copper alloys with the diffusion coefficient calculated from equation (36) (taken from ref. 39)

Noting also that:

$$\alpha q_a = -\beta q_b \qquad (40)$$

and using the relationship:

$$\mu_i = \mu_i^\circ + kT \ln(C_i/C_i^\circ) \qquad (41)$$

enables $\nabla \phi$ to be eliminated from equation (38) when $i = a,b$ and the flux of species 'a' can then be calculated to be:

$$J_a = -D_a \nabla C_a [D_b(\alpha + \beta)/(\beta D_a + \alpha D_b)] \qquad (42)$$

For the case where $\alpha = \beta = 1$ this reduces to:

$$J_a = -[2D_a D_b/(D_a + D_b)]\nabla C_a \qquad (43)$$

and if $D_b \gg D_a$ this further reduces to:

$$J_a = 2D_a \nabla C_a \qquad (44)$$

Thus the overall diffusion process is in this case controlled by the slower-moving species.

When a stress is applied to the material, the local concentrations of A- and B-type vacancies change. Thus for a tensile stress:

$$C_a = C_{ao} \exp[\sigma \Omega_M/(\alpha + \beta)kT] \qquad (45)$$

where Ω_M is the molecular volume. The creep rate can be arrived at by following similar reasoning to that used in developing equation (4) and can be shown to be:

$$\dot{\varepsilon} = \frac{B\sigma \Omega_M D_a^T}{(\alpha f_a)d^2 kT} \qquad (46)$$

where D_a^T is the tracer diffusion coefficient of species 'a' and $f_a \sim 1$ is the correlation factor for tracer motion of the A-atoms. It should be noted that equation (46) differs from the original creep expression in that the molecular volume is used and the factor (αf_a) also appears. The equations become parametrically identical, however, if the molecular diffusion coefficient, D_M, is used given by:

$$D_M = D_a^T/f_a \alpha \qquad (47)$$

Diffusional creep has been studied extensively in ceramics and some examples showing the parametric dependence of creep rate upon stress and grain size have already been shown in Figs. 2 and 3. A common feature of these studies is that although reproducibility of results on any particular batch of material is usually good, there is often large batch to batch variations. This arises because impurities in materials having a degree of ionic bonding have considerably more influence upon diffusivities than in metallic systems. For example in many oxide ceramics, the defect formation energy is typically ~ 6 eV, such that the intrinsic defect concentration at high temperatures is of the order of 10^{-8} vacancies/lattice site, several orders of magnitude lower than impurity levels in even the purest samples.[48] In contrast, for metals the equilibrium defect concentration is of the order 10^{-4} vacancies/lattice site, some hundreds of times greater than impurity levels in the purest samples. Consequently in impure or doped ceramic materials, the vacancy concentration, C_v, may be controlled by composition and the effective diffusion coefficient for creep may then also depend strongly upon this variable. Consequently it is often difficult to compare experimental results with theoretical predictions.

Nevertheless, the general trend is for experimental rates to exceed those predicted using the diffusion coefficient of the slowest-moving species. A typical example is shown in Fig. 12 for UO_2, a material which has been extensively studied because of its importance as a reactor fuel. In this material, uranium is the slower-moving species. There is wide scatter in tracer diffusion measurements (in this material measurements are further complicated since UO_2 can exist in a range of non-stoichiometric compositions); however, the diffusion coefficients estimated from creep are substantially higher than those of uranium ions.

Other oxide systems exhibit similar behaviour. Magnesia (MgO) creeps about two orders of magnitude faster than predicted;[61] in this case oxygen is the least mobile species and hematite (Fe_2O_3) creeps between one and two orders faster.[62] For hematite, iron and oxygen have roughly similar mobilities. Alumina (Al_2O_3) also creeps faster than predicted using the oxygen ion diffusion coefficient, and rates coincide with those predicted using the (faster) diffusion coefficient of aluminium. This has led Coble et al.[63,64] to suggest that this enhancement may occur because oxygen diffusion could be enhanced by the presence of grain boundaries, such that creep is controlled by the faster-moving aluminium ions. Enhanced diffusion rates of oxygen ions

in polycrystals over those in single crystals have in fact been observed.[65] However, since many ceramic materials show enhanced rates whichever ion has the lower mobility, this specific explanation for Al_2O_3 may not be generally relevant.

CONDITIONS UNDER WHICH DIFFUSION CREEP PREDOMINATES

The rate of diffusional creep depends linearly upon stress whereas creep by dislocation movement is much more strongly stress dependent. These two mechanisms are independent and the faster one controls the overall rate. Thus diffusional creep predominates at lower stress levels. As an illustration of the relative importance of diffusional creep over other creep processes, a deformation map reproduced from Ashby[66] is shown in Fig. 13. This type of diagram, shown here for silver, represents fields of predominance of particular creep mechanisms in stress/temperature space. The diagram clearly shows the important range of predominance of diffusional creep, especially that which occurs by grain-boundary diffusion. The elastic regime shown, that is where creep rate is too small to be measurable under normal conditions, is set in this diagram at 10^{-8} per second. It is to be noted that lowering this arbitrary limit serves to extend the range of dominance of diffusional creep, especially that controlled by grain-boundary diffusion.

The deformation map shown is for material of constant grain size. Because diffusional creep depends strongly upon grain size and dislocation creep is relatively insensitive, it follows that diffusional creep is predominant up to higher stress levels for fine-grained material. Thus the diffusional creep field in the deformation map is correspondingly extended. It also follows that any factor which slows down dislocation creep will similarly extend the diffusional creep field. In metals, solute additions affect factors such as diffusivity, shear modulus, stacking fault energy and may also slow down dislocation movement by solute hardening or by segregation to dislocations. Changes in diffusivity affect both diffusional and dislocation creep by about the same extent and thus do not appreciably change the stress range of predominance of either mechanism. However, since diffusional creep is not sensitive to any of the other factors, it can predominate over a wider range if solutes impede dislocation movement.

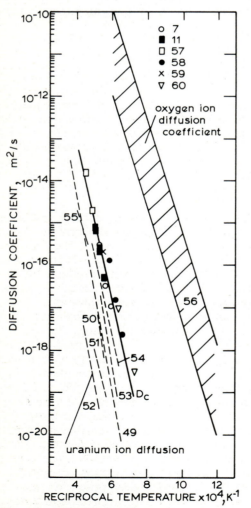

12 **Effective diffusion coefficient for creep of UO_2 compared with the tracer diffusivities of uranium and oxygen ions**

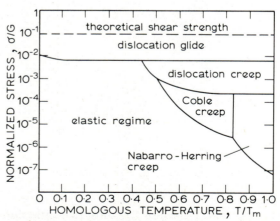

13 **A deformation mechanism map for pure silver calculated for a grain size of 32 μm and a strain rate of 10^{-8}/s (taken from Ashby[66])**

In ionic and covalently bonded materials, dislocation movement is often more difficult than in metallic systems owing to their higher elastic moduli, the geometrical complexity of moving dislocations through a structure containing more than one species and the possible influence of electrostatic charge effects. Furthermore, there may not be sufficient independent slip systems to allow coherent deformation of polycrystalline aggregates. Consequently, diffusional creep is correspondingly more important as a deformation mode in ceramics than in metals. Ceramic materials are often prepared by the sintering of powders and the residual porosity remaining from this manufacturing process tends to pin grain boundaries. Thus a fine stable grain size can be maintained and this factor also increases the importance of diffusional creep in these materials. Figure 2 shows clearly the extended stress range of predominance of diffusional creep of ceramics compared to metals. The behaviour of SiC is to be particularly noted. This material shows no evidence of a stress index greater than unity even at stresses of ~ 200 MN/m^2.

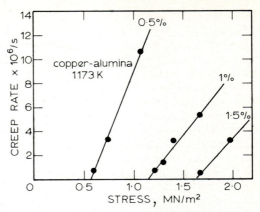

14 Variation of creep rate with stress for Cu–Al$_2$O$_3$;[70] a threshold stress for diffusional creep is revealed which varies linearly with volume fraction of second phase

CONDITIONS UNDER WHICH DIFFUSIONAL CREEP IS SUPPRESSED

Early theories of diffusional creep considered grain boundaries to be perfect sinks and sources for vacancies. If they are, then all of the applied stress is available to drive the diffusive flux and creep occurs at a rate which varies linearly with stress. If, however, boundaries are not perfect sources and sinks, then some of the applied stress must drive the interfacial reaction responsible for vacancy creation and annihilation and reduced creep rates will be obtained. Some of the earliest evidence that diffusional creep could be inhibited was obtained on a sintered magnesium alloy.[67,68] In this alloy, diffusional creep was totally suppressed compared to material produced in the conventional manner and this difference has been attributed to the presence of MgO particles on grain boundaries which are present in the sintered material.[69] In alloys containing lesser quantities of second phase, diffusional creep rates are suppressed at low stresses, but may still provide the greatest contribution to strain. The way in which this suppression occurs is in two forms. First, creep rates continually diminish as particles build up on boundaries parallel to the tensile stress axis. This build-up occurs since particles act only as inert markers during diffusional creep and consequently collect on boundaries where vacancies are being absorbed, while particle-free zones form on emitter boundaries.

The second feature of the diffusional creep of many particle-containing materials is that a threshold stress exists below which diffusional creep does not contribute to the strain. As an illustration, the creep behaviour of some sintered Cu–Al$_2$O$_3$ alloys is shown in Fig. 14 taken from Burton.[70] To avoid problems due to the concurrent decrease in creep rate with strain, initial creep rates are plotted in this figure. The relationship, $\dot{\varepsilon} \propto \sigma - \sigma_0$ is clearly revealed, with the threshold stress σ_0 varying approximately linearly with volume fraction of second phase. Similar behaviour has been observed in Au–Al$_2$O$_3$ alloys.[71] This variation of σ_0 with volume fraction clearly explains the observed decrease in creep rate with strain, since the effective volume fraction on emitter boundaries increases with strain.

Several theoretical attempts have been made to explain the threshold stress in materials containing particles. Particles inside grains are evidently unimportant at small volume fractions and the theories consider the role of those in grain boundaries. Ashby[72] considers grain-boundary line defects having edge character to be responsible for the emission and absorption of vacancies. These defects are envisaged to move with a climb-like motion along the boundary in a sense which depends on whether they are emitting or absorbing vacancies.

In order to bypass an array of particles of spacing λ, the defect is forced to adopt a local curvature $\sim 2/\lambda$ such that a critical stress $2E/b\lambda$ must be exceeded. Noting that the interparticle spacing $\lambda \sim 1/\sqrt{n_A}$ and the volume fraction $V \sim r^2 n_A$, where n_A is the number of particles intersecting unit area of boundary, gives the critical or threshold stress due to this model as $\sigma_0 \sim (2E/br)\sqrt{V}$. This is not in accord with the experimental observations of the linear dependence between σ_0 and V.

Other models have considered the situation where the particle/matrix interface is not a good vacancy source and sink while the particle-free regions maintain this ability. Harris et al.[69] suggest metals containing high melting point particles as an example. In this case the source sink ability might be impaired at the particle/matrix interface owing to low atomic mobility in an interface across which exists a phase at a low fraction of its melting point. Vacancy condensation (taking a boundary under compression as an example) will cause a progressive increase in local stress at the particle. If the particle interface remains inactive as a vacancy source and sink, creep can only continue if the stress concentration at the particle is relieved by some irreversible process. Harris[73] suggests the punching of dislocation loops as a means of relaxing the stress. If there are n_A particles per unit area each of radius r then for punching to be energetically favourable it is required that the work done by the stress on condensing a single layer of vacancies on unit area of boundary (σb) exceeds the energy of the dislocation loops ($2\pi r n_A E$). The line energy $E \sim 0.5Gb^2$. Thus a threshold stress $\sigma_0 \sim 2\pi r n_A E/b$ is predicted, or in terms of the volume fraction $\sigma_0 \sim (2E/br)V$. An alternative way in which the stress can be relaxed is by the nucleation of interfacial defect loops in the particle/matrix interface.[74] From nucleation theory, the critical loop size is $E/\sigma_p b$ where

σ_p is the stress at the particle and the activation energy ΔG for nucleating loops of this size is $\pi E^2/\sigma_p b$. The nucleation rate is proportional to $\exp -\Delta G/kT$, which is a very strong function of σ_p. It can be shown that the rate changes from a very low to a very high value over a narrow range of stress and behaviour approximates to threshold behaviour with $\sigma_0 \sim (\pi E^2/55 b k T) V$.

These latter two models give the correct parametric dependence of σ_0 upon V; however, one feature which is not explained is the strong temperature dependence of σ_0 which is observed experimentally.[70]

While the inhibition of diffusional creep by the presence of particles may to some extent be anticipated, one surprising feature has been the observation of suppressed rates even in single-phase materials. The interpretation of these observations falls into two main categories. Hay *et al.*[75] interpret data on Fe_2O_3 in terms of diffusional creep with some interdependent process operating sequentially. This process is not specified, but is considered to have a stress dependence higher than unity such that it predominates at low stress levels. This has the effect of causing the stress index ($\mathrm{d}\log\dot\epsilon/\mathrm{d}\log\sigma$) to increase from unity at low stresses to a value which characterizes the sequential process. The other interpretation of suppressed rates in Mg, Ag, Ni, Cd, Cu,[20] UO_2,[58] and UC[76] has been in terms of a threshold stress σ_0, such that $\dot\epsilon\alpha(\sigma-\sigma_0)$. It should be noted that it is often difficult to distinguish experimentally the difference between a power law process and a threshold process because of the similarity of the curves $\dot\epsilon\alpha\sigma^{n>1}$ and $\dot\epsilon\alpha(\sigma-\sigma_0)$ at low values of σ. Both give rise to values of ($\mathrm{d}\log\dot\epsilon/\mathrm{d}\log\sigma$) greater than unity at low stress levels.

The data available on metals which exhibit threshold behaviour have been reviewed by Crossland[20] and a systematic dependence upon temperature, shear modulus, and grain size is revealed such that:

$$\sigma_0 \propto \frac{Gb}{d}\exp\left(\frac{U}{kT}\right) \tag{48}$$

This empirical expression also fits the data on UO_2 and UC. In Figs. 15 and 16, the variation of σ_0 with temperature and grain size is demonstrated.

The work of Hondros[78] is not easily included in either of the two previous categories. These experiments were performed on bamboo structures of γ-iron. In a series of alloys containing small amounts of Mn, Ni, Si, and Cr, little difference was detected from the behaviour of the pure material except for slightly enhanced rates in the Fe–Cr and Fe–Si alloys. These could, however, be attributed to solute-enhanced diffusion. However, in Fe–Sn and the system Cu–O, considerable retardation of creep was observed. Hondros notes that the alloying constituents in these latter two systems, unlike the former, have a very high interfacial activity and are thus likely to segregate at boundaries. The retardation was attributed to the inhibition of vacancy source and sink activity by the presence of these segrated elements. The behaviour differs to all other reports of inhibited diffusional creep behaviour, however, in that the linear stress dependence was maintained. They, however, represent the only data on inhibition obtained at a temperature near to the melting point.

At present there has been only one theoretical attempt to explain the threshold stress in single phase materials. Ashby and Verrall[37] note that during grain-

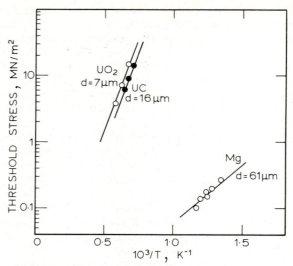

15   Variation of the threshold stress for diffusional creep for various materials; the threshold stress increases with decreasing temperature and grain size; ceramics show higher σ_0 values than metals

switching events as depicted in Fig. 10, the total boundary area fluctuates. The area increases as the cluster of four grains moves from the initial to the intermediate state. This increase in area can be calculated to be $0\cdot26d^2$ and if the boundary energy per unit area is γ, then the result is that a threshold stress is necessary to enable deformation to continue. Ashby and Verrall estimate this threshold stress to be $\sim 0\cdot7\gamma/d$. The value of the threshold stress is, however, smaller than the experimentally observed values, neither can it predict an appreciable temperature dependence.

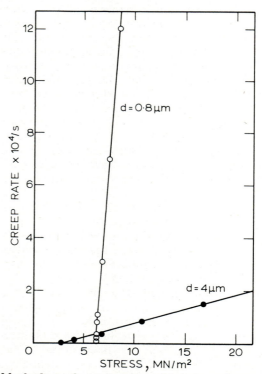

16   Variation of creep rate with stress for two hydrided Mg–Zr alloys;[77] the threshold stress varies with (1/d) where d is the grain size

OTHER DIFFUSIONAL CREEP CONTRIBUTIONS

In addition to grain boundaries, surfaces and dislocations can also act as vacancy sources and sinks. Consequently, in materials containing dislocations or cavities, an additional diffusional creep contribution can arise if these defects are situated such that vacancy diffusion between them allows the applied stress to perform work.

Nabarro[79] has presented a detailed model of the diffusional creep contribution which can arise if vacancies diffuse from source dislocation links to sink links. The model considers strain to arise by the climb of links in the dislocation network by a Bardeen–Herring[80] mechanism with the resulting increase in dislocation density being balanced by the mutual elastic attraction and annihilation of dislocations of opposite sign. The creep rate is predicted to be:

$$\dot{\epsilon} = \frac{Db\sigma^3}{\pi k T G^2} \Big/ \ln\left(\frac{4G}{\pi\sigma}\right) \tag{49}$$

The logarithmic factor changes only slowly with stress (it is typically ~ 10 for metals) and thus the creep rate varies essentially as σ^3.

For creep by dislocation climb to exceed ordinary diffusional creep, the material must have a large grain size. The creep rate predicted by equation (49) exceeds the Nabarro–Herring creep rate only above a certain stress σ_t which can be calculated by equating the two rates to be:

$$\sigma_t \sim 10 Gb/d \tag{50}$$

Thus even for a relatively coarse-grained material having $d = 100 \ \mu$m, climb creep will only predominate at stresses greater than $\sim 1 \ MN/m^2$. It should be noted that in metallic systems the transition from diffusional creep to usual recovery creep processes often occurs around this stress level and climb creep may often be masked by these processes. For example, it never appears as the predominating mechanism in the deformation map of silver in Fig. 13. It may, however, be important at lower temperatures when dislocation core diffusion can predominate over lattice diffusion. Nabarro[79] has calculated this contribution to be:

$$\dot{\epsilon}_c = 4 D_c b\sigma^5 / \pi^4 k T G^4 \tag{51}$$

where D_c is the dislocation core diffusivity. The calculation follows similar lines to that shown above. The creation and annihilation fluxes, this time occurring by core diffusion, are calculated. The rates of creation and annihilation are then equated to give the steady-state spacing. The creep rate can then be calculated from the density and speed of the dislocation array. The higher stress dependence arises because diffusion is now restricted to dislocation cores, such that over any plane the area across which matter can diffuse is reduced by the factor $\sim (b/\lambda)^2$ and noting that $\lambda \propto (\sigma)^{-1}$ gives the five power law.

While climb creep is generally unimportant in many metallic systems it may well be much more important as a deformation mode in coarse-grained covalent or ionically bonded materials where creep by dislocation glide may be difficult. Nabarro has suggested its possible importance in single-crystal turbine blades and in planetary matter.

Although steady-state deformation by dislocation climb is not generally of importance, transient creep by climb can be. This can arise if the dislocation network is strongly pinned. Then, since the angular distribution of links is generally such that some can act as vacancy sources and others as sinks, creep occurs as dislocations bow out between their pinning points. At stresses below that required for the operation of a Bardeen–Herring source, the creep component saturates out as dislocations approach the equilibrium radius of curvature given by:

$$r = Eb^2/kT \ln(C_v/C_0) = E/b\sigma \tag{52}$$

(C_v/C_0) is the equilibrium vacancy supersaturation at the dislocation line. It is also implicit that this transient component should be reversible on stress removal as dislocations return to their new equilibrium configuration. This reversible creep component has been shown to decay exponentially with time[81] such that:

$$\epsilon = \epsilon_m \left\{ 1 - \exp - \frac{\pi^2 GDb^3 t}{2\lambda^2 kT} \right\} \tag{53}$$

where λ is the link spacing and ϵ_m is the maximum transient strain which has a value approximately equal to (σ/G). This transient and reversible diffusional creep has been detected experimentally in UO_2.[82] ϵ_m has been shown to vary linearly with stress and the strain decays exponentially with a rate constant which has an activation energy close to that for lattice diffusion of uranium ions.

The other source of diffusional strain, predicted by Harris et al.,[83] involves vacancy diffusion between grain-boundary cavities and the grain boundary. The vacancy potential at a cavity is $2\gamma\Omega/r$, where γ is the surface energy and r the cavity radius and in the absence of applied stress the potential at the boundary is zero. Consequently the cavity can sinter by the diffusion which is driven by this potential gradient. If, however, a tensile stress σ is applied across the boundary, the potential at the boundary is $\sigma\Omega$ and the potential gradient driving diffusion is now $\sim [\sigma - (2\gamma/r)]\Omega/\lambda$ where λ is the inter-cavity spacing. Thus sintering can now only occur if $(2\gamma/r) > \sigma$. Under the opposite conditions cavities will open by diffusion at a rate calculated by Hull and Rimmer[84] to be:

$$\frac{dr}{dt} = \frac{2\pi\sigma\Omega\omega}{\lambda r k T} D_g \tag{54}$$

where ω is the grain-boundary width and D_g the grain-boundary self-diffusion coefficient. It should be noted that the growth of cavities involves the plating out of material from the growing cavity onto adjacent regions of grain boundary and thus gives rise to a tensile creep strain. Harris et al.[83] have suggested the possible technological importance in creep-resisting alloys, since these often fail by grain-boundary cavity growth. They relate the incremental change in cavity volume to changes in strain by the equation:

$$4\pi r^2 \, dr = \lambda^2 dg \tag{55}$$

where dg is the incremental increase in the thickness of the layer of atoms plated out on the adjacent grain boundary. To calculate the contribution to creep strain by cavity growth ϵ_g, equations (54) and (55) are integrated and combined to give:

$$\epsilon_g = \frac{4\pi}{3\lambda^2 d}\left(\frac{4\pi\sigma\Omega\omega}{\lambda kT}D_g t\right)^{\frac{3}{2}} \qquad (56)$$

where d is the grain size. (It should be noted that the $(2\lambda/r)$ term has been omitted since in most practical cases it is much smaller than σ.)

If the material is in compression, the stress-enhanced sintering of cavities can similarly give rise to a diffusional creep contribution. This is calculated as before, but the boundary conditions $r = r_0$ at $t = 0$ are used in the integration since the material is assumed to have pre-existing cavities of radius r_0. This contribution by sintering is:

$$\epsilon_s = \frac{4\pi}{3\lambda^2 d}\left\{r_0^3 - \left(r_0^2 - \frac{4\pi\sigma\Omega\omega}{\lambda kT}D_g t\right)^{\frac{3}{2}}\right\} \qquad (57)$$

Both these treatments are at best only illustrative. In real materials, cavities have a range of sizes so that different sized cavities sinter to closure at different times in compression[59] and they are continually nucleated under tensile stresses.[85] Nevertheless this contribution to creep strain is likely to be significant in certain cases and has been shown to be an important source of compressive creep strain in the ceramic material uranium dioxide.[59] This material contains sintering cavities remaining from the manufacturing process.

CONCLUSIONS

The original theory of diffusional creep predicts the creep rate of bamboo-structured materials very accurately. These experiments, however, represent very idealized conditions. The grain structure is very simple, ensuring that all the grain boundaries are normal to the stress axis and temperatures used in the experiments are so high that contributions by short-circuiting diffusion are not likely.

In the diffusional creep of materials of technological importance, several additional important features have to be considered. The geometrical aspects of the creep process in polycrystals is more complex. Boundaries are not generally normal to the stress axis and must therefore carry both shear and normal contributions. Grain-boundary sliding must also occur to preserve specimen coherency during creep, and at large strains grain switching events can occur. Account must be taken of contributions by grain-boundary diffusion which may give the largest creep contribution in many cases.

The effective diffusion coefficient to use in the creep equation may be a complex coefficient in multi-component systems and small impurity additions, although relatively unimportant in metallic systems, may substantially change the behaviour of ionic solids.

One important feature, not considered in the early theories, is the fact that grain boundaries do not always act as perfect sources and sinks for vacancies. While the theoretical understanding of this feature is at an early stage, enough empirical data are usually available to indicate the necessary conditions for the inhibition of creep.

Two additional diffusional creep contributions due to the presence of dislocations or cavities may sometimes be important although at present only a small amount of experimental evidence is available.

ACKNOWLEDGMENT
This paper is published by permission of the Central Electricity Generating Board.

REFERENCES
1 F. R. N. NABARRO: 'Strength of solids', 75, 1948, London, The Physical Society
2 C. HERRING: J. Appl. Phys., 1950, 21, 437
3 G. B. GIBBS: Mém. Sci. Rev. Mét., 1965, 62, 781
4 H. JONES: Met. Sci. Eng., 1969, 4, 106
5 B. BURTON AND G. W. GREENWOOD: Acta Met., 1970, 18, 1237
6 C. K. L. DAVIES AND S. K. SINHA RAY: Special Ceramics, 1972, 5, 193
7 L. E. POTEAT AND C. S. YUST: 'Ceramic microstructures', (Eds. R. M. FULRATH AND J. A. PASK), 646, 1968, New York, John Wiley and Sons
8 P. L. FARNSWORTH AND R. L. COBLE: J. Am. Ceram. Soc., 1966, 49, 264
9 W. D. KINGERY AND E. D. MONTRONE: J. Appl. Phys., 1965, 36, 2412
10 J. L. ROUTBORT et al.: J. Nucl. Mat., 1972, 44, 247
11 P. E. BOHABOY et al.: 1969, GEAP 100 54
12 R. T. TREMPER et al.: J. Am. Ceram. Soc., 1974, 57, 421
13 P. E. EVANS: ibid., 1970, 53, 365
14 E. M. PASSMORE et al.: ibid., 1966, 49, 594
15 R. L. COBLE: J. Appl. Phys., 1963, 34, 1679
16 R. B. JONES: Nature, 1963, 207, 70
17 P. G. SHEWMON AND F. N. RHINES: J. Met., 1954, 6, 1021
18 I. M. BERNSTEIN: Trans. AIME, 1967, 239, 1518
19 B. BURTON AND G. W. GREENWOOD: Met. Sci. J., 1970, 4, 215
20 I. G. CROSSLAND: 'Physical metallurgy of reactor fuel elements', 66, 1975, London, The Metals Society
21 I. G. CROSSLAND: Physica Status Solidi, 1974, 23, 231
22 R. B. JONES: paper presented at NPL Conference, Trends in Diffusion, 1970
23 R. B. JONES: 'Quantitative relationship between properties and microstructure', (Eds. D. G. BRANDON AND A. ROSEN), 343, 1969, Haifa
24 R. N. STEVENS: Phil. Mag., 1971, 23, 265
25 R. N. STEVENS: Surface Sci., 1972, 31, 543
26 W. R. CANNON: Phil. Mag., 1972, 25, 1489
27 R. S. GATES: ibid., 1975, 31, 367
28 A. E. AIGELTINGER AND R. C. GIFKINS: J. Mat. Sci., 1975, 10, 1889
29 R. RAJ AND M. F. ASHBY: Metall. Trans., 1971, 2, 1113
30 M. V. SPEIGHT: Acta Met., 1975, 23, 779
31 W. B. BEERÉ: Met. Sci., 1976, 10, 133
32 M. F. ASHBY: Surface Sci., 1972, 31 498
33 M. F. ASHBY et al.: Scripta Met., 1970, 4, 737
34 G. L. REYNOLDS et al.: Acta Met., 1975, 23, 573
35 J. H. HENSLER AND G. V. CULLEN: J. Am. Ceram. Soc., 1967, 50, 584
36 W. HOTZ et al.: J. Mat. Sci., 1975, 10, 2003
37 M. F. ASHBY AND R. A. VERRALL: Acta Met., 1973, 21, 148
38 J. WEERTMAN: Trans. ASM, 1968, 61, 681
39 H. JONES AND G. M. LEAK: ibid., 1966, 14, 21
40 E. D. HONDROS: Physica Status Solidi., 1967, 21, 375
41 B. YA. PINES AND A. F. SIRENKO; Fiz. Met. Metallored., 1959, 7,(5), 766
42 B. YA. PINES et al.: ibid., 1967, 23,(1), 179
43 B. YA. PINES AND V. P. KHIZHKOVYY: ibid., 1960, 22,(1), 82
44 B. BURTON AND B. D. BASTOW: Acta Met., 1973, 21, 13
45 B. BURTON AND G. W. GREENWOOD: ibid., 1970, 18, 1237
46 A. L. RUOFF; J. Appl. Phys., 1965, 36, 2903
47 Y. OISHI AND W. D. KINGERY: J. Chem. Phys., 1960, 33, 480
48 W. D. KINGERY: J. Am. Ceram. Soc., 1974, 57, 1

49 P. NAGELS *et al.*: paper SM-66/46, IAEA Symp. Thermodynamics, Vienna, 1965

50 P. McNAMARA: PhD. Thesis, London, 1963

51 G. B. ALCOCK *et al.*: paper SM-66/36, IAEA Symp. Thermodynamics, Vienna, 1965

52 D. K. REIMANN AND T. S. LUNDY: *J. Nucl. Mat.*, 1968, **20**, 54

53 R. LINDNER AND F. SCHMITZ: *Z. Naturforsch*, 1961, **169**, 1373

54 A. B. AUSKERN AND J. BELLE: *J. Nucl. Mat.*, 1961, **3**, 267

55 S. YAJIMA *et al.*: *ibid.*, 1969, **20**, 162

56 J. BELLE: *ibid.*, 1969, **30**, 3

57 R. A. WOLFE AND S. F. KAUFMAN: WAPD-TM-587, 1967

58 B. BURTON AND G. L. REYNOLDS: *Acta Met.*, 1973, **21**, 1073

59 B. BURTON AND J. P. BARNES: *Met. Sci. J.*, 1975, **9**, 18

60 M. S. SELTZER *et al.*: *J. Nucl. Mat.*, 1972, **44**, 331

61 W. B. BEERÉ: *Acta Met.*, 1977, to be published

62 A. G. CROUCH: *Trans. Brit. Ceram. Soc.*, 1973, **72**, 307

63 A. E. PALADINO AND R. L. COBLE: *J. Am. Ceram. Soc.*, 1963, **46**, 133

64 R. L. COBLE AND Y. H. CUERARD: *ibid.*, 1963, **46**, 353

65 Y. OISHI AND W. D. KINGERY: *J. Chem. Phys.*, 1960, **33**, 480

66 M. F. ASHBY; *Acta Met.*, 1972, **20**, 887

67 P. GREENFIELD AND W. VICKERS: *J. Nucl. Mat.*, 1967, **22**, 77

68 W. VICKERS AND P. GREENFIELD: *ibid.*, 1968, **27**, 73

69 J. E. HARRIS *et al.*: *J. Australian Inst. Met.*, 1969, **14**, 154

70 B. BURTON: *Met. Sci. J.*, 1971, **5**, 11

71 F. K. SAUTTER AND E. S. CHEM: 'Oxide dispersion strengthening', (Ed. C. S. ANSELL), 495, 1968, New York, Gordon and Breach

72 M. F. ASHBY: *Scripta Met.*, 1969, **3**, 837

73 J. E. HARRIS: *Met. Sci. J.*, 1972, **7**, 1

74 B. BURTON: *Met. Sci. Eng.*, 1973, **11**, 337

75 K. A. HAY *et al.*: 'Physical metallurgy of reactor fuel elements', 95, 1975, London, The Metals Society

76 B. BURTON: CEGB Report No. RD/B/N3225, 1974

77 A. U. KARIM AND W. A. BACKOFEN; *Met. Trans. ASM*, 1972, **3**, 709

78 E. D. HONDROS: 'Physical metallurgy of reactor fuel elements', 79, 1975, London, The Metals Society

79 F. R. N. NABARRO: *Phil. Mag.*, 1967, **16**, 231

80 J. BARDEEN AND C. HERRING: 'Imperfections in nearly perfect crystals' (Ed. W. SCHOCKLEY), 1952, New York

81 G. L. REYNOLDS *et al.*: CEGB Report No. RD/B/N3646, 1976

82 B. BURTON AND G. L. REYNOLDS: *Phil. Mag.*, 1974, **29**, 1359

83 J. E. HARRIS *et al.*: *Met. Sci. J.*, 1974, **8**, 310

84 D. HULL AND D. E. RIMMER: *Phil. Mag.*, 1959, **4**, 673

85 R. P. SKELTON: *Met. Sci. J.*, 1975, **9**, 192

Diffusional growth of creep voids

J. E. Harris

The development of the theory of diffusional growth of creep voids is reviewed. It is pointed out that the process of growth can be considered as the removal of atoms from the surface of the voids and the plating of these as an even layer along those grain boundaries which are normal to the stress axis. The plating process jacks apart the two hemispheres of the void either side of the boundary and the significance of this is explored. Plating also leads to overall creep strain and this is an important deformation mode; it is a form of diffusional creep which has been termed 'Hull–Rimmer creep' and its magnitude and kinetics determined. It is suggested that a suitable array of precipitate particles prevents the formation of a plated layer of even thickness and this inhibits void growth. The active nucleation sites for voids are considered to be precipitate particles in sliding grain boundaries. It is usually not feasible to completely remove all inclusions from materials and the most effective means of avoiding high-temperature creep embrittlement is to arrange to have copious intergranular precipitates in the microstructure of the alloy. These precipitates may suppress grain-boundary sliding and hence minimize void nucleation. As an added safeguard the diffusional growth of any void which did form would be inhibited by the presence of the precipitates.

The author is with the CEGB, Berkeley Nuclear Laboratories

The phenomenon of brittle fracture in creep has been known for a very long time since engineers first strained metal components slowly at elevated temperatures. Such brittle behaviour must have been puzzling for it has been known for millenia that working and fabrication processes are facilitated by raising the temperature of the metal or alloy which is to be formed. The fact that ductility varies with strain rate at high temperatures probably implies that the fracturing process is thermally activated.

On the other hand, over a wide range of creep strain rates (i.e. the fastest of which is much slower than those imposed during hot working) the product of creep rate and time-to-fracture is approximately constant.[1-5] Hence if an alloy is strengthened by, say, precipitation it creeps more slowly for a given stress, but the fracturing process is slowed down correspondingly. It has been argued,[3-5] very strongly, that this is evidence that failure is a direct consequence of plastic deformation and not an independent diffusional process. The contrary view, which will be the one developed in this paper, is that deformation and the growth of creep voids are normally independent processes but both may be affected approximately equally by changes in microstructure. The subject has been extensively reviewed[6-9] in recent years so a comprehensive survey is not attempted in this paper.

Attention will be focused on the fracturing process caused by the growth and impingement of approximately spherical (r type voids)[1] on high-angle grain boundaries located normal to a tensile stress axis. Diffusional growth only will be considered although it has already been noted that there are many who prefer models based on growth being due to grain-boundary sliding and other deformation processes. The controversy is not entered here except to observe that the diffusional growth model is very well based on a simple thermodynamic argument and where such growth does not occur it is perhaps more fruitful to determine the factors preventing diffusion rather than to attack the basis of the theory.

ORIGINS OF THE DIFFUSIONAL THEORY

It was the similarity in the appearance of creep cavities to Kirkendall voids which first led to the suggestion that they form by vacancy condensation.[10,11] The majority of the early workers considered that the vacancies were generated by plastic deformation in the interior of the grains. However, in 1957, Balluffi and Seigle[12] made the important advance when they pointed out that the more likely source of vacancies was the boundary itself. A simplified form of the Balluffi and Seigle analysis is now presented for it is the starting point for all subsequent analyses of diffusional growth.

Consider an isolated void of radius r on a high-angle grain boundary which is normal to a tensile stress σ. If the void receives a vacancy from the boundary its surface energy increases by an amount $2\gamma b^3/r$ (but see below) where γ is the surface energy per unit area and b^3 the atomic volume. The emission of a vacancy from the boundary is equivalent to the plating out of a new atom so that a force σb^2 is moved through a distance b. If this positive work increment exceeds the corresponding increase in surface energy then void growth must occur. That is to say the condition for void growth is

$$\sigma \geqslant \frac{2\gamma}{r} \tag{1}$$

When the growing voids impinge with their neighbours final failure will occur. This is the most efficient of all forms of fracture for it converts a very high fraction of the work done by the applied stress into surface energy of fracture. For this reason it can occur at very low stresses, lower in fact than those needed to cause intragranular deformation by dislocation movement.

Vacancies may be transported from the grain-boundary source to the void either along the boundary itself or through the lattice, the relative importance of the two routes depending on the ratio of the product of diffusion coefficients and diffusing areas. In most practical situations $D_g w > D_l d$ so that grain-boundary diffusion predominates (D_g and D_l are grain boundary and lattice self-diffusion coefficients respectively, w is the width of the grain boundary, and d is the diameter of the void).

The void growth process can then be thought of as the removal of atoms from the surface of the voids and the plating of an equivalent number of atoms along the grain boundary. Much can be learned of the mechanism of fracture by concentrating attention upon the plating process itself. Three aspects are considered: (a) the fact that plating forces apart the two hemispheres of the void either side of the boundary (the jacking action) and this gives an additional small increase in void volume; (b) the plating process produces significant overall creep strain and is in fact a form of diffusional creep deformation; and (c) the plated layer must be of even thickness along the boundary otherwise back stresses will be generated opposing further void growth; such even plating may not be possible if precipitate particles exist along the boundary.

Jacking caused by plating

An expression for the critical stress for void growth will be derived once again, but this time taking into account the jacking action and the fact that the net stress increases as the voids grow and the remaining area of load-bearing grain boundary decreases.

This time consider an array of voids of radius r and spacing $2x$. The emission of a vacancy from the grain-boundary area associated with a void jacks apart the two hemispheres of the void by a vertical distance $b^3/\pi(x^2-r^2)$ with a corresponding increase in surface energy of $2rb^3\gamma/(x^2-r^2)$. When the vacancy condenses into the void there is an additional increase in void surface energy of $2b^3\gamma/r$ so the total change in surface energy is

$$\frac{2\gamma b^3}{r}\left[\frac{x^2}{x^2-r^2}\right]$$

Equating this with the work done when a vacancy is emitted and the *net* force, $\sigma b^2 x^2/x^2-r^2$, moves through a distance b, the condition for void growth remains as $\sigma \geqslant 2\gamma/r$. However, the driving force for subsequent growth now becomes

$$\left[\sigma-\frac{2\gamma}{r}\right]\frac{x^2}{x^2-r^2}$$

This correction, which becomes important in the later stages of the fracture process, should be included in any expression for the rate of void growth. This was not done in the Hull and Rimmer[13] derivation nor that due to Speight and Harris.[14] In later developments of these analyses Vitovec[15] and Harris et al.[16] incorporated the net stress but Weertman[17] drew attention to shortcomings in the Hull and Rimmer treatment and the fact that Speight and Harris had used incorrect boundary conditions. However, Weertman's analysis itself neglected the jacking action. The only treatment which is physically sound and has the correct boundary conditions is that due to Speight and Beeré[18] and this is the analysis which should be used when accurate estimates of the void growth rates are required.

Creep strain due to plating

The progressive plating of atoms along the boundary achieves strain in the direction of the applied stress and the process is a type of diffusional creep deformation.[16,19] For convenience it will be termed Hull–Rimmer (HR) creep. The practical importance of the HR creep strain was first pointed out by Cottrell in 1961[20] but he overestimated its magnitude by up to a factor 2. An estimate is now made of the maximum creep strain due to void growth.

Consider a volume $\pi x^2 l$ associated with each void where l is the mean grain diameter. Growth involves simply the removal of atoms from the surface of the void and deposition of these into its associated volume, the number of atoms remaining constant; that is:

$$\pi x^2 l - \tfrac{4}{3}\pi r^3 = \text{constant} \tag{2}$$

By differentiating:

$$x^2\, dl = 4r^2\, dr \tag{3}$$

Where $r = x$ adjacent voids impinge and this can be taken as the final fracture criterion. Hence the total strain at failure can be derived as follows

$$x^2\int_{l_0}^{l} dl = 4\int_0^x r^2\, dr \tag{4}$$

$$\Delta l = \tfrac{4}{3}x \tag{5}$$

The maximum HR strain $\epsilon_{fm} = \Delta l/l = 4x/3l$. Typically $x \sim l/20$ so $\epsilon_{fm} \sim 7\%$. In fact many metals and alloys deformed slowly at elevated temperatures fail with total rupture ductilities less than this value so that HR creep can be a significant, if not dominant, mode of deformation in these cases.

From above, $\epsilon_f \propto \Delta l \propto r^3$. During the early stages of void growth $r \propto t^{\frac{1}{3}}$ whereas during a later stage $r \propto t^{\frac{1}{3}}$.[14] It follows that initially the void strain will vary linearly with time, $\epsilon_f \propto t$, whereas subsequently the deformation rate will increase with increasing time, $\epsilon_f \propto t^{\frac{3}{2}}$. This gives some idea of the shape of the HR creep curve as illustrated in Fig. 1. In this diagram HR creep is shown

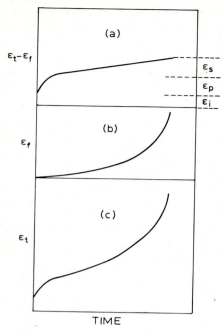

1 The creep strains due to dislocation movement (instantaneous ϵ_i, primary ϵ_p, and secondary ϵ_s) are added to the HR creep strain ϵ_f to give the total strain ϵ_t

occurring independently from, and additional to, deformation involving dislocation movement. The model has, of course, been somewhat oversimplified because it assumes that all voids are nucleated at the very commencement of the creep test. Analyses involving progressive nucleation of voids with increasing strain are available.[21-23]

It is surprising that little attention has been paid to HR creep. It is a form of diffusional creep where the average diffusion distance is only $\sim x/2$. This diffusion distance may be up to two orders of magnitude shorter than that operative during conventional Nabarro–Herring or Coble diffusional creep where material has to be transported across a complete grain. For this reason HR creep could prove to be of far greater practical significance than the other, more familiar, diffusion creep processes.

HR creep is linearly dependent upon stress and the activation energy for the process is that for grain-boundary self diffusion. This contrasts with dislocation creep where the stress exponent is often ~ 5 and the relevant activation energy is that for lattice self diffusion. It follows that HR creep will dominate at low temperatures and low stresses, i.e. those conditions which give rise to low rates of deformation. Hence HR creep could readily become the dominant deformation mode in lightly stressed engineering components with long design lifetimes and operating at low or intermediate temperatures. It is certainly a deformation mode which should be taken into account when long-term extrapolation of creep data is attempted.

Inhibition of plating

It appears to be the case that vacancies cannot be created or destroyed at a two-phase interface without the co-operative movement of atoms either side of the boundary, i.e. those that comprise both phases.[24,25] Thus, if a high-angle grain boundary is decorated with precipitates with a higher melting temperature than the matrix

then its vacancy emitting or absorbing capability may be adversely affected.[26-28] In the case of diffusional growth of voids in a decorated boundary the laying down of an even plated layer may be inhibited close to the precipitates. This will lead to the generation of back stresses which will oppose further void growth. This may explain why it is that the stress rupture lives of two-phase alloys are often so much longer, for a given stress, than those of single-phase alloys or pure metals.[5]

The best illustration that it is the intergranular and not the intragranular precipitates which are important in prolonging rupture life is provided by the results of Betteridge as reported in a review article by McLean.[5] Creep rupture tests were carried out on Nimonic 80A at 750°C and 17 tons in^{-2} in two conditions of heat treatment which produced none and many grain-boundary precipitates respectively. The following results were obtained:

Condition produced by heat treatment	Rupture lifetime, h	Rupture elongation, %
No grain-boundary precipitates	6	0·6
Many grain-boundary precipitates	130	78

In this section it will be assumed that the beneficial influence of intergranular precipitates is due to the inhibition of void growth; in a later part of the paper the alternative, equally valid, approach will be adopted that it is due to inhibition of void nucleation.

As already mentioned, during growth of the creep voids, stress will be concentrated at the intergranular precipitates and this may be relieved by plastic deformation involving the generation of dislocations at the precipitate/matrix interface and the movement of these into the matrix. A simple case to analyse is where this local deformation is achieved by the punching out of vacancy loops;[29] for each monolayer of atoms plated along the boundary a single loop would have to be generated at each precipitate. The work done in the direction of the applied stress must now not only be sufficient to compensate for the increase in surface energy of the voids, but also for the total dislocation line energy of the loops. Such considerations lead to the following expression for the critical stress for void growth in a two-phase alloy:

$$\sigma \geqslant \frac{2\gamma}{r} + \frac{N_p G b r_p}{2} \ln \frac{r_p}{b} \tag{6}$$

where N_p is the number of precipitates per unit area of grain boundary, G the elastic shear modulus of the matrix, and r_p the precipitate radius.

Work is currently in hand to determine the influence of precipitates on the kinetics of HR creep. Three approaches are being attempted; (a) the LHS minus the RHS of equation (6) is substituted for the $(\sigma - 2\gamma/r)$ term in the Harris et al.[16] analysis; (b) it is assumed that it is the annealing of the vacancy loops which is rate controlling, i.e. applying the model for inhibition of Coble creep due to Crossland and Clay[30] to the case of diffusional fracture; (c) ascribing to the regions adjacent to the precipitates the macroscopic creep properties of the matrix and thereby working out the back stress opposing void growth.

NUCLEATION

Reversing the normal order, this paper will conclude with a discussion and brief historical survey of the theories of nucleation of creep voids.

For typical creep stresses, the critical radii for void nucleation are calculated from equation (1) to lie within the range 0.1–$1.0\ \mu$m, i.e. sizes readily observable in the electron microscope. The fact that such embryonic voids are not normally found in undeformed metals and alloys suggests that they do not exist *ab initio* but result from the deformation process. There is in fact fairly general agreement that they nucleate by the rupturing of some obstacle to grain-boundary sliding such as ledges,[31-34] intersections with sub-boundaries[35] or precipitates.[8,20,36-38]

An immediate difficulty is that the length of free sliding boundary between obstacles may be too short to permit sufficient stress concentration to occur to cause loss of cohesion. (Creep cavitation of the type considered in this paper occurs at stresses lower than those necessary to nucleate grain-boundary triple-point cracks where the whole of the grain boundary is available to concentrate the stress.[39,40]) In an attempt to overcome this difficulty nucleation at very small ledges, only several atomic diameters high where very high stress concentrations could occur, has been proposed.[34] However, it was subsequently demonstrated[38] that the rate of radial sintering of such small voids would greatly exceed the rate of grain-boundary sliding so a viable size could never be achieved. On recognition of these difficulties it was suggested[20] that the nucleation sites were precipitates with such a low work of adhesion with the matrix that separation could readily occur at relatively low stress concentrations. In support of this idea it was first proposed,[41] and later demonstrated,[42] that the stress concentration at a particle of finite dimensions which allowed flow to occur around it would be far higher than the stress concentrated at the obstacle of infinite size incorporated in the Stroh[43] analysis. A new difficulty then appeared for it has been shown[44] that if the diffusional properties of a high-angle grain boundary are ascribed to a precipitate matrix interface then diffusional flow can occur around the precipitate at such a rapid rate that it could never be an effective obstacle to grain-boundary sliding. Precipitates return as probable nucleation sites, however, as soon as it is accepted, for reasons mentioned earlier, that their interfaces are not good sources or sinks for vacancies.[28] In these circumstances diffusional flow around the precipitate is not possible so that the stress concentration can rise to a level sufficient to cause loss of cohesion across the interface.

If precipitates are such effective barriers to sliding then, presumably, if a sufficiently high density were present then sliding could be completely suppressed and cavitation prevented. This is the alternative explanation for the beneficial influence of intergranular precipitates on creep rupture properties to which reference has already been made. Unlike diffusional creep deformation, when creep is occurring by dislocation movements within the grains, boundary sliding is not essential to retain continuity between neighbouring grains. A recent analysis due to Speight[45] permits an estimate to be made of the increase in stress necessary to maintain a given deformation rate when sliding is completely suppressed. If σ_s is the stress with infinitely easy grain-boundary sliding and if σ_{ns} is the stress to maintain the same deformation rate in the absence of sliding, the following relationship applies when the stress exponent takes a value of 5:

$$\frac{\sigma_{ns}}{\sigma_s} = (1.63)^{\frac{1}{5}} \tag{7}$$

That is to say σ_{ns} exceeds σ_s by only 10% indicating that the complete suppression of sliding may be quite a practical possibility. This is an important result for it indicates that during a constant strain rate test, as more and more obstacles to sliding are introduced into a grain boundary, the shear stress does not rise inexorably but reaches some limiting value which is well below the theoretical fracture stress. At constant applied stress, prevention of grain-boundary sliding will lead to a reduction in creep rate of only 39%. It is, of course, not usually possible to confine precipitation to the grain boundary and concurrent intragranular precipitation will cause additional strengthening. It can then be seen that a change in microstructure leading to strengthening can also prevent the formation of voids and this is in no way at variance with the theoretical basis of the diffusional void growth model. If voids cannot be nucleated they cannot grow by any of the proposed mechanisms.

In conclusion, Cottrell's suggestion[20] that the most probable sites for the nucleation of creep voids is precipitate particles, has been supported. It is now possible to draw up a fairly tight specification for the properties, size, and distribution for such nucleating precipitates. First, there must obviously be sufficient precipitates to account for the observed cavity density with a few to spare but they must not be too numerous otherwise they may suppress completely grain-boundary sliding. Secondly, the precipitates must not be much smaller than the critical void radius (defined as $r = 2\gamma/\sigma$) or sintering will exceed sliding rates preventing the formation of a viable cavity; on the other hand the precipitates must not be too large for then insufficient stress would be generated to cause loss of cohesion at the interface. Thirdly, the precipitates should have a low work of adhesion with the matrix but not so low that the interface takes on the diffusional characteristics of a high-angle grain boundary thereby allowing rapid diffusional flow around the precipitate. Finally, also to inhibit local diffusional flow, the precipitate should have a higher melting temperature than the matrix.

In this paper attention has been drawn to the important roles played by intergranular precipitates in the nucleation and growth of creep voids. Creep cavitation is seen as an extrinsic process and in theory, could be eliminated if a single-phase alloy or metal is made completely free from second-phase particles. Alternatively, for alloys intended for high-temperature service where low creep rates are also required copious intergranular precipitates can be introduced. The precipitates, as well as being numerous, should have a high melting temperature and be strongly bonded to the matrix. In this way sliding and hence cavity nucleation may be suppressed completely, or even if this fails theory indicates that subsequent diffusional growth of the voids may be greatly inhibited.

ACKNOWLEDGMENT

This paper is published by permission of the Central Electricity Generating Board.

REFERENCES

1 F. GAROFALO: 'Fundamentals of creep and creep rupture in metals', 1965, New York, Macmillan

2 F. C. MONKMAN AND N. J. GRANT: *Proc. ASTM*, 1956, **56**, 593

3 P. W. DAVIES AND B. WILSHIRE: 'Structural processes in creep', 34, 1961, London, ISI/Inst. Metals

4 P. FELTHAM: *ibid.*, p. 82

5 D. MCLEAN: *Rep. Prog. Phys.*, 1966, **29**, 1

6 G. W. GREENWOOD: 'Interfaces', (Ed. R. C. GIFKINS), 223, 1969, Melbourne, Butterworths

7 A. J. PERRY: *J. Mater. Sci.*, 1974, **9**, 1016

8 R. RAJ AND M. F. ASHBY: *Acta Met.*, 1975, **23**, 653

9 J. A. WILLIAMS: 'The mechanics and physics of fracture', 159, 1976, London, The Metals Society

10 J. N. GREENWOOD et al.: *Acta Met.*, 1954, **2**, 250

11 L. SEIGLE AND R. RESNICK: *Acta Met.*, 1955, **3**, 605

12 R. W. BALLUFFI AND L. SEIGLE: *Acta Met.*, 1957, **5**, 449

13 D. HULL AND D. E. RIMMER: *Phil. Mag.*, 1959, **4**, 673

14 M. V. SPEIGHT AND J. E. HARRIS: *Metal Sci. J.*, 1967, **1**, 83

15 F. H. VITOVEC, *J. Mater. Sci.*, 1972, **7**, 615

16 J. E. HARRIS, et al.: *Metal Sci.*, 1974, **8**, 311

17 J. WEERTMAN: *Scr. Metall.*, 1973, **7**, 1129

18 M. V. SPEIGHT AND W. BEERÉ: *Metal Sci.*, 1975, **9**, 190

19 B. BURTON AND J. P. BARNES: *Metal Sci.*, 1975, **9**, 18

20 A. H. COTTRELL: 'Structural processes in creep', 1, 1961, London ISI/Inst. Metals

21 G. W. GREENWOOD: *Phil. Mag.*, 1963, **8**, 707

22 R. T. RATCLIFFE AND G. W. GREENWOOD: *Phil. Mag.*, 1965, **12**, 59

23 R. P. SKELTON: *Metal Sci.*, 1975, **9**, 192

24 J. E. HARRIS AND B. C. MASTERS: *Proc. Roy. Soc. A*, 1966, **292**, 240

25 M. F. ASHBY AND R. M. A. CENTAMORE: *Acta. Met.*, 1968, **16**, 1081

26 J. G. EARLY et al.: *Trans. AIME*, 1964, **230**, 1641

27 J. E. HARRIS: *Metal Sci. J.*, 1973, **7**, 1

28 G. B. GIBBS, AND J. E. HARRIS: 'Interfaces', (Ed. R. C. GIFKINS), 53, 1969, Melbourne, Butterworths

29 J. E. HARRIS: *J. Nucl. Mat.*, 1976, **59**, 303

30 I. G. CROSSLAND AND B. D. CLAY: to be published

31 R. C. GIFKINS: *Acta Met.*, 1956, **4**, 98

32 C. W. CHEN AND E. S. MACHLIN: *Acta Met.*, 1956, **4**, 655

33 P. W. DAVIES AND B. WILSHIRE: *J. Inst. Metals*, 1961–62, **90**, 470

34 D. MCLEAN: *J. Australian Inst. Metals*, 1963, **8**, 45

35 A. E. B. PRESLAND AND R. I. HUTCHINSON: *J. Inst. Metals*, 1963–64, **92**, 264

36 R. RESNICK AND L. SEIGLE: *Trans. AIME*, 1957, **209**, 87

37 C. W. WEAVER: *J. Inst. Metals*, 1959–60, **88**, 296

38 J. E. HARRIS: *Trans. AIME*, 1965, **233**, 1509

39 C. ZENER: 'Fracturing of metals', 1, 1948, Cleveland, Ohio, ASM

40 R. EBORALL: Proc. NPL Symp. 'Creep and fracture of metals at high temperatures', 229, 1956, London, National Physical Laboratory

41 R. EBORALL: 'Structural processes in creep', 75, 1961, London, The Iron and Steel Institute

42 E. SMITH AND J. T. BARNBY: *Metal Sci. J.*, 1967, **1**, 1

43 A. N. STROH: *Proc. Roy. Soc.*, 1954, **A223**, 404

44 G. B. GIBBS: *Mem. Sci. Rev. Met.*, 1965, **LX11** (10), 781

45 M. V. SPEIGHT: *Acta Met.*, 1976, **24**, 725

Lattice defects in splat-quenched metals

H. Jones

Current research effort involving solidification at high cooling rates ($>10^5/10^6 K/s$) started with the availability of the gun technique of splat quenching, first described by Duwez et al. in 1960. Subsequent developments have been such that microstructural refinement, solubility extension, and new phases can now be produced in continuous or discontinuous ribbon, wire or wool as well as in flake made by earlier techniques. Pure metals so formed show small grain sizes (typically $0 \cdot 1 - 10\ \mu m$) and can show increased vacancy concentration retained as clusters, compared with samples quenched or at equilibrium in the solid state. Splat-quenched alloys show reduced solute clustering compared with quenching from high temperature in the solid state but clustering during quenching is a major effect in the new highly extended solid solutions obtained. This, together with heterogeneous precipitation on the many grain boundaries or stacking faults present, limits age-hardening, but dispersion hardening is very effective in such alloys showing low solid solubility under equilibrium conditions. Plasticity is also enhanced as a result of refined matrix grain size and finer dispersions of second phases and improvements in endurance under a variety of conditions can be expected to result from further employment of techniques available. While formation of non-crystalline alloys is limited to certain composition ranges at accessible cooling rates, the effects associated with refinement of crystalline microstructure occur over the entire spectrum of alloy composition and cooling rates and should be less susceptible to degradation during consolidation.

The author is at the University of Sheffield

Methods of quenching from the liquid into the solid state have been available at least from the beginning of the century (e.g. refs. 1, 2), but their potential for achieving non-equilibrium metallurgical effects was hardly recognized until the 1950s when Falkenhagen and Hofmann,[3] following Hofmann[4] and Hofmann and Wiehr,[5] reported that terminal solid solubility extension was produced in thin sections of a number of alloys by injection chill casting. This was followed by reports of systematic shifts in composition for coupled eutectic growth in quenched liquid droplets with decreased quenching bath temperature;[6,7] of dispersion hardening due to refinement of second phases by atomization;[8] and of *complete* solid solubility extension and *new* non-equilibrium phases, one of them non-crystalline, by novel methods of quenching the melt devised by Duwez et al.[9,10] The novelty was to employ a shock wave to impel a *small* quantity of melt ($<0 \cdot 1$ g) at the inside surface of a rotating cylinder or (Fig. 1a) at an inclined ski-slide substrate[11] to produce flake specimens $<0 \cdot 1$–$100\ \mu m$ thick. Cooling rates were estimated to be typically $\sim 10^6$ K/s[12-14] and are considered to reach 10^{10} K/s in the thinnest areas.[14-16] These values

substantially exceed upper limits $\sim 10^4$ K/s for normal chill casting[3,17] or atomization[18] and 10^5 K/s for conventional quenching entirely within the solid state,[13] and so made more accessible a whole new regime of cooling rates and their effects.

Developments and variants of the basic techniques proliferated in the 1960s so that it is now possible to produce discontinuous or continuous ribbon, wire or wool as well as flake by direct quenching from the melt. Comprehensive surveys of the literature up to the end of 1972 are available[19-21] so the present review is directed towards aspects relevant to the formation of crystal lattice defects in splats, in keeping with the general emphasis of the present conference and the subject of one earlier review.[22] Lattice defect formation in splats appears to be influenced by the choice of quenching technique and the conditions of quenching characterized in the next section. In subsequent sections, results on lattice defect generation are discussed firstly for pure metals and then for alloys. Finally, established effects on properties related to the high volume densities of defects produced are briefly considered.

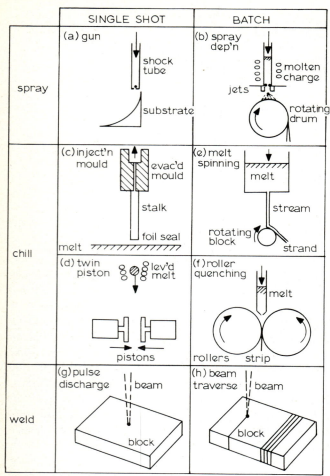

	SINGLE SHOT	BATCH
spray	(a) gun — shock tube, substrate	(b) spray dep'n — molten charge, jets, rotating drum
chill	(c) inject'n mould — evac'd mould, stalk, foil seal, melt	(e) melt spinning — melt, stream, rotating block, strand
	(d) twin piston — lev'd melt, pistons	(f) roller quenching — melt, rollers, strip
weld	(g) pulse discharge — beam, block	(h) beam traverse — beam, block

a gun technique (after ref. 11); b spray deposition (after ref. 28); c injection mould (after refs. 3, 17); d two-piston technique (after ref. 132); e melt spinning (after ref. 38); f roller quenching (after refs. 39, 40); g single-shot power discharge (after refs. 42–44); h beam traverse quenching (after ref. 45)

1   Basic methods of rapid quenching from the melt

CONDITIONS AND METHODS FOR RAPID QUENCHING FROM THE MELT

Calculations[23] assuming one-dimensional heat flow normal to the splat-substrate interface show that high cooling rates depend on the obtaining of a thin layer of sample in good thermal contact with the heat sink. Methods for achieving these conditions divide most naturally (e.g. ref. 24) into three categories. One class of methods employs impulsion of a spray of droplets at the quenching substrate so that freezing may take place on the substrate *during* deposition, typically producing layer structures in the deposit. Another group includes methods in which a single slug or relatively coherent stream of melt is cast between two (sometimes on one) substrate(s). The remaining approach involves the application transiently of a high-intensity heat source to a block of material so that rapid melting followed immediately by rapid freezing occurs in a thin layer at the surface. All three groups now include variants capable of continuous production of quenched material

as well as their basic single-shot versions still adequate for many experimental purposes.

Spray deposition

The gun technique (Fig. 1a) pioneered by Duwez *et al*.[9–12] has established itself as the standard method and is capable of the highest cooling rates, especially when used in a controlled atmosphere.[16] Continuous versions (e.g. Fig. 1b) employing deposition on a substrate by rotary[25,26] or gas blast[27,28] atomization, or by plasma jet[29,30] or arc[31] spraying have been developed to overcome the drawback of a limited charge per shot. Control of rate, area, and thickness of deposition is achieved, as required, by scanning[32] the source of spraying relative to the substrate which may be a rotating disc,[25,27] drum[28] or rollers,[32,33] a travelling belt[34] or simply a stationary surface.[26,32] Deposits from all these techniques are characterized by inclusions of porosity and surface films and by non-uniformities of metal macrostructure and microstructure. Porosity can be closed by subsequent pressing or working and surface films can be controlled by employing inert atmospheres in deposition, while metallic structural non-uniformities in deposits are an intrinsic feature that can be controlled but not eliminated at the deposition stage.

Chill-casting

The injection chill technique[3] (Fig. 1c), itself limited to thicknesses above ~1 mm for useful penetrations of metal into the mould, may be regarded as the forerunner of a larger class of related methods. These include: piston-and-anvil[35] and kindred techniques (e.g. Fig. 1d) in which a slug of melt is spread into a disc; melt-spinning[36] (e.g. Fig. 1e) and its variants in which extensive lengths of ribbon,[36] wire[37] or wool[38] are made by solidifying liquid metal on the surface or periphery of a rotating cup,[36] block[38] or wheel;[37] and roller-quenching[39,40] (e.g. Fig. 1f) in which lengths of sheet some several millimetres wide result from the solidifying of liquid metal directly in the nip of a precision-made rolling mill with the rolls essentially in contact. All these methods produce relatively sound and uniform material typically ~50–100 μm thick solidified at cooling rates ~10^5–10^6 K/s and are under intensive development at present.

Weld-casting

The technique of traversing the surface of a piece of parent material with a conventional welding arc to achieve refinement of solidification structure, as pioneered by Brown and Adams,[41] is limited to cooling rates up to ~100 K/s. Capacitance discharge[42–44] (e.g. Fig. 1g), electron beam,[45] laser[44,46,47] and explosive[48] welding have all been employed to melt and resolidify to a depth of ~100 μm at cooling rates as high as 10^6 K/s.[44,46,47] All are adaptable (e.g. Fig. 1h) to continuous treatment of a surface and again relatively uniform microstructures are produced, though banding attributable to input energy spikes and included spherical pores as large as 0·1 mm in diameter are sometimes a feature of capacitance discharge and laser methods.

DEFECTS IN PURE METALS

Information is available on the effect of rapid quenching from the melt for some 30 pure metals, usually limited as part of a study of effects in an alloy system. Substantial studies have been reported however for Be,[25] B,[49,50] Al,[27,31,32,51–69] V and Nb,[70] Co,[71–73] Ni,[74–76] Ag,[60,77] Zn,[78] Cd,[79–81] Ge,[16,82] Te,[16] Sb,[57] and Pb.[83]

Grain structure

Grain structures of gun splats range from elongated in the plane of the flake to columnar and equiaxed through its thickness.[20,25,51,52] Grain sizes for pure metals are typically in the range $0.1–10 \ \mu m$,[25,51,52,56,60,69,70,76,77,79] although values of $0.01 \ \mu m$ and below are attainable in certain alloys (e.g. ref. 15). Preferred orientations are sometimes evident, notably for particular samples of Al,[60] Pb,[60] Ag,[84] Cu,[84] and Cd.[79,81] Twins are also scarce especially in alloys, although occurrence of evident twinning was reported for Ag and to a lesser extent for Cu.[85]

Wood and Sare[76] proposed that grains elongated in the plane of the sample were an indication of heat flow parallel to this plane towards areas of better thermal contact, that is, departing from the assumption of heat flow everywhere normal to the splat/substrate interface implicit in conventional heat flow models. Such growth structures may thus be characteristic of 'lift-off' areas[86] although the possibility of substantial grain-boundary migration following solidification in splats (e.g. ref. 60) cannot be ruled out. There is evidence (e.g. ref. 15) that grain size in the plane of the foil decreases with decreasing local thickness in thin areas of gun splats resembling the observation of a limiting grain size related to the sheet thickness resulting from solid-state growth in thin sheets (e.g. ref. 87). Otherwise the absence of preferred crystallographic orientation except in certain samples is consistent with the observation that protracted columnar growth is necessary to develop this by natural selection in solidification of cast metals. The observation of more twinning in Ag than in Cu may be attributed to its lower stacking fault energy, although any contribution of twinning or stacking faults to X-ray line broadening of splat-cooled Ag was reported[60,77] to be negligible. This broadening was consistent, however, with a diffracting domain size $\sim 0.11 \ \mu m$, at the lower end of the range of grain sizes observed in pure metal splats by electron microscopy. No significant broadening was detected for Al and Pb in this study[60] on gun splats. Broadening attributed to quenching strains of $1.5–2.9 \times 10^{-4}$ was detected, however, in a later work on Al,[69] reported to be in good agreement with estimates from dislocation density by electron microscopy. That study employed two-piston and roller-quenched samples, however, expected to be more prone than gun splats to thermal stress or deformation during or following quenching.

Dislocation structure

Despite these possibilities during preparation, samples quenched from the melt show no evidence of dislocation arrays that could be associated with major plastic deformation. Dislocation loops characteristic of condensation of excess vacancies quenched-in from higher temperatures, however, were observed in some of the earliest studies.[12,51,52,56] Thomas and Willens[51] reported uniform distributions of $1.3–1.6 \times 10^{13}$ loops per mm^3 except for denuded zones $0.05–0.1 \ \mu m$ wide aside grain boundaries for electron-transparent areas of pure aluminium gun splats with a grain size of $0.3–1 \ \mu m$ (see Fig. 2). Quenched-in vacancy concentration C_v^q derived

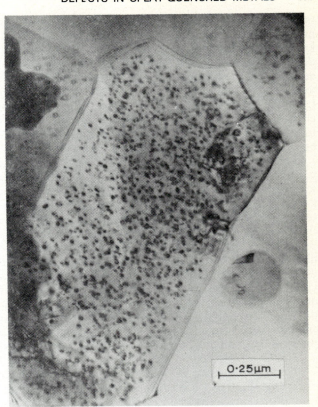

2 Vacancy loops and vacancy denuded zones in splat-quenched aluminium (from ref. 51) ×60 000

from the number and radius of loops increased from $1.0 \times 10^{-3}–1.6 \times 10^{-3}$ with increase of melt superheat from 65 to 505 K. Extrapolation gave a value of 9.5×10^{-4} at the melting point T_m compared with 6×10^{-4} (for example ref. 88) for quenching entirely within the solid state. McComb et al.[52] made similar electron microscope observations on aluminium gun splats but, possibly because they studied thinned samples, could detect no effect of superheat in the range 90 to 290 K, and obtained a smaller C_v^q of approximately 2×10^{-4} from loop density and radius with appreciable variations from point to point not entirely attributable to thickness fluctuations. McComb's observations probably reflect the lower cooling rates and complex thermal histories of thicker and layered regions of gun splats where a larger and more variable proportion of quenched-in vacancies could be lost to available sinks.

The observations by Thomas and Willens were a subject of some discussion,[53–56] especially their implication that a retained vacancy concentration higher than the equilibrium value for solid at T_m might be attainable by quenching from the melt. Subsequent studies on Cd[79–81] and Al[61,63,65] samples made by piston or roller quenching methods at least support the observations that vacancy clustering does occur in splat quenching. Indications from small angle X-ray scattering (SAXS) studies[63,65,81] were reflected in decreased lattice parameters,[61,79,80]* attributed to the effect of quenched-in point defects in dilating the crystal lattice. For both

* The dissenting results of Suryanarayana[66] for Al (no significant departure from normal lattice parameter with superheat) possibly again reflect the increased opportunity for loss of excess vacancies to sinks in thicker regions of gun splats.

metals the lattice parameters were reported to increase on annealing to their equilibrium values for 298 K and, for Al,[61] a systematic decrease in lattice parameter with decreasing splat thickness was explained as an effect of associated increased cooling rate in increasing C_v^q. The value of C_v^q of 4×10^{-3} was estimated[80] for Cd splats from the fractional change in unit cell volume of the crystal lattice divided by the estimated fractional relaxation of atomic volume around a vacancy, in reasonable agreement with values estimated by other methods.

Babić et al.[64] observed a linear increase of residual resistivity with increase of this change in lattice parameter in splat-cooled Al. This relationship corresponded to a level of C_v^q (7×10^{-4}–1.8×10^{-3}) as indicated by resistivity measurements, of 1.4–1.6 times that indicated by the lattice parameter. Resistivity measurements on Pb water-quenched from a series of temperatures in the liquid and solid states,[83] appear to support the more direct observations of Thomas and Willens on Al, i.e. a value of C_v^q increasing with increasing temperature of quenching whether in the liquid or solid states, a discontinuous increase ($\times 5$) in C_v^q for liquid compared with solid quenched from T_m and values of C_v^q by quenching the melt exceeding the equilibrium value C_v^e for solid at T_m (see Fig. 3).

Such observations of C_v^q by quenching the melt exceeding C_v^e at T_m, if confirmed, imply trapping in the solid of vacancies derived from the liquid state or from the moving solid/liquid interface (e.g. ref. 89), corresponding to solute trapping giving rise to solid solubility extension in alloys. Otherwise the increase in C_v^q with increasing melt superheat could be ascribed for gun or piston splats to increasing cooling rate resulting from decreasing splat thickness and better contact with the substrate, allowing C_v^q to attain a larger fraction of the value of C_v^e at T_m. An increment in C_v^q for quenching from the melt again is expected for the higher cooling rate through the solid state associated with splat quenching, though neither explanation accounts adequately for the same effects obtained in the water quenching study on Pb in which every effort was made to quench from the melt at the same rate as for solid samples.

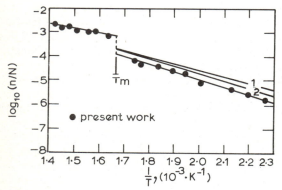

Curves *1* and *2* are from equilibrium and from quenching plus equilibrium measurements for the solid state

3 Retained vacancy concentration as a function of sample temperature prior to quench for lead quenched from liquid and solid initial states (from ref. 83)

Phase constitution

Evidence of X-ray reflections not attributable to the normal crystal forms of the element have been reported for Al,[57] Co,[71] and Zn[78] quenched from the melt. These reflections for Al[58] and Co[72] were later attributed to formation of hexagonal nitrides by nitrogen absorption, although this interpretation was questioned for Co.[73] The only pure elements apparently forming noncrystalline phases by solidification at cooling rates less than 10^6 K/s are S and Se. The first report of elemental glass formation by splat cooling was for B[49] and more recently glassy Ge,[16,82] Te,[16] and Ni[74–76] have been reported to form in electron-transparent areas of gunsplats made in low oxygen inert atmospheres. Calculations (e.g. ref. 90) suggest however that critical cooling rates as high as 10^{10} K/s, not attainable in sections thicker than 1 μm,[20] must be exceeded to form glasses in normally close-packed metals such as Ni and probably depend on impurities for their stabilization.*

DEFECTS IN ALLOYS

Splat-quenching effects from the beginning have been viewed primarily as alloying phenomena and studies for some 200 binary and many higher alloy systems had already been reported by the end of 1972.[20] The possibilities of non-equilibrium phase formation and stabilization are considerably increased and the new effects of solid solubility extension and associated solute distribution within these solid solutions have to be taken into account.

Grain and dislocation structure

Grain structure shows the same features as for pure metals but with the possibility of forming grain sizes as small as 0.01 μm (e.g. ref. 91). Twinning appears to be even less evident than in pure metals,[12,85] but stacking faults are a prominent feature in certain alloys such as Al–10 at.-% Ag,[92] Ag–5 to 20 at.-% Ge,[77,93,94] and Pu–15 at.-% Ti.[95] The Ag–Ge alloys form an extended fcc terminal solid solution and, at high enough Ge levels, a non-equilibrium cph Hume–Rothery phase. Both phases exhibit profuse faulting in the form of randomly arranged stacking faults separating the fcc and cph structures coexisting as a lamellar mixture. Absence of dislocation loops associated with collapsed vacancy clusters in solid solutions made by splat cooling for Al–0.5 to 5 at.-% Au,[96] Al–17.3 at.-% Cu[97] and Al–3 to 5 at.-% Fe[91] was attributed to solute–vacancy pairing. Dislocation loops were observed by electron microscopy, however, in splat-quenched solid solutions of Al–6 at.-% Ag,[98,99] Al up to 5 at.-% Cr[100] in thicker areas (see Fig. 4a), Al–2.5 at.-% Cu,[101] Al up to 0.5 at.-% Fe[102] and Al–1 at.-% Si[103] in certain regions. It has been suggested[100] that the lower limit of composition for loop formation in Al–Fe compared with Al–Cr reflects a higher binding energy of solute to vacancies in the former system.

Solidification substructure

Disc-shaped predendritic areas have been identified[15,104] as the sources of dendritic solidification in splats. Corresponding evidence for chill-cast alloys[105]

* The formation, structure, and properties of glassy metals is now a field of activity in itself and outside the scope of the present review; for two recent general surveys see refs. 166 and 167.

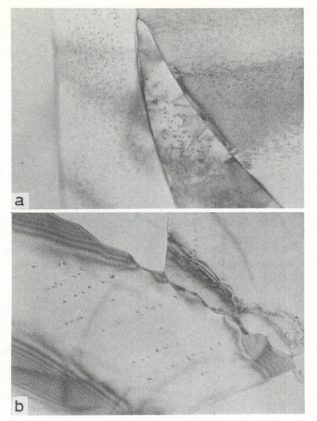

a random loops in Al–3·6 at.-% Cr, ×21 700;
b aligned loops in Al–1·2 at.-% Cr, ×14 400

**4   Vacancy loops in splat-cooled Al–Cr alloys
(from ref. 100)**

suggests that these growth centres are points of good local contact with the chill or substrate surface. Predendritic growth occurs radially at high undercooling without change of composition until the smooth growth front unstabilizes in favour of cellular or dendritic growth. The power relationship between secondary dendrite arm spacings down to ~10 μm and cooling rates up to ~30 K/s first reported for series of Al–Cu[106] and commercial[107] Al casting alloys was extended via collected data[108] for Al–4·5 wt.-% Cu into the regime of splat cooling by Dean and Spear[109] and by Matyja *et al.*[110] Dendrite arm spacings range from as small as 0·01 μm (e.g. ref. 111) to as large as 3 μm (e.g. ref. 112) in splats reflecting effects of composition as well as variations in thickness, thermal contact, and resultant cooling rate. Solute partitioning during solidification may give rise to intercellular dislocation arrays[86,101,113] (e.g. Fig. 4*b*) or

networks of second phase (e.g. ref. 114). Fully eutectic lamellar microstructures with spacings reported to be as small as 0·008 μm[16] are obtainable, becoming degenerate at high enough cooling rates (e.g. ref. 115). Extrapolation of the established power relation with eutectic spacing yields estimates of solidification front velocities in the range 3–2 000 mm/s (e.g. refs. 16, 116, 117).

Solute clustering in the solid state

Results of SAXS studies[98,99,118,119] for as-quenched Al–4 to 6 at.-% Ag alloys that can be fully solution-treated in the solid state clearly show (Table 1) that solute clusters were either undetectable or smaller in average size for a quench from the melt (LQ) than for one entirely within the solid state (SQ). This smaller size for LQ was maintained during growth of GP zones on aging at least up to 413 K for Al–4·5 at.-% Ag[119] with a much narrower size distribution than for SQ specimens.[98,119] Maximum hardness on aging was reached at a lower temperature for LQ.[119] Addition of 5 at.-% Zn to the Al–5 at.-% Ag alloy, however, was reported[120] to be sufficient for GP zones not to be suppressed by LQ, in spite of the effectiveness of LQ in the binary. Similarly, solute clustering is known to be a significant effect at high enough solute levels in the considerably extended solid solutions obtainable by LQ. For example, although lattice parameter measurements indicate that solid solubility of Fe in Al can be increased from an equilibrium maximum of 0·026 at.-% to as much as 4 at.-% (e.g. ref. 121), residual resistivity,[122] Mössbauer[67,123,124] and electron microscope[91,102,125] studies indicate solute clustering at concentrations above 0·7 at.-%. High angle X-ray line broadening,[69] SAXS,[65,126] and electron microscope[123,125,126] studies indicated similar clustering in Al–Sn[69,126] and Al–Ni[65,69,126] but none was detectable in Al–Mn[65,125] alloys. This stronger tendency for clustering in Al–Fe and Al–Ni compared with Al–Mn is also reflected in a higher lattice distortion (as indicated for example by the extent of deviation from Vegard's Law) and a higher hardness increment per unit alloy addition in Al–Fe, –Co, and –Ni compared with Al–V, –Cr and –Mn.[127]

Studies of the initial stages of aging of Al–Si alloys[103,128,129] quenched from the melt are not in agreement. Agarwal *et al.*[103,128] reported that resistivity measurements indicated a more sluggish reaction in LQ compared with SQ samples, attributing this to *fewer* retained vacancies in the splat resulting in fewer nuclei for precipitation. They accounted for a reduced decomposition wavelength for spinodal decomposition in LQ samples of Al–15 at.-% Ag in the same way.[103] Matyja *et al.*,[129] on the other hand, attributed the lower activa-

TABLE 1   Cluster size from SAXS studies on Al–Ag alloys quenched from the melt (LQ) compared with quenching from within the solid state (SQ)

Alloy composition at-% Ag	Wt-% Ag	Quenching technique from the melt	Cluster radius, Å LQ	SQ	Ref.
4·2, 5·9	15, 20	Piston and anvil	7, 9	—	Kahkönen[118]
4·5	16	Two-piston or roller quench	7	16	Kranjc and Stubičar[119]
5·9	20	Torsion catapult	5·5–8·1	7·4	Roberge and Herman[98]
5·9	20	Duwez gun	ND(<8)	14	Morinaga *et al.*[99]

ND = not detectable

tion energy for precipitation they found by electron microscopy of LQ samples of Al–Si alloys to *increased* retained vacancy concentration compared with SQ samples. A possible rationalization is that Agarwal *et al.* made their resistivity measurements on LQ samples made by the torsion catapult technique which could give rise to the same losses of quenched-in vacancies as for thicker layered regions of gun splats ('Dislocation structure', p. 177). Matyja *et al.* in contrast studied electron-transparent regions of unthinned gun splats known to be effective in retaining excess vacancies.

The presence of high volume densities of heterogeneous nucleation sites in LQ samples in the form of grain boundaries and stacking faults can result in a significant discontinuous decomposition at the expense of continuous transformations within the matrix (e.g. ref. 84). Thus spinodal decomposition evident (Fig. 5) in splat-quenched Al–22 at.% Zn[130] was displaced by discontinuous decomposition from grain boundaries for Al–28 at.-% Zn.[131] Precipitation along grain boundaries is also a feature of aging of LQ Al–14 at.-% Ag[98] while growth of non-equilibrium Al_6Fe at grain boundaries occurs alongside precipitation of equilibrium Al_3Fe within grains of aged Al–Fe extended solid solutions made by splat cooling.[91,114] These effects together with solute clustering during quenching limit the amount of further hardening that can occur as a result of aging. Age-hardening has been detected, however, for splat-cooled extended solid solutions of Al–0.25 to 2.5 at.-% Au,[96] Al–5 wt.-% Cr,[132] Al–6 and 12 at.-% Cu,[30] Al–0.7 at.-% Fe,[62] Al–1 and 2 at.-% Fe,[102] Al–3 at.-% Fe,[91,125,133] Al–4 at.-% Fe,[134] Al–4 at.-% Fe–0.3 to 1.5 at.-% Mn,[132,135,114] Al–4 at.-% Fe–0.3 at.-% Zr,[114,135] Al–3 at.-% Mn,[125] Al–3.6 at.-% Ni,[136] Al–Ni–Fe[32] and Al–1.3 at.-% V,[137] the effect being substantial in a number of cases (e.g. refs. 30, 102, 132, 135) (*see* Fig. 6). It is reported, however, that sequences of precipitation are not significantly affected for LQ Al–17 at.-% Cu[97] and Al up to 11 at.-% Si[129,138] compared with more dilute SQ alloys and, in the matrix, to be analogous to equally dilute SQ Al–Cu for LQ Al–0.2 to 3 at.-% Fe.[91,102,114,125,133]

Martensitic and massive transformations

Martensite formation via rapid quenching from the melt compared with normal solid-state quenching has been

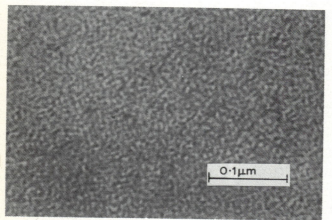

5   Spinodal decomposition in splat-cooled Al–22 at.-% Zn after 15 min in the microscope (from refs. 103, 130) ×200 000

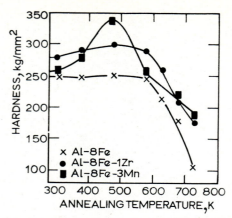

6   Age-hardening in spray-deposited Al–8 wt.-% Fe base alloys; hardness measured at room temperature after isothermal holding at elevated temperature for 1 h (from refs. 114, 135)

studied for Ni–35 at.-% Al,[139] Cu–17 to 30 at.-% Al,[140,141] Cu–12 at.-% Sn,[142] Cu–39 to 42 wt.-% Zn,[143] Ag–21 to 25 at.-% Al,[144,145] Ag–5 to 22 at.-% Ge,[140,146] (Au, Pd or Pt), Ti,[147] AuZn[148] and Au–47 to 50 at.-% Cd.[149,150] Chandrasekaran and Mukherjee[139] showed a refinement of twinned martensite spacing in Ni–35 at.-% Al from 0.03 μm when bulk-quenched within the solid state to 0.004 μm when splat-quenched from the liquid state (*see* Fig. 7). This alloy is typical in showing a martensitic or faulted structure already on quenching, raising the possibility that it formed directly by solidification rather than indirectly during quenching from a precursor phase formed from the melt. Warlimont and Furrer,[140] however, concluded that the latter two-stage process is operative in Cu–Al, Cu–Sn, Ag–Al and Au–Cd to which may be added Cu–Sn,[142] Cu–Zn[143] and Au–Zn,[148] for all of which the precursor is a bcc–B2 CsCl-type or A2-type Hume–Rothery 3/2 electron-compound. The transformation to martensite is slow enough in Au–Cd in fact to allow retention of untransformed β on quenching.[149,150] For Au–50 at.-% Cd, this phase decomposed at room temperature to three distinct martensitic phases.[149] One exhibited an unusually fine structure containing numerous planar defects. Another was characterized by ~1 μm 'packets' of transformation twins of spacing ~0.1 μm. The third structure was identified as lamellae of spacing 0.02 μm of two close-packed phases of different stacking sequences. For Au–47.5 at.-% Cd, transformation to martensite was induced by *in situ* beam heating in an electron microscope. Martensite laths nucleated heterogeneously in β from grain boundaries and occasionally at dislocations within grains and grew lengthwise while thickening on {110} planes to form a wedge shape propagating into the β-phase. For Ag–Al alloys, martensitic morphologies were already evident on quenching and three variants α', β', and γ' were distinguished in terms of stacking sequences.[144] The α'-phase (stacking ABCABC) was dominant at 21 at.-% Al as plate-like (thickness ~0.3 μm) with internal striations (spacing ~0.008 μm) due to stacking faults. The β'-phase (stacking ABCBCACAB), mainly at 22.5 at.-% Al, formed larger plates (thickness ~1 μm) with finer striations (spacing ~0.004 μm). The γ'-phase (stacking ABABAB) characteristic of 23.5 at.-% Al, occurred either as

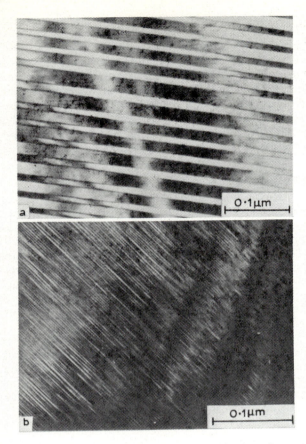

a conventional solid-state quench; *b* splat-quenched (from ref. 139) ×160 000

7 Refinement of martensitic substructure in Ni–35 at.-% Al by splat quenching

striated plates not really distinguishable from α' and β', or as alternating twin-related bands further subdivided by fine striations due to twins and stacking faults. Massive phases were found alongside the martensitic products in the same alloy samples.[145] Again three forms were distinguished: μ_m (22·5 at.-% Al alloy) had a domain size ∼0·02 μm (growing to 0·25 μm on beam heating) and matched the β–Mn structure; ζ_m (21–23·5 at.-% Al) formed as fine striated regions at grain boundaries and within grains of β: α_m (21·5 at.-% Al) was often duplex with ζ_m as striated lamellae ∼0·01 μm thick. It is notable that regular and randomly faulted close-packed structures with long period (up to 24 layer) stacking sequences observed by electron microscopy in the Ag–Ge alloys[140,146] confirmed earlier X-ray evidence for light to heavy faulting.[93]

EFFECT OF DEFECTS ON PROPERTIES
Enhanced mechanical properties attributable to defects produced by quenching from the melt have been reported even for pure metals. Kaufmann and Muller[25] obtained improved strength and ductility in hot-pressed Be attributable to the small grain size (7 μm in this case) derived from splat-quenching. Microhardness values for splat-quenched Al nearly double those of single-crystal Al have been reported,[63] decreasing with decreasing cooling rate.[68] This effect of cooling rate was associated with an increasing grain size for splat-cooled Nb and V.[70] Increased corrosion rate of Al gun-splats in air

polluted with SO_2 however was attributed to high vacancy concentrations, large surface areas, and possible absence of a continuous protective oxide coating.[59]

For alloys, examples of enhancement in properties attributable to grain refinement by quenching the melt include superplastic behaviour (in this case 600% elongation to fracture at 673 K when strained at 0·3 min^{-1}) in Al–8 at.-% Cu made by a two piston technique[151] and (*see* Fig. 8) increased elongation to fracture (at 623 to 773 K and 0·02 min^{-1}) in two commercial Al alloys made via melt spinning.[152] Similarly increased strength[27,33,38,45,132,134,135,137,153–155] maintained (e.g. Fig. 9) at increasing test temperature[38,132,134,135,137,153,154] is mainly due to more effective dispersion hardening in susceptible alloys such as Al–Si,[45,132] Al–V,[137] Al–Mn[45,132] and Al–Fe.[132,134,135,154] Increased fatigue and stress rupture lives resulting from processing via splat cooling[155] were attributed to refinement and a more even distribution of metallic inclusions (Fig. 10). Promising results were also obtained for Al–Nb alloys made via quenching the melt.[156] Increased fracture toughness,[157] wear resistance,[158,159] resistance to stress corrosion and exfoliation corrosion,[160] creep life[161] and conductivity[162] combined with high strength have been demonstrated even for relatively modest increases in cooling rates during solidification and even better results can be expected from splat cooling.

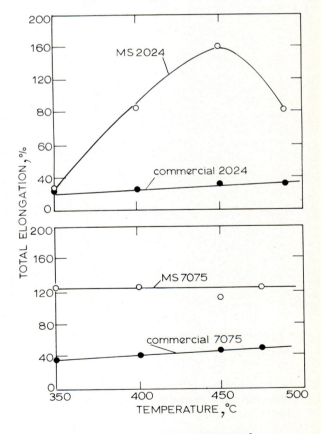

8 Elongation to fracture at 0·02 min^{-1} as a function of test temperature for 2024 and 7075 wrought aluminium alloys processed by melt spinning (MS) and extrusion compared with conventional product from ingot (from ref. 152)

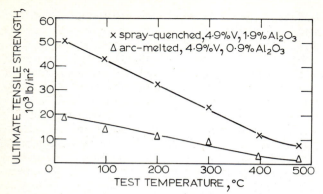

9 UTS as a function of test temperature for Al–4·9 wt.-% V made by plasma spray deposition compared with arc melting; products hot-pressed in the same manner before testing (after ref. 137)

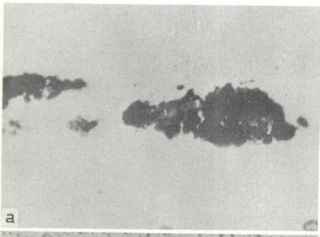

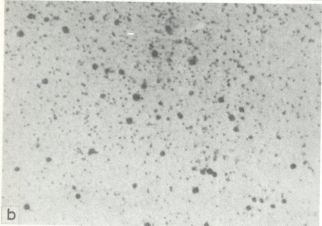

a conventionally processed from cast ingot; *b* splat-quenched and consolidated by extrusion (from ref. 155) ×1275

10 Refinement by splat-quenching of metallic inclusion size in commercial 2024-T4 wrought aluminium alloy

The refinement of crystalline microstructure originating these effects thus has the feature of being obtainable over the spectrum of available cooling rates and alloy compositions and it has been shown that consolidation of samples can be achieved without unacceptable losses of inherent property benefits.[38,135,137,152,155] The forma-

tion at accessible cooling rates of non-crystalline phases is, in contrast, severely restricted in alloy composition (e.g. ref. 163) and their lower stability (e.g. ref. 164) compared with crystalline counterparts raises the question of whether these phases can survive conventional consolidation. In the present context, however, the possibility of enhancing properties by forming refined duplex crystalline/non-crystalline microstructures, either by controlled heat treatment or during controlled consolidation, is intriguing (e.g. ref. 165).

CONCLUSIONS

1 Increased volume densities of vacancy defects can be attained by the technique of quenching from the melt provided that loss of vacancies to increased numbers of grain boundaries and other sinks is not too effective.

2 Corresponding solute clustering during quenching through the solid state can be more effectively suppressed by the higher cooling rates attainable, although clustering is a major effect in the new families of extended solid solutions that cannot be stabilized normally in the solid state.

3 This clustering and the high volume density of heterogeneous nucleation sites afforded by grain boundaries or stacking faults restricts age-hardening but dispersion-strengthening is very effective and the fine distribution of second phases enhances plasticity, endurance, and other properties.

REFERENCES

1 P. F. COWING: US Patent 809671, 1906
2 M. U. SCHOOP: British Patent 5712, 1910; 21066, 1911
3 G. FALKENHAGEN AND W. HOFMANN: *Z. Metallkunde*, 1952, **43**, 69
4 W. HOFMANN: *Aluminium*, 1938, **20**, 865
5 W. HOFMANN AND H. WIEHR: *Z. Metallkunde*, 1941, **33**, 369
6 E. SCHEIL AND Y. MASUDA: *Aluminium*, 1955, **31**, 51
7 E. SCHEIL AND R. ZIMMERMANN: *Z. Metallkunde*, 1957, **48**, 509
8 R. J. TOWNER: *Met. Progr.*, 1958, **73**(5), 70
9 P. DUWEZ *et al.*: *J. Appl. Phys.*, 1960, **31**, 1136, 1137, 1500
10 W. KLEMENT *et al.*: *Nature*, 1960, **187**, 869
11 P. DUWEZ AND R. H. WILLENS: *Trans. Met. Soc. AIME*, 1963, **227**, 362
12 R. H. WILLENS: Proc. 5th Int. Congr. on Electron Microscopy, Vol. 1, EE-6 (Ed. S. S. BREEZE), 1962, New York, Academic Press
13 P. DUWEZ: 'Progress in solid state chemistry' (Ed. H. REISS), Vol. 3, 377, 1966, Oxford, Pergamon Press
14 R. C. RUHL: *Mater. Sci. Eng.*, 1967, **1**, 313
15 P. RAMACHANDRARAO *et. al.*: *Phil. Mag.*, 1972, **25**, 961
16 H. A. DAVIES AND J. B. HULL: *J. Mat. Sci.*, 1974, **9**, 707
17 C. P. HINESLEY AND J. G. MORRIS: *Metall. Trans.*, 1970, **1**, 1476
18 N. J. GRANT: in 'Powder Metallurgy for High Performance Applications' (Ed. J. J. BURKE AND V. WEISS), Proc. 18th Sagamore Army Materials Conference, Syracuse University Press, 1972, p. 85
19 T. R. ANANTHARAMAN AND C. SURYANARAYANA: *J. Mat. Sci.*, 1971, **6**, 1111
20 H. JONES: *Rep. Progress Physics*, 1973, **36**, 1425
21 H. JONES AND C. SURYANARAYANA: *J. Mat. Sci.*, 1973, **8**, 705
22 T. R. ANATHARAMAN AND P. RAMACHANDRARAO: *Banaras Metallurgist*, 1973, **5**, 37
23 R. C. RUHL: *Mater. Sci. Eng.*, 1967, **1**, 313

24 H. JONES: in 'Rapidly quenched metals' (Eds. N. J. GRANT AND B. C. GIESSEN), 1, 1976, Cambridge, Mass., M.I.T. Press

25 A. R. KAUFMANN AND W. C. MULLER: in 'Beryllium technology', Vol. 1, 629, 1966, New York, Gordon and Breach

26 H. JONES AND M. H. BURDEN: J. Physics E. (Sci. Instruments), 1971, 4, 671

27 G. D. LAWRENCE AND G. S. FOERSTER: Int. J. Powder Metall., 1970, 6, 45

28 G. T. THURSFIELD AND H. JONES: J. Physics E. (Sci. Instruments), 1971, 4, 675

29 M. MOSS et al.: Appl. Phys. Lett., 1964, 5, 120

30 K. D. KRISHNANAND AND R. W. CAHN: as ref. 24, p. 67

31 H. WARLIMONT AND P. KUNZMANN: as ref. 24, p. 197

32 A. R. E. SINGER: Metals and Materials, 1970, 4, 246; J. Inst. Metals, 1972, 100, 185; Light Metal Age, 1974, 32, (9, 10), 5

33 J. P. H. A. DURAND et al.: Mater. Sci. Eng., 1976, 23, 247

34 A. R. KAUFMANN: J. Met., 1968, 20, 109A

35 P. PIETROKOWSKY: Rev. Sci. Instrum., 1963, 34, 445

36 R. POND AND R. MADDIN: Trans. Met. Soc. AIME, 1969, 245, 2475

37 R. E. MARINGER AND C. E. MOBLEY: J. Vac. Sci. Technol., 1974, 11, 1067

38 C. E. MOBLEY et al.: J. Inst. Metals, 1972, 100, 142

39 E. BABIĆ et al.: J. Physics E. (Sci. Instruments), 1970, 3, 1014; Fizika, 1970, 2, suppl. 2, paper 2

40 H. S. CHEN AND C. E. MILLER: Rev. Sci. Instrum., 1970, 41, 1237

41 P. E. BROWN AND C. M. ADAMS: Weld. J. Suppl., 1960, 39, 520–s; Trans. Amer. Foundrymen's Soc., 1961, 69, 879

42 K. MUTSUZAKI et al.: J. Japan Inst. Metals, 1963, 27, 424

43 H. KANEKO AND J. IKEUCHI: Proc. 1st Int. Conf. on Electrodischarge Machining, Tokyo, 1965, p. 23

44 H. JONES: Mater. Sci. Eng., 1969, 5, 1

45 B. LUX AND W. HILLER: Praktische Metallographie, 1971, 8, 218

46 M. LARIDJANI et al.: J. Mat. Sci., 1972, 7, 627

47 W. A. ELLIOTT et al.: Appl. Phys. Lett., 1972, 21, 23

48 B. CROSSLAND AND J. D. WILLIAMS: Metallurgical Rev., 1970, 15, 79

49 F. GALASSO et al.: Appl. Phys. Lett., 1966, 8, 331

50 F. GALASSO AND R. VASLET: Rev. Sci. Instrum., 1966, 37, 525

51 G. THOMAS AND R. H. WILLENS: Acta Met., 1964, 12, 191

52 J. A. McCOMB et al.: J. Phys. Soc. Japan, 1964, 19, 1691

53 R. W. BALLUFFI AND R. D. SIMMONS: Acta Met., 1964, 12, 957

54 G. THOMAS AND R. H. WILLENS: Acta Met., 1965, 13, 139

55 K. A. JACKSON: Acta Met., 1965, 13, 1081; see also 'Solidification', 121, 1971, American Society for Metals

56 G. THOMAS AND R. H. WILLENS: Acta Met., 1966, 14, 1385

57 R. KUMAR AND A. N. SINHA: Trans. Indian Inst. Met., 1968, 21, 9

58 C. SURYANARAYANA AND T. R. ANANTHARAMAN: ibid., 1968, 21, 67

59 C. JANSEN, B. C. GIESSEN AND N. J. GRANT: J. Metals, 1968, 20, 10

60 P. RAMACHANDRARAO AND R. R. ANANTHARAMAN: Trans. Met. Soc. AIME, 1969, 245, 892

61 A. KIRIN et al.: Scripta Met., 1969, 3, 943

62 D. ROČÁK et al.: Fizika, 1970, 2, suppl. 2, paper 21

63 A. PRODAN AND A. BONEFAČIĆ: ibid., paper 29

64 E. BABIĆ et al.: ibid., paper 30; Phys. Lett., 1970, 33A, 368

65 E. DARTYGE et al.: Acta Met., 1972, 20, 233

66 C. SURYANARAYANA: Physica Status Solidi, 1973, a18, K135

67 P. SHINGU et al.: J. Japan Inst. Met., 1973, 37, 433

68 P. FURRER: 'The microstructure and design of alloys', Vol. 1, 229, 1974, London, The Metals Society

69 A. KIRIN AND A. BONEFAČIĆ: J. Physics F (Metal Physics), 1974, 4, 1608

70 E. M. SAVITSKII et al.: Doklady Akad. Nauk. SSSR, 1973, 210, 405

71 E. KRAINER AND J. ROBITSCH: Z. Metallkunde, 1970, 61, 350

72 T. R. ANANTHARAMAN: Z. Metallkunde, 1970, 61, 760

73 E. KRAINER AND J. ROBITSCH: Metall., 1971, 25, 1361

74 H. A. DAVIES et al.: Nature Phys. Sci., 1973, 246, 13

75 H. A. DAVIES AND J. B. HULL: J. Mat. Sci., 1976, 11, 215; Mater. Sci. Eng., 1976, 23, 193

76 J. WOOD AND I. SARE: Metall. Trans., 1975, 6A, 2153

77 P. RAMACHANDRARAO et al.: Z. Metallkunde, 1970, 61, 471

78 A. V. ROMANOVA AND V. V. BUKHALENKO: Phys. Met. Metallogr., 1973, 35(6), 185

79 E. LAINE AND I. LAHTEENMAKI: J. Mat. Sci., 1971, 6, 1418

80 E. LAINE: Physica Status Solidi, 1972, 13, K27

81 H. KAHKÖNEN AND E. LAINE: J. Mat. Sci., 1974, 9, 1190

82 H. A. DAVIES AND J. B. HULL: Scripta Met., 1973, 7, 637

83 P. K. RASTOGI AND K. MUKHERJEE: Metall. Trans., 1970, 1, 2115

84 R. K. LINDE: Trans. Met. Soc. AIME, 1968, 236, 58

85 R. STOERING AND H. CONRAD: Acta Met., 1969, 17, 933

86 J. V. WOOD AND R. W. K. HONEYCOMBE: J. Mat. Sci., 1974, 9, 1183

87 C. G. DUNN AND J. L. WALTER: in 'Recrystallisation, grain growth and textures', 461, 1966, Cleveland, American Society for Metals

88 F. J. BRADSHAW AND S. PEARSON: Phil. Mag., 1957, 2, 570

89 G. F. BOLLING AND D. FAINSTEIN: Phil. Mag., 1972, 25, 45

90 H. A. DAVIES et al.: Scripta Met., 1974, 8, 1179

91 P. FURRER AND H. WARLIMONT: Proc. 7th Int. Congr. on Electron Microscopy, Grenoble, 1970, 507; Z. Metallkunde, 1973, 64, 236

92 B. JOUFFREY et al.: J. Physique, 1966, 27, C3-114 and Proc. 6th Int. Congr. on Electron Microscopy, Kyoto, 1966, Vol. 1, 81

93 P. RAMACHANDRARAO AND T. R. ANANTHARAMAN: Phil. Mag., 1969, 20, 201

94 P. FURRER et al.: Phil. mag., 1970, 21, 873

95 R. B. ROOF AND R. O. ELLIOTT: J. Mat. Sci., 1975, 10, 101

96 T. TODA AND R. MADDIN: Trans. Met. Soc. AIME, 1969, 245, 1045

97 M. G. SCOTT AND J. A. LEAKE: Acta met., 1975, 23, 503

98 R. ROBERGE AND H. HERMAN: Fizika, 1970, 2, suppl. 2, paper 6; J. Mat. Sci., 1973, 8, 1482; 1974, 9, 1123

99 M. MORINAGA et al.: J. Mat. Sci., 1974, 9, 1385

100 H. WARLIMONT et al.: Mater. Sci. Eng., 1976, 23, 101

101 Y. FUJINAGA et al.: J. Japan Inst. Met., 1968, 32, 1210

102 E. BLANK: Fizika, 1970, 2, suppl. 2, paper 24; Z. Metallkunde, 1972, 63, 315, 324

103 S. C. AGARWAL AND H. HERMAN: in 'Phase transitions' (Ed. L. E. CROSS), 207, 1973, Oxford, Pergamon Press

104 P. RAMACHANDRARAO AND T. R. ANANTHARAMAN: Trans. Met. Soc. AIME, 1969, 245, 890

105 M. PRATES AND H. BILONI: Metall. Trans., 1972, 3, 1501

106 J. A. HORWATH AND L. F. MONDOLFO: Acta Met., 1962, 10, 1037

107 R. E. SPEAR AND G. R. GARDNER: *Trans. Amer. Foundrymen's Soc.*, **71**, or *Modern Casting*, 1963, **43**, 209

108 B. P. BARDES AND M. C. FLEMINGS: *Trans. Amer. Foundrymen's Soc.*, **74**, 406, or *Modern Casting*, 1966, **50**, 100

109 W. A. DEAN AND R. E. SPEAR: in 'Strengthening mechanisms: metals and ceramics', Proc. 12th Sagamore Army Materials Conference (Eds., J. J. BURKE *et al.*), 268, 1966, Syracuse University Press

110 H. MATYJA *et al.*: *J. Inst. Metals*, 1968, **96**, 30

111 M. H. BURDEN AND H. JONES: *Fizika*, 1970, **2**, suppl. 2, paper 17

112 T. Z. KATTAMIS *et al.*: *J. Cryst. Growth*, 1973, **19**, 229

113 A. NAGATA *et al.*: *Trans. Japan Inst. Met.*, 1969, **10**, 52

114 M. H. JACOBS *et al.*: Proc. 25th Anniversary Meeting of EMAG, Inst. of Physics, 1971, 284; *J. Mat. Sci.*, 1974, **9**, 1631

115 I. S. MIROSHNICHENKO: *Russ. Metall. (Metal)*, 1968, No. 5, 128

116 M. H. BURDEN AND H. JONES: *J. Inst. Metals*, 1970, **98**, 249

117 M. G. SCOTT: *J. Mat. Sci.*, 1974, **9**, 1373; 1975, **10**, 269

118 H. A. KAHKÖNEN: *Metall. Trans.*, 1972, **3**, 739; *Ann. Acad. Sci. Fennica*, 1972, **A7**, No. 4

119 K. KRANJC AND M. STUBIČAR: *Fizika*, 1970, **2**, suppl. 2, paper 31; *Metall. Trans.*, 1973, **4**, 2631

120 H. A. KÄHKÖNEN AND M. T. YLI-PENTTILÄ: *J. Appl. Cryst.*, 1973, **6**, 412

121 A. TONEJC AND A. BONEFAČIĆ: *J. Appl. Phys.*, 1969, **40**, 419

122 E. BABIĆ *et. al.*: *Phys. Lett.*, 1970, **32A**, 5; *Fizika*, 1970, **2**, suppl. 2, paper 28

123 R. MISHIMA *et al.*: *Seisan-Kenkya*, 1971, **23**, 85: *J. Japan Inst. Light Metals*, 1974, **24**, 211

124 S. NASU *et al.*: *J. Physics F (Metal Physics)*, 1974, **4**, L24

125 A. FONTAINE *et al.*: Compt. rend., 1970, **B271**, 231; *Fizika*, 1970, **2**, suppl. 2, paper 23

126 A. BONEFAČIĆ *et al.*: *J. Mat. Sci.*, 1975, **10**, 243

127 E. BABIĆ *et al.*: *Physica Status Solidi*, 1973, **16**, K21

128 S. C. AGARWAL *et al.*: *Scripta Met.*, 1973, **7**, 365

129 H. MATYJA *et al.*: *Metall. Trans.*, 1975, **6A**, 2249

130 S. C. AGARWAL AND H. HERMAN: *Scripta Met.*, 1973, **7**, 503

131 S. AGARWAL *et al.*: *ibid.*, 401

132 P. ESSLINGER: *Z. Metallkunde*, 1966, **57**, 109

133 A. FONTAINE AND A. GUINIER: *Phil. Mag.*, 1975, **31**, 839

134 H. AHLBORN AND D. MERZ: *Aluminium*, 1971, **47**, 730

135 G. THURSFIELD AND M. J. STOWELL: *J. Mat. Sci.*, 1974, **9**, 1644

136 A. TONEJC *et al.*: *Fizika*, 1970, **2**, Suppl. 2, paper 7; *Acta Met.*, 1971, **19**, 311

137 M. MOSS: *Acta Met.*, 1968, **16**, 321; *Trans. Amer. Soc. Met.*, 1969, **62**, 201

138 M. ITAGAKI *et al.*: *Trans. Amer. Soc. Metals*, 1968, **61**, 330

139 M. CHANDRASEKARAN AND K. MUKHERJEE: *Mater. Sci. Eng.*, 1974, **14**, 97

140 H. WARLIMONT AND P. FURRER: *Fizika*, 1970, suppl. 2, paper 26

141 P. FURRER AND H. WARLIMONT: *Z. Metallkunde*, 1973, **64**, 626

142 W. VANDERMEULEN AND A. DERUYTERRE: in 'The mechanism of phase transformations in crystalline solids', Institute of Metals Monograph No. 33, 1969, p. 294; *Fizika*, 1970, **2**, suppl. 2, paper 8

143 S. SRINIVASA RAO *et al.*: *Indian J. Technol.*, 1971, **9**(1), 11

144 S. P. GUPTA: *J. Phys. Soc. Japan*, 1972, **32**, 1682; *Mater. Sci. Eng.*, 1972, **10**, 341

145 S. P. GUPTA: *Mater. Sci. Eng.*, 1973, **12**, 67

146 P. FURRER *et al.*: *Proc. Indian Acad. Sci.*, 1972, **75**, 103

147 H. C. DONKERSLOOT AND J. H. N. VAN VUCHT: *J. Less-Common Metals*, 1970, **20**, 83

148 P. FERRAGLIO *et al.*: U.S. Clearinghouse Report AD710807, 1971

149 K. MUKHERJEE AND P. FERRAGLIO: *Fizika*, 1970, **2**, suppl. 2, paper 27; *Acta Met.*, 1974, **22**, 835

150 P. FERRAGLIO *et al.*: *Acta Met.*, 1970, **18**, 1067

151 G. BEGHI *et al.*: *J. Mat. Sci.*, 1970, **5**, 820

152 B. A. WILCOX AND A. H. CLAUER: 'The microstructure and design of alloys', Vol. 1, 227, 1974, London, The Metals Society

153 D. M. SCHUSTER AND M. MOSS: *J. Metals*, 1968, **20**(10), 63

154 G. FANINGER *et al.*: as ref. 24, p. 483

155 M. LEBO AND N. J. GRANT: *Metall. Trans.*, 1974, **5**, 1547

156 G. BEGHI *et al.*: *J. Nucl. Mat.*, 1968, **26**, 219; 1968, **31**, 259

157 H. W. ANTES AND H. MARKUS: *Met. Eng. Q.*, 1970, **10**(4), 9

158 C. PANSERI AND M. PAGANELLI: *Aluminio*, 1966, **35**, 325; 1968, **37**, 387

159 H. M. SKELLY AND C. F. DIXON: *Int. J. Powder Metall.*, 1965, **1**(4), 28; 1971, **7**(3), 47; *Metal Progr.*, 1968, **94**(5), 103

160 J. P. LYLE AND W. S. CEBULAK: in 'Powder metallurgy for high performance applications' (Eds. J. J. BURKE *et al.*), 231 1972, Syracuse University Press; *Met. Eng. Q.*, 1974, **14**(1), 52; *Metall. Trans.*, 1975, **6A**, 685

161 G. BEGHI *et al.*: *Met. Ital.*, 1967, **59**(5), 333

162 P. K. ROHATGI AND K. V. PRABHAKAR, *Metall. Trans.*, 1975, **6A**, 1003

163 S. TAKAYAMA: *J. Mat. Sci.*, 1976, **11**, 164

164 H. S. CHEN: *Acta Met.*, 1974, **22**, 1505

165 S. TAKAYAMA AND R. MADDIN: *J. Mat. Sci.*, 1976, **11**, 22

166 J. J. GILMAN: *Phys. Today*, 1975, **28**(5), 46

167 P. DUWEZ: *Ann. Review Mater. Sci.*, 1976, **6**, 83

Session IV
Oxidation

Chairman: D. R. Holmes (Central Electricity
Research Laboratories)
Technical Secretary: J. Stringer (University of
Liverpool)

High-temperature oxidation
of metals

J. Stringer

The overall processes involved in the reactions of a metal with an oxidant to form a solid reaction product are described. Transport in the oxide is summarized, both by the movement of lattice defects and by short-circuit processes, and the location of the oxide growth step is discussed in terms of the transport. The problem of maintaining contact between oxide and metal in the case where the oxide grows at the oxide/oxidant interface is analysed, and the nature of oxide/metal adhesion, the structure of the oxide/metal interface, and the general problem of the transfer of atoms from the metal to the oxide are discussed. It is shown that vacancy injection is not a necessary requirement of any oxidation process, but may occur in an attempt to accommodate loss of metal volume without having metal/oxide contact; in this case the vacancies will tend to condense as voids within the metal.

The author is at the University of Liverpool

At elevated temperatures, metals react with gaseous constituents of their environments, and this can destroy structural integrity in a number of ways. First, and most obviously, useful structural metal is removed to form the reaction product. Secondly, the reaction will occur more readily with some constituents of the alloy than with others, and the effect of this may be to remove useful strengthening elements from a depth much greater than the thickness of the superficial product. Thirdly, the reactant can penetrate rapidly along grain boundaries and similar structural discontinuities, again affecting the strength much more than would be anticipated.

The important species that can participate in reactions of this kind include oxygen, nitrogen, chlorine, bromine, sulphur, and carbon among others; but the net effect of the reaction is to remove electrons from the metal atoms, and this process is referred to as oxidation. For convenience, in this paper the reaction product will be referred to as the oxide but the principles are quite general.

TRANSPORT IN THE OXIDE

In the simplest and most desirable situation the oxide forms a continuous layer over the metal surface and acts as a barrier between the reactants. Further reaction then involves transport of one or both reactants across the oxide. The transport may be by bulk diffusion or by a variety of short-circuit processes, and clearly the slowest and thus the most desirable process is bulk diffusion.

Most oxides have a limited stoichiometry range, and transport involves the movement of one or more defect species. In the simplest situation the defects can be treated as unassociated point defects, and if the oxide is regarded as an ionic solid the possible defects are anion or cation vacancies, or interstitials of either species. If there is, for example, a cation vacancy then the lattice in the vicinity of the defect will have a net negative charge, and it is a premise of the theory that this is not permitted; local electrical neutrality must be restored in one way or another. Clearly, this can be done by introducing an interstitial cation or an ion vacancy, both of which conserve not only electrical neutrality but also stoichiometry. However, if matter is to be transported this implies a concentration gradient and hence a departure from stoichiometry. The alternate method is to balance the charge with electronic defects: in this case by an appropriate number of positive charges. Depending on the nature of bonding, these charges may be located on nearby atoms or in extended lattice states.

The mobility of the ionic defect and the electronic defect will in general be different, and thus in an oxide growing as a result of the chemical activity gradient a charge difference will develop between the inside and outside of the scale. The sign of the charge difference will tend to slow down the faster-moving species and accelerate the slower-moving, until at steady state the fluxes will be equal. Because the oxidation represents a removal of electrons from the metal, and an addition to the oxidant, the process always occurs so that there is a net flux of electrons outwards through the scale, and thus a net movement of negatively charged ionic defects inwards. If the electronic mobility is higher than the ionic, therefore, the charge difference will be such that the oxide/oxidant interface is negative with respect to the oxide/metal interface; the reverse is true if the ionic mobility is higher than the electronic. This is true irrespective of the detailed nature of the defect species. An oxide in which the electronic species is more mobile is called a semiconductor; the reverse case is termed an ionic conductor.

The defects are injected by reactions at the interfaces. Thus, at the metal/oxide interface the reactions

$$M_m = M_{i,o}$$
$$M_{i,o} = M_{i,o}^{p+} + pe_o \qquad (1)$$

may occur, where the metal is represented as M, the subscript m signifies the bulk metal, and o the oxide at the metal/oxide interface; the subscript i indicates an interstitial. The injection and the ionization reactions are here separated, but will probably occur together. Equivalent reactions can be written for the injection of an anion vacancy at the metal/oxide interface, and for the injection of anion interstitials or cation vacancies at the oxide/oxidant interface. The defects are also eliminated by interface reactions; thus, at the metal/oxide interface

$$M_{v,o}^{p+} + pe_{v,o} + M_m = nil \qquad (2)$$

where the subscript v signifies a vacancy.

Finally, the oxide is created somewhere. It is usual to assume that one ionic defect moves much more rapidly than any of the others, and in this case it is effectively the only one to move: the oxide is then created at one or other interface. The scale-forming reaction may inject defects:

$$pM_m = M_pO_q + qO_{v,o}^= + 2qe_0 \qquad (3)$$

or eliminate them:

$$M_{i,x}^{2+} + \tfrac{1}{2} O_2 + 2e = MO \qquad (4)$$

In this equation the monoxide has been written to avoid needless overcomplexity.

Now, these concepts of the injection of defects, the movement of defects, and the creation of oxide are quite general, and do not depend on the details of the process. To discuss kinetics does require some further modelling, which becomes less general. Before doing that, consider some of the simplifications introduced above. The concept of local electrical neutrality clearly falls down if the oxide is very thin, since it will be difficult to develop a charge difference without transporting all the mobile electronic defects right through the film. An example is the tunnelling of electrons through thin films. For thick films however the assumption can be justified in some detail, at least for the bulk of the film. However, in the immediate vicinity of both interfaces there will be a region of charge imbalance called the space–charge region, resembling (but not identical to) the well-known space–charge region at a metal/semiconductor junction, or the double layer at electrode surfaces. The thickness of the space–charge regions will depend on a number of factors: for example, in the case of the oxidation of copper to form Cu_2O the thickness of the space–charge

region at the metal/oxide interface may be written[1]

$$\lambda_0 = \frac{\chi kT}{24\pi e^2 \eta_0} \qquad (5)$$

where χ is the dielectric constant of the oxide, k is Boltzmann's constant, e is the electronic charge, and η_0 is the concentration of cation vacancies at the metal/oxide interface.

For the oxidation of copper in oxygen at 1 000°C this has a value of 15 Å. In other cases, the space–charge region may be as much as a few hundred Angstroms.

However, the assumption of local electrical neutrality is not all that important in relation to the defect reactions written above: it is of more importance in discussing reaction kinetics, or the influence of impurities on the oxidation.

A more important assumption is that the lattice defect species are unassociated point defects. It is well known that this is not so: very early in the development of oxidation theory Wagner[2] showed that in the case of the formation of thick Cu_2O films on copper, assuming independent point defects, the parabolic rate constant k_p should depend on the $\tfrac{1}{8}$th power of the oxygen pressure,

$$k_p \propto p_{O_2}^{\frac{1}{8}} \qquad (6)$$

whereas experimentally it was shown that[3]

$$k_p \propto p_{O_2}^{\frac{1}{7}} \qquad (7)$$

Wagner was able satisfactorily to explain this discrepancy by taking account of the interaction of the point defects using the Debye–Huckel theory of strong electrolytes, although this too is now recognized as an oversimplification.

More recently, high-resolution electron microscopy has demonstrated the existence of very extended defects in some non-stoichiometric oxides, notably Nb_2O_5 and TiO_2;[4] and calculations of defect interactions have suggested the likelihood of rather smaller defect aggregates in other oxides. It is not yet clear to what extent the existence of associated defects modifies the simple equations 1–4; since the extended defects must be in equilibrium with unassociated point defects it is possible that these equations may still be conceptually valid, although determination of equilibrium constants from experimental data will not usually be possible. Transport theory, and hence detailed reaction kinetics, will almost certainly have to be modified.

The degree of ionization of defects also changes with increasing departure from stoichiometry, as might be expected; this also does not change the qualitative validity of the simple equations, but makes estimation of the effect of impurities difficult. The experimental data have been reviewed in some detail by Kofstad.[5]

Much more important is the assumption of transport by the motion of lattice defects of whatever sort. There is increasing evidence that oxidation can proceed by the transport of material down short-circuit paths of one sort or another; the data have recently been reviewed by Stringer.[6] An important example is the oxidation of Ni–Al alloys to form Al_2O_3: Pettit and Tien[7] showed by means of marker experiments that at 1 100°C Al_2O_3 grew by the movement of oxygen inwards (small platinum markers were at the scale/oxygen interface after the reaction). However, experiments on single crystal samples of Al_2O_3 have shown that the cation diffuses several orders of magnitude faster than the oxygen, and that neither diffuse rapidly enough to explain the observed oxidation rate. Oishi and Kingery[8] demonstrated that in a polycrystalline compact of Al_2O_3 with a mean grain size of $100 \mu m$, the aluminium diffusion rate was essentially the same as in the single crystal, but the oxygen transport was much faster, with a considerably reduced activation energy; in fact the oxygen transport rate was substantially the same as that of the aluminium. The grain size of the Al_2O_3 formed in the oxidation of alloys at elevated temperatures is much smaller still; typically of the order of $1 \mu m$, and it seems certain therefore that the dominant transport process in the oxidation is the grain-boundary diffusion of oxygen. For this, the defect reactions written above are wholly inappropriate, and the calculation of reaction rates or of the effect of impurities is not possible without a detailed model which does not presently exist. How common is short-circuit diffusion? It is difficult to be sure, since the estimation of reaction rates from diffusion data is not simple; the interpretation of marker experiments is not straightforward, and in few other cases are data quite as clear as they are for Al_2O_3. However, there is increasing evidence that *sometimes* the growth of Cr_2O_3 may involve the transport of oxygen, probably by a short-circuit mechanism: arguments have been presented to suggest that the more usual chromium transport is also by a short-circuit process, but this is certainly a minority view.[9] The growth of NiO on nickel almost certainly involves short circuit transport; in this case Wolf and Rhines[10] have argued that oxygen is transported along the columnar grain boundaries, and nickel through the bulk of the oxide (by a cation vacancy process) with *some* of the oxide forming with the bulk of the scale layer at the oxide grain boundaries. The major portion of the oxide forms at the scale/oxygen interface in the normal way. Gibbs[11] has discussed some related mechanisms, but whatever the details of the models, the kinetic data indicating the existence of a short-circuit contribution remain valid. Similar data have been presented for the oxidation of iron.

OXIDE FORMATION

Metal transport outwards

However the reactants are transported, at some point oxide is formed, perhaps by a reaction such as (3) or (4). If the oxide is formed at the oxide/oxygen interface there would seem to be little problem in accommodating the newly formed oxide: the situation is essentially the same as the precipitation of a solid from the liquid or gas, and one would expect growth on existing surface steps such as emergent screw dislocation steps. However, if this happens, all the oxide is formed outside the first-formed layer, which presumably was at the original metal surface. For a finite-sized specimen, and ignoring the problem of transfer at the metal/oxide interface, this would eventually result in a hollow block of oxide with a hole in the centre the size and shape of the original specimen. Experiments seldom come out as cleanly as this, but in essence one can obtain this result. However, sometimes even with oxides which are believed to form at the oxide/oxygen interface, complete oxidation results in a fairly compact oxide block, with little or no internal porosity. There has been a great deal of dis-

cussion of this behaviour, without any wholly satisfactory explanation having been proffered.

Commonly, the scale has neither of these two appearances. Instead, there is an outer compact layer of scale, often formed of large columnar crystals, and an inner fine-grained oxide layer, which is often porous. It has been suggested that this results from a dissociation reaction[12] in which first the metal loses contact with the oxide. Then, the innermost oxide dissociates, liberating oxygen which diffuses across the gap to the metal, there reacting to form fresh oxide; and metal which diffuses out to the scale/oxygen interface. Brückman and Mrowec and their co-workers[13] have developed this model, pointing out that dissociation will take place more readily at the columnar oxide grain boundaries, thus opening paths towards the surface. When these paths finally reach the surface of the columnar oxide, gaseous oxygen is admitted to the inner layer.

On coupon specimens of pure metals, the scale appears often to be able to maintain contact with the centre of the broad faces for a considerable time, losing contact early at the edges and corners where the fine-grained inner layer appears, producing a characteristic 'dog-bone' appearance in cross-section. Brückman and Mrowec suggest that the scale is able to relax after the receding metal interface by a form of plastic deformation, but this is impossible at the edges and corners. In support of this, specimens containing small amounts of a more reactive element which oxidized internally broke away uniformly over the whole specimen surface so that the boundary between the outer and inner layers was the same as the original shape of the specimen. Brückman[14] was even able to restrain the scale from relaxing by the use of pins normal to the specimen surface. Most of the work of Mrowec, Brückman, and their co-workers has been concerned with sulphidation rather than oxidation, but the morphologies and mechanisms appear to be fairly similar.

Attractive as this explanation is, it implies a very high degree of plasticity in the oxide. Stringer[15] has reviewed the plastic deformation of oxides and has expressed the opinion that plastic strains of greater than 1 or 2% are extremely improbable; and Hancock and co-workers[16] have shown that oxides on metals do indeed fracture at tensile strains essentially equal to the bulk fracture strain, which is normally very small. It is of course possible that an oxide may be capable of undergoing compressive plastic strains of larger magnitude, and an oxide may possibly deform by creep at a rate comparable to that at which it is growing to very high strains. Stringer[17] has introduced the concept of 'stress-directed growth' to account for these and similar observations. Borie et al.[18] have observed a variation in dislocation density through an oxide film to accommodate differences in lattice parameter between the outside and inside, and discussed the possibility that oxide growth might take place by the addition of atoms to the extra half-plane in suitable dislocations, a process clearly analogous to creep by dislocation climb. They concluded that such a process could account for only a few percent of the oxide growth in the particular case they examined, but clearly it is a possibility that must be examined for each reaction.

If the process is one of plastic deformation, it is worth asking where the stress comes from to produce the plastic flow. The only possibility is the oxide/metal adhesion. Adhesion between metal and oxide is very imperfectly understood: it is possibly the least well-understood aspect of oxidation reactions. Adhesion can result from the following causes:

(i) close contact of two surfaces; if two surfaces are in intimate contact there is an adhesion between them which becomes quite large if the distances are of atomic dimensions

(ii) electrostatic interactions; the existence of the space–charge layer will result in an attraction between the oxide and the metal

(iii) chemical bonding; if there exists a bond between the innermost anions of the oxide and the metal, there will be an attractive force analogous to the chemical bonding responsible for cohesion in the oxide itself.

It is not possible quantitatively to estimate the magnitudes of these effects, but probably the adhesion between the oxide and the metal is at the maximum one half of the cohesion of the oxide. Adhesion has been discussed in a little more detail elsewhere.[17]

An alternative approach is to look at the so-called growth stresses in the oxide itself. If the adhesion is compelling the oxide to contract plastically onto a receding metal surface, the observed growth stress in the oxide will be compressive and of the order of the plastic flow stress. Notice that this is quite different to the Pilling–Bedworth model for growth stresses, which would predict no growth stresses for an oxide growing by cation transport.[15] Again, there is little experimental evidence, but it seems likely that the growth stresses are usually significantly smaller than would be expected unless the flow stress is much lower in a growing oxide.

Furthermore, in some cases it has been possible to oxidize metals completely virtually without any voids being formed, and this implies a very large plastic strain for the first-formed oxide layer.[15] On balance, while plastic deformation of growing oxides may well be possible, the ability of the oxide to follow the receding metal layer most probably involves some other process, probably involving inward transport of oxygen.

Oxygen transport inwards

If the oxide-forming reaction takes place at the metal/oxide interface, the situation is clearly different. To describe it in terms similar to those given above, if the metal is oxidized to completion, all the oxide formed must be contained within the envelope of the first-formed oxide layer. If the oxide has a larger volume than the metal from which it was formed, clearly the effect will be to dilate the first-formed envelope, which will thus be under tension, and compress the newly forming oxide. Conversely, if the oxide has a smaller volume than the metal from which it was formed, either voids will appear or interface forces will attempt to collapse the first-formed envelope: the newly formed oxide will thus be in tension. This is the point first made by Pilling and Bedworth,[19] who further suggested that since oxides tend to be stronger in compression than tension, the first situation would result in a protectively forming scale, whereas in the second situation the scale would tend to fail in tension on forming and thus be non-protective. They recognized that in the first case the outer layers of

the scale would be in tension, on a finite specimen, and thus limited their remarks to concave surfaces.

For the majority of metals, the oxide exceeds the volume of the metal from which it was formed, so that most should form in compression. Often, it is implied in discussions of the Pilling–Bedworth model that the growth stresses in oxides arise from the elastic isotropic deformation of the oxide to equal the original metal volume, but this is plainly absurd: the volume change can be as much as 200%, and even in what is regarded as a nearly ideal ratio is normally over 20%. The main part of the volume change must be accommodated in some other way, and it is usual (again) to suggest that plastic flow of the oxide occurs. This and other possibilities have been discussed by Stringer.[15,17]

Atom transfer processes at the metal/oxide interface

In relation to the problem of the injection of vacancies, the nature of the processes by which an atom is transferred from the metal to the oxide is clearly important, and this requires a model for the metal/oxide interface. In the case of some oxides in the thin film region there appears to be a good correspondence between the metal atoms in the oxide and those in the metal: the film is then referred to as epitaxial. Thick oxide films however do not in general appear to be epitaxial, although in a number of cases the orientation of the oxide grains is clearly non-random: this seems to be related to anisotropy in the growth process rather than to interface constraints. It is probable that the metal/oxide interface most closely resembles an incoherent phase boundary, although it may possess some degree of coherency in some cases.

Vermilyea[20] was the first to point out that metal removal from the metal surface might well take place at a step, and that the oxide might accommodate to that step by bending rather than by matching it with a corresponding step. This model was for the growth of a scale in which the metal atoms were mobile, but Stringer[17] made the logical extension that steps might also exist on the oxide surface at the interface, and new oxide forms at these positions in the case where oxygen was mobile. The effect of these suggestions is that there need be no intrinsic difference in the metal removal process depending on the mobile species in the scale or the location of the scale-forming reaction.

Oxidation of alloys

The discussion so far has, by implication, confined itself to the oxidation of pure metals. The oxidation of alloys is a much more complex problem, lying outside the scope of this paper; but in the practically important situations an alloy will form a protective scale consisting of the oxide of one of its constituents. This introduces a new variable, because the constituent which is oxidizing must diffuse from the body of the alloy to the metal/oxide interface, and there is thus necessarily a requirement for the diffusion of the non-reacting elements away from the interface. If there is an inequality in the rates of these processes there will be a net vacancy flux, with the possibility of the formation of Kirkendall voids, and indeed gross void formation in alloys can be produced by suitable oxidation—for example, in the oxidation of Ni–20 Cr alloys in high-velocity air.[21]

Growth stresses and substrate creep

It has already been pointed out that it is observed experimentally that a stress system is established in the vicinity of the surface of an oxidizing metal, although there is still considerable argument about the origin of the stresses.[15,17] Nevertheless, there is very clear evidence that the stress can be sufficiently high to cause macroscopic creep of components. Noden,[22] for example, showed that steel fuel element cans extended significantly as a result of oxidation; Jones[23] reported a contraction in magnesium, although others have reported extensions: Antill[24] oxidized Magnox coated with carbon in a CO_2 atmosphere containing 1% decane at 9 atm. pressure at 500°C, and at the end of 500 h extensions of nearly 20% were observed.

VACANCY INJECTION DURING OXIDATION

Other papers will discuss the injection of vacancies during oxidation in more detail. The earliest work suggesting that the loss of adhesion between scale and metal was due to the condensation of vacancies at the interface appears to have been that of Dunnington, Beck and Fontana;[25] Tylecote and Mitchell[26] were able to maintain adhesion for longer periods by drilling holes in specimens near to the metal surface to act as artificial vacancy sinks. Hancock and Fletcher[27] observed large grain-boundary voids in nickel after oxidation, which they attributed to the condensation of vacancies injected by the oxidation. Stringer[28] suggested that the improvement in the adhesion of Cr_2O_3 scales produced by rare-earth additions to heater alloys could be attributed to their acting as vacancy sinks, preventing the nucleation of voids at the scale/metal interface. More recently, several investigators have used transmission electron microscopy in elegant demonstrations that vacancies can indeed be injected during oxidation; some of these will be referred to later.

The point of the preceding analysis has been to show that in macroscopic terms systems in which the oxide grows by metal transport outwards have to accommodate the reducing metal volume by:

(i) losing contact between metal and oxide
(ii) forcing the oxide to contract round the shrinking metal, presumably by plastic flow
(iii) forming voids within the metal
(iv) initiating a process allowing transport of oxygen inwards.

In the last analysis, vacancy injection provides a mechanism to allow some of these processes to occur. The equilibrium concentration of vacancies in solution in the metal will always be too low to make any significant contribution to the volume change. In addition, vacancy injection may encourage substrate creep, but this is not of particular significance in terms of the approach used here, which considers the oxidation of a finite specimen. Clearly, the most important process for which vacancy injection would play a significant role is that in which voids are to be formed in the metal.

Which process will be favoured in any particular system? Various authors have discussed the different situations of 'strong interface, weak oxide' (thus encouraging plastic contraction of the oxide) or 'weak interface, strong oxide' (encouraging void formation at the interface, leading to scale separation); but these imply that gross plastic deformation of the oxide produced by adhesion stresses is possible: this is at least arguable. However, when it comes to void formation, it is perhaps legitimate to conclude that the most important criterion will be surface energy, and if a void within the metal has a lower total surface energy than the stable void at the metal/oxide interface having the same volume, then void formation in the metal (and thus vacancy injection) will be favoured. In this calculation, there will be a strain energy term since the formation of a void at the metal/oxide interface will relax to some extent the growth stresses in the immediate vicinity of the void: the effect of this term will be to tend to favour interface voids.

Since vacancy injection appears to be a consequence of the relative volume changes, it would appear that no vacancy injection should take place if an oxide forms by oxygen transport provided the Pilling–Bedworth ratio is greater than one. However, if one of the other methods of accommodating the oxide volume change discussed by Stringer takes place, and recalling the 'stepped interface' model for atom transfer, vacancy injection is possible: compressive growth stresses are observed for scales growing by both methods of transport, so there seems little reason for believing there to be significant differences in the interface processes.

On balance therefore it must be concluded that the incorporation of a metal atom into the oxide does not necessarily result in the injection of a vacancy, any more than does the evaporation of a metal atom from a metal surface. Vacancy injection, if it occurs, does so because the oxidizing system is attempting to maintain oxide/metal cohesion under circumstances where the metal would otherwise shrink away from the oxide, and the function of the injected vacancy is ultimately to condense at a void within the metal. Whether or not this happens depends on the mechanisms responsible for the shape changes associated with the formation of the oxide and the adhesion at the metal/oxide interface. Since no satisfactory models exist for either of these, detailed predictions of behaviour are not possible.

REFERENCES

1 T. B. GRIMLEY: in 'Chemistry of the solid state', (Ed. W. E. GARNER), 336, 1955, London, Butterworths
2 C. WAGNER: Z. Phys. Chem., 1933, B21, 25; 1936, B32, 447
3 C. WAGNER AND K. GRÜNEWALD: Z. Phys. Chem., 1938, B40, 455
4 J. S. ANDERSON: in 'Defects and transport in oxides' (Eds. M. S. SELTZER AND R. I. JAFFEE), 25, 1974, New York, Plenum Press
5 P. KOFSTAD: 'Stoichiometry, diffusion, and electrical conductivity in binary metal oxides', 1972, New York, Wiley
6 J. STRINGER: in 'Defects and transport in oxides' (Eds. M. S. SELTZER AND R. I. JAFFEE), 495, 1974, New York, Plenum Press
7 J. K. TIEN AND F. S. PETTIT: Metall. Trans., 1972, 3, 1587
8 Y. OISHI AND W. D. KINGERY: J. Chem. Phys., 1960, 33, 480
9 J. STRINGER et al.: Oxid. Metals, 1972, 5, 11
10 F. N. RHINES AND J. S. WOLF: Metall. Trans., 1970, 1, 1701
11 G. B. GIBBS: Corrosion Sci., 1967, 7, 165
12 N. BIRKS AND H. RICKERT: J. Inst. Metals, 1962–63, 91, 308

13 S. MROWEC: *Corrosion Sci.*, 1967, **7**, 563; A. BRÜCKMAN: *ibid.*, 51; A. BRÜCKMAN *et al.*: *Fiz. Metal. i Metalloved.*, 1963, **15**, 362; 1965, **20**, 702; A. BRÜCKMAN *et al.*: *Oxid. Metals*, 1969, **1**, 241

14 A. BRÜCKMAN: personal communication

15 J. STRINGER: *Corrosion Sci.*, 1970, **10**, 513

16 G. WARD *et al.*: *Metall. Trans.*, 1974, **5**, 1451

17 J. STRINGER: *Werkstoffe u. Korrosion*, 1972, **23**, 747

18 B. BORIE *et al.*: *Acta Met.*, 1962, **10**, 691

19 N. B. PILLING AND R. E. BEDWORTH: *J. Inst. Metals*, 1923, **29**, 529

20 D. A. VERMILYEA: *Acta Met.*, 1957, **5**, 492

21 F. J. CENTOLANZI: 'Hypervelocity oxidation tests of thoria dispersed nickel–chromium alloys', NASA-TM-X62,015, Feb. 11, 1971

22 J. D. NODEN *et al.*: *Brit. Corrosion J.*, 1968, **3**, 47

23 R. B. JONES: private communication (*see* ref. 14 in ref. 22)

24 J. E. ANTILL *et al.*: *J. Nucl. Mat.*, 1970, **36**, 1

25 B. W. DUNNINGTON *et al.*: *Corrosion*, 1952, **8**, 2t

26 R. F. TYLECOTE AND T. E. MITCHELL: *J. Iron Steel Inst.*, 1960, **196**, 445

27 P. HANCOCK AND R. FLETCHER: *Journées Int. d'Etude sur l'Ox. des Metaux*, 1965, 70, Brussels

28 J. STRINGER: *Metallurgical Rev.*, 1966, **11**, 113

Microscopic observations of oxidation/point defect behaviour

R. E. Smallman and P. S. Dobson

The growth of an oxide film on a metallic surface involves mass transport of either metal ions leading to oxide growth at the oxide–oxygen interface or oxygen ions leading to growth at the oxide–metal interface. The importance of point defects in the oxide on these diffusion processes is well understood, but the interrelationship between point defects in the oxide and the metal has received less attention. When vacancies are the predominant point defects in the metal then the transfer of metal atoms into the oxide and their subsequent diffusion must leave behind vacant lattice sites in the metal. Conversely, the substitutional diffusion of oxygen through the oxide and its eventual incorporation into the metal near the interface must result in a flow of vacancies from metal to oxide. Thus an oxide growing by cation diffusion will produce vacancies in the metal whereas an oxide growing by anion diffusion will continually absorb metallic vacancies (or produce interstitials). This paper discusses the experimental evidence for point defect injection during oxidation.

The authors are at the University of Birmingham

The growth of an oxide film on a metallic surface involves mass transport of either metal ions leading to oxide growth at the oxide/oxygen interface or oxygen ions leading to growth at the oxide/metal interface. When vacancies are the predominant point defects in the metal then the former process gives rise to vacant lattice sites in the metal while the latter, involving the substitutional diffusion of oxygen through the oxide and its eventual incorporation into the metal near the interface, will result in a flow of vacancies from metal to oxide. The vacancies injected into the metal during oxidation may be absorbed by the climb of dislocations or by the nucleation and growth of voids at grain boundaries and at the oxide/metal interface, while for the converse situation vacancy absorption by the oxide has been found to result in the growth of defects of interstitial character.

The most convincing and direct evidence for the generation of excess vacancies in a metal by surface oxidation has come from direct observations of either the growth or shrinkage of lattice defects, such as dislocation loops or voids, in the vicinity of an oxidizing interface. The two techniques which have been used are: (*a*) transmission electron microscopy and (*b*) X-ray topography and this paper briefly reviews the observations made from the injection of point defects during oxidation.

ELECTRON MICROSCOPE OBSERVATIONS

If a thin foil contains a dislocation loop then, provided that the temperature is sufficiently high to enable diffusion to occur, the loop will either grow or shrink depending on the sense of the vacancy flux between the loop and the foil surface. The rate of change of the loop radius, r, with time is given by:

$$\frac{dr}{dt} = AD_s \left[\frac{c_s}{c_0} - \frac{c_L}{c_0} \right] \tag{1}$$

where A is the geometrical factor depending on the diffusion geometry and the Burgers vector of the dislocation, D_s is the coefficient of self-diffusion, c_0 is the equilibrium vacancy concentration, and c_s is the vacancy concentration in equilibrium with the foil surface. The vacancy concentration in equilibrium with the loop, c_L, is given by $c_L = c_0 \exp(\Delta F / kT)$ where ΔF is the change in the loop energy per vacancy emitted ($= dE/dn$).

If the dislocation loop is of vacancy type then ΔF is positive and providing that the surface acts as a perfect vacancy sink, i.e. $c_s = c_0$, the loop will shrink on annealing. This type of behaviour is observed for dislocation loops enclosing a stacking fault in pure aluminium when annealed at 140°C and above. For this faulted defect the self-energy of the loop is dominated by the stacking fault energy γ and thus $\Delta F \simeq \gamma B^2$ where B^2 is the cross-sectional area of a vacancy in the (111) plane. For large loops the diffusion geometry approximates to cylindrical diffusion and a solution of the time independent diffusion equation ($\nabla^2 C = 0$) gives for the annealing rate,

$$dr/dt = -\{2\pi D/b \ln(L/b)\}[\exp(\gamma B^2/kT)-1] \quad (2)$$

where $D = D_0 \exp(-U/kT)$ is the coefficient of self-diffusion and L is half the foil thickness. A linear dependence of loop radius on annealing time is observed[1,2] for loop radii greater than 500 Å. In accordance with this equation an increased shrinkage rate is observed at small radii due to the contribution of the dislocation line energy to the total loop energy. Prismatic dislocation loops, i.e. not enclosing a fault, have also been observed to shrink in aluminium and in this case the driving force is provided by the self-energy of the dislocation itself. For a prismatic loop $\exp(\Delta F/kT)$ can be approximated to $\{1+(\alpha b/r)\}$ where α is a constant determined by the elastic constants of the material and the annealing rate becomes:

$$dr/dt = -\{2\pi D/b \ln(L/b)\}(\alpha b/r) \quad (3)$$

A parabolic relationship is observed between loop radius and annealing time[2] which confirms that (c_s/c_0) is very close to unity. From a comparison of the annealing rates of faulted and prismatic loops at the same temperature, the stacking fault energy γ and the coefficient of self-diffusion D have been determined.[2]

When thin foils of materials other than aluminium are annealed either outside the electron microscope or *in situ* it is often observed that dislocation loops in the metal grow rather than shrink. This phenomena can occur in a wide variety of materials, both pure metals and alloys, with totally different structures.

Hexagonal metals

The growth of dislocation loops in zinc at room temperature was first observed by Berghezan *et al.*[3] who suggested that the excess vacancies absorbed by the growing loops were injected into the foil as a consequence of surface oxidation. Further experiments on this effect were carried out by Price[4] who suggested that the excess vacancies were created by ion damage of the foil during observation in the electron microscope. This controversy was resolved by Dobson and Smallman[5] who showed that loop growth occurred irrespective of whether the foils were annealed inside or outside the microscope and they concluded that surface oxidation was responsible for loop growth. The oxidation of zinc occurs by the creation of interstitial–vacancy pairs at the oxide/metal interface and the subsequent transport of zinc interstitials to the oxide/oxygen interface. Excess vacancies are thus created in the metal at the metal/oxide interface and these may diffuse away from the interface to be finally absorbed on dislocation loops and other vacancy sinks in the metal.

The conclusion that the excess vacancies, which give rise to loop growth in zinc, are created as a consequence of oxidation was confirmed[6] by experiments which correlated the growth rate of the dislocation loops with those parameters which are known to affect the oxidation rate. Figure 1a shows a plot of loop radius versus annealing time for loops in zinc annealed at room temperature under different oxygen partial pressures. It can be seen that the rate of growth increases as the oxygen partial pressure decreases and that the effect is reversi-

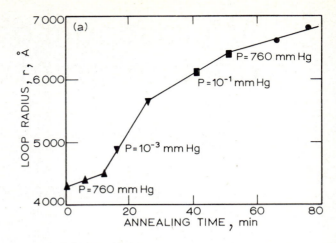

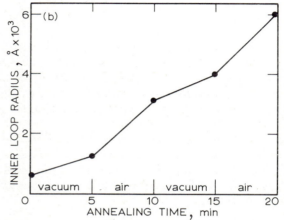

1 Radius *v.* time plot for a loop in (*a*) zinc annealed at 23°C under different partial pressures of oxygen, and (*b*) magnesium annealed at 180°C alternatively in air and in vacuum

ble from one pressure to another. A similar experiment carried out on loops in magnesium at 180°C showed, Fig. 1b, that the rate of loop growth increased with increasing oxygen pressure. It is not entirely clear why the dependence of the loop growth rate on oxygen pressure is of the opposite sense in the two cases but, nevertheless, the fact that such a dependence is observed is strong evidence that oxidation is responsible for vacancy injection. The oxidation rate of zinc is also known to be strongly affected by the addition of small impurity concentrations, the oxidation rate being increased by the addition of monovalent impurities and decreased by the addition of trivalent impurities. The effect of small alloying additions of lithium and aluminium on the loop growth rate was studied by comparing the average rate of loop growth in the alloys with that in pure zinc. The results are shown in Fig. 2. The growth rate was considerably enhanced in the zinc–lithium alloys and decreased in the zinc–aluminium alloys showing a direct correlation between the loop growth rate, vacancy injection, and oxidation rate.

Work by Rozhanskii *et al.*[7] on the climb of loops in zinc has shown that the nucleation and growth of an oxide film and its related vacancy production can be stopped by chemical treatment of the foil surface. The method consists of removing any oxide film remaining after electropolishing by immersing the specimen in

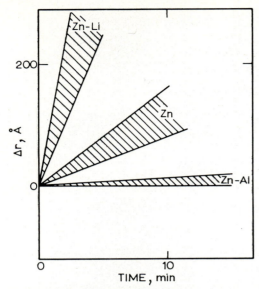

2 **A plot of the change in radius with time for loops in zinc, zinc–0·1% lithium and zinc–0·24% aluminium annealed in air at 23°C; the growth rate varied from loop to loop in the same material and each shaded region shows the range of values obtained**

ammonia and then wetting the surface with a weak solution of mercurous nitrate. Using this treatment it was found that dislocation loops contracted at rates corresponding to the surface vacancy concentration being equal to c_0. This technique has also been successfully applied[8] to the climb of dislocation loops in cadmium.

In metals with hexagonal structure the dislocation structure sometimes takes the form of double-faulted loops.[9] In this case the precipitation of a second layer of vacancies removes the stacking fault created by the precipitation of the first layer and the defect therefore consists of an annulus of stacking fault surrounding a loop of perfect material. Under the growth conditions produced by oxidation it is observed (Fig. 3) that stacking fault is created by vacancy absorption at the growing outer perimeter of the loop and is destroyed at the growing inner perfect loop. The perfect region expands faster than the outer stacking fault since the addition of a vacancy to the inner loop decreases the energy of the defect by γB^2 whereas the addition of a vacancy to the outer loop increases the energy by the same amount. This effect is further enhanced as the two loops approach each other due to vacancy transfer from the outer to inner loops. The growth of these double loops is affected by oxygen pressure in a similar manner to the single faulted loops (*see* Fig. 3) and analysis of the relative growth rates of the individual loops and the perfect prismatic loop when the two loops coalesce shows that the γ for magnesium is 90 mJ/m^2 and that the supersaturation (c/c_0) is about 7.

Aluminium alloys

The electron microscope observations of loop kinetics indicate that neither vacancy injection nor any barrier effects[10] occur at temperatures of the order of 150°C. This ideal behaviour is considerably modified in aluminium alloys containing a few percent of mag-

nesium. When these alloys are annealed in the temperature range 150°–200°C it was observed[11] that the faulted loops shrank and the prismatic loops grew. The driving force for loop shrinkage is greater for faulted loops ($\Delta F \sim \gamma B^2$) than for prismatic loops which clearly accounts for the opposite sense of climb in the two cases. The growth rate of the prismatic loops was again dependent on the oxygen partial pressure, the rate increasing with increasing oxygen pressure and it was concluded that vacancy injection was occurring during the anneal as a result of the growth of a magnesium-rich oxide on the foil surface. Evidence for the formation of magnesium oxide on the oxide on the surface of aluminium–magnesium alloys is to be found in electron diffraction studies of the oxide[12] and from observations that the magnesium content in the alloy is depleted during annealing.[13]

The growth behaviour of the prismatic loops was more complex than that in pure magnesium where a constant rate of growth was observed. Figure 4 shows typical growth curves for alloys containing 0·65%, 3·3% and 6·6% magnesium. The rate of growth generally increased continuously during the anneal for the Al–0·65% Mg alloy whereas, apart from the early stages, a constant growth rate was observed for the more concentrated alloys. The driving force for prismatic loop shrinkage is dependent on loop radius, i.e. $\exp(\Delta F/kT) \simeq \{1 + (\alpha b/r)\}$, and it is to be expected therefore that the growth rate will increase as the loops grow, but calculations of the magnitude of this effect indicated that it was too small to account for the accelerating growth rate in the Al–0·65% Mg alloy. It was therefore concluded that the vacancy concentration c_s in equilibrium with the oxide is increasing during the anneal. The surface vacancy concentration remained constant throughout the anneal for the more concentrated alloys. Further evidence for the build-up of the vacancy concentration with annealing time for Al–0·65% Mg is shown in Fig. 5 which shows that, after interrupting an anneal at 190°C by an intermediate anneal at 140°C, the rate of growth is approximately equal to that found at the beginning of the first 190°C anneal. No appreciable loop growth occurred during the 140°C anneal and therefore a negligible amount of vacancy production must occur at this temperature, although the temperature is sufficiently high to allow any vacancies already present in the foil to diffuse to sinks. The vacancy supersaturation built up during the first 190°C anneal is therefore dissipated during the intermediate anneal and the conditions in the foil at the start of the second 190°C anneal will be similar to those at the beginning of the first 190°C anneal. In a dilute alloy the rate of vacancy production will be determined by the rate of diffusion of magnesium atoms to the surface. These vacancies will then migrate into the foil and enhance diffusion of the magnesium atoms to the surface with consequent further increased vacancy production. Because of this co-operative process the vacancy concentration at the foil surface will increase with annealing time. This process cannot go on indefinitely, however, because as the chemical stress due to the vacancy supersaturation increases it will progressively become more difficult to produce further vacancies and ultimately the rate of vacancy production will be determined by the rate of oxidation in the presence of a chemical stress. This situation appears to be present in the more

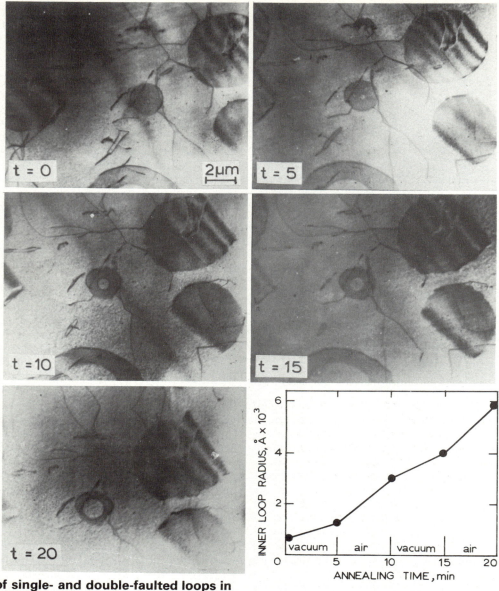

3   Growth of single- and double-faulted loops in magnesium on annealing alternately in air and vacuum at 180°C

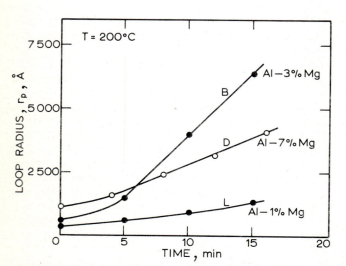

4   Growth behaviour of prismatic loops in aluminium–magnesium alloys at 200°C

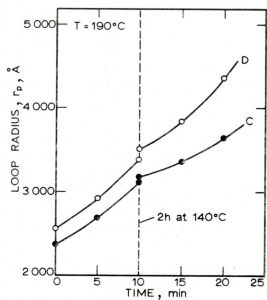

5   Growth of prismatic loops in Al–0·65% Mg at 190° before and after an intermediate anneal of 2 h at 140°C

concentrated alloys where, except possibly in the early stages of loop growth, the vacancy concentration remains constant during the anneal.

On the basis of this model it is to be expected that the loop growth rate will increase with increasing magnesium concentration. This was found to be the case for the alloys containing 0·65% Mg and 3·3% Mg but the loop growth rate was found to be less in an alloy containing 6·6% Mg than in the alloy containing 3·3% Mg. The reason for this is not absolutely clear but may be associated with competition for the excess magnesium between the oxide and the growing precipitates of Mg$_2$Al$_3$ in the supersaturated alloy.

Intermetallic compounds

Vacancy generation by surface oxidation can also occur in intermetallic compounds and recent results have shown the effect to occur in NiAl. It has been observed[14,15] that single crystals of NiAl which have been slowly cooled from the melt contain voids and these voids have been observed to grow in thin foils annealed at 900°C as shown in Fig. 6. Vacancy-type dislocation loops produced by deformation before thinning were also observed to grow on annealing (see Fig. 7). Since the Burgers vector of the loops is a lattice vector of a type $a\langle 100\rangle$ it is clear that prismatic loop growth requires the condensation of equal numbers of both nickel and aluminium vacancies and it is reasonable to expect that both species of vacancy are involved in void growth. Auger electron spectroscopy has recently been used by Hutchings[16] to study the high-temperature oxidation of NiAl in situ. A duplex oxide is formed with α-alumina at the gas interface and a mixed oxide at the metal interface. The inner layer has not been unambiguously identified but is similar to the oxide formed on NiAl at room temperature containing both nickel and aluminium. This oxide is therefore very probably NiO·Al$_2$O$_3$ and almost certainly a metal defective oxide. With the formation of such an oxide both nickel and aluminium vacancies will be injected into the specimen during oxidation.

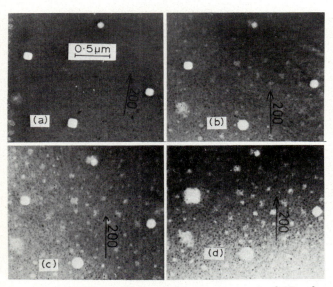

a before annealing; b after an anneal of 15 min at 890°C; c after a further 15 min anneal at 920°C; d after a further 60 min anneal at 920°C

6 Growth of voids in NiAl

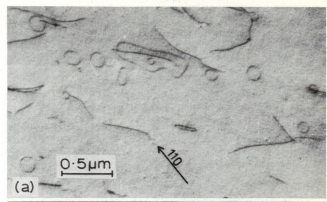
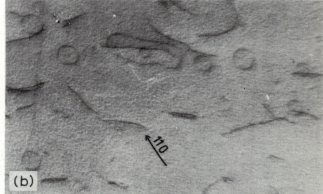
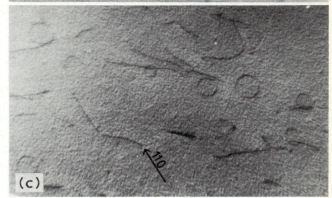

a t = 0 min; b t = 10 min; c t = 20 min

7 Growth of dislocation loops in NiAl at 640°C; T = 640°C

It is not necessary that all the vacancies which are absorbed by the prismatic loops are injected into the specimen at the oxide/metal interface since a large concentration of constitutional vacancies are already present in the NiAl. The material used by Fraser et al.[14] was aluminium-rich ((50·4 at.-%) and thus contained nickel vacancies. The oxide formed is aluminium-rich, and hence as aluminium is removed from the non-stoichiometric NiAl the composition of the NiAl moves closer to the stoichiometric composition. Thus removing one aluminium atom by oxidation produces one aluminium vacancy and, additionally, frees one nickel vacancy.

Silicon

In contrast to the cases described above, the oxidation of silicon takes place at the oxide/silicon interface by the diffusion of the oxygen anions through the oxide.[17] In order to accommodate the inward-growing oxide there is a flow of vacancies towards the oxidizing interface or an equivalent flow of interstitials from the interface. Thus the interstitial concentration is enhanced and the vacancy concentration decreased in the vicinity of the oxide/silicon interface.

The excess interstitials injected into silicon by oxidation has been shown to result in the growth of extrinsic stacking faults.[18] A number of mechanisms have been proposed for the nucleation of these faults; one possible mechanism involves the dissociation of a prismatic dislocation into a Frank and Shockley partial dislocation bounding an area of extrinsic stacking fault according to the reaction:

$$a/2[1\bar{1}0] \rightarrow a/3[1\bar{1}1] + a/6[1\bar{1}\bar{2}]$$

Providing the Frank dislocation is on the side away from the specimen surface the fault can grow by absorbing interstitials and thereby advance into the crystal. Investigation of the behaviour of these faults by thin foil annealing experiments showed that the faults grew when the foils were annealed in an oxidizing atmosphere and shrank when annealed in vacuum (10^{-5} torr).

During the last few years there has been considerable interest in replacing the current diffusion processes by ion-implantation as a highly reproducible technique to incorporate doping impurities in silicon. After implantation the slices are annealed at about 1000°C to produce the required junction depth and to anneal the damage associated with the high energy implantation. Damage is, in general, deleterious to the device performance; for example, dislocation loops may act as recombination centres and give rise to excess leakage currents under reverse bias or reduced gain.

Nicholas et al.[19] investigated the replacement of the base diffusion process in bipolar transistors with a boron ion-implantation. The boron base was driven-in at 1 180° under oxidizing conditions and all the other processes were as used for the standard diffused transistors. The gain of these transistors was only half that of the all-diffused transistors. It was also found that this reduction in gain did not occur if the boron was implanted through 1 000 Å of oxide and driven-in without the oxide being removed or if the boron was implanted in bare silicon and driven-in using an inert atmosphere. It was concluded that these effects were associated with lattice defects produced by the implantation whose density and size were affected by the nature of the annealing environment.

Silicon slices have been implanted[20] with boron doses of $1 \cdot 10^{14} - 1 \cdot 10^{15}$ ion cm^{-2} and energies of 10 and 40 kV. When these slices were annealed at 1 100°C using steam or dry oxygen large stacking faults were created which were observed by both optical and electron microscopy. In general the size and density of these faults increased with increasing severity of the implantation. When the slices were annealed in argon the surface appeared perfect under the optical microscope and electron microscopy revealed a large density of small dislocation loops. This is illustrated in Fig. 8 which compares micrographs obtained from slices implanted with 5.10^{14} boron ion cm^{-2} at 40 kV and annealed for 15 min at 1 100°C in oxygen (Fig. 8a) and argon (Fig. 8b).

Figure 8b is typical of an implanted and annealed specimen. The excess point defects produced during the implantation condense out during the subsequent anneal to form prismatic and faulted dislocation loops

which are of interstitial character. This process is strongly influenced by the excess interstitials generated at the oxide/silicon interface when the anneal is carried out in an oxidizing atmosphere (Fig. 8a). Many of the prismatic loops grow and intersect the surface and then presumably dissociate according to the dislocation

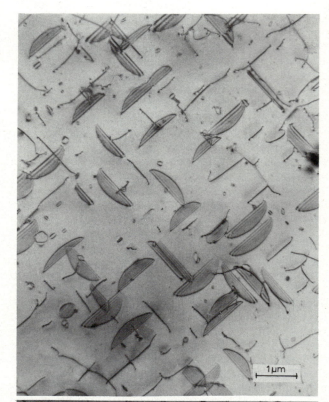

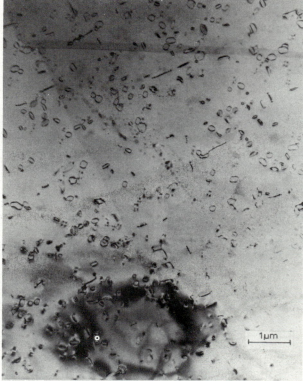

8 Electron micrographs from silicon slices implanted with $5 \cdot 10^{14}$ boron ion cm^{-2} at 40 K and annealed for 15 min at 1100°C in; a oxygen, and b argon

reaction above to form a stacking fault which continues to grow as oxidation proceeds. The damage produced during an oxidizing anneal thus extends to greater depths and penetrates the emitter-base junction causing the reduced gain or in some cases excessive leakage currents.

X-RAY TOPOGRAPHY OBSERVATIONS

A disadvantage of electron microscopy for oxidation and vacancy injection studies is the limited thickness of the specimens and oxide which can be studied. The high-voltage electron microscope has additional disadvantages in that radiation damage, consisting of vacancies and interstitials, occurs in many of the metals of interest at relatively low operating voltages. A complementary technique to electron microscopy which overcomes the above disadvantages and is receiving increasing attention is that of X-ray topography.[21] In this technique, monochromatic X-rays are used to examine single crystals of low dislocation density set at the Bragg angle for a particular family of lattice planes. Lattice defects are revealed by producing differences in diffracted intensity from the perfect crystals.

Lang topography has been used by Michell and Ogilvie[22] to investigate zinc and cadmium crystals grown from the vapour phase. These crystals initially contained few dislocations but after exposure to air for several days small unresolvable centres of strain contrast were nucleated which with increasing exposure became resolvable as loops. After this initial period of nucleation the dislocation loops grew with increasing exposure, but their number remained essentially constant. Other line defects appeared in the crystal apart from the loops so that the crystal became increasingly imperfect with time of exposure in air. X-ray topographic analysis of cadmium samples produced similar observations. The cadmium crystal became increasingly imperfect with exposure to air. Unlike zinc, fresh loops were continually nucleated and moreover the growth rate of these loops was faster than observed for the zinc crystal. This was reflected in the loop diameter finally observed which was 80–100 μm for zinc and 600 μm or greater for cadmium. Loops with $b = \langle 20\bar{2}3 \rangle$ were identified in addition to those with $\langle 0001 \rangle$ as found in zinc. More recently G'Sell and Champier[23] have also studied the nucleation and growth of loops in zinc crystals exposed to air for approximately four months. Control experiments with crystals maintained in vacuum and an inert atmosphere showed no evidence of loop formation. In further work on cadmium by Michell and Smith[24] CdO_2 was observed to form on the vapour grown crystals which decoheres after several weeks' exposure in air and this has been attributed to a high Pilling–Bedworth ratio (1·84). Badrick and Puttick[25] have also studied Cd crystals grown by a modified Bridgman technique. Unfaulted loops of about 50 μm diameter could be produced on cooling which on exposure to air were observed to shrink in some areas and grow in others.

Until recently very little work had been carried out on magnesium because of the difficulty of obtaining crystals with low dislocation contents. However, Michell and Smith[24] have made a preliminary study of vapour-grown crystals. The as-grown crystals were covered with a very thin highly textured polycrystalline layer of MgO but no dislocation loops were revealed. After some weeks exposure in air the initially adherent oxide developed blisters and extensive cracking was observed by electron microscopy but again no dislocation loops were observed by X-ray topography.

Recently Vale[26] has produced magnesium crystals with a relatively low dislocation density using a modified Bridgman method. The initial defect structure was of low dislocation content but contained some small 'centres of strain' from the growth process which were thought to be unresolvable loops. Exposure to air for several days at room temperature and indeed at temperatures up to 500°C failed to produce any desirable change in the as-grown defect structure. At 500°C simultaneous weight-gain measurements showed an increase of 5 μg which, if distributed evenly, would produce an oxide film approximately 270 Å thick. Above 500°C oxidation becomes extensive and is accompanied by the nucleation and growth of a large number of dislocation loops and in the temperature range 500°–550°C the rate of oxidation was found to be very dependent on surface finish. Typical X-ray topographs are given in Fig. 9 which show unfaulted dislocation loops lying on the basal plane. The initial stages of oxidation are thought to occur by the diffusion of Mg^{2+} ions through the oxide causing vacancy loops to grow to diameters of up to 135 μm. With increasing oxidation the oxide begins to decohere and the number of loops decreases until eventually virtually all the dislocation loops disappear; this is shown in Fig. 10a and b for a crystal oxidized more severely at the outside regions. Figure 11 shows a plot of weight gain during oxidation at 550°C. This curve was produced in four separate runs, as labelled, and Fig. 12a–d shows corresponding topographs taken in each of the four stages. The characteristic feature of loop nucleation and growth is evident in the early stages but when the oxide becomes incoherent and decoheres the loop diameter and density decrease.

These results clearly show the relative advantages and disadvantages of the technique of electron microscopy and X-ray topography and particularly their complementary nature.

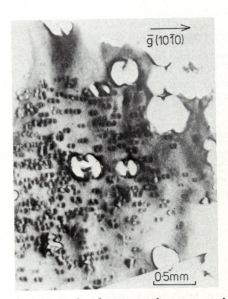

9 X-ray topograph of magnesium crystal oxidized at 550°C; the white areas are thought to be formed from internal oxidation of impurities

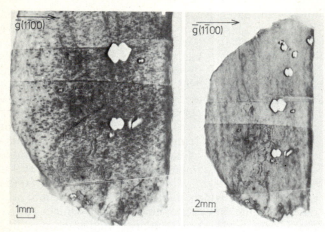

10 X-ray topographs of oxidized magnesium showing; *a* loops present prior to heavy oxidation; at this stage the region near the curved edge is the most heavily oxidized; *b* the defect structure after heavy oxidation when nearly all the loops have disappeared

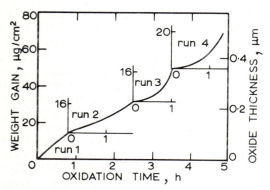

11 A plot of the weight gain during four different oxidation runs at 550°C

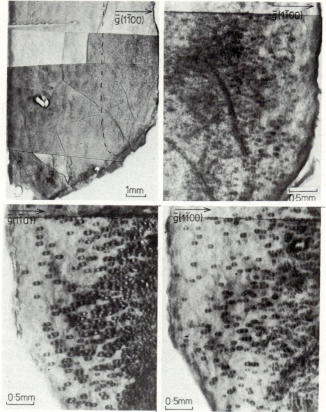

a no significant loop growth has occurred; the dotted region is more heavily oxidized; *b* taken after run 2 when loop growth is evident; *c* taken after run 3 when unfaulted loops lying on the basal plane are clearly evident; *d* after run 4 when the size and density of loops is smaller than in *c* probably due to decohesion of the oxide scale

12 X-ray topographs taken after each of the four runs shown in Fig. 11

ACKNOWLEDGMENTS
The authors would like to thank Mr R. Hutchings, Mr J. Lambert, and Dr R. Vale for permission to quote their unpublished results, and Dr M. H. Loretto for helpful discussions.

REFERENCES
1 J. W. EDINGTON AND R. E. SMALLMAN: *Phil. Mag.*, 1965, **11**, 1109
2 P. S. DOBSON *et al.*: *Phil. Mag.*, 1967, **16**, 9
3 A. BERGHEZAN *et al.*: *Acta Met.*, 1961, **9**, 464
4 F. PRICE: 'Electron microscopy and strength of crystals', 1963, **103**, Interscience Publication
5 P. S. DOBSON AND R. E. SMALLMAN: *Proc. Roy. Soc.*, 1966, **A293**, 423
6 R. HALES *et al.*: *Metal Sci. J.*, 1968, **2**, 224
7 V. N. ROZHANSKII, *et al.*: *Sov. Phys. Solid State*, 1967, **9**, 593
8 K. H. WESTMACOTT: *T.A.I.M.E.*, (1970), **1**,
9 R. HALES *et al.*: *Proc. Roy. Soc.*, 1968, **A207**, 71
10 J. E. HARRIS AND B. C. MASTERS: *Proc. Roy. Soc.*, 1966, **A292**, 240
11 P. S. DOBSON *et al.*: *Phil. Mag.*, 1968, **17**, 269

12 L. DE BROUCKERE, *J. Inst. Metals*, 1945, **71**, 131
13 K. FØRSVOLL AND D. FOSS: *Phil. Mag.*, 1967, **15**, 329
14 H. L. FRASER, *et al.*: Conf. on 'Voids formed by irradiation of reactor materials', (Eds. S. PUGH, *et al.*), 1971, 177, British Nuclear Eng. Soc.
15 H. L. FRASER, *et al.*: *Phil. Mag.*, 1973, **28**, 639
16 R. HUTCHINGS: PhD thesis, University of Birmingham, 1976
17 M. M. ATALLA: 'Properties of elemented and compound semiconductors', (Ed. H. GATOS), 1960, **5**, 163, New York, Interscience Publishers, Inc.
18 I. R. SANDERS AND P. S. DOBSON: *Phil. Mag.*, 1969, **20**, 881
19 K. H. NICHOLAS *et al.*: *Appl. Phys. Lett.*, 1975, **26**, 320
20 J. LAMBERT AND P. S. DOBSON: to be published
21 A. R. LANG: 'Modern diffraction and imaging techniques in materials science', Eds. S. AMELINCKX, *et al.*), 1970, North-Holland Publ. Co.
22 D. MICHELL AND A. J. OGILVIE: *Physica Status Solidi*, 1966, **15**, 83
23 C. G'SELL AND G. CHAMPIER: *Phil. Mag.*, 1975, **32**, 283
24 D. MICHELL AND A. SMITH: *Physica Status Solidi*, 1968, **27**, 291
25 A. BADRICK AND K. PUTTICK: *Phil. Mag.*, 1971, **23**, 585
26 R. VALE: PhD thesis, University of Birmingham, 1976

Influence of metal lattice vacancies on the oxidation of high temperature materials

G. B. Gibbs and R. Hales

When an oxide scale grows on a metal by the outward diffusion of cations, it is anticipated theoretically that lattice vacancies will be injected into the metal substrate. There is a large body of evidence supporting this proposition including electron microscope observations on the climb of secondary defects, the optical microscopy of voids, and some tracer diffusion effects. The influence of a vacancy supersaturation on the kinetics of oxidation is considered in detail and an analogy is drawn between the effect of vacancy concentrations and variations in oxygen potential on scale growth rates. However, for most practical situations the effects are shown to be small. Injected lattice vacancies are annihilated at various internal sinks such as the scale/metal interface or defects within the matrix. Factors which determine where vacancies are likely to be annihilated are discussed. Vacancy annihilation produces dimensional changes in the metal substrate which favour the eventual breakdown of the scale/metal interface and loss of scale adhesion. Thus a compact scale growing by the outward diffusion of vacancies has inherent long-term instability. Previously compact scale overlying an interfacial void is able to dissociate to produce microchannels which allow gas access to the underlying metal. This does not necessarily result in a loss of corrosion protection; an inner layer of oxide can form at a parabolic rate in the vacancy condensation zone made available by the outward diffusion of cations. The formation of an inner layer can restore contact between metal and the scale thereby improving scale/metal adhesion. However, if the oxidant is CO_2 or H_2O a redox mixture can form in the pores at the scale/metal interface. This can modify the underlying metal by carburization, decarburization or the formation of hydrides. The oxidation of iron in $CO-CO_2$ mixtures can result in very high oxidation rates due to the deposition of carbon in the inner layers via a catalysed Boudouard reaction. The oxidation of alloys in which one constituent is preferentially oxidized can be adversely affected if the initially protective scale breaks down above interfacial voidage to expose a depleted alloy on which faster-growing oxide may form. In view of the possibly adverse consequences of loss of scale/metal adhesion it may be considered prudent to delay this transition. Possible approaches to achieve this are considered.

The authors are with the CEGB, Berkeley Nuclear Laboratories

Alloys chosen for high-temperature application in oxidizing environments generally form protective oxide scales which grow by solid-state diffusion of metal ions outwards. The protection afforded by the surface scale is of two kinds. First, a corrosion rate which is controlled by diffusion across a scale of increasing thickness decreases with increasing time, and quickly falls to a low level corresponding to an acceptable rate of metal loss. Secondly, the metal substrate is protected from adverse effects of the environment such as carburization or decarburization which affect the mechanical properties. A large class of corrosion-resistant alloys relies on the selective oxidation of one component which forms an oxide in which cation mobility is low; the chromium-bearing stainless steels all fall into this category.

When metal ions enter a growing oxide film as part of the outward cation flux, they necessarily leave behind metal lattice vacancies which are in excess of the thermal equilibrium number. These vacancies condense out to produce voidage, or are annihilated at the scale/metal interface and at internal grain-boundary and dislocation sinks. The eventual result of the vacancy precipitation processes is some loss of corrosion protection, although the service performance of an alloy does not necessarily become inadequate. This paper reviews the effects of injected metal lattice vacancies on the kinetics of oxidation and on the structure of oxide scales.

VACANCY INJECTION AND DIFFUSION TO SINKS

If the oxide/metal interface is an incoherent boundary, it will have some of the properties of a high-angle grain boundary and injected lattice vacancies may be annihilated at the interface. Electron microscopy of oxidizing thin foils indicates that only a fraction of the total number of vacancies is injected into the underlying metal. These excess vacancies will initially sink out:

  (i) at grain boundaries and dislocations which are existing 'ideal' sinks
  (ii) at vacancy-type dislocation loops formed by collapse of discs formed by coalescence of excess vacancies within grains, Fig. 1a
  (iii) at voids also formed by vacancy coalescence, Fig. 1b.

The first two processes cause the metal sample to shrink, whereas with the third the external dimensions of the metal are preserved.

The existence of the various vacancy precipitation or annihilation processes means that no sample can develop a very high vacancy supersaturation during oxidation. This result is already well known from studies of diffusion couples[1] and processes such as dezincification of brass,[2] where vacancy generation rates are comparable with those during oxidation.

For example, consider a vacancy diffusion flux j between an oxidizing surface where the concentration is $[V_L]$ and grain-boundary or dislocation sinks, at distance d, holding the local concentration at the equilibrium value $[V_L]_e$. If D is the metal self-diffusion coefficient and Ω' is its atomic volume, straightforward application of Fick's law gives

$$\frac{[V_L]}{[V_L]_e} = 1 + \frac{j\Omega'd}{D} \qquad (1)$$

Substituting typical values of $j \sim 10^{14}$ vacancies cm^{-2} s^{-1}, $D \sim 10^{-12}$ cm^2 s^{-1}, $d \sim 10^{-4}$ cm, and $\Omega' \sim 10^{-23}$ cm^3 gives an excess vacancy concentration only 10% higher than the equilibrium concentration.

In laboratory experiments where coarse-grained thin foils have been used, it has sometimes been suggested that much higher supersaturations can develop.[3] This is not so on account of process (ii) above. Simple random walk calculations indicate that vacancy–vacancy interactions will rapidly produce vacancy loops or clusters which will act as dispersed vacancy sinks. Vacancy loops generated by *in situ* oxidation of thin foils in the electron microscope are an example of this, Fig. 1a.

The influence of a surface vacancy supersaturation on the kinetics of initial oxidation is examined in the next section; it is small since the predicted supersaturation is small. The main effects of injected vacancies arise from the way in which they condense out, and the microstructure of a metal or alloy can therefore have a profound influence on its corrosion behaviour.

Annealed coarse-grained pure metals, with a low density of pre-existing vacancy sinks and void nuclei, can develop a coarse dispersion of large grain-boundary condensation voids,[4] Fig. 1b. Clearly, these can affect the structural integrity of the sample.

Fine-grained materials, particularly commercial steels with dispersed phases, have numerous pre-existing sinks where excess vacancies can be annihilated. The materials tend not to show internal void formation. Rather, the metal core shrinks inside its oxide case resulting eventually in loss of scale/metal adhesion and a change in the corrosion process (*see* p. 204).

It is possible to preserve the external dimensions of an oxidizing sample by condensing out injected vacancies as a fine dispersion of voids which do not seriously affect structural integrity. A dispersion of (rare earth) oxide particles in the matrix may provide appropriate nuclei and this may partly explain the beneficial effects of rare earth additions on scale/metal adhesion.[5]

SINGLE-LAYER GROWTH: INFLUENCE OF VACANCIES ON KINETICS

The initial oxidation of a pure metal or alloy frequently produces a single-layer oxide which is non-porous and grows entirely by solid-state diffusion of cations outwards. Eventually, there may be loss of scale/metal adhesion as a result of vacancy condensation (or other processes generating stress at the scale/metal interface). Before this, there will be a small vacancy supersaturation at the scale/metal interface associated with a flux of vacancies to internal sinks. The supersaturation can

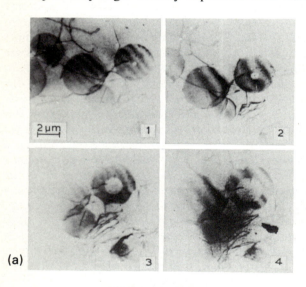

(a)

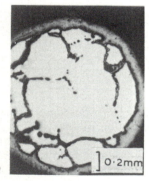

(b)

1 t = 0 min; *2* t = 5 min; *3* t = 15 min; *4* t = 25 min

1   Formation and growth of vacancy clusters by annihilation of injected vacancies at *a* dislocation loops in magnesium, and *b* voids in nickel

influence the kinetics of oxidation of a pure metal. For an alloy with one component oxidizing preferentially, the vacancy diffusion flux may exert an additional influence by (a) determining the moving boundary condition for diffusion of the alloying element out of the metal, and (b) influencing the diffusion rate of the alloying atoms. In all practical cases the effects are small.

Pure metals

Figure 2a illustrates the defect gradient across the scale of an n-type oxide in which the mobile defect is a cation interstitial. If Ω is the volume of oxide per metal ion, the rate of scale thickening is

$$dX/dt = j\Omega,$$

and noting that $[M_i]_g \ll [M_i]_m$, the diffusion flux across the scale is given by

$$j = [M_i]_m \frac{D_i}{X}$$

For interstitials at the scale/metal interface it is usual to write

$$[M_i]_m = B(p_{O_2})^{-1/n}$$

where p_{O_2} is the oxygen partial pressure that would be in equilibrium with both oxide and metal at the interface. Combining these equations the corrosion rate may be written as

$$dX/dt = B\Omega(p_{O_2})^{-1/n}(D_i/X) \qquad (2)$$

If the same analysis is applied to the p-type oxide of Fig. 2b, the corrosion rate becomes

$$dX/dt = A\Omega(p_{O_2})_g^{1/n}(D_v/X) \qquad (3)$$

since the vacancy concentration in equilibrium with the external gas at pressure $(p_{O_2})_g$ is $[V_m]_g \gg [V_m]_m$.

Equations (2) and (3) contain the well known result that external partial pressure of oxygen has a significant effect on corrosion rate only if the mobile defect is a cation vacancy.[6] It also follows from these equations that a modest vacancy supersaturation in the metal substrate will influence the corrosion rate only of the n-type oxide, by affecting $[M_i]_m$ or p_{O_2} at the scale/metal interface.

The effective oxygen partial pressure at the scale/metal interface p_{O_2} is obtained by applying the law of mass action to the equation

$$MO \overset{K_a}{\rightleftarrows} M + \tfrac{1}{2}O_2$$

and is given by

$$p_{O_2}^{\frac{1}{2}} = K_a/a_m$$

where a_m is the thermodynamic activity of the metal. It is readily shown[7] that if $a_{m,e}$ refers to metal containing the equilibrium concentration of vacancies $[V_L]_e$ and with a redox potential $p_{O_2,e}$, then

$$\frac{p_{O_2}}{p_{O_2,e}} = \left\{\frac{a_{m,e}}{a_m}\right\}^2 = \left\{\frac{[V_L]}{[V_L]_e}\right\}^2$$

On substituting into equation (2), we have

$$\frac{dX}{dt} = B\Omega(p_{O_2})_e^{-1/n}\left\{\frac{[V_L]_e}{[V_L]}\right\}^{2/n}(D_i/X) \qquad (4)$$

The effect of varying the vacancy supersaturation at the scale/metal interface is thus seen to be analogous to that of varying the oxygen partial pressure at the scale/gas interface of a p-type oxide. As anticipated from Le Chatelier's principle, increasing the vacancy supersaturation decreases the corrosion rate.

It is possible to derive various kinetic laws for the growth of n-type oxides by substituting into equation (4) a relationship for the time-dependence of $[V_L]$ derived from the assumed distribution of vacancy sinks.[8,9] However, it is simpler to argue that for a typical value of $n \sim 6$ and $[V_L]/[V_L]_e < 10$, the quantity $\{[V_L]_e/[V_L]\}^{2/n}$ never departs greatly from unity. That is, oxidation kinetics should remain almost parabolic with hardly any change in rate constant.

Alloys

The analysis of the previous section may be applied, with minor modifications but the same general conclusions, for the selective oxidation of one component of a corrosion-resistant alloy. In this latter case it is necessary to enquire whether other effects of the vacancy flux can have a more significant effect on the corrosion rate, and this requires more detailed examination of diffusion of the oxidizing element in the substrate.

If the diffusion coefficient of the alloying element is independent of concentration and the scale/metal interface moves slowly by comparison with diffusion in the alloy (stationary boundary condition), the depletion profile is described by[10]

$$\frac{C_x - C_I}{C_0 - C_I} = \text{erf}(x/2\sqrt{Dt})$$

where the number of solute ions per unit volume C has a value C_I at the interface and C_0 in bulk material. The corresponding flux of solute ions at the metal surface is

$$j = (C_0 - C_I)D^{\frac{1}{2}}/\sqrt{\pi t}$$

A constant value of C_I at the interface therefore gives the correct time dependence of flux to match parabolic oxidation. If the parabolic rate constant k_p is defined by

$$X^2 = k_p t$$

then

$$\frac{dX}{dt} = j\Omega = \frac{1}{2}\sqrt{\frac{k_p}{t}}$$

It follows that the magnitude of solute depletion in

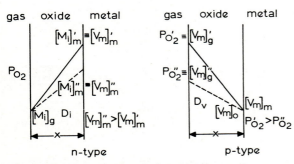

2 Schematic diagram showing the oxide defect concentration gradients and the influence of oxygen pressure and metal lattice vacancy concentrations

dynamic equilibrium at the interface is

$$(C_0 - C_I) = \frac{1}{2\Omega}\sqrt{\frac{\pi k_p}{D}} \qquad (5)$$

The parabolic rate constant k_p may be related to the activity of the solute in the alloy and to the defect parameters of the oxide as in the previous subsection (see Wagner[11] and Hales[12]). The maximum possible value of k_p is that which corresponds to the outward diffusion flux when C_I is zero:

$$(k_p)_{max} = \frac{4C_0^2 \Omega D}{\pi} \qquad (6)$$

In some cases the defect parameters of an oxide would tend to give a larger value of k_p, if there was no restriction on the rate of diffusion of solute out of the alloy. In that case the surface concentration of solute will fall essentially to zero and the rate-controlling process in oxidation will be solute diffusion in the substrate. If k_p is the parabolic rate constant dictated by oxide defect parameters, this transition to metal lattice diffusion control occurs when the amount of solute dissolved in the virgin alloy is less than

$$C_0 = \left(\frac{\pi k_p}{4\Omega D}\right)^{\frac{1}{2}} \qquad (7)$$

(This critical concentration is often wrongly regarded as a condition for scale instability;[11,13] it is merely a condition for change in the principal process determining the rate of oxide scale growth.)

There are two ways in which a vacancy diffusion flux can modify the present analysis. If injected vacancies are all annihilated in a zone which is small compared with the solute diffusion zone $\sim \sqrt{Dt}$, then the diffusion problem should be solved for a moving interface. In practice, this also affects the final corrosion rate. Figure 3 shows the calculated effect for Ni–Pt alloys.[14,12]

The second possible effect of injected vacancies is that the countercurrent of excess vacancies will speed the diffusion of solute atoms to the surface. If excess vacancies are all annihilated in a zone which is small compared with the size of the solute depletion zone, any local enhancement of the diffusion coefficient will clearly have little influence. If the vacancies are diffusing to more remote sinks the problem may be tackled by the methods of irreversible thermodynamics.[15] Depending on the magnitude of the solute-vacancy binding energy, the excess vacancies will not necessarily enhance diffusion to the surface. However, in practical cases, the evidence from Kirkendall diffusion couples is that effects on solute diffusion are negligible in the zone $\sim \sqrt{Dt}$ for the levels of vacancy supersaturation that can be sustained.[1,2]

LOSS OF SCALE/METAL ADHESION

Probably the most important consequence of vacancy injection during oxidation is that it gives an inherent long-term instability to single-layer oxide scales formed by outward diffusion of cations. In a normal commercial material containing a variety of vacancy sinks, injected vacancies will tend to be annihilated so that the metal core will shrink inside its oxide case. Vacancy condensation (to give local loss of adhesion) at the scale/metal interface will be favoured initially only if

$$kT \ln\{[V_L]/[V_L]_e\} \geqslant \Omega'^{\frac{2}{3}} W \qquad (8)$$

Assuming a typical value of the work of adhesion $W \sim 10^3 \ \mathrm{erg \ cm^{-2}}$, this requires supersaturations much greater than anticipated (section on 'Vacancy injection and diffusion to sinks'). However, with continued vacancy annihilation within the metal, elastic strains will be generated at the scale/metal interface and W in equation (8) must be replaced by $(W - E)$. The elastic energy, E, quickly becomes of the order of W near local discontinuities (Fig. 4) when voidage will grow along the scale/metal interface.

It must be pointed out that other processes which give rise to scale stresses, including differential thermal expansion of oxide and metal under thermal cycling conditions and mechanical loading of a component,[16] may cause an oxide scale to separate from the metal substrate. Vacancy injection ensures that local separation will eventually occur even in the absence of these effects.

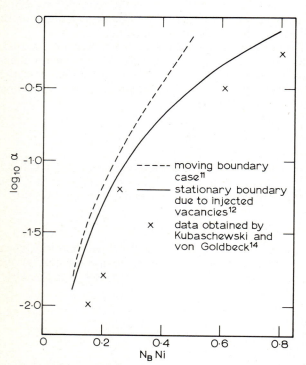

3 Variation of oxidation rate as a function of composition for Ni–Pt alloys oxidized at 1 100°C

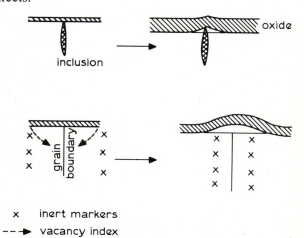

4 Influence of stress raising features on scale/metal interface stability

Pure metals

For pure metals, the major consequence of local loss of scale/metal adhesion (assuming that the scale does not become completely detached) is that the first-formed protective oxide will become porous. The development of microporosity as a consequence of the thermodynamic instability of the oxide once it becomes detached from the metal was first discussed by Mrowec and co-workers,[17–20] and the ideas have been elaborated elsewhere.[21,7]

Once microchannels have been formed, oxidant can reach the scale/metal interface and cause oxide growth there. Outer layer growth may also continue at the scale/gas interface by solid-state diffusion of cations through oxide crystallites that are still in contact. This produces a duplex oxide scale described more fully in the following section. The rate-controlling process remains the solid-state diffusion of cations, and protective corrosion rates are not changed by more than a factor of two.* However, the fact that gas now has access to the scale/metal interface may permit other reactions with the substrate such as carburization, decarburization, and hydriding.

Oxidation of alloys

The section on single-layer growth in alloys described selective oxidation of one component of a corrosion-resistant alloy. Often this constituent will be chromium; for example, in the 18 Cr series of austenitic stainless steels. Loss of scale/metal adhesion for mechanical reasons or on account of the vacancy condensation processes described above will expose the surface of the alloy now depleted of the protective element. Further oxide growth at this surface may therefore produce a much less protective oxide.

In the case of the chromium-bearing stainless steels, the surface concentration of chromium required to make the very protective Cr_2O_3 the thermodynamically preferred oxide is only $\sim 10^{-6}\%$. Equation (5) indicates that such low levels are not generally attained. However, experiments on a range of alloys demonstrate that at chromium levels $\lesssim 9$ wt-% there are kinetic reasons why the less-protective oxide Fe_3O_4 may be formed in preference to Cr_2O_3. It is therefore quite common for 18 Cr austenitic steels, which initially develop a tarnish film of Cr_2O_3, to exhibit a change to more rapid protective oxidation kinetics (Fig. 5) when this scale becomes detached.[22] In fact, a duplex scale forms with a mixed iron–chromium spinel underlying the Fe_3O_4 and providing the new rate-controlling barrier for oxidation.

DUPLEX SCALE GROWTH
Pure metals

A model for duplex scale growth has been described elsewhere.[21,7] Metal ions leaving the scale/metal interface to produce outer-layer growth create local voidage in which the inner layer can grow. Inner-layer growth is therefore coupled to outer-layer growth and controlled by the outward diffusion of cations. Mrowec and co-workers[19,20] describe the inner layer as forming in the 'metal consumption zone'. Figure 6 provides a descrip-

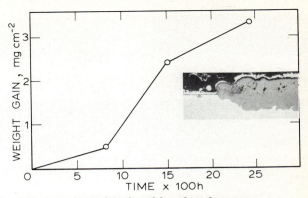

5   Increase in oxidation kinetics due to breakdown of the protective oxide and the formation of duplex oxide on 18–8 stainless steel

tion of inner layer growth based on a lattice vacancy model. Once voidage has formed at the scale/metal interface, these voids will be the nearest ideal sinks to the points of oxide-metal contact where vacancies are injected. Vacancies will therefore continually sink out at the interface providing the volume for inner-layer growth.

An interesting feature of the vacancy model for duplex scale growth has been pointed out by Gibbs and Wootton:[23] the vacancies initially injected during single-layer growth will produce elastic strain energy as a result of the strains generated by dimensional changes when they are annihilated at dislocations and grain boundaries. Once some voidage has formed at the scale/metal interface, this strain energy can be completely relieved by re-creation of the vacancies at the extended defects and their diffusion to the surface voidage—processes analogous to vacancy loop annealing and diffusion creep. This enables inner-layer growth to catch up outer-layer growth before both continue at the same rate (Fig. 7); after prolonged exposure, inner and outer layer thicknesses become identical. This would also explain some features of the experiments reported by Cox et al.[3] without the need to invoke very high vacancy supersaturations.

Duplex scales on alloys

As pointed out in the previous section, duplex scales often form on corrosion-resistant alloys once the initial selective oxidation film has failed. Although this leads to an increase in corrosion rate (Fig. 5), the new

*In duplex scale growth, an increment of oxide is added at the scale/gas interface *and* in the equivalent vacancy volume left behind at the scale/metal interface, for each cation diffusing outwards.

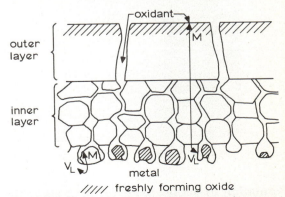

6   A schematic representation of the growth of a duplex oxide scale

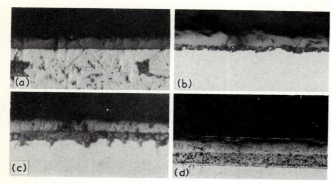

a a single compact layer of oxide is first formed ×1 500; *b* an inner layer nucleates at the scale/metal interface ×1 500; *c* it grows to the same thickness as the outer layer ×1 500; *d* inner and outer layers then grow at the same rate ×250

7   Duplex scale on mild steel in CO_2 at 500°C

protective corrosion kinetics may still be acceptable. There are two reasons for this.

If the new outer layer is Fe_3O_4, most alloying elements in commercial steels will not dissolve appreciably in the outer oxide and the rate-controlling process for duplex scale growth will be the outward solid-state diffusion of iron ions. However, at the scale/metal interface, continually created voidage will expose all alloying elements to oxidant entering via microchannels and all oxidizable species are expected in the inner-layer crystallites. For iron–chromium alloys these become a mixed iron–chromium spinel, and since iron diffusion in the spinel is slower than iron diffusion in Fe_3O_4 it provides the rate-controlling barrier. It is possible for this form of duplex scale growth, with essentially parabolic kinetics, to continue indefinitely. The composition of the inner layer adjusts so that the duplex scale as a whole consumes all alloying constituents in the fractions in which they are represented in the alloy. There is neither build-up nor depletion of one particular element below the scale, Fig. 8.

Often, the corrosion resistance of the alloy is superior to that described above. A 'healing layer' of chromium-rich oxide may form at the scale/metal interface and

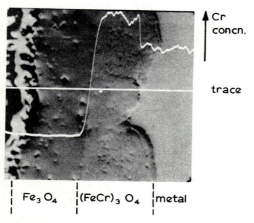

8   Formation of duplex oxide on 18 Cr austenitic steel without preferential oxidation of chromium

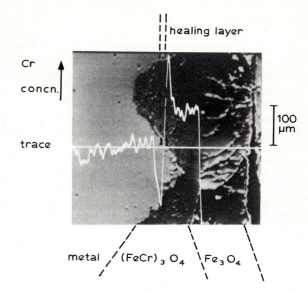

9   Formation of a chromium-rich 'healing layer' at the interface between metal and inner oxide by the selective oxidation of chromium in the alloy.

provide a more effective diffusion barrier. This is favoured by high chromium levels and high chromium diffusivity, especially the availability of short-circuit diffusion paths provided by fine grains or cold work. In the 18 Cr series of steels, there is therefore a tendency for the 'healing layer' to develop after a period of duplex scale formation (Fig. 9).

CONSEQUENCES OF DUPLEX SCALE GROWTH

There are a number of consequences of duplex scale growth arising from vacancy injection and condensation which are commercially important.

Oxide adhesion

It has been noted already that duplex oxide on commercial slabs may provide adequate corrosion resistance for high-temperature application. The somewhat irregular nature of the scale/metal interface means that the scale is well keyed onto the metal, and since injected vacancies are now consumed at the scale/metal interface it is stable in relation to vacancy injection effects. When a 'healing layer' forms at the base of the scale, this is now in an ideal position protected from external damage—in contrast to an initial Cr_2O_3 layer.

A disadvantage of duplex oxide scales is that the boundary between inner and outer layers often contains clearly defined remnants of the initial interfacial voidage and is therefore a plane of weakness. The outer layer may spall off under T-cycling or mechanical loading conditions. When the inner layer is a mixed spinel and provides the rate-controlling barrier, the metal attack rate is unchanged. However, the spalled oxide may be a nuisance, for example, in the steam pipe circuit of a boiler.

Other gas–metal reactions

The principal disadvantage of duplex scale growth is that the external gas now has access to metal in voids at the scale/metal interface and metal-gas reactions other than oxidation may take place. Depending on the composition of the corroding atmosphere, possible reactions are carburization,[24] decarburization,[25] sulphidation[26] and hydriding.[27] In high-pressure CO_2, catalysed carbon deposition at the scale/metal interface may produce a rapid 'breakaway' corrosion attack of some ferritic steels.[21]

These adverse reactions tend to be inhibited by the formation of a chromium-rich 'healing layer' at the base of duplex scale.

CONCLUSIONS

Metal lattice vacancies injected by oxidation can have a profound effect on the corrosion behaviour of metals and alloys in high-temperature service. Direct effects of a vacancy supersaturation on oxidation kinetics are small, but vacancy condensation effects can change the reaction product from a single-layer of impervious oxide to a duplex porous scale often growing at a somewhat faster rate.

Duplex scale growth rates are often acceptable. Reactions such as carburization, decarburization, and hydriding may also be acceptable since the reaction area (voidage at the scale/metal interface) is small in relation to the total surface area, and penetration depths are diffusion limited. However, this is not always true and there may be some incentive to avoid initial single-layer breakdown. Alloying to improve scale plasticity and the provision of a fine dispersion of void condensation nuclei are obvious approaches.

If a duplex scale does develop on an alloy which relies on chromium for corrosion resistance, formation of a 'healing layer' of chromium-rich oxide at the base of the scale is an advantage. This is favoured by high chromium content and high chromium diffusivity. The latter increases with decreasing grain size, the presence of cold work, and increasing temperature.

REFERENCES

1 R. S. BARNES AND D. J. MAZEY: *Acta. Met.*, 1958, **6**, 1
2 R. W. BALLUFFI AND L. L. SEIGLE: *Acta. Met.*, 1955, **3**, 170
3 M. G. C. COX *et al.*: *Phil. Mag.*, 1973, **28**, 309
4 R. HALES AND A. C. HILL: *Corrosion Sci.*, 1972, **12**, 843
5 J. STRINGER *et al.*: *Oxid. Met.*, 1972, **5**, 11
6 P. KOFSTAD: 'High temperature oxidation of metals', 1966, New York, Wiley
7 G. B. GIBBS AND R. HALES: *Corrosion Sci.*, in press
8 G. B. GIBBS: *Phil. Mag.*, 1968, **18**, 1175
9 M. J. MINDEL AND S. R. POLLACK: *J. Phys. Chem. Solids*, 1969, **30**, 993
10 W. JOST: 'Diffusion in solids, liquids and gases', 1960, New York, Academic Press
11 C. WAGNER: *J. Electrochem. Soc.* 1952, **99**, 369
12 R. HALES: CEGB report No. RD/B/N3727, 1976
13 D. P. WHITTLE: *Oxid. Met.* 1972, **84**, 171
14 O. KUBASCHEWSKI AND O. VON GOLDBECK: *J. Inst. Metals*, 1949, **76**, 740
15 P. S. DOBSON: *Phil. Mag.*, 1972, **26**, 1307
16 P. HANCOCK AND R. C. HURST: Proc. BNES Conf. on Corrosion of Steels in CO_2, Reading University, Sep. 1973, 320
17 S. MROWEC AND T. WERBER: *Corrosion Sci.*, 1965, **5**, 717
18 S. MROWEC: *Corrosion Sci.*, 1967, **7**, 563
19 A. BRUCKMAN *et al.*: *Oxid. Met.*, 1972, **5**, 137
20 A. DRAVNIEKS AND H. McDONALD: *J. Electrochem. Soc.* 1948, **94**, 139
21 G. B. GIBBS: *Oxid. Met.*, 1973, **7**, 173
22 S. J. ALLEN *et al.*: Proc. BNES Conf. on Corrosion of Steels in CO_2, Reading University, Sep. 1973, 284
23 G. B. GIBBS AND M. R. WOOTTON: to be published
24 J. E. ANTILL *et al.*: *Corrosion Sci.* 1968, **8**, 689
25 G. J. BILLINGS *et al.*, *J. Electrochem. Soc.*, 1970, **117**, 111
26 S. MROWEC *et al.*: *Oxid. Met.*, 1969, **1**, 93
27 E. C. POTTER: *Research*, 1955, **8**, 450

Point defects in the formation of duplex oxide layers on metals

V. D. Scott

Duplex oxide scales, where the constituent layers are different in composition and/or structure, may be associated with a number of experimental situations, some connected with the oxidizing conditions and some with the nature of the metal. Many data, however, can be related to a rate-controlling outward movement of metal cations through the oxide via a solid-state diffusion process, together with a simultaneous inward transport of the oxidant to the metal/oxide interface. As a result of the outward movement of cations, metal atom vacancies are left behind. This paper focuses attention on the role of point defects, both cations and vacancies, in the growth and structure of duplex oxide scales formed on metals subjected to gaseous oxidation. The occurrence of segregation within the layers of any addition elements to the metal is described. An analysis of diffusion path networks in close-packed oxide lattices and of interaction energies between different cations and available interstitial sites is shown to account for much of the published data on element partitioning effects in duplex oxide layers. The characteristic presence of voids within oxide scales is discussed in terms of vacancies generated during the oxidation process, and a figure for the concentration of injected vacancies is derived. The discussion includes some results on oxidation kinetics in the thin film regime from which similar vacancy concentrations have been deduced. The implication of these ideas in the development of alloys which give reduced oxidation is discussed, of particular interest being the possibility that vacancies may be trapped in the lattice by adding solute atoms to the metal which have a high binding energy with the vacancy.

The author is at the University of Bath

OCCURRENCE OF DUPLEX OXIDE SCALES

Duplex oxide scales, where the constituent layers are different in composition and/or in structure, are commonly observed on oxidized metals; a typical example of a duplex layer is illustrated in Fig. 1 taken from an iron–chromium alloy oxidized at 600°C for 50 h. The occurrence of this type of scale depends upon oxidizing conditions, such as gas composition and temperature, upon the properties of the particular oxide–metal system, and also upon metal composition.

With regard to oxidizing conditions, it is known that the structure of oxide is determined by the oxygen potential of the gaseous environment, i.e. the highest oxidation state which is stable in a particular gas atmosphere will form on the surface. However, between this oxide and the metal an intermediate layer (or layers) may form whose structure will be controlled by the metal–oxide redox potential. Hence the layers are unlikely to have the same structure, although by altering the oxygen potential the nature of the outer layer may be changed and, at a sufficiently low oxygen potential, a single-phase oxide may result. In addition, other elements from the gas phase may be incorporated into the oxide scale to form multiphase reaction products. The presence of carbon in the scale developed on metals oxidized in carbon-containing gases is a well known example of this phenomenon,[1-3] although in most of the cases reported the carbon is present in the inner layer and its effect appears to be an alteration in the nature of the duplex scale rather than promotion of the initial formation of a duplex layer.

The mechanical properties of the oxide are also important in determining whether or not multilayer oxides form. In the case of tungsten, for example, where the oxide grows by anion diffusion, considerable compressive stresses are developed in the oxide. These increase as the film thickens and eventually fragmentation of the oxide occurs at the outer surface to form a duplex scale, both layers being referred to as tungstic oxide.[4] Mechanisms which are based on stresses in oxide

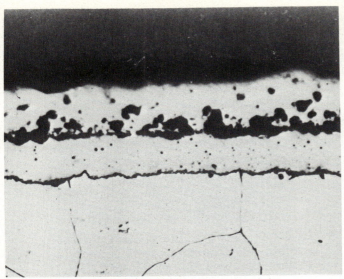

1   Typical duplex oxide scale: Fe–5% alloy oxidized at 500°C for 50 h; optical micrograph of taper (4 : 1) section ×1 000

scales have been invoked to explain a number of features of duplex layers, such as spalling, porosity, etc.;[5,6] and the numerous studies carried out on the oxidation of iron are especially relevant here. Normally, three oxide layers are formed in air or oxygen: wustite (FeO) adjacent to the metal with a thin outer layer of hematite (Fe_2O_3) and an intermediate layer of magnetite (Fe_3O_4). At oxidation above 1 000°C, the oxide is sufficiently plastic to maintain contact with the metal and this three-layer oxide persists until the metal is consumed.[7] At lower temperatures, however, the oxide is much less plastic and, as the metal cations move outwards through the layers, space is likely to be created next to the metal interface. What happens next can be affected by the geometry and properties of an oxide shell and two limiting situations may be envisaged. If oxidant is prevented from reaching the metal the void increases in size until complete oxidation leaves an oxide shell with a central hole of about the same size as the starting metal; this phenomenon is well illustrated in the literature by oxidation studies on samples of metal wire, etc.[8,9] Alternatively, where oxidant can gain access to the metal, the space may become filled with oxide and another layer of wustite will be formed.[10]

The nature of the oxide scale is known to depend sensitively also upon the composition of the metal. A good example of this is provided by oxidation studies on iron which contains chromium additions, where the formation on the metal of an inner layer of wustite is suppressed and a chromium-containing spinel phase is produced instead.[11] Duplex scales are frequently a characteristic of alloy oxidation where one of the constituent elements oxidizes preferentially. Hence on an iron–chromium–aluminium alloy, aluminium oxide forms first on the metal and this is subsequently covered by chromium oxide;[12] the amount of iron taking part in the reaction is small owing to its low diffusivity in alumina.

Clearly it will be evident from the examples referred to above that the formation of duplex oxide scales may be associated with a number of different experimental situations, some connected with the oxidizing conditions and some with the nature of the material. Hence it is not surprising that no single unified theory exists which can account for the phenomenon, although as discussed below many of the data on duplex scales can be related to the occurrence of a two-way diffusion process, with the outer oxide being formed as a result of the outward movement of metal cations through the oxide via a solid-state diffusion process and the inner layer by the simultaneous inward transport of oxidant to the metal/oxide interface.

It is not the intention in the present paper to catalogue the considerable amount of experimental data which has now been accumulated on duplex oxide scales but rather to discuss mechanistic aspects involved in their formation, and in particular the role of lattice point defects. There are, unfortunately, no comprehensive review articles on this subject, but the paper by Wood[13] deals with a number of relevant aspects as does the rather older treatise by Kubaschewski and Hopkins.[14] For information on specific systems the reader is referred to the numerous articles published in the literature and to some recent conference proceedings (*see*, for example, the International Conferences on Metallic Corrosion held at intervals under the auspices of NACE; ASM Seminar on Oxidation of Metals and Alloys published in 1971; BNES Conference on Corrosion of Steels in CO_2 published in 1975, etc.).

The present treatment is, for convenience, divided into two main sections. The first deals particularly with interstitial point defects, i.e. solute atoms, in the oxide lattice and examines their diffusion and distribution during oxidation. The following section is concerned with vacancy point defects and the way in which they may contribute to the structure of oxide scales. Although data on the effect of duplex oxide layers on the oxidation behaviour of numerous metal systems are not reviewed here, nor are the many, frequently speculative, explanations put forward to explain oxidation kinetics discussed, some comments on the possible way in which vacancy point defects may affect the properties of oxide layers are reserved for a concluding section.

MASS TRANSPORT PROCESSES

We shall begin by noting a number of microstructural observations which may be regarded as characteristic of many duplex oxide scales.

The outer oxide consists of large columnar crystals which appear to develop as a result of unconstrained growth away from the substrate (*see* Fig. 2); the crystals may exhibit facets associated with low index crystal faces or sometimes show a 'spaghetti-like' morphology.[15] Porosity is a common feature of this columnar structure, with voids tending to lie along grain boundaries of the oxide crystals.

The inner layer of oxide shows a different morphology from the outer layer, the oxide crystals being very much finer. Pores are present and cases have been reported[16] where the voids tend to grow in a platelike manner parallel to the metal/oxide interface; this feature is discussed in more detail in the following section on vacancies.

Sometimes the crystal structure of the two layers is different. For example when a nickel–chromium alloy is oxidized, a doped layer of Cr_2O_3 is formed on the metal with an outer layer of NiO; between the two a spinel, $NiCr_2O_4$, tends to grow.[17] Oxidation of iron in low

2   As Fig. 1; scanning electron micrograph of fracture section ×4 500

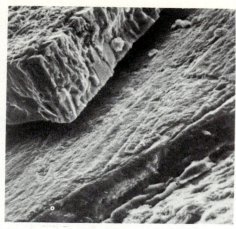

3   Fe–5% Cr alloy oxidized at 600°C for 100 h; scanning electron micrograph of fracture section (courtesy *Phil. Mag.*) ×300

partial pressures of oxygen will also give a duplex scale, an outer layer of Fe_3O_4 and an inner layer of FeO.[1,2] However, neither of these examples is primarily the product of a two-way diffusion process, the oxides growing mainly by outward movement of cations. They will not therefore be discussed further here except to note that the two-layer scale on the nickel-chromium alloy results from the slower diffusion of chromium in oxide lattices compared with nickel, while on iron the oxygen potential gradient is responsible.

In many other cases the crystal structure of the two layers is found to be the same. For example, on mild steel oxidized in CO_2-based gas, both oxide layers consist of magnetite, Fe_3O_4. On similarly oxidized iron–chromium alloys the layers are also isomorphous and adopt spinel-type crystal structures,[1] while multilayer scales of NiO have been observed on oxidized nickel.[18] These latter data can be related to the outward movement of metal cations through the oxide via a solid-state diffusion process, which many other workers[19,20,2] also believe to be the rate-controlling process. The inner oxide layer is produced by simultaneous inward transport of oxidant to the metal/oxide interface. Evidence to support this view is available from experiments using markers and tracers. For example, Fig. 3 shows that polishing scratches present originally on the metal surface are, after oxidation, situated at the interface between the two oxide layers. There have been two mechanisms proposed to explain the formation of the inner oxide. One idea assumes that the oxidant passes through the oxide via pores or cracks to reach the metal,[2,16,19,21] while the other suggests that the oxidant gains access to the metal by continual dissociation of the oxide.[21] Of the two proposed mechanisms, it is unlikely that oxide dissociation could account for the observed growth rate of the inner oxide on ferrous alloys, although this mechanism may be involved in pore formation. The evidence here is more consistent with the passage of oxidant gas through the oxide to form the inner layer.

In general the relative thicknesses of the oxide comprising a multilayer scale vary widely and are a function of temperature.[22] However, where duplex scales are formed by a two-way diffusion process the thickness of the two layers is approximately equal and almost temperature invariant.[18] Thickness ratios have been reported which tend to lie within the range 0·5–2; usually the outer layer is the thicker in the earlier stages of oxidation and, as the reaction proceeds, the difference becomes relatively less.

A notable characteristic of duplex layers formed on oxidized alloys is a difference in their chemical composition as well as morphology. The alloying elements may be concentrated almost entirely in the inner layer or present to different levels in both layers. This occurrence was noted many years ago by Pfeil[10] who showed that most of the transition elements with the exception of manganese were concentrated within the inner layer of oxide formed on alloy steels heated in air at 1 000°C. The findings have since been substantiated and extended by a large number of workers. For example, Brenner[23] found molybdenum concentrated in the inner scale on iron–molybdenum alloys oxidized at 1 000°C, while Moreau[24] observed a similar effect on iron–chromium alloys and Rahmel[25] reported that vanadium, silicon, chromium, and molybdenum were confined to the inner layer on the respective binary alloys oxidized at 1 000°C in oxygen. Similar partitioning behaviour has been observed on more complex iron alloys, both ferritic and austenitic, and also in a variety of oxidizing environments. Antill *et al.*[1] noted that duplex scales were formed on steels oxidized in CO_2 gas, with chromium and molybdenum concentrated in the inner oxide and with iron and manganese distributed throughout the double layer. Segregation of chromium in the inner oxide was also found by Surman and Castle[26] on iron–chromium alloys after oxidizing them in steam. These workers postulated a vapour-phase transport mechanism to explain the morphology of the outer oxide and used it to account also for element partitioning in the double layer. They suggested that a volatile iron hydroxide was produced by reacting steam with the metal which was then decomposed at the gas/oxide interface to oxide. The absence of chromium in the outer layer was attributed to the fact that chromium forms no known volatile hydroxide. As Cox *et al.*[20] have pointed out, however, such a mechanism cannot be generally applied to account for observed element partitioning effects and these workers have recently put forward an alternative hypothesis based on analysis of interstitial cation diffusivity through oxide lattices. This chemical diffusion model is described below.

Cox et al.[20] began by considering the distribution of interstitial positions in close-packed oxide ion lattices of the spinel structure commonly found on oxidized transition metals and then proceeded to examine the energetics involved as cations move through the structure by way of these different interstices. Following Azaroff,[27] the available cation diffusion path through the oxide lattices was assumed to consist of a succession of jumps from an octahedral interstice to an adjacent unoccupied tetrahedral position and then to the next unoccupied octahedral interstice; neighbouring sites are separated by a trigonal constriction through which the cation must squeeze. Since most cations of the first-row transition elements are of similar size, the strain energy associated with the passage of a cation through a trigonal constriction was assumed to be constant and cation mobility to be essentially a function of the difference between octahedral and tetrahedral site energies for the diffusing ion. Site energies for transition metal cation species were then calculated using crystal field theory and a ranking order for cation mobility was drawn up based upon the site energy difference:

$$Fe^{3+} > V^{3+} > Mn^{3+} > Cr^{3+} \quad \text{(trivalent ions)}$$

$$Mn^{2+} > Fe^{2+} > CO^{2+} > Ni^{2+} \quad \text{(divalent ions)}.$$

Support for the proposed model was obtained from a comparison of the parabolic rate constants observed upon oxidizing the transition metal elements. Hence it should follow that Cr^{3+}, with its octahedral preference energy, would diffuse much more slowly than Fe^{3+} through close-packed oxide lattices. This is known to be the case with iron–chromium alloys oxidized at temperatures below 1 000°C, where only iron reaches the gas/oxide interface and the oxide in the outer layer is essentially Fe_3O_4. The inner oxide, formed by gas penetration through the porous oxide as referred to above, will thus contain both chromium and iron, i.e. the mixed spinel $(Fe_{3-x}Cr_x)O_4$, where x was shown by Cox to be related to the chromium content of the alloy. It is noteworthy that the above ranking order for cation mobility would appear to account for most element partitioning effects in oxide scales on transition metal alloys which have been reported in the literature.

ROLE OF VACANCIES

The frequent occurrence of void dispersions in oxide scales was attributed as long ago as 1952[8] to the precipitation of vacancies generated during the oxidation process. Indeed, the production of a tube of oxide from the oxidation of metal wire can be interpreted in terms of a vacancy generation and condensation mechanism.[9] Other experiments involving the oxidation of tubular specimens[17,28] have also provided indirect evidence for the generation of vacancies. Mindel and Pollack[29] too have attributed voids in oxide scales to the condensation of vacancies. However, in general, it is difficult to obtain quantitative data on vacancy concentrations from such observations due to pull-out of oxide particles during sectioning of the sample. Scanning electron microscope studies on fracture sections are regarded by many workers as being a more satisfactory method of evaluating pore structures, although even here the results may be biased due to the fracture path following the voids.

Probably the most convincing evidence for vacancy generation in the metal substrate during oxidation has been provided by observations on dislocation loop growth in metal foils oxidized in the electron microscope.[30–32] This type of experiment can clearly give important information on the effect of alloy additions on the rate of loop growth, i.e. on the behaviour of vacancies during oxidation but this will not be discussed further since the subject is dealt with in another paper in this volume.[33] Instead we shall examine in some detail results obtained in the author's laboratory[16,34] which have also given information on vacancies and from which quantitative data have been extracted.

Reference has earlier been made to the formation of lamellar voids in the inner oxide scales on high purity, annealed iron–chromium alloys. These voids are regularly spaced (see Fig. 4), the spacings being larger with thicker specimens. It should be noted that such regular arrangements of voids are not commonly observed on commercial materials which have been oxidized, due presumably to the large number of vacancy sinks present. Also the metallurgical condition of the metal sample is critical and the presence of cold work can affect the regularity with which voids may form. The occurrence of lamellar voids was explained by invoking a periodic vacancy process as follows. As outer oxide is formed by the outward migration of iron cations, vacancies are injected into the metal. Provided that the rate of vacancy injection is greater than the rate of vacancy loss at sinks then initially some will be retained in solution in the metal. Eventually a critical concentration is reached when vacancies may condense into voids at suitable precipitation sites. The regular alignment of voids parallel to the metal/oxide interface indicates that this is the preferred site in these high-purity alloys. It follows that after vacancy condensation has occurred, the vacancy excess in the metal would have to increase to the critical fraction before further vacancy rejection can take place. By this time the metal/oxide interface would have advanced a discreet distance into the metal. Thus the periodic character of lamellar voids was explained. The authors were able to discount the likelihood of stress-induced cracking being responsible for void formation from the fact that rapid cooling of the specimen did not encourage interfacial voids.

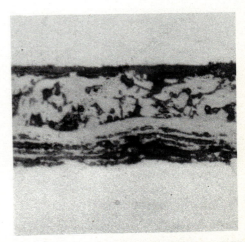

4 Fe–9% Cr alloy oxidized at 600°C for 142 h; optical micrograph (courtesy *Phil. Mag.*) ×500

From measurements of the interlamellar void spacings, Cox estimated the critical fraction of vacancies in the metal required for condensation to take place at the metal/oxide interface as 6×10^{-3}. The maximum energy, E_v, required to induce a vacancy to condense at the interface was estimated from the vacancy supersaturation to be $\approx 2\cdot7 \times 10^{-19}$ J. This value, computed from the void distribution in oxide scales, was then compared with that obtained independently by considering the energy required to create a new internal surface by separation of the oxide/metal interface. Using typical values for the specific surface free energies for metal/vapour, oxide/vapour and metal/oxide interfaces respectively gave $E_V \approx 1\cdot3 \times 10^{-19}$ J. Taking into account the approximations involved in both calculations for E_V the values are seen to be in close agreement, which lends support to Cox's proposed vacancy generation and condensation model. Of course, E_V calculated from the vacancy condensation model would be expected to be the higher value, since in principle more energy would be required to precipitate the first vacancy than would be needed to enlarge the cluster by further vacancy absorption. Furthermore, according to the argument, with thicker specimens more vacancies would have to be produced to reach the critical level, and therefore a correspondingly thicker oxide film would have to be formed on the metal surface. Hence a direct relationship between specimen thickness and interlamellar void spacing should occur, as was noted by Cox. It also follows that at intersecting surfaces a higher local vacancy flux would be injected into the metal than at a planar surface; this would lead to a greater incidence of voids at specimen corners, as has been observed by a number of workers (see, for example, ref. 19).

Additional information on the role of vacancies in the formation of duplex oxide layers has been provided in a more recent paper. It was noted[34] that the thin protective film of chromium-rich α-M_2O_3 which may form first on oxidized iron–chromium alloys appeared to break down in localized regions, usually close to grain boundaries of the underlying metal. Associated with this breakdown was the development of a number of nuclei of magnetite which, once formed, spread laterally across the surface as well as increasing in thickness, Fig. 5. Beneath each island of magnetite was an inner layer of

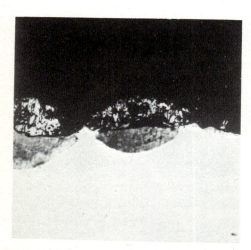

5   **Fe–15% Cr alloy oxidized at 600°C for 100 h; optical micrograph (courtesy *Phil. Mag.*) ×350**

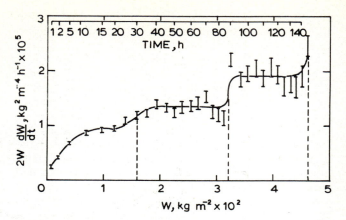

6   **Kinetic data for Fe–20% Cr alloy oxidized at 600°C (courtesy *Phil. Mag.*)**

chromium-containing spinel oxide and, as oxidation proceeded, these patches of duplex structure eventually impinged upon one another to give complete surface coverage as illustrated earlier, Fig. 1. Detailed analysis of the oxidation kinetics during this process revealed a series of abrupt discontinuities in the weight-gain curve which appeared to occur at approximately equal weight-grain increments, Fig. 6. Once, however, the metal was entirely covered by duplex oxide, the discontinuities ceased and parabolic kinetics were recorded. Cox measured the weight-gain increment and calculated that the associated number of vacancies injected into the metal was $\sim 9 \times 10^{-3}$. Since this figure was closely similar to his value calculated from lamellar void distributions, Cox considered that each discontinuity in the kinetic curves corresponded to situations in which vacancies injected into the metal reach supersaturation level and then condense at suitable sites in the specimen. It was suggested that the metal/oxide interface acted again as the most favourable site for condensation and that when the free energy accompanying the vacancy supersaturation is sufficient to exceed the work of adhesion at the oxide/metal interface a void is nucleated. The oxide above the void would then be expected to become mechanically unstable and could collapse under the external pressure of gas, as suggested by Howes.[35] Subsequent breakdown of the the thin protective oxide may then allow the direct access of gaseous oxidant to the metal. Protective oxide would not form to heal the damage, since the metal would now be depleted in chromium due to the earlier formation of chromium-rich M_2O_3, and M_3O_4 would be produced instead, giving characteristic duplex oxide growth.

While accepting that the above ideas based on vacancy generation and condensation have many attractive features, they have been criticized by Gibbs on the basis of the large vacancy supersaturations which were derived. He considers that other vacancy sinks such as dislocations and grain boundaries would restrict the development of such high fractions. However, calculations by Kuhlmann–Wilsdorf[36] would indicate that dislocations are inefficient sinks for vacancies, especially at high temperatures and, as Cox has pointed out, similarly high vacancy levels can be inferred from measurements on irradiated materials despite the presence of dislocations, interstitials, and grain boundaries.

CONCLUDING REMARKS

From the foregoing it is apparent that when the formation of a duplex oxide scale can be clearly and simply related to a two-way diffusion process it is possible to describe the growth in terms of particular diffusion mechanisms. Where the diffusants are interstitial atoms or oxidant species their behaviour may be followed, using available techniques, as oxidation proceeds and structural observations may be explained in quantitative or semiquantitative terms and the rate-controlling process identified. The application of such data to the interpretation of oxidation kinetics has, however, not generally met with a comparable degree of success and it may be argued that additional factors are sometimes involved here to complicate interpretation, for example effects caused by stress being developed in the growing oxide, etc.

It is perhaps in connection with quantitative descriptions of the role of vacancies in the oxidation processes that information is most needed. The problem arises from a deficiency of experimental data giving information directly on vacancy behaviour, arising in turn from a shortage of suitable techniques which can be applied to such studies. Apart from the observations on dislocation loop growth in foils oxidized in the electron microscope and some X-ray topography experiments, most published data on vacancies in oxidation have been derived from indirect methods such as the work described in the previous section. While such conjectural approaches have some value they must expect to receive criticism. However, there does appear to be a measure of agreement that the inner layer in duplex oxide scales is formed in space made available beneath the outer oxide by the movement of cations to the gas/oxide interface. Hence, while little doubt remains that vacancies are produced in the metals surface regions as a result of this diffusion, the main argument appears to centre around the fate of vacancies. Three possibilities may be envisaged:

(i)   vacancies may dissolve in the metal and then precipitate into voids at suitable nucleation sites such as interfaces
(ii) vacancies may dissolve in the metal and then become annihilated at dislocations, etc. and/or form vacancy loops
(iii) the vacancies may be annihilated at the metal/oxide interface as fast as they form.

In all three cases, loss of metal will occur and, where the oxide is rigid, voidage would be expected. In such situations filling of voids with an oxide may take place and, since this process will occur by a different oxidation mechanism, the inner oxide is readily distinguished by its different morphology and/or composition from the outer layer. This difference allows, of course, the total vacancy production to be assessed but may not give information on the vacancy concentration in the metal. It may be argued, therefore, that the observations referred to in the previous section which infer a cyclic vacancy condensation process are especially valuable in this respect.

Once formed, voids present in the oxide scale may alter the course of oxidation reaction in a number of ways. In most cases discussed in the literature it would appear that voids are deleterious, although the extent to which they may contribute to an increased oxidation rate will depend upon the precise oxidation conditions, for example whether the metal is subjected to thermal cycling. The ways in which voids may affect oxidation can be listed as follows.

1 Voids present in the scale may act as regions of mechanical weakness, cases having been reported where substantial voidage has been associated with ready detachment of the oxide scale. Spalling is then accentuated by external stresses, imposed for example by an impact or by thermal cycling, and the protective nature of the oxide scales suffers accordingly

2 Voids may help to provide a path for gaseous oxidant to reach the metal where the gas reacts to form an inner layer. Since the inner oxide is also generally porous, gas will permeate the layer. In this way the passage of gas to the metal will be restricted and, if the oxidant consists of a mixture of gases, its composition may change and so also the oxidation behaviour. For example, with steels oxidized in $CO-CO_2$ mixtures, the gas adjacent to the metal would become increasingly richer in CO gas as a result of the reaction

$$xM + CO_2 \rightleftharpoons M_xO + CO$$

Hence the increased carburizing potential of the gas may eventually result in carbon deposition. The presence of carbon formed in this manner in the inner oxide is believed[2,3] to play an important role in breakaway oxidation behaviour.

3 It may also be envisaged that voids block some of the lattice diffusion paths for cations and hence reduce the rate of growth of the outer layer. The importance of this effect will, of course, depend upon void structure although the possible contribution of surface diffusion to cation migration cannot usually be ignored.

In general then, porosity in oxide scales may be considered undesirable and its prevention or minimization a worthwhile objective in the development of improved corrosion-resistant alloys. It would hence follow that, if voids are intimately associated with vacancy condensation, any way of stabilizing vacancies in solution in the metal may be beneficial. This may be achieved by adding to the metal solute atoms with a large vacancy binding energy. The addition is more likely to be one which can be substitutionally incorporated into the metal lattice since interstitial atoms, while having a high vacancy binding energy[37] are probably too mobile to inhibit vacancy migration; this is particularly important with ferritic steels.[38] Certainly, the addition of yttrium or scandium atoms to an Fe–25Cr–4Al alloy[39] has been shown to improve greatly oxide scale adhesion, and one mechanism put forward to account for this result involved the association of the vacancy with the large yttrium or scandium solute atom. However, it should also be noted that an alternative explanation suggested that yttrium and scandium improved mechanical keying of the scale by forming oxide 'roots' into the metal. Such ambiguities of interpretation may be removed by carrying out studies on model systems where one of the mechanisms can be eliminated. For example, work in the author's laboratory is focused upon iron–platinum and iron–nickel alloys where mechanical keying of the oxide is unlikely. Finally, it would be valuable if measurements of vacancy–solute binding energies for a range of alloy systems were available in order to obtain further information on the role of vacancies in the oxidation of metals.

ACKNOWLEDGMENTS

The author wishes to acknowledge the many stimulating discussions on metal oxidation he has enjoyed with his colleagues, Dr M. G. C. Cox and Dr B. McEnaney, and which have led to the preparation of this paper.

REFERENCES

1 J. E. ANTILL et al.: Corrosion Sci., 1968, **8**, 689
2 G. B. GIBBS et al.: Proc. BNES Conf. on Corrosion of Steels in CO₂, 1975, 59
3 P. L. SURMAN AND A. M. BROWN: Proc. BNES Conf. Corrosion in Steels in CO₂, 1975, 85
4 W. B. JEPSON AND D. W. AYLMORE: J. Electrochem. Soc., 1961, **108**, 942
5 D. L. DOUGLASS: Oxid. Metals, 1969, **1**, 127
6 J. STRINGER: Corrosion Sci., 1970, **10**, 513
7 J. PAIDASSI: J. Metals, 1952, **4**, 546
8 B. W. DUNNINGTON et al.: Corrosion, 1952, **8**, 2
9 H. J. ENGELL AND F. WEVER: Acta Met., 1957, **5**, 695
10 L. B. PFEIL; J. Iron Steel Inst., 1929, **119**, 501
11 C. T. FUJII AND R. A. MEUSSNER: Trans. AIME, 1968, **242**, 1259
12 F. H. STOTT et al.: Oxid. Met., 1971, **3**, 103
13 C. C. WOOD: 'Oxidation of metals and alloys', 1971, 201, ASM
14 O. KUBASCHEWSKI AND B. E. HOPKINS: 'Oxidation of metals and alloys', 1962, London, Butterworths.
15 G. B. GIBBS: Proc. BNES Conf. on Corrosion of Steels in CO₂, 1975, 281
16 M. G. C. COX et al.: Phil Mag., 1973, **28**, 309
17 D. L. DOUGLASS: Corrosion Sci., 1968, **8**, 665
18 J. A. SARTELL AND C. H. LI: J. Inst. Metals, 1961, **90**, 92
19 S. MROWEC: Corrosion Sci., 1967, **7**, 563
20 M. G. C. COX et al.: Phil Mag., 1972, **26**, 839
21 A. DRAVNIEKS AND H. J. McDONALD: J. Electrochem. Soc., 1948, **94**, 139
22 G. VALENSI: Proc. Internat. Conf. Surf. React., 1948, 156
23 L. S. BRENNER: J. Electrochem. Soc., 1955, **102**, 7
24 J. MOREAU: Compt. Rend., 1953, **236**, 85
25 A. RAHMEL: Z. Electrokem., 1962, **66**, 363
26 P. L. SURMAN AND J. E. CASTLE: Corrosion Sci., 1969, **9**, 771
27 R. F. TYLECOTE AND T. E. MITCHELL: J. Iron Steel Inst., 1960, **196**, 445
28 L. V. AZAROFF: J. Appl. Phys., 1961, **32**, 1638
29 M. J. MINDELL AND S. R. POLLACK: J. Phys. Chem. Solids, 1969, **30**, 993
30 D. MICHELL AND G. J. OGILVIE: Physica Status Solidi, 1966, **15**, 83
31 R. HALES et al.: Metal. Sci., 1968, **2**, 224
32 H. L. FRAZER et al.: Phil. Mag., 1973, **28**, 639
33 R. E. SMALLMAN AND P. S. DOBSON: this volume
34 M. G. C. COX et al.: Phil. Mag., 1975, **31**, 331
35 V. R. HOWES: Corrosion Sci., 1968, **8**, 221
36 D. KUHLMANN-WILSDORF: Scripta Met., 1973, **7**, 1059
37 R. A. ARNDT AND A. C. DAMASK: Acta Meta., 1964, **12**, 341
38 M. G. C. COX et al.: Metal Sci., 1976, **10**, 379
39 J. K. TIEN AND F. S. PETTIT: Metall. Trans., 1972, **3**, 1587

Influence of vacancies produced by oxidation on the mechanical properties of nickel and nickel–chromium alloys

P. Hancock

When metals oxidize by cationic diffusion, vacancies are generated at the metal/scale interface which have a significant influence on the mechanical properties of the substrate metal. Thin uniform oxide films, produced by oxidation, of 4–40 μm, are shown to cause a pronounced reduction in creep-rupture life and appreciable reduction in tensile ductility. These effects are shown to be associated with voids precipitated by vacancy injection due to the oxidation reaction. Vacancy injection by oxidation is also shown to cause strengthening at low strains and the mechanisms by which this effect occurs are considered. The influence of chromium content on vacancy generation by oxidation is examined and discussed in terms of its influence on the mechanical properties of the substrate metal. Precipitation and dispersion hardening are shown to have a significant influence on the process of vacancy injection due to oxidation. The effects of cold work on nickel and nickel–chromium alloys are investigated; it is shown that nickel becomes more vulnerable to deterioration at high temperatures whereas nickel–chromium alloys become more resistant. The reasons for this behaviour are explained.

The author is at the Cranfield Institute of Technology

When the mechanical properties of metals and alloys are measured at high temperatures the influence of the ambient atmosphere is invariably ignored, especially if the atmosphere is mainly oxidative and is not seriously contaminated with known aggressive species such as chlorides, vanadates or sulphates. If internal oxidation or corrosion occurs, either because of the presence of such contaminants,[1,2] or because of unwise alloy selection, then it is obvious that the environment will influence the mechanical properties of the component. However, if the surface oxide forms as a thin uniform coating on the testpiece, or component, and this coating does not change the load-carrying section appreciably, then the effect of the environment is invariably dismissed as being of no importance. It may be argued that the test environment is usually ambient air and, therefore, the environmental effect is included in the test results. This approach can lead to trouble if the tests are used to interpret the high-temperature deformation mechanisms due to stress, or if the results are used to predict longer-term behaviour or the influence of multiaxial stressing.

In this paper the influence of vacancies produced by surface oxidation on the mechanical properties of metals is discussed and the nickel–chromium system, which is widely used for alloys intended for high-temperature service, has been chosen to illustrate the interactive effects of environmental and deformation mechanisms.

VACANCY PRODUCTION DURING OXIDATION OF NICKEL

Metals which are used successfully for high-temperature service in oxidizing environments form surface oxides which afford a degree of protection to the underlying metal. Further reaction occurs either by cationic diffusion of metal ions outwards through the scale to react at the scale/gas interface, or by anionic diffusion of oxygen ions through the scale to react at the metal/scale interface. Most commercial alloys for high-temperature service are based on the nickel, cobalt or iron systems[3] and oxidize by cationic diffusion of metal ions. Nickel and the nickel–chromium alloys are typical of this class of materials.

When metal ion diffusion is the dominant mechanism of transport through the oxide scale, this should result in vacancy formation in the metal at the metal/scale interface. Early work on the oxidation of nickel[4] showed that the oxidation was accompanied by the formation of voids in the metal at the grain boundaries. It was suggested that the mechanism of void formation was by the precipitation of vacancies generated according to a reaction of the form:

$$Ni \rightarrow Ni^{2+} + 2e + \square_{Ni}$$

where Ni^{2+} represents a metal ion moving into the oxide leaving a vacancy ($\square_{Ni}$) in the metal lattice. The number of voids was shown to increase with exposure time and temperature, as required by vacancy precipitation produced by the oxidation process. A typical result for nickel oxidized for 100 h at 1 000°C is shown in Figure 1a.[4] At lower temperatures and shorter times the voids are less numerous but they can be easily detected after 24 h at 700°C.[4] This effect in nickel has been confirmed by Hales and Hill[5] and similar effects have been reported in other metals oxidizing by cation diffusion as reviewed recently.[6] The early experiments merely postulated that the voids were formed by vacancy injection but the elegant experiments by Smallman and co-workers, who oxidized specimens in the electron microscope, showed that vacancy dislocation loops are formed as a consequence of surface oxidation.[7-10] Further work showed that their rate of growth could be controlled both by alloying additions and by oxygen pressure, when the metals oxidized by cationic diffusion, as predicted from oxidation theory.[11-14] Similarly, it has been shown that for silicon, which oxidized by anionic diffusion of oxygen through the scale, the reverse effect happened, i.e. existing vacancy dislocation loops shrank as oxidation proceeded.[15,16]

The above experiments clearly demonstrate that vacancy injection into the metal can occur when cationic diffusion is the principal means of diffusion through the oxide scale. However, consideration of the structure of the oxide and metal lattices suggests the metal/oxide interface should be highly disordered; it might be expected that this would be a natural site for void precipitation. In fact, Howes[17] has shown that for Fe–Cr alloys oxidized at 950°C vacancies precipitate voids at the metal/scale interface. It can be shown that, if the same nickel shown in Fig. 1a is first vacuum annealed for long periods, about 48 h at 1 000°C, before being exposed in air at 1 000°C, then the vacancies precipitate at the scale/metal interface rather than in the grain boundaries, as shown in Fig. 1b. It should be noted that the amount of oxide formed on both Figs. 1a and b is about the same (actually the amount of scale in the annealed material is somewhat smaller), but of more importance is the fact that the site of void precipitation has been altered. It was thought that perhaps gas solubility in the nickel might have acted as void nucleation sites but this was disproved by cold rolling by 25% after vacuum annealing. On subsequent oxidation the vacancies again precipitated in the grain boundaries as shown in Fig. 1a. Harris and Masters[18] claimed that the oxide scale on aluminium alloys at temperatures up to 385°C, far from acting as a site for precipitation of vacancies, prevented vacancy escape at the metal/scale interface and Gibbs and Harris suggest that a coherent or semicoherent interface associated with epitaxial growth

a

b

1   a Grain-boundary voids produced by oxidizing cold-rolled nickel for 100 h at 1 000°C ×150; b metal/oxide interface voids produced by oxidizing vacuum annealed nickel for 100 h at 1 000°C ×150

is not able to absorb vacancies in the same way as an incoherent boundary.

The structure of the metal/oxide interface and that of the underlying metal are therefore very important in determining whether voids will migrate into the metal or remain at the metal/scale interface.

The results on nickel show that the effect of cold work in the metal can alter the site for void nucleation and it

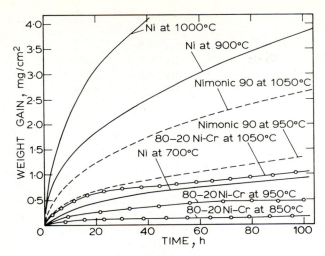

2 Influence of chromium and temperature on the oxidation of nickel

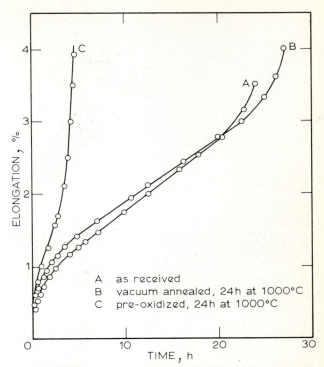

A   as received
B   vacuum annealed, 24h at 1000°C
C   pre-oxidized, 24h at 1000°C

3 Influence of preoxidation on creep of nickel at 725°C

seems reasonable that the increased dislocation density should aid vacancy migration from the metal/oxide interface. This effect should also increase the oxidation rate and this has been shown to be true for cold-worked nickel,[19–22] copper[22] and iron.[20,22] Thus, for nickel, cold work increases the oxidation rate but, more importantly, relative to the mechanical properties of the underlying metal, the rate of vacancy diffusion into the metal is also increased and this will be shown to have a significant influence on the mechanical behaviour of the material.

As the temperature increases the amount of oxidation increases as shown in Fig. 2 and therefore the number of injected vacancies should also increase and this has been observed.[4] However, it is obvious from a comparison of the amount of oxide formed on Figs. 1a and b with the number of voids present in the metal, or at the interface, that not every metal-ion incorporated into the scale has produced a vacancy which has migrated into the metal. Some 'collapse' of the oxide must occur to follow the metal as it retreats. This means that the plasticity of the oxide scale is important in determining whether the scale can accommodate the deformation necessary to follow the retreating metal and this has been examined in some detail in the author's laboratory.[23–27] It has been shown that the oxide scale formed on nickel is one of the most mechanically sound and highly adherent oxides.[24,27] Therefore, as cation diffusion occurs and the metal retreats, the oxide should be able to follow across a planar interface,[5,25] but on a cylindrical specimen the oxide would have to exhibit considerable plastic deformation to remain adherent to the metal.[5,25] This effect has been discussed in detail elsewhere[4,5,25] and the results show that the interfacial adhesion is dependent upon specimen geometry. This means that, as the interfacial properties are critical in deciding whether vacancies precipitate at the metal/scale interface or in the grain boundaries, the effect of specimen geometry must not be ignored.

INFLUENCE OF OXIDATION ON THE MECHANICAL PROPERTIES OF NICKEL

The influence of preoxidation on the creep properties of nickel was demonstrated in 1966[4] where it was shown that preoxidation had a pronounced effect on the creep behaviour of nickel tested at 23·2 MN/m²

(1·5 Tonf/in²) at 725°C; these results are shown in Fig. 3.[4] The results for as-received material are shown in curve A, and curve C shows the influence of preoxidizing in air for 24 h at 1 000°C. This treatment produced a thin uniform oxide scale on the metal with no internal oxidation. The specimens were 6·35 mm in diameter (0·25in) and the loss in section due to oxidation was less than 0·01%. To demonstrate that the effect was not due to prior annealing at 1 000°C, curve B shows the results of specimens also treated for 24 h at 1 000°C but in vacuum. It is apparent that heat treatment in vacuum had no effect, but preoxidation caused a drastic reduction in creep life. Later study of the fracture surfaces, Fig. 4, showed that the initial prior oxidation had a pronounced influence on the fracture mechanism. The as-received (Fig. 4a) and pre-vacuum annealed fractured surfaces (Fig. 4b) show no evidence of voiding and no evidence of secondary creep cracks, whereas the preoxidized specimen (Fig. 4c) showed pronounced secondary creep cracking. Later work[28] showed that smaller amounts of preoxidation, e.g. 24 h at 950°C, also reduced the creep life at test temperatures of 600°–950°C but the primary creep rate of the preoxidized specimens was always lower than that observed for the untreated specimens. This strengthening effect was only noticeable in the initial stages of creep because the influence of the grain-boundary voids on the fracture process soon became the dominant deformation mechanism. If the void growth during oxidation caused appreciable grain-boundary voiding before testing, as in Fig. 3, then the strengthening effect became masked. However, with small amounts of oxidation, it was quite pronounced and reproducible. Creep strengthening in the diffusion creep or Herring–Nabarro creep region at very high temperatures, due to oxidation, has been observed by Hales et al.[29] They tested magnesium at 450°C with a surface oxide and suggested that vacancy

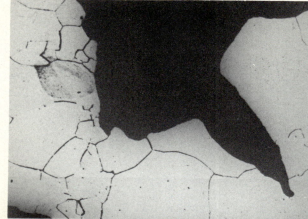

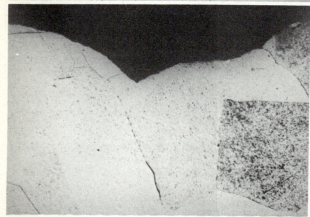

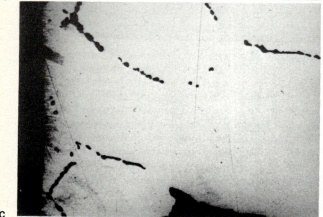

a as-received; *b* pre-vacuum annealed 24 h at 1 000°C; *c* preoxidized 24 h at 1 000°C

4   Fracture surfaces of creep specimens shown in Fig. 3

gradients produced by oxidation opposed those set up by the applied stress system. Hondros and Clark[30] found that oxide films gave a strengthening effect to copper at 927°C and suggested that the film acted as a barrier to vacancy flow, as first suggested by Harris and Masters[18] to account for the stability of vacancy dislocation loops in aluminium alloys at 385°C. Although these mechanisms are reasonable for the diffusion creep régimes, the creep experiments with nickel are not at sufficiently high temperatures for diffusion creep to be the predominant mechanism. However, the creep tests span a range of temperatures from 600° to 900°C and it is suggested that the initial strengthening effect is due to the strengthen-

ing produced by vacancy clusters[31,32] formed during oxidation. However, there is also a secondary strengthening effect which is particularly relevant at the higher temperatures, when grain-boundary sliding mechanisms would be expected to occur. This second effect is due to the restraining of grain-boundary movement by void precipitation. This inhibits boundary sliding and rotation, hence causing strengthening in the regions where grain-boundary sliding is operating. These effects of boundary pinning have been demonstrated by studying the annealing of cold-worked nickel. In these experiments half of the specimens were given a prior vacuum anneal for 25 h at 1 000°C. Then, both vacuum-annealed and cold-worked specimens were oxidized for times up to 100 h in air at 1 000°C. The specimens were sectioned normal to the surface and the angle that the surface grains made with the normal to the metal/oxide interface was measured. The frequency of occurrence of grains whose angle with the surface normal was within ±20° is plotted in Fig. 5. This shows that after prior annealing at 1 000° for 24 h the surface grain boundaries are mainly normal to the surface, as expected from surface energy considerations. However, the surface grain boundaries of the cold-worked specimens annealed in air are inhibited from moving to positions normal to the metal/oxide interface, even after times of 100 h. Grain size counts on the two sets of specimens also showed that grain growth of the cold-worked material, annealed only in air, was appreciably inhibited. These experiments show that voids produced on the grain boundaries inhibit boundary movement and this inhibition can lead to strengthening in creep conditions where grain-boundary sliding is in operation.

The influence of vacancies produced by oxidation on the deformation and fracture of nickel can also be

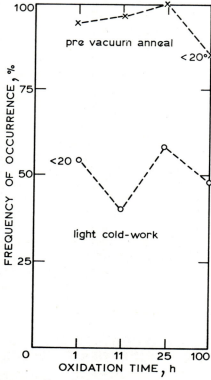

5   Influence of grain-boundary voids in restraining the grain movement in cold-worked nickel annealed at 1 000°C in air[19]

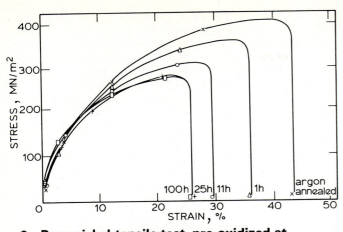

6 Pure nickel tensile test, pre-oxidized at 1 000°C for various times

investigated by considering the effects of preoxidation on the tensile properties of the nickel at 20°C. Sheet specimens 1·25 mm thick were preoxidized in air at 1 000°C for varous periods and then tensile tested at 20°C. The results, Fig. 6,[19] show that the effect of preoxidation is to:

(i) raise the yield point slightly
(ii) reduce the maximum load dramatically
(iii) reduce the ductility significantly.

Comparable tests annealed in argon instead of air showed no change in properties. As these tests are conducted at room temperature, the strengthening effect on the yield point cannot be due to restrictions in grain-boundary movement, but can reasonably be assumed to be due to vacancy cluster hardening. It is certainly not due to any contribution to strength that the thin oxide might contribute, for the effect is observed even when the oxide is removed by careful polishing. The pronounced influence of the grain-boundary voids on fracture is shown by the marked change in maximum load and ductility observed after preoxidation. In fact, examination of the specimens preoxidized for more than 11 h showed large amounts of secondary cracking with very little grain deformation and little local necking, while specimens annealed in inert atmosphere fractured with large amounts of grain deformation as shown in Fig. 7.[19] After testing, the fracture appearance of the preoxidized specimens was similar to that normally observed

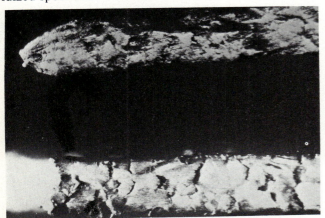

7 Fracture of annealed (top) and preoxidized (bottom) (24 h at 1 000°C) nickel after tensile fracture at 20°C

in high-temperature low-stress creep, the fracture being intergranular in all cases.

These experiments show that the vacancies produced during oxidation have a pronounced influence on the mechanical properties of the underlying metal. The effects produced are exactly as predicted from the oxidation behaviour, assuming that cationic diffusion causes vacancy injection into the material. The influence of chromium additions to the nickel matrix must now be considered.

VACANCY PRODUCTION DURING OXIDATION OF NICKEL–CHROMIUM AND ITS INFLUENCE ON MECHANICAL PROPERTIES

As many high-temperature-service alloys are based on the nickel–chromium system, the effect of the additions of chromium on the vacancy injection problem must be considered. The influence of chromium on the oxidation behaviour of the nickel–chromium alloys has been extensively researched and well reviewed.[3,28] In the steady state, chromic oxide or nickel–chromium spinel forms an oxide barrier through which outward diffusion of cations is still the predominant diffusion mechanism. However, as shown in Fig. 2 the rate of oxidation is reduced considerably compared to nickel. For instance, 80 Ni–20 Cr at 1 050°C oxidizes at about the same rate as nickel at 700°C and, if compared at the same temperature, the oxidation rate of nickel at 700°C is more than an order of magnitude greater than that of 80 Ni–20 Cr at the same temperature. This means that if all other factors remained constant, the amount of vacancy injection for 80 Ni–20 Cr should be considerably less than that experienced with nickel and exposure times and temperatures would have to increase considerably to obtain comparable effects. The same can be said for the commercial strengthened high-temperature alloys based on the nickel–chromium system and a typical example, Nimonic 90* (approximate composition 18–21 Cr, 15–21 Co, 1·8–3·0 Ti, 1·35 Al, bal Ni), is included in Fig. 2. The oxidation rate is higher than the 80 Ni–20 Cr alloys, but not nearly as high as the oxidation rate observed with nickel and, for many of the superalloys, the observed oxidation rate is comparable to that of the 80 Ni–20 Cr system.[34]

However, apart from the reduction in oxidation rate there are other basic differences between the oxidation of nickel and nickel–chromium alloys. In the second section it was stated that the oxide formed on nickel is mechanically sound and highly adherent.[24,27] However, the oxide scales formed on the nickel–chromium alloys do not have the same degree of mechanical stability,[6,25,26,27] and it has been shown recently that the nickel-based superalloys can be ranked in terms of the mechanical stability of their oxide scales.[34] Briefly, the method used to measure mechanical scale stability relies on measurement of the natural resonant frequency of freely suspended cylindrical specimens during oxidation.[23,24,26] As the oxide film grows, the modulus of the oxide is different from the modulus of the metal and the growth of the oxide can be monitored by measuring the change in resonant frequency of the oxidizing rod. However, any mechanical discontinuity, such as a crack forming in the oxide, will manifest itself as a sharp change in the frequency response. So, an oxide which

* Trade Name Henry Wiggin & Co. Ltd.

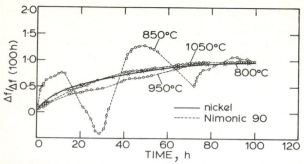

8 Normalized change in frequency for isothermal oxidation of nickel and Nimonic 90[34]

grows as a coherent, adherent scale, shows a smooth curve for frequency v. time, whereas an oxide which cracks during growth will exhibit discontinuities. Typical results for nickel and Nimonic 90 are shown in Fig. 8. The results have been normalized by plotting frequency change divided by the total frequency change in 100 h v. time. In this way results of different materials at different temperatures can be compared.[34] The oxide film on nickel gave a smooth continuous response, whereas the response from Nimonic 90 showed that there is pronounced mechanical instability at 850°C but, as the temperature increases, the scale became more mechanically coherent and this is due to the increased scale plasticity at the higher temperatures.[27,34] Results for a higher strength nickel–chromium superalloy, Nimonic 105*[27,34] (composition 13·5–15·75 Cr, 18–22 Co, 0·9–1·5 Ti, 4·5–4·9 Al–5·01 Mo, bal Ni) showed that the mechanical instability of this oxide was more pronounced than Nimonic 90, although the oxidation characteristics were similar to the 80 Ni–20 Cr alloys. This shows that although the oxidation properties of the superalloys may be similar, the mechanical integrity of the oxide scales can be appreciably different.

However, the oxide films become more coherent and adherent as the temperature increases and, therefore, vacancy injection would only be expected at very high temperatures where coherent scales were present.

As the effective barrier oxide is usually chromic oxide in the nickel–chromium system, when any mechanical crack or discontinuity occurs, it is essential that this should be repaired by chromium diffusing from the body of the alloy to aid repair and reduce further attack. Any process which speeds the diffusion of chromium to the oxide scale should therefore reduce the rate of oxidation. Increasing the number of grain boundaries and the dislocation content of the metal by cold work enhances chromium diffusion and consequently Giggins and Pettit[35] and Wysiekierski et al.[36] have shown that cold work reduces the rate of oxidation in these alloys. A similar effect of cold work was observed with 18 Cr–10 Ni and 25 Cr–20 Ni steels and this was explained[1] in terms of enhanced chromium diffusion to allow a protective film to be maintained. It should be noted that the influence of cold work in materials which form a protective layer by selective oxidation of one component (as in the Ni–Cr alloys) is invariably to reduce the rate of oxidation. In single-component systems such as nickel, copper, and iron the effect of cold work is to increase the rate of oxidation. Both are diffusion-controlled conditions operating as described above.

From these arguments it would appear that the influence of vacancy injection by oxidation should only become apparent at high temperatures and at long exposure times. When vacancy ingestion does occur there is a further factor to consider. Voids form in both the grains and grain boundary so their influence on mechanical properties should not be as pronounced as in nickel.

The effect of oxidation on the tensile properties of 80 Ni–20 Cr has been examined by tensile tests at 1 000°C. A comparison of specimens oxidized in air with equivalent vacuum-annealed specimens showed that maximum stress and ductility were reduced slightly by preoxidation at 1 000°C. Values of the reduction in properties expressed as a percentage of the equivalent vacuum-annealed specimens are shown in Table 1.

TABLE 1 Reduction in tensile properties of 80 Ni–20 Cr at 1 000°C due to oxidation (values of reduction expressed as a percentage of the equivalent vacuum-annealed specimens)

Oxidation time, h	Maximum stress, %	Elongation to max. stress, %	Elongation to fracture, %
1	97	94	94
10	96	91	90
25	94	81	81

As expected, the results are not as dramatic as observed with nickel but they do show a significant effect which increases with time of exposure. Hence with these simple single-phase Ni–Cr alloys, there is still an effect of surface oxidation on the mechanical properties of the metal, although the oxide scale thickness is much smaller than found on nickel at these temperatures.

The high-temperature high-strength commercial alloys are strengthened by appreciable amounts of second-phase precipitates, and the interfaces in the metal matrix produced by these precipitates should act as vacancy sinks, restraining vacancies from migrating to appreciable depths into the metal. This is shown in Fig. 9,[36] which shows Nimonic 105 with a cold-worked surface layer oxidized for 100 h at 950°C. The barrier film

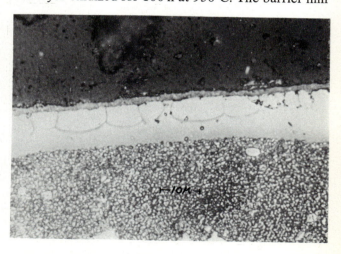

9 Nimonic 105 oxidized for 100 h at 950°C showing no voiding[36]

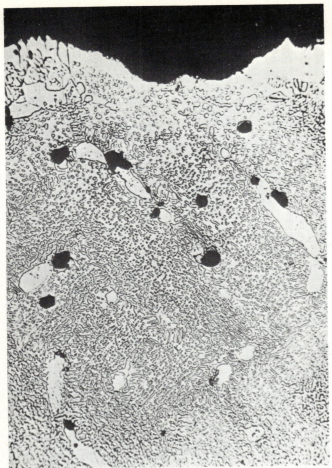

10 High-strength Ni–Cr alloy oxidized for 420 h at 1 100°C, showing internal voiding[37]

in this case is alumina but the complex matrix structure does not allow appreciable vacancy penetration. However, with this type of alloy, which has lower chromium contents, which means greater oxidation, voidage can occur. Figure 10[37] shows such voidage after oxidation of an experimental alloy (composition 6·5 Cr, 11·6 Co, 8·5 W, 8·5 Ta, 5·9 Al, bal Ni) for 400 h exposure at 1 100°C. However, at this temperature most of the precipitates would be in solid solution. Only the carbides would remain and, as shown, voids have nucleated on the carbide phases. This demonstrates that the influence of second phases in inhibiting vacancy injection is only significant if precipitate phases are not in solution. Oxide dispersion strengthened alloys do not have the problem of phase solution at operating temperatures and the oxidation of nickel thoria[38] and nickel–chrome–thoria[39] has been studied at temperatures from 900° to 1 400°C. The oxide scales were found to be highly adherent, even at the high temperatures, but no void injection was found, which is in direct contrast to the behaviour of nickel at even lower temperatures. It should therefore be expected that the effect of vacancy injection produced by oxidation should be markedly reduced in dispersion-strengthened alloys by vacancy adsorption at the phase interfaces.

CONCLUSIONS

1 Vacancies can be injected into the metal substrate when metals oxidize by cationic diffusion through the oxide.

2 Thin uniform oxide scales of the order of 4–40 μm can cause creep lives in 10 mm diameter specimens to be reduced dramatically. Void precipitation, formed by vacancy ingestion, due to the oxidation process, is responsible for acceleration of the creep fracture process.

3 The voids are shown to increase with temperature and exposure time and hasten tensile fracture at ambient temperatures. Their effect is to cause the fracture surface of room-temperature tensile specimens to become indistinguishable from high-temperature long-time creep rupture tests. This suggests that many of the effects of secondary creep cracks observed during high-temperature creep may be related to the test environment.

4 The oxidation-induced vacancy injection also causes strengthening at low strains. The effect is attributable to cluster hardening at low temperatures and also to restriction of grain-boundary movement due to void pinning at higher temperatures. These effects of surface oxidation are shown to influence creep, yield stress, and recrystallization.

5 The effect of oxidation on mechanical properties is reduced for the more oxidation-resistant alloys such as the 80 Ni–20 Cr system, as the oxidation rate is lower and voids precipitate within both the grain and grain boundaries. However, the effect of oxidation reappears at higher temperatures and longer exposure times.

6 The influence of second-phase particles reduces the effect of vacancy injection by offering a multitude of sinks below the metal/oxide interface, so preventing void migration deep into the underlying metal.

7 The effect of cold work on single-component systems such as nickel is shown to increase the oxidation rate and encourage vacancy migration into the metal. Cold-worked material is therefore shown to exhibit the greatest effect of environment on the mechanical properties of the underlying metal.

8 Cold work on systems that rely for protection on selective oxidation of one element, such as the Ni–Cr system, gives exactly the reverse effect to that observed with the single-element systems. The cold work increases the diffusion of chromium to the surface of the alloy to allow repair of the oxide scale and therefore cold-worked material shows reduced oxidation rates.

9 This paper has only dealt with the effect of vacancies produced by oxidation on mechanical properties. It is an extremely important effect but it must be mentioned that vacancy injection is not the only way in which oxidation will influence the mechanical properties of components. The effect of stress causing rupture of the oxide film is equally important and, for a full understanding of the effects of oxidation on the mechanical properties of components, both aspects must be considered.

REFERENCES

1 P. HANCOCK: 'Corrosion of alloys at high temperatures in atmospheres consisting of fuel combustion products and associated impurities', 1968, London, HMSO
2 J. STRINGER: 'Hot corrosion in gas turbines', Metal and Ceramics Information Centre, Batelle, MCIC Report 1972
3 P. KOFSTAD; 'High temperature oxidation of metals', 1966, Wiley
4 P. HANCOCK AND R. FLETCHER: *Métallurgie*, 1966, **6**, 1
5 R. HALES AND A. C. HILL: *Corrosion Sci.*, 1972, **12**, 843

6 P. HANCOCK AND R. C. HURST: 'Advances in corrosion science and technology', Vol. 4, 1, 1974, Plenum
7 P. S. DOBSON AND R. E. SMALLMAN: *Proc. Roy. Soc.*, 1966, **A293**, 423
8 R. HALES *et al.*: *Proc. Roy. Soc.*, 1968, **A307**, 71
9 R. HALES *et al.*: *Metal Sci. J.*, 1968, **2**, 224
10 H. L. FRASER *et al.*: *Phil. Mag.*, 1973, **28**, 639
11 S. KRITZINGER *et al.*: *Phil. Mag.*, 1967, **16**, 217
12 P. S. DOBSON *et al.*: *Phil. Mag.*, 1968, **17**, 769
13 S. KRITZINGER *et al.*: *Phil. Mag.*, 1967, **16**, 217
14 M. E. WHITEHEAD *et al.*: *Acta Met.*, 1975, **23**, 911
15 I. R. SANDERS AND P. S. DOBSON: *Phil. Mag.*, 1969, **20**, 881
16 R. E. SMALLMAN AND P. S. DOBSON: 'Surface and defect properties of solids', Vol. 4, 103, 1975, Chemical Society
17 V. R. HOWES: *Corrosion Sci.*, 1970, **10**, 99
18 J. E. HARRIS AND B. C. MASTERS: *Proc. Roy. Soc.*, 1966, **A292**, 240
19 P. HANCOCK AND J. McLAVERTY: Internal Report, January 1969
20 D. CAPLAN AND M. COHEN: *Corrosion Sci.*, 1966, **6**, 321
21 D. CAPLAN *et al.*: *Corrosion Sci.*, 1970, **10**, 9
22 S. I. ALI AND G. C. WOOD: *J. Inst. Metals*, 1969, **97**, 6
23 D. BRUCE AND P. HANCOCK: *J. Inst. Metals*, 1969, **97**, 140
24 D. BRUCE AND P. HANCOCK: *J. Inst. Metals*, 1969, **97**, 148
25 D. BRUCE AND P. HANCOCK: *J. Iron Steel Inst.*, 1970, **208**, 1021
26 R. C. HURST AND P. HANCOCK: *Werkstoffe u. Korrosion*, 1972, **9**, 773
27 P. HANCOCK: 'Mechanical properties of growing surface oxide scales', Paper to AIME Conference, Detroit 1974, in press
28 P. HANCOCK AND R. FLETCHER: unpublished work
29 R. HALES *et al.*: *Acta Met.*, 1969, **17**, 1323
30 E. D. HONDROS AND C. R. CLARK: *J. Mat. Sci.*, 1970, **5**, 374
31 R. MADDIN AND A. H. COTTRELL: *Phil. Mag.*, 1955, **46**, 735
32 P. B. HIRSH: this volume
33 G. C. WOOD: *Oxid. Met.*, 1970, **2**, 11
34 J. B. JOHNSON *et al.*: to be published
35 S. A. GIGGINS AND F. S. PETTIT: *Metall. Trans.*, 1969, **245**, 2509
36 A. G. WYSIEKIERSKI *et al.*: US–UK Conference on Marine Gas Turbines, Bath, Sept. 1976
37 J. D. W. RAWSON: private communication
38 F. S. PETTIT AND E. J. FELTON: *J. Electrochem. Soc.*, 1964, **111**, 135
39 H. H. DAVIS AND H. C. GRAHAM: *Oxid. Met.*, 1971, **5**, 431

Author index

Subject index